STATISTICS
An Introduction

STATISTICS

ALBERT D. RICKMERS
and
HOLLIS N. TODD
of the
ROCHESTER INSTITUTE OF TECHNOLOGY

An Introduction

McGraw-Hill Publishing Company

New York   St. Louis   San Francisco   Auckland   Bogotá
Caracas   Hamburg   Lisbon   London   Madrid   Mexico   Milan
Montreal   New Delhi   Oklahoma City   Paris   San Juan
São Paulo   Singapore   Sydney   Tokyo   Toronto

STATISTICS: An Introduction

Preface

The field of statistics comprises a body of theory and practice which includes data collection methods, analysis methods, and interpretation. The field has grown to the point where it has application to every kind of situation in which difficult decisions must be made, based on observations.

On the one hand, pure and applied research is dependent for its success on the appropriate analysis of data. On the other hand, statistical methods are essential to the initiation and maintenance of complete quality control. In these two different kinds of situations, there is no real alternative to statistical methods. The truth of a scientific law can be established only when the data can be demonstrated to confirm theory. This confirmation can come only from the successful completion of a well-designed experiment, part of which design is necessarily statistical in basis. Statistical methods, when combined with other techniques and other sources of information, are essential to maximize the quality of an industrial process and to minimize cost.

Statistical methods are so powerful that they should be introduced early in the educational process. From a study of statistical concepts, the student grasps the fundamental understanding that variability is encountered in all scientific, research, and industrial activities. From this understanding he sees the need for adequate sampling in every measurement process. The study of statistics sheds light on the scientific method. The student learns the basis for the approach to experimentation which permits him to make an analysis of the factors affecting data variability.

This book is addressed to an undergraduate audience and to others with comparable academic backgrounds. It is intended as a text in two, three, and four-year programs in science and engineering. The authors believe that the principles of statistics should be taught to those intending to do even limited research in medicine, biology, agriculture, and engineering fields, as well as in physics and chemistry and other sciences. The contents of this book have been taught successfully to students of widely varying educational preparation. It has been our experience that the principles of statistics are received by most students with astonishment and enthusiasm.

Only arithmetic and elementary algebra are needed as mathematical prerequisites. The calculus is used only once in the development of a concept. In general, we have omitted formal proofs and complex mathematical analyses. We have tried, however, to justify if only on an intuitive basis, the most important concepts. Usually new material is introduced with examples which illustrate situations that are likely to be familiar to the student. Principles and techniques are applied by means of examples and problems.

The number of symbols is kept to a minimum. Unfortunately, there is as yet no universally accepted set of symbols. Those used in this book are in accordance with the usage of many statisticians.

Depending on the objectives of a course in elementary statistics, this book may be used in its entirety for a full year, or in part for a one-semester or two-quarter course. Emphasis may be placed, in the second case, either on research and development applications or on process and product control.

The first seven chapters discuss fundamentals. They deal with the concept of variation, measures of variability and central tendency, the normal distribution, and tests of significance and interval estimates for the mean and variance.

If the text is to serve those who are interested in the applications of statistics to research, the book may be followed through Chap. 17. In these chapters we discuss analysis of variance, finding sample size, regression and correlation, and planning and execution of basic experiment designs, including response surface techniques.

For those primarily interested in process and product control, the sequence may be Chaps. 1 to 7, followed by 18 to 24. In the latter chapters, we discuss nonparametric methods, the elements of probability, sampling plans, and types of control charts, including those most recently developed.

We do not pretend that this is other than an introduction to a fascinating and multifarious branch of applied mathematics. Students who master this book will have acquired a new and useful vocabulary; they will have acquired some insight into a fruitful method of thinking and working; they will at least know that an efficient set of tools is available. Perhaps they will be stimulated to learn more.

Albert D. Rickmers and *Hollis N. Todd*

Contents

STATISTICS
An Introduction

1 Variability

Within any group of supposedly similar objects, or among the data from a repeated experiment, we can always find differences if we examine the data closely enough. If we make observations on a repetitive process and apparently find no variation, this seeming consistency indicates that our examination has been inadequate to detect differences of a smaller magnitude. If these differences are too small to be of practical importance, we may choose to ignore them and continue to take measurements as we have in the past; but at least we should be aware that these differences do exist.

A central theme of the statistical approach to data analysis is this: variability always exists. This statement applies to research, to engineering development, and to quality control. No experiment can be repeated exactly. Similar experiments never truly yield identical results if measured on a fine enough scale. Common experience supplies illustrations: No two persons are exactly alike, the phrase "identical twins" notwithstanding. No two snowflakes are identical (though for most purposes we may consider that they are identical). Similarly, no two items from the same production line are identical; no two samples from the same lot of material are identical.

Variability in a scientific experiment or in an industrial process is often thought to be a curse, something to be eliminated. It can never be wholly eliminated. In fact, one of the keystones of a statistical approach is to make *use* of variability.

One of the basic questions we shall deal with is of this general type: Is the observed variability small enough to be neglected? The answer to such a question involves at least three aspects:

1 The specification of the magnitude of the variation that can be tolerated.
2 The measurement of the variability actually occurring.
3 The comparison of the size of the permissible variability with that actually being found.

The first of these aspects is not primarily a statistical matter. It is instead a problem requiring a managerial decision, necessitating a study of the ultimate use of the production item or of the experimental data. A decision may be required about the tolerable failure rate of a part or about the supportable number of customer complaints. In a scientific experiment, a judgment must be made about the level of reliability that should be attached to the outcome of the experiment. These decisions are made on other than statistical grounds.

The statistical methods we are about to discuss relate to the other aspects of the question. They involve the detection of the pattern of variability present and analysis of its sources; the measurement of this variability; a comparative evaluation of the variability with the desired characteristics of the process.

The owner of a metalworking firm wishes to make blades for saber saws; he must compete with other manufacturers of similar products. The owner should set specifications for the quality of the product in terms of sharpness, resistance to breaking, durability, etc. These specifications should be based on a study of the requirements of the market, and are basically not statistical. Statistical quality control in this example would involve a study of a process for making the blades, to determine the characteristics of the product, and to compare these characteristics with the specifications. Perhaps the manufacturer must change his production methods to enter the market, or he may easily be in a position to compete when both cost and quality are considered.

Experiments are conducted on occasions when the sun is eclipsed in order to measure the small deviation of starlight passing near the sun. From theory the expected deviation can be estimated; this is a mathematical problem, not a statistical one, involving only the equations derived from scientific theory. Statistics is, however, involved in the actual measurements. There is variability of the observations associated with differences in apparatus and in observers, and this variability enters into the comparison of the measured results with those to be expected from theory.

Examples like the two above could be extended indefinitely; many will be added in the discussion which follows. It is enough for now if we have shown that we necessarily begin any study of statistical methods of data analysis with the methods used in studying the variability of a repetitive process.

1·1 KINDS OF DATA

In studying a process, we necessarily acquire information about it by making observations on the set of items of interest. We thus accumulate a set of data. Depending on the nature of the process, the raw data may be of different kinds.

1 The data may merely be *qualitative*, as when we judge the item as satisfactory or unsatisfactory, in which case the data are counts of the numbers of items falling into either one of two (or perhaps more) classes. Characteristics such as these are called *attributes*.

2 The data may be *quantitative*, resulting from a measurement or other numerical estimation. Measurements yield *variables*. Here we may find either of two kinds of data appropriate. One kind is that which we would obtain if we were studying the performance of a machine which fills bottles with pills. If we make a count of the number of pills put in each of many bottles, the values we obtain can vary only by whole units. Such data are called *discrete*, or *discontinuous*. If, on the other hand, we were studying the variation in diameter of a set of machine parts, we could measure the diameter of each part to any degree of precision we pleased, limited by the characteristics of the measuring instrument. Such data are called *continuous*; there is no natural discontinuity between parts of one size and parts of another.

PROBLEMS

1·1 Classify the data that would be obtained in each of the following situations as (1) attributes or (2) variables. If the data are variables, indicate whether they would be continuous or discrete.
   (a) The number of typographical errors in a newspaper
   (b) Judgments of a taste panel of samples of tomato soup
   (c) The prices of a given stock
   (d) Time to failure of each of a lot of radio tubes
   (e) Diameters of ball bearings measured with a micrometer
   (f) Percentage of sugar in fruit juice
   (g) The glossiness of different varnishes
   (h) Stock market averages
   (i) Tests of ball bearings with a go, no-go gauge
   (j) Measurements of the speed of light

1·2 Using a ruler graduated in the metric system, have 5 different people measure the length of this textbook.
   (a) What type of measurement was made?
   (b) Did all 5 measurements agree?
   (c) If each person had measured the length of the book to the nearest centimeter, would all the answers have agreed?

(d) If, using a ruler graduated in millimeters, each length is recorded to the nearest centimeter, what type of observation is being recorded?

1·3 List 5 types of data which would be qualitative.
1·4 List 5 types of data which would be quantitative.
1·5 The length of time one works on a job is recorded in several ways.
(a) Explain how "time" might be a continuous measurement.
(b) Explain how "time" might be considered as a discrete value.

1·2 FREQUENCY DISTRIBUTIONS

Consider Table 1·1, in which we present the raw data obtained when we found the deviations from the aim point for a lot of 200 items. Such data could arise from a process intended to put 100 sheets of paper into a package; the data represent the results of an examination of 200 packages and indicate the number of sheets over or under the desired count of 100.

Table 1·1   Study of Packaging Process

0	2	1	0	1	3	−1	0	0	0	−2	0	2	−3
2	−1	0	−1	0	−1	0	−1	1	−3	3	0	2	2
−2	0	0	2	−2	2	0	0	1	2	−2	0	0	0
0	−1	2	0	2	1	−2	3	1	−1	−2	−1	0	1
0	1	0	2	1	0	3	−1	−1	0	1	2	1	−1
−2	−1	1	0	1	1	3	3	−3	−3	2	−3	1	−4
0	−1	−2	−2	2	−2	2	−1	3	0	−2	0	−1	0
3	1	−1	−1	0	1	3	0	−1	1	−1	0	−1	0
−1	0	1	−1	1	4	−2	0	1	1	2	−1	−2	1
−2	−1	0	1	−1	−2	−1	0	2	0	2	4	0	−1
2	1	0	−4	3	−3	1	−1	1	−4	2	5	−2	1
−1	−5	−1	−1	0	−2	2	−2	0	1	1	0	1	4
1	−3	0	−2	−3	−1	−3	1	−1	−1	0	1	−1	−2
−1	−3	−2	1	1	2	0	0	−1	1	−2	−1	1	−2
2	−1	1	0										

As presented, these data exemplify a discrete variable. There is no organization in this table, other than sequence. In certain problems involving a production sequence, the order might well have considerable significance; here, we will suppose that order is of no importance. In this form of data presentation, it is difficult to interpret the observations.

In Table 1·2, we present the same data, but now we have organized them into a classification according to the magnitude of the deviations. We have grouped all measurements of the same magnitude together. What we have now is a *frequency histogram*, "frequency" meaning the number of items in a class. Such a plot is valuable in the following ways:

1   The data are condensed into a graph which is pictorially meaningful.
2   The pattern, if any, into which the data fall can be observed. Clustering

Table 1·2   Frequency Histogram (Data from Table 1·1)

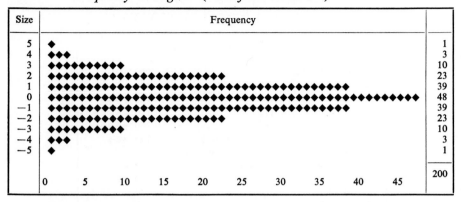

Size	Frequency	
5	◆	1
4	◆◆◆	3
3	◆◆◆◆◆◆◆◆◆	10
2	◆◆◆◆◆◆◆◆◆◆◆◆◆◆◆◆◆◆◆◆◆◆	23
1	◆◆◆◆◆◆◆◆◆◆◆◆◆◆◆◆◆◆◆◆◆◆◆◆◆◆◆◆◆◆◆◆◆◆◆◆◆	39
0	◆◆	48
−1	◆◆◆◆◆◆◆◆◆◆◆◆◆◆◆◆◆◆◆◆◆◆◆◆◆◆◆◆◆◆◆◆◆◆◆◆◆◆◆	39
−2	◆◆◆◆◆◆◆◆◆◆◆◆◆◆◆◆◆◆◆◆◆◆◆	23
−3	◆◆◆◆◆◆◆◆◆◆	10
−4	◆◆◆	3
−5	◆	1
		200

0   5   10   15   20   25   30   35   40   45

of the data, symmetry or lack of symmetry, and the range of the data can be seen.

3   We can calculate the fraction of the observations which fall between any specified limits.

Thus frequency tables or histograms alone often provide the basis for considerable insight into the characteristics of a repetitive process. The study of such a process should usually begin with this kind of treatment of the data.

Patterns of the type shown in Table 1·2 (though perhaps of different shape) can be found wherever fine enough measurements are made. There is a mathematically derived theoretical distribution which is associated with this kind of variability. We need relatively large numbers of observations if a specific frequency histogram is to be a good approximation of the underlying frequency distribution, the theoretical curve which applies to that particular situation.

Very often, mere inspection of a frequency distribution in itself permits useful inferences to be made. Three paper mills supplied the same type of paper to a newspaper press. One of the major causes of downtime on the press was that some rolls of paper tended to tear more often than others. Lots of 20 rolls from each of the three suppliers were tested for tear resistance, with results as in Table 1·3. The frequency tabulations show: The paper supplied by A was extremely variable in quality; the paper supplied by B was consistent in quality; the paper supplied by C was somewhat less consistent than that from B but was at a higher level of quality (more tear resistance) generally. On the assumption that the lots tested were representative, and supposing that a tear strength of 130 is the minimum desirable, it would be wise to confine the purchases of the paper to supplier B; suppliers A and C should be encouraged to reduce the variability of their product; supplier C could supply a very desirable paper if the variability could be reduced.

Table 1·3   Frequency Histograms of Tear Strength of
Paper from Three Different Suppliers

Mill A	Mill B	Mill C
200 ◆	200	200
190	190	190
180 ◆◆	180	180 ◆◆◆
170	170	170 ◆◆◆◆◆
160 ◆◆◆	160	160 ◆◆◆◆◆
150 ◆	150 ◆◆◆◆◆◆◆	150 ◆◆◆
140 ◆◆◆◆◆	140 ◆◆◆◆◆◆◆◆◆	140
130 ◆◆◆◆	130 ◆◆◆◆	130 ◆◆
120 ◆	120	120 ◆◆
110 ◆◆◆	110	110
100	100	100

In addition to merely inspecting the pattern displayed by a frequency histogram, we usually make calculations from the data which give numbers descriptive of the process and which permit us to make comparisons of processes of the same kind in a meaningful way. Such calculated numbers describe, among others, two important characteristics of a frequency distribution: (1) the point about which the observations tend to cluster; (2) the variation among the observations. The subject of this chapter is the preparation of frequency-distribution plots and the calculation of the descriptive numbers derived from quantitative data.

1·3 POPULATIONS AND SAMPLES

Occasionally, though rarely, we examine all the items of a set or all the items coming from a process. We then say that we are dealing with the entire *population* or *universe*. A population includes the whole set of items of a specified kind at a specified time. Examples are the heights of all the persons in the United States now; the maximum speed of all the MG automobiles now in operation; the sharpness of all the razor blades produced in 1965 by a single specified machine; the purity of the entire production of sulfuric acid from an industrial plant.

Numbers which describe populations are called *parameters*. One example is the arithmetic average of the characteristic of interest, which we symbolize by the Greek letter μ, read "mu." Only where *every* member of the population has been examined and the value of the characteristic determined can we actually find μ.

Usually we must be content with examining a part of a population, a *sample*. Numbers which describe samples are called *statistics*. We will symbolize as X values measurements on the individual items making up a sample. If we wish to specify an X value, we use subscripts, such as X_1, X_5, etc., the generalized symbol being X_i. The arithmetic mean of the items making up a sample is symbolized by $\bar{X}$, read "X bar." The arithmetic mean is the sum of all the items divided by the number of items:

$$\bar{X} = \frac{X_1 + X_2 + X_3 + \cdots + X_n}{n}$$

We think of $\bar{X}$ as *estimating* μ; that is, the arithmetic average of any single sample of a population only approximates the arithmetic average of the population itself. In general, statistics determined from samples are estimates of parameters, which in turn are descriptive of populations. The distinction between population and sample is especially important. We usually would like to say something about populations when we have only a sample for study. We use our sample data to make *inferences* about the population. *Statistical* inference is one of the basic themes of this book. If we know the population characteristics, we can *deduce* from these the probable characteristics of a sample; this is usually an easier task. But, in general, we can get only a sample estimate of a population parameter. Statistics plays a major role in showing how the estimate (the statistic) is related to the population parameter.

PROBLEMS

1·6   Define clearly the populations implied in each of the following:
(a) A study of the variability of a machine bottle-filling process
(b) The cost of living in New York City
(c) A comparison of the miles per gallon of gasoline required by each of two makes of automobiles
(d) The shear strength of a batch of cold-rolled steel
(e) Water hardness of a specific water supply
(f) The efficiency of a motor generator

1·7   Consider the following population of numbers:

3, 5, 4, 6, 5, 7, 3, 5, 4, 6, 4, 3, 5, 3, 5, 4, 4, 5, 2, 6, 5

(a) Calculate the mean of the entire population.
(b) Calculate the mean of the odd values.
(c) Calculate the mean of the even values.
(d) Does the value of the mean of either sample approximate the mean value of the population?

1·8   Measure the thickness of 100 pages of your text, and then by dividing by the number of sheets involved, calculate the thickness of each sheet.
(a) Is the answer to the above calculation a sample value, an average value, or a population mean?

(b) Repeat the instructions given above, and calculate another value of the thickness of each sheet. Do the two thickness values agree?

(c) Why should the two calculated thickness values be averaged to obtain a better measure of the sheet thickness?

1·4 SELECTION OF SAMPLES

If we intend to evaluate the population from a sample, or from several samples, it is essential that the sample be representative of the population. Thus, once we have defined the population of interest, the method of selecting the sample must be our next consideration.

If we were to sample a sheet of plastic by examining patches only from the four corners, we would quite possibly (in fact, probably) be using a nonrepresentative sample, since it is reasonable to think that the center of the sheet would be subject to different manufacturing stresses from those at the corners. Similarly, the surface of a carload of coal is unlikely to be representative of the entire car because stratification may have occurred during loading or shipping. In a continuous process, samples may be non-representative if the samples are taken always at the beginning of operations each day or each shift.

To avoid the mistakes inherent in nonrepresentative sampling, we often use *random* sampling procedures. If it is practicable to assign a different number to each item in a lot, or to each position within a sheet or container, chance devices should be used to determine the selection of the sample. If the lot size is small, a deck of cards or a group of numbered tags drawn from a box can serve.

An alternative method of randomization uses a table of random numbers, available from a variety of sources. Numbers are assigned in advance to the items in the lot. We enter the table at any place we please and take numbers from the table by moving in any prearranged way (up, down, diagonally, etc.). The sequence of numbers dictates the sample selection.

Often we want to randomize the order in which a set of tests will be made, when we fear that over the time of testing some factor may alter the results. By randomizing the sequence, we hope to reduce the possibly systematic effect associated with time.

We can never be completely confident that we have successfully random-ized a sampling procedure. Rather, we take the opposite approach and look for signs of nonrandomness. It is often useful to record the observations in the sequence in which they occur, and to look for patterns. When we find no systematic arrangement of the observations, we assume that no important nonrandomness is occurring.

PROBLEMS

1·9 Consider the following population of values:

6, 6, 4, 7, 5, 6, 4, 7, 8, 4, 6, 6, 5, 4, 7, 7, 4, 5, 5, 7, 4

(a) Calculate the population mean.

(b) Randomly select 5 of the values from the population and calculate the sample mean.

(c) If a series of randomly selected samples of 5 values were taken, and their averages were then averaged, do you think the resulting value would be a better approximation of the population mean? Explain.

1·10   Drill bits, all the same size, are stored in a container in 5 columns and 10 rows. How would you randomly select 2 of the drill bits from storage?

1·11   One member of a group of 10 people is to be sent on an errand that no one wishes to go on. Explain how the person should be selected on a random basis.

1·12   You are to call 5 people on the telephone, and the names are to be selected at random from a telephone book. Explain how the 5 names could be randomly selected.

1·13   When the value of π is carried out to a great number of decimal places, the digits are said to be random.

(a) Could the digits from π be used to make random selections?

(b) If the digits are random, should they start to repeat themselves in the same order at some point? Explain.

1·5   FREQUENCY HISTOGRAMS USING UNGROUPED DATA

Some sets of raw data comprise only a relatively small number of different values. This is often true for discrete data and is occasionally true for continuous data. In these situations, inspection of the data often suggests how they can be conveniently classified. We can then prepare a frequency table and a frequency plot by making a count of the number of observations falling into each class. We followed this procedure in preparing Table 1·2.

In Table 1·4 we give the results of height measurements (to the nearest inch) of 100 male college students. The largest value in the table is 78; the smallest is 60. The *range* of the observation is 18; this is the difference between the smallest and largest values. We can therefore group the data into 18 different classes; this we have done in Table 1·5, using tally marks which form the frequency histogram.

Because we are dealing here with continuous data, we may make a somewhat different kind of graph—a line graph—called a *frequency polygon*. We draw straight lines connecting the midpoints of adjacent classes, as in Fig. 1·1.

Table 1·4   Heights of Male College Students (Inches)

66	72	69	67	70	69	72	64	70	69	71	62	68	70	75	78	76	71	69	70	76	65	72
66	68	74	71	73	67	70	68	68	70	65	69	64	73	75	69	73	70	75	68	69	64	70
72	70	67	71	69	70	69	72	66	70	74	67	74	64	70	74	73	70	71	69	73	68	66
70	65	70	71	61	68	72	67	70	63	72	70	66	65	69	68	70	67	71	69	60	72	69
71	68	71	69	70	68	71	77															

Table 1·5 Tally Sheet Showing Frequency Distribution (*Data from Table 1·4*)

Height	Tally	f
78	◆	1
77	◆	1
76	◆◆	2
75	◆◆◆	3
74	◆◆◆◆	4
73	◆◆◆◆◆	5
72	◆◆◆◆◆◆◆◆	8
71	◆◆◆◆◆◆◆◆◆◆	10
70	◆◆◆◆◆◆◆◆◆◆◆◆◆◆◆◆◆◆◆	19
69	◆◆◆◆◆◆◆◆◆◆◆◆◆	13
68	◆◆◆◆◆◆◆◆◆◆◆	11
67	◆◆◆◆◆◆	6
66	◆◆◆◆◆	5
65	◆◆◆◆	4
64	◆◆◆◆	4
63	◆	1
62	◆	1
61	◆	1
60	◆	1
		Total = 100

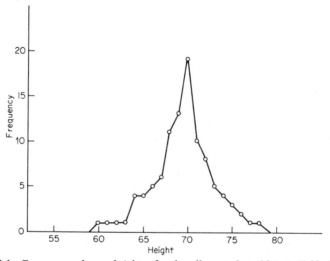

Fig. 1·1 Frequency polygon: heights of male college students (data in Table 1·2).

Note especially these points concerning the frequency polygon:

1   We have the right to draw lines connecting the midpoints of the cells because of the continuous nature of the data. Although we placed all numbers recorded as 71 in. in the same cell, we realize that some of these heights, if measured more carefully, would have been, for example, 71.3, 71.2, and 69.7 in. There is therefore real meaning to the line: it expresses our best estimate, on the basis of the data we collected, of where the data would have fallen had we made more precise measurements.
2   We assume, in any calculations, that the midpoint of the cell is the average value of all the observations placed in that cell.
3   The frequency polygon is a closed curve. It falls to a frequency of zero at the first cell interval on either side of our recorded data.
4   The area under the polygon represents the total collected data.
5   If the number of cells were greatly increased (by making more precise measurements) and if the sample size were also greatly increased, the polygon would approximate a smooth curved line. Although this smooth curve is a limiting shape for the polygon, we can never actually find it precisely, since we can never have an "infinite" number of cells or an "infinite" sample size. Nevertheless, this concept of a smooth curve as the limit of a frequency polygon is an important one. Data from many real situations appear to approximate a particular smooth curve known as the *normal curve of distribution*, or *normal curve* for short. Because of the importance of the normal curve, the following chapter is devoted to its analysis. Here it is enough to say that the normal curve is not only smooth but symmetrically placed about its highest point. The frequency polygon in Fig. 1·1 suggests that this kind of pattern could be the underlying distribution for this example.

A gross-weight filling machine was supposed to place 14 oz of a chemical product in each container. To see how well the filler operated, and to obtain data about the variability of the packaging operation, samples were collected at random from the process and net weights determined to the nearest 0.1 oz. The data are shown in Table 1·6; the values range from 13.5 to 14.4. An ungrouped frequency histogram was prepared, as in Fig. 1·2.

A quick evaluation of the data using the histogram suggested that the average was approximately correct. There was, however, an indication of

Table 1·6   Weights of Filled Bags (Ounces)

14.1	14.3	13.8	13.9	14.0	14.0	13.8	14.1	14.1	14.2	14.2
13.7	13.9	14.0	14.4	13.8	13.8	14.1	13.5	13.7	14.0	14.2
13.9	14.0									

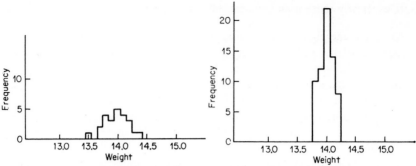

Fig. 1·2 Frequency histogram: weights of filled bags before use of sorter (data in Table 3·3).

Fig. 1·3 Frequency histogram: weights of filled bags after use of sorter.

1.6

many underfills and overfills. A platform weight sorter was placed in the packaging line to permit 100 percent inspection of all packages. Bags which were underfilled or overfilled were sidetracked from the line. A subsequent frequency histogram of packages passed by the checker is given in Fig. 1·3.

Thus the preparation of a frequency-distribution plot often leads to insight into a production situation. In simple cases, judgments can be based on such plots with considerable assurance of validity.

PROBLEMS

1·14   Place the following sets of data into cells, using ungrouped data:
(a) 4, 5, 4, 7, 5, 8, 4, 5, 6, 3, 6, 4, 3, 7, 8, 5, 3, 5, 6, 3, 2
(b) 0.5, 0.6, 0.4, 0.5, 0.4, 0.6, 0.5, 0.4, 0.2, 0.5, 0.9, 0.5, 0.6
(c) 163, 159, 150, 164, 158, 158, 155, 159, 154, 160, 162, 163, 164, 155, 158, 160, 153, 155, 158, 156, 157, 155, 153, 152, 150, 160

1·15   Select 16 pennies and after shaking them in your hands until they are well mixed, toss them onto a blanket or rug, and record the number of heads that you find. Toss the coins, in this manner, 100 times and construct a frequency distribution of the results.

1·16   Forty randomly selected families that live on a given Air Force Base included the following number of members per family:
6, 4, 4, 3, 9, 3, 5, 3, 2, 4, 5, 3, 5, 3, 11, 3, 4, 4, 3, 5, 4, 3, 6, 3, 5, 4, 4, 3, 5, 3, 6, 5, 4, 5, 5, 6, 3, 4, 3, 2

(a) Make a frequency histogram, showing the frequency of the different-sized families.
(b) Construct a frequency polygon from the above data.
(c) Does a histogram or a polygon give the better picture of the various family sizes? Why?
(d) What type of data was being tabulated?

1·17   List five situations where the underlying distribution for the frequency polygons might be the normal curve.

1·18   Select at random 25 people and ask them their height correct to the nearest inch.
  (a) Identify the population from which you have taken your sample.
  (b) Construct a frequency polygon of the heights recorded.
  (c) Calculate the mean of the sample.
  (d) Estimate the mean of the population of heights.
  (e) Could the normal distribution be the underlying distribution for this population? Explain.

1·6   FREQUENCY HISTOGRAMS USING GROUPED DATA

In most cases where we are dealing with continuous data, or where the different values of discrete data are large in number, we cannot assign a single class to a single value without defeating the whole purpose of drawing a histogram. The frequency of occurrence in any given class would be too small to show any useful pattern.

In such situations, we group the data into a reasonable number of *cells*, usually about 15. A cell is a classification into which we put all values that are closely alike. Based on the range of the data we choose a convenient *cell interval*, or cell width, with cell *boundaries* which define the upper and lower limits for each cell. We specify the cell boundaries to have one significant digit (usually 5) *more* than in the recorded observations. We do this so that there will never be any doubt about the cell to which any observation should be assigned.

Note especially that the cell *midpoint* is the average of the two cell boundaries. In calculations which are based on grouped data, we assign the value of the cell midpoint to all the observations placed in that cell.

By using cells covering a specified small range of the data, we can treat continuous data in the same way we treat discrete data. The grouping process may cause us to lose some of the detail present in the original data, but the advantages of grouping are great, and the loss is usually small.

Consider, for example, a set of data representing the average score per game of all league bowlers in the United States. A tabulation of every one of the possible different averages would fill many pages. The data were actually grouped, and reported as in Table 1·7. Ignoring the catchall cells for the largest and smallest scores, the cell interval for these grouped data is 5. That 8,585 scores are reported for the cell 195–199 means that any score lying between these two values is included in the count for that cell. We cannot tell how many of these were for a specific value of, say, 196.

We may criticize this tabulation for failing to indicate the cell midpoints, for having an indefinite cell interval for the largest and smallest cells, and especially for not having precisely defined cell boundaries: it is not clear, for example, where an average score of 194.5 is placed, since the lower boundary of one cell is given as 195 and the upper boundary of the next is given as 194.

Table 1·7   A.B.C. Averages for League Bowling in 1961

Average per Game	Number of Bowlers
200 and over	4,632
195–199	8,585
190–194	22,567
185–189	56,610
180–184	124,312
175–179	221,552
170–174	327,377
165–169	423,512
160–164	481,057
155–159	502,817
150–154	479,527
145–149	428,145
140–144	354,450
135–139	273,402
130–134	198,815
125–129	133,450
120–124	86,487
115–119	51,977
110–114	30,515
105–109	16,405
100–104	9,095
Under 100	11,730

Nevertheless, this tabulation does provide us with the possibility of making some interpretations: (1) the distribution appears to be approximately symmetrical, although poorer bowlers seem to have a little more spread in their scores than do the better ones; (2) a reasonable guess would be that the overall average of the bowlers is near the cell 155–159 (the actual average was reported to be 153.9); (3) more than 75 percent had scores between 135 and 175. The point here is that a tabulation of large numbers of data in a grouped frequency-distribution array enables us to arrive at general statements about the data which could hardly be made without such treatment. For this reason, such a tabulation is the basis of further statistical analysis.

We now illustrate the methods of making a correct tabulation without the defects in the example of the bowling scores. The marks of 100 students on a final examination in statistics are shown in Table 1·8.

Table 1·8   Scores of 100 Students on a Final Examination in Statistics

78	96	52	77	74	66	78	76	83	88	81	73	74	92	91	77
91	75	89	75	82	94	93	80	60	87	76	79	79	58	76	90
79	71	89	78	70	68	79	87	85	65	76	73	83	71	84	73
69	64	73	79	77	83	84	82	82	83	81	82	80	90	86	67
63	92	83	86	70	55	66	84	82	98	80	79	72	78	62	77
80	85	76	67	85	70	69	86	81	88	75	87	80	81	87	81
77	74	75	80												

To determine the cell interval, we first find the *range* of the data; this is the difference between the largest and the smallest scores in the raw data The largest score is 98, and the smallest is 52; the range is therefore 98–52 or 46. We wish to use about 15 cells. We find the necessary cell interval by dividing the range by the number of cells. Thus, $46 \div 15$ is a little more than 3; we shall therefore take 3 as the cell interval.

In the tabulation, the uppermost cell must contain the largest observation (98), must be 3 units wide (since this is the cell interval), and must have cell boundaries specified to *one more digit* than the raw data. Since the test scores were given to the nearest integer, we specify the cell boundaries to the nearest $\frac{1}{2}$ unit; we do this so that there will never be any doubt about which cell should contain any specific score—by always defining the cell boundaries to one more digit than the raw data. To ensure that the 98 will fall in the first cell, we make the value 98 the midpoint of that cell and define the cell boundaries as $1\frac{1}{2}$ units on either side of the midpoint. The first cell will then have boundaries of $96\frac{1}{2}$ and $99\frac{1}{2}$. (We might have chosen 97 or 99 as the midpoint of the first cell and still included 98 in that cell. It is probably a little more convenient to choose 98 unless there is a specific reason to do otherwise. Only a very small change in the histogram will result if we do change the value for the midpoint—it will not affect the general picture.)

Once we have defined the cell boundaries and midpoints of the first cell, we need only subtract successively the constant cell interval of 3 units from each of the three values for the first cell (cell midpoint and cell boundaries) to define all the other cells. We continue thus until the final cell contains the smallest test score 52.

After we have defined the cells, we make as before a tally of the raw data. This procedure gives us the frequency plot in Table 1·9.

By examining this plot, we can see that we have again a rough symmetry, with apparently an unduly large number of rather poor scores. The average appears to be near the cell having a midpoint of 80.

In the shipping department of a sheet-metal plant, metal shelving units were shipped with nuts and bolts for their assembly. Since the nuts and bolts shipped with each order were not counted but were weighed out roughly, it was suspected that more than the required pieces were being shipped. Shipping personnel were asked to count the number of bolts they found each time they weighed out 1 lb for an order. The results for 1 day were as shown in Table 1·10. The raw data indicate a range of 60, from 130 to 70. A tally was made using a cell interval of 5, as in Table 1·11.

The frequency plot indicated that an average of about 90 bolts per package was being shipped. Since the original shipping schedule assumed 85 bolts per pound, an average overshipment was clearly indicated by the frequency plot, along with considerable variation. A better system of weighing was thus considered worth investigating. Careful weighing on a new scale which was checked for accuracy indicated that about 85 bolts to the pound was indeed a good estimate. New scales were placed in the shipping room and packers

Table 1·9　Frequency Distribution of 100 Test Scores on a Final Examination

Cell boundaries	Cell midpoint	Tally	Frequency
96.5–99.5	98	/	1
93.5–96.5	95	//	2
90.5–93.5	92	/////	5
87.5–90.5	89	///// /	6
84.5–87.5	86	///// /////	10
81.5–84.5	83	///// ///// ///	13
78.5–81.5	80	///// ///// ///// //	17
75.5–78.5	77	///// ///// ////	14
72.5–75.5	74	///// ///// /	11
69.5–72.5	71	///// /	6
66.5–69.5	68	/////	5
63.5–66.5	65	////	4
60.5–63.5	62	//	2
57.5–60.5	59	//	2
54.5–57.5	56	/	1
51.5–54.5	53	/	1
			Total = 100

Table 1·10　Number of Bolts per Pound (56 Samples)

112	101	105	87	70	106	75	130	98	87	100	76	89	104
88	109	82	78	90	73	92	100	79	88	83	85	79	94
121	93	122	99	115	92	81	101	82	90	77	106	93	90
89	94	88	90	96	83	124	87	89	117	92	91	85	80

Table 1·11　Frequency Distribution of Packaged Bolts

Cell limits	Cell midpoint	Tally	Frequency
127.5–132.5	130	◆	1
122.5–127.5	125	◆	1
117.5–122.5	120	◆◆	2
112.5–117.5	115	◆◆	2
107.5–112.5	110	◆◆	2
102.5–107.5	105	◆◆◆◆	4
97.5–102.5	100	◆◆◆◆◆◆	6
92.5–97.5	95	◆◆◆◆	4
87.5–92.5	90	◆◆◆◆◆◆◆◆◆◆◆◆◆◆◆◆	16
82.5–87.5	85	◆◆◆◆◆◆◆	7
77.5–82.5	80	◆◆◆◆◆◆	6
72.5–77.5	75	◆◆◆◆	4
67.5–72.5	70	◆	1
			Total = 56

were instructed to make weighings with more care. A little training, together with better equipment, produced a frequency distribution with a smaller range and with a better approximation to the desired number of bolts.

PROBLEMS

1·19 Make a frequency histogram and a frequency polygon of the data in Table 1·11.

1·20 Place the following data into cells; use a cell interval of 0.1 in.:
8.37, 7.56, 7.39, 7.73, 7.49, 7.22, 7.40, 7.39
7.44, 7.00, 6.99, 6.89, 6.84, 6.45, 6.43, 6.48
6.66, 6.60, 7.01, 6.49, 7.02, 6.68

1·21 200 measurements were recorded to the nearest 0.01 in. The largest value was 22.34 and the smallest 22.01.
(a) What is the range of the data?
(b) Using grouped frequencies, what should be the cell interval?
(c) For the uppermost cell, what should be the cell midpoint and cell boundaries?

1·22 The smallest of 150 measurements was recorded as 5.18 in.; the largest recorded value was 7.44 in.
(a) What size cell interval should be used to group the data?
(b) Determine a suitable cell interval, giving the cell midpoint and cell limits for the cell containing the largest values.
(c) List the cell boundaries for all the required cells.

1·23 The smallest of 2,000 measurements was 5.18 in. and the largest was 7.44 in. What size cell interval should be used when grouping the data for a frequency polygon? (Compare your answer with the one obtained in Prob. 1·22.)

1·7 RELATIVE FREQUENCY

For the comparison of frequency plots of different numbers of observations, it is useful to plot *relative frequencies* instead of actual frequencies. If n is the total number of observations in a *sample*, the relative frequency for any cell is found by dividing the actual frequency by n; the result is expressed usually as a decimal fraction, occasionally as a percent. For example, we could convert the frequencies in Table 1·11 to relative frequencies by dividing each of the values in the right-hand column by 56. This would give, for the uppermost cell, a relative frequency of 1/56, or about 0.018 (1.8 percent). Similarly, the cell having a midpoint of 90 has a relative frequency of 16/56, or about 0.286.

When frequency histograms or frequency polygons are plotted in terms of relative frequencies, a comparison is facilitated because differences in sample size are ignored, and shapes of the plots are more easily seen.

More important, perhaps, relative frequencies are related to the *probability* of occurrence of a specific value in the set. This statement means that if the relative frequency of a value is 0.10, in a *large number of observations* from a given set of data, we would expect to encounter this value 10 percent of the time. The relative frequencies calculated from small sample sizes are likely to be rather poor approximations of the actual probabilities of occurrence; the relative frequencies approach the probabilities as the sample size is increased.

1·8 CALCULATIONS FROM FREQUENCY TABLES

Many populations and samples appear to be symmetrically distributed; many of these appear to be normal, i.e., to have a particular kind of symmetrical pattern. For such populations and samples, numbers may be derived from frequency tables which are completely descriptive of the data. We can, by making these calculations, replace a large set of raw data with a few derived numbers.

We need two kinds of derived numbers: (1) an indication of the general location of the data on a scale—often called a measure of *central tendency;* (2) an indication of the spread of the data—their dispersion or scatter about the central value.

Measures of Central Tendency

1 The most frequently occurring value in a set of data is the *mode.* In a frequency polygon having one maximum (which will be the case in most data from a single population) the mode is the X value for the highest point. In the set of observations 4, 4, 5, 5, 5, 6, 6, 7, 8 the mode is 5. We sometimes encounter data for which the plot gives two peaks; these are called *bimodal.* Bimodal distributions suggest two sources of data, either because we are dealing with two populations or because some factor in the accumulation of the data gives two sets of results.

The mode is the simplest of the various measures of central tendency to obtain; like many other things that come cheaply, it has limited use. There will be occasions for which it is a sufficient measure, and others where it is inadequate. The mode is most useful for data in which similar values occur often, with relatively few scattered values far from these, as in the set 4, 6, 8, 8, 8, 8, 8, 12, 20. The mode is not useful for rectangular distributions, that is, those having practically the same frequency for any value of X. Furthermore, the value of the mode is unrelated to the size of the observations lying away from the region about which the data cluster; by ignoring these, we may fail to make use of much of the data we have collected.

2 If the data are arranged in order of magnitudes, the middle measurement is the *median.* The median is that value of the data dividing the observations into two lots containing equal numbers of observations. The median

of the set of data (4, 5, 5, 6, 6, 7, 9) is 6, since three observations lie to the right of 6 and three to the left of 6. If n is an even number, the median is taken to lie midway between the two middle observations; thus in the set of observations (5, 5, 6, 7, 9, 10) the median would be 6.5. Since the median is merely the middle value, it is unaffected by the extreme values: the median would be the same if the largest value were (instead of 10) 20 or 100. Therefore, the median (like the mode) does not take into account the distances between the various observations and the central value. There will be times when this property is advantageous and others when it is clearly undesirable. For example, if we want to discover the selling potential for a product we have to offer, the median income of our potential customers would probably be a useful guide (especially if there is one and only one millionaire in the group). If, on the other hand, we earned $1 on each of four occasions and $21 on a fifth, the median would badly underestimate our earning power.

3   The most generally useful, and most common, measure of central tendency is the arithmetic *mean*, which is the arithmetic average of all the data. The formula for the mean of a *population* may be written compactly:

$$\mu = \frac{\sum_{i=1}^{N} X_i}{N}$$

where Σ, the Greek capital letter sigma, is the sign of summation, meaning "add." The subscript ($i = 1$) for Σ means "begin with the first observation"; the superscript N for Σ means "conclude the summation with the Nth observation." Since in calculating the population mean we always use all the members of the population, the formula is often simplified to

$$\mu = \frac{\sum X}{N} \tag{1·1}$$

We distinguish between the mean of the *population*, which requires an observation from every member of the population, and the mean of a *sample* (taken from a population) which requires an observation only from each member of the sample of size n. The sample mean $\bar{X}$ is found in the same manner as the population mean μ (but note the changes in symbols):

$$\bar{X} = \frac{\sum_{i=1}^{n} X_i}{n} \tag{1·2}$$

Since in most situations it is impractical to examine every member of the population of size N, we usually use a sample of size n. If the sample is properly selected and is representative of the population, we believe that the relative frequency distribution of the sample approximates that of the population and that $\bar{X}$ approximates μ. We say that the statistic $\bar{X}$ is a *point estimate* of the parameter μ.

PROBLEMS

1·24　In what ways are mean, median, and mode alike? In what ways are they different?

1·25　Find the arithmetic mean and the median of the following populations:

(a)　4, 5, 4, 7, 3, 8
(b)　3, 15, 5, 4, -5, -3, -5
(c)　3.6, 7.4, 8.4, 5.2, 6.4
(d)　0.0034, 0.0037, 0.0039, 0.0039,
　　　0.0034, 0.0036

1·26　Using the following data, set up boundaries, cell midpoints, tallies, frequencies, and relative frequencies to show the distribution of the data:

　　　117, 114, 118, 114, 115, 116, 113, 117, 118, 116, 115
　　　115, 117, 113, 115, 113, 115, 118, 116, 116, 119, 113
　　　115, 113, 115, 115, 113, 113, 111, 114, 114, 117, 114
　　　115, 116, 116, 119, 112, 114, 115, 113, 118, 114, 117

1·27　From the data in Prob. 1·26:
(a)　What is the mode?
(b)　What is the median?
(c)　By merely looking at the plot obtained in answer to Prob. 1·26, estimate the value of the mean.
(d)　Compute the mean of the data.

1·28　Using the data as recorded in Table 1·9:
(a)　Calculate the mean of the data.
(b)　Calculate the median of the data.
(c)　Find the mode, and compare its value with that of the mean and median.
(d)　What is the general shape of the distribution of the data?

1·29　Using the data as recorded in Table 1·11:
(a)　Calculate the mean of the data.
(b)　Calculate the median and the mode of the data.
(c)　Compare the values of mean, median, and mode, and note the general shape of the distribution of the data.

1·9　MEASURES OF DISPERSION

1　The *range* of a set of observations is the absolute difference (regardless of mathematical sign) between the largest and smallest numbers in the set. The range of the set of data $(-4, -2, 3, 5, 8, 10)$ is 14.

The value of the range is primarily determined by the presence in the observations of exceptional values, and is totally unaffected by even large numbers of observations occurring in the central area of the distribution. Large sample sizes may contain extreme values, causing the range to be great. Furthermore, the range makes use of only two pieces of data out of all those collected.

The average of a set of ranges from a number of samples (obtained by randomly subdividing a large set of data) is more reliable than a single value of the range from the entire set. We will make use of the average range when we discuss the preparation of control charts.

2   The *average deviation* is the mean of the absolute deviations of the observations from the mean (or median). This measure of dispersion has the advantage of using all the data, and of being relatively easy to compute. For these reasons it is sometimes used for simple applications. It is, however, not generally applicable in more complex situations because of its difficult mathematical properties.

3   The *variance* is a measure of dispersion that we will find to be of exceptional importance. It has the property of giving due weight to every piece of data in terms of the square of the deviation of each observation from the mean.

For a population the variance is given by

$$\sigma^2 = \frac{1}{N} \sum_{i=1}^{N} (X_i - \mu)^2 \qquad (1\cdot3)$$

where σ^2 is the variance (σ is the small Greek letter sigma); the other symbols were identified in connection with the formula for the population mean. From Eq. (1·3) we see that the variance is the average of the squares of the deviations of the X_i values from the population mean μ.

An important property of variance as a measure of dispersion is that of *additivity:* that is, if several different factors contribute to the dispersion in a distribution, the total variance is the sum of the variances due to the individual factors. This rule holds only where the factors are *independent* of each other. The additivity of variances is of use in planning parts assembly, in studying testing methods, and in evaluating processes where the final product is affected by a series of independently operating factors. Additivity of variance plays a part in many statistical methods; one of the more important of these, which we discuss later in this book, is called *analysis of variance.*

If, as is most often the case, we must estimate the variance from a sample, we use

$$s^2 = \frac{1}{n-1} \sum_{i=1}^{n} (X_i - \bar{X})^2 \qquad (1\cdot4)$$

where s^2 is the sample variance. Note the changes from Eq. (1·3): in Eq. (1·4) $\bar{X}$ replaces μ, since $\bar{X}$ symbolizes the sample mean; $n-1$

replaces N. The substitution of $n - 1$ for N corrects for the consistent tendency to underestimate the variance for small sample sizes. Some texts define the sample variance with n as the denominator. Most modern statisticians, and the leading statistical societies, recommend the use of $(n - 1)$.

4 The *standard deviation* is the measure of dispersion we will use most often. It is the square root of the variance. The reason for taking the square root is that thereby we restore the unit of measurement to that of the original data. Thus, if the observations are in inches, the standard deviation is also in inches.

For a population, the standard deviation is

$$\sigma = \sqrt{\frac{\sum_{i=1}^{N} (X_i - \mu)^2}{N}} \qquad (1 \cdot 5)$$

and for a sample the standard deviation is

$$s = \sqrt{\frac{\sum_{i=1}^{n} (X_i - \bar{X})^2}{n - 1}} \qquad (1 \cdot 6)$$

Many statistical procedures are based on the standard deviation. The properties of this measure of dispersion will be discussed in the text which follows.

1·10 ALTERNATIVE METHOD FOR CALCULATING STANDARD DEVIATION

Especially for large sample sizes a more convenient formula for the standard deviation, and one less liable to rounding-off errors (and calculating mistakes), is

$$s = \sqrt{\frac{n \sum X_i^2 - (\sum X_i)^2}{n(n - 1)}} \qquad (1 \cdot 7)$$

Table 1·12

Raw Data: 4, 3, 6, 2, 6, 5, 3

X_i	X_i^2
4	16
3	9
6	36
2	4
6	36
5	25
3	9
$\Sigma X_i = 29$	$\Sigma X_i^2 = 135$

$$s = \sqrt{\frac{7 \times 135 - (29)^2}{7 \times 6}}$$

$$= \sqrt{\frac{945 - 841}{42}}$$

$$= \sqrt{2.48} = 1.57$$

Note that this formula does not require the calculation of the mean, since the only values that enter into the formula are X_i and n. Observe especially that the first term of the numerator specifies the *sum of the squares* of the observations, and that the second term specifies the *square of the sum* of the observations. We illustrate the use of Eq. (1·7) for the data in Table 1·12.

PROBLEMS

1·30 Calculate the standard deviation for the data in Table 1·12 by the use of Eq. (1.6). What conclusions do you draw?

1·31 Calculate the standard deviation from the following sample data:
34 ml, 32 ml, 30 ml, 29 ml, 32 ml, 32 ml, 31 ml

1·32 Calculate the standard deviation of the following sample data (given in inches):
0.29, 0.25, 0.27, 0.26, 0.29

1·33 Calculate the standard deviation of the following sample data: (weights to the nearest ounce) 23, 26, 24, 28, 24

1·11 CODING DATA

Calculations of the mean and standard deviations can be facilitated by the process of *coding*. This process transforms raw data, which may be awkward to work with because they are either large values or small decimal numbers, into more convenient numbers which are easier to handle. At the final stage, a reverse process of uncoding then restores the calculated statistics to appropriate values.

We have a set of raw data: 57, 59, 53, 55. We wish to know the mean and standard deviation. We can code by subtracting 50 from each X value, thus producing the coded numbers 7, 9, 3, 5.

1 We find the mean of the coded values to be 6. To uncode we add 50 to this mean, and find the sample mean to be 56. Note that the coding process has not changed the relationship of the observations to each other but has merely shifted their location.
Coding Rule 1: If, in coding, a constant is subtracted from (or added to) every observation, that constant must be added to (or subtracted from) the coded mean to find the mean of the original data.

2 We find the sample standard deviation of the coded data from Eq. (1·7) to be 2.6. This is also the standard deviation of the original data, since the dispersion of the set of data is independent of the location of the values.
Coding Rule 2: If, in coding, a constant is subtracted from (or added to) every observation, the standard deviation is unchanged; no uncoding is necessary.

Assume that our set of observations was 57,000, 59,000, 53,000, and 55,000. We could code these data by first dividing each of the observations by 1,000. The mean of the coded data would then also be $1/1,000$ of that of the original observations; we would uncode by multiplying the mean of the coded data to find the mean of the original data. In this case, we would have changed the scale of measuring our data; the distribution has been shrunk until we uncode it. Because of this scale change, the need to uncode would now also apply to the calculation of the standard deviation.

Coding Rule 3: If, in coding, each observation is divided by (or multiplied by) a constant, the mean and the standard deviation of the coded data must be multiplied by (or divided by) that constant to find the mean and standard deviation of the original data.

Coding Rule 4: If, in coding, a sequence of operations is applied to the data, then in uncoding, the reverse operations must be used, and in an order *opposite* to that of the coding order.

We wish to find the sample mean and standard deviation of these data: 0.0047, 0.0045, 0.0048, 0.0048, 0.0044, 0.0047. We code the data in two steps: (1) We multiply each observation by 10,000, giving 47, 45, 48, 48, 44, 47; (2) from each of these numbers we subtract 40, giving 7, 5, 8, 8, 4, 7.

For the coded data the mean is 6.5 and the standard deviation 1.64. To uncode, for the mean we reverse the coding operations: by adding 40 we obtain 46.5; then by dividing by 10,000 we obtain 0.00465. To find the uncoded standard deviation no addition is necessary; we therefore divide by 10,000 and obtain 0.000164.

Note that we could have coded the raw data by subtracting in the second step 46 instead of 40. The coded values would then have been 1, -1, 2, -2, and 1. The average of the coded values would have been 0.5. Uncoding would have required the addition of 46, and the complete uncoding process would have given the same mean as before. Similarly, the standard deviation could have been computed by the use of these coded data, and it would have been found more easily than before.

PROBLEMS

1·34 An antibiotic is placed in capsules by an automatic filler. The weight in milligrams of 36 sample capsules was recorded as follows:

44.0, 43.1, 45.2, 44.2, 44.5, 43.9, 44.1, 42.3
45.9, 45.9, 43.7, 43.8, 45.4, 43.9, 46.7, 44.2
47.5, 44.8, 44.1, 44.4, 43.5, 44.6, 43.0, 45.0
43.9, 43.9, 43.8, 44.0, 43.1, 44.2, 43.9, 42.3
43.7, 43.9, 46.7, 44.5

Find the standard deviation of these sample data, using coded data, and then uncode the answer.

1·35   Code the following observations to single digits:
(a) 0.0034, 0.0037, 0.0032, 0.0031
(b) 142.2, 142.5, 142.0, 141.8, 142.3
(c) 16.234, 16.230, 16.229, 16.233, 16.235

1·36   The volumes in milliliters of five randomly selected bottles from an automatic filling operation were recorded as follows:
124.5, 123.9, 124.0, 124.2, 124.4
(a) Code the values to small whole numbers.
(b) Calculate the value of the coded mean.
(c) Calculate the value of the coded standard deviation.
(d) List the values of the uncoded mean and standard deviation.

1·37   Using coded values, find the mean and the standard deviations of the following samples and give answers as uncoded values:
(a) 2.34, 2.33, 2.30, 2.38, 2.35, 2.33,
      2.32, 2.30 in.
(b) 0.45 lb, 0.49 lb, 0.47 lb, 0.39 lb,
      0.41 lb
(c) 123.8 cc, 123.3 cc, 123.5 cc, 123.5 cc,
      123.4 cc, 123.3 cc

1·12   CALCULATION OF MEAN AND STANDARD DEVIATION

We return to the data in Table 1·9 to demonstrate the methods of finding the sample mean and standard deviation for a large number of observations which have been grouped into cells. The sequence of steps should be followed by reference to Table 1·13.

1   Estimate the cell which is likely to contain the mean value. Here we assume that it will lie in the cell having a midpoint of 80. Assign to that cell the value 0 in the column headed d (signifying deviation from the assumed mean). It is not important that the cell we choose actually contain the mean, since the calculating method to be described will correct for any incorrect assumption. Note that in effect we are coding the data by subtracting 80 from every cell.
2   Assign positive or negative deviation values (in terms of cells) to the other cells: positive, if the cell lies above the assumed mean position; negative, if the cell lies below the assumed mean. Now we are coding by dividing by the cell interval. This means that each cell is just 1 unit farther from or closer to the center than the preceding one.
3   Find the algebraic product of the pairs of numbers in the columns headed f and d, thus generating the numbers in the column headed fd. Find the algebraic sum of the numbers in this column. This value is Σfd, which is here -50. If this value were 0, the position of the actual mean would be the midpoint of the cell containing the assumed mean.

Table 1·13 Frequency Distribution of 100 Test Scores, with Assumed Mean at 80

Cell boundaries	Cell midpoint	f	d	fd	fd^2
96.5–99.5	98	1	+6	6	36
93.5–96.5	95	2	+5	10	50
90.5–93.5	92	5	+4	20	80
87.5–90.5	89	6	+3	18	54
84.5–87.5	86	10	+2	20	40
81.5–84.5	83	13	+1	13	13
78.5–81.5	80	17	0	0	0
75.5–78.5	77	14	−1	−14	14
72.5–75.5	74	11	−2	−22	44
69.5–72.5	71	6	−3	−18	54
66.5–69.5	68	5	−4	−20	80
63.5–66.5	65	4	−5	−20	100
60.5–63.5	62	2	−6	−12	72
57.5–60.5	59	2	−7	−14	98
54.5–57.5	56	1	−8	−8	64
51.5–54.5	53	1	−9	−9	81
Total		$\Sigma f = 100$		$\Sigma fd = -50$	$\Sigma fd^2 = 880$

That Σfd is negative means that the actual mean is slightly less than the assumed mean. We calculate the actual mean by finding the required correction to be applied to the assumed mean, thus

$$\bar{X} = A + \frac{i \Sigma fd}{n} \qquad (1\cdot8)$$

where A = midpoint of cell containing assumed mean
$\quad i$ = cell interval used in grouping
Σfd = total of fd column
$\quad n$ = sample size.

Substituting in Eq. (1·8) the data from Table 1·13, we obtain

$$\bar{X} = 80 + \frac{(3)(-50)}{100} = 78.5$$

This method of finding the mean will probably give a value slightly different from that which we would have obtained by working with the raw data. To get exactly the same value would require that each of the sets of observations placed in a cell be grouped symmetrically and similarly about the midpoint of every cell; this is probably not exactly true, hence the slight difference in the two calculated means. The method just described, however, saves much computational time, and the difference in the mean is (especially for large sample sizes) usually insignificant.

4 To find the standard deviation, we need one more column of data: this column is headed fd^2; the numbers it contains are the products of the terms in the f column and the squares of the deviations in the d column. In practice, it is easier to find the fd^2 values by multiplying the terms in the fd column by those in the d column. Note that the signs of these terms are all positive. We find the sum of the terms in this column, and symbolize the sum as Σfd^2.

5 The standard deviation is now found from

$$s = i\sqrt{\frac{n\,\Sigma fd^2 - (\Sigma fd)^2}{n(n-1)}} \qquad (1 \cdot 9)$$

where the symbols are as defined before, with this reminder: Σfd^2 is the sum of the terms in the last column in Table 1·13; $(\Sigma fd)^2$ is the square of the sum of the terms in the fd column, the fifth column in Table 1·13. We substitute in Eq. (1·9):

$$s = 3\sqrt{\frac{(100)(880) - (-50)^2}{(100)(99)}}$$

$$= 3\sqrt{\frac{88{,}000 - 2{,}500}{9{,}900}}$$

$$= 3\sqrt{8.636} = (3)(2.9) \text{ or } 8.7$$

PROBLEMS

1·38 A company made *only* 10 racing engines of a kind, which had the following weights when assembled (weights in pounds):

   1,283, 1,260, 1,277, 1,289, 1,284, 1,278, 1,275
   1,280, 1,278, 1,276

(a) What was the average weight per engine when assembled?
(b) Find the standard deviation of the 10 racing engines when assembled.

1·39 Calculate the standard deviation of the sample data contained in Table 1·5.

1·40 Calculate the sample standard deviation from the data in Table 1·11.

1·13 SIGNIFICANT FIGURES

The process of coding data requires an understanding of the reliability of computed statistics or parameters. We discuss here the concept of "significant figures" as applied to the kinds of calculations involved in statistics.

The reliability of measured numbers may be estimated by counting the digits without regard to the location of the decimal point. The greater the number of significant figures, the more the reliability of the number.

The observation 2.03 has three significant figures. The observation 3.412 has four significant figures and is therefore a better measurement than the preceding. The observation 0.00674 has three significant figures, the zeros merely establishing the position of the decimal point. The observation 6.0023 has five significant figures.

When measured data are multiplied or divided, the following rule usually holds: Retain in the result as many significant figures as in the *least* reliable of the data entering into the computation. Thus, in the result of the calculation 2.34×3.7, the product is written 8.7, retaining only two significant figures because the poorer of the original data contains two significant figures.

When measured data are added or subtracted, the following rule usually holds: Retain in the result only that decimal place (reading from the left) that appears in the worst of the original data. Thus, in the calculation $2.34 + 3.724$, only the hundredths place would be retained in the result, since the poorer of the original data is given only to hundredths.

Modifications of these basic rules are needed in statistical calculations. The following are guides:

1 Carry out the calculation of the uncoded mean to one more significant figure than in the raw data. Thus the mean of the set of observations 0.26, 0.27, 0.25, 0.28, 0.26 would be written 0.264. The averaging process gives a value for the mean that is more reliable than any single observation.

2 Carry out the calculation of the uncoded standard deviation to the same number of decimal places as in the mean. Thus the estimate of the standard deviation of the data in rule 1 above would be written 0.011. The mean enters into the calculation of s; hence the value of s is as well known as is the mean.

3 Carry out the calculation of the variance to twice as many total decimal places as in the standard deviation. Thus the variance of the data in rule 1 above would be written 0.000130. Since the variance is the square of the standard deviation, two decimal places are generated in the variance for every decimal place in the standard deviation.

1·14 CUMULATIVE FREQUENCY DISTRIBUTION

It often happens that we are interested in finding the number of items in a lot which are smaller than, or greater than, a specified value. If machine parts are usable if they are at least 2.000 in. long, we would want to know what fraction of the product would be 2.000 in. or greater in length. This kind of information is not easily obtainable from a conventional plot of frequency distribution; it is best displayed by a plot of cumulative frequency.

To illustrate, we use the data from Table 1·9, and from the original data calculate the following numbers, working from the bottom of the table upward (see Table 1·14):

Table 1·14   Cumulative Distribution of 100 Test Scores on Final Exam in Statistics

Cell midpoint	Absolute frequency	Cumulative frequency	Relative frequency	Cumulative relative frequency
98	1	100	0.01	1.00
95	2	99	0.02	0.99
92	5	97	0.05	0.97
89	6	92	0.06	0.92
86	10	86	0.10	0.86
83	13	76	0.13	0.76
80	17	63	0.17	0.63
77	14	46	0.14	0.46
74	11	32	0.11	0.32
71	6	21	0.06	0.21
68	5	15	0.05	0.15
65	4	10	0.04	0.10
62	2	6	0.02	0.06
59	2	4	0.02	0.04
56	1	2	0.01	0.02
53	1	1	0.01	0.01

1   Cumulative frequency: the successive sums of the numbers in the frequency column, thus: Opposite each cell midpoint we write the sum of all the frequencies for that cell and all cells below. Opposite 65, for example, we write 10; this is the sum of the frequencies 4, 2, 2, 1, and 1, which are opposite 65 and below. The topmost number in this column will be equal to n.

2   Relative frequency: the quotient found by dividing each of the absolute frequency values by the total number of observations n.

3   Cumulative relative frequency: Numbers are generated in a manner similar to that for cumulative frequency, except that we use the numbers in the relative frequency column. The numbers in this column will always lie between 0.0 and 1.0, inclusive.

We now plot a graph, as in Fig. 1·4, plotting cumulative relative frequency against the *upper cell boundary* appropriate for each cell. We use the upper cell boundary for each cumulative relative frequency because we are interested in all the observations included in each cell, and these have values up to and including the upper cell boundary.

For discontinuous data, we connect the points by a broken straight line. For continuous data, we draw a smooth curve approximating the points. In Fig. 1·4, we have assumed that the test scores are in reality continuous in nature.

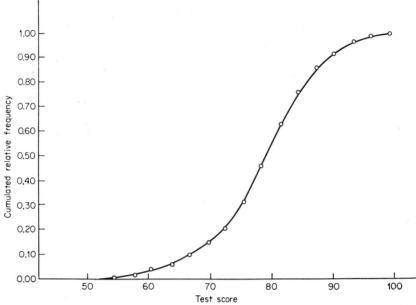

Fig. 1·4 Cumulative frequency distribution of examination scores (data from Tables 1·6 and 1·8).

From a cumulative-frequency-distribution plot, we can readily find

1   The median: the observation corresponding to the 0.50 value of the ordinate, since the median is defined as that value which divided the observations into two equal lots. Here, the median is at approximately 79.
2   The fraction of observations falling below any chosen value. For example, here 0.10 of the scores lie at 66.5 and below.
3   By difference, the fraction of observations lying above any chosen value. For instance, here 0.10 of the scores lie at 89.5 and above, since 0.90 lie at 89.5 and below.
4   Conversely, the value of the observation that will include any desired fraction of the data. For example, if we desire to assign an F grade to 15 percent of the class, we would assign such a grade to all scores of 69.5 and below.

PROBLEMS

1·41   Using the sample data in Table 1·11:
(a) Add a column of relative frequencies.
(b) Add a column of cumulative relative frequencies.
(c) Construct a graph of the cumulative relative frequencies.

1·42 The final grades in a chemistry course in college A are recorded as follows:

       68, 84, 75, 82, 68, 90, 62, 88, 76, 93, 73, 79, 88
       73, 60, 93, 71, 59, 85, 75, 61, 65, 75, 87, 74, 62
       86, 67, 73, 81, 72, 63, 76, 75, 85, 77, 96, 78, 89

(a) Construct a frequency distribution of the grades.
(b) Plot a cumulative-relative-frequency graph of the data.
(c) From the graph in answer to (b), determine the median score.
(d) From the graph, determine the value of the grade that separated the top 30 percent of the class from the rest.

1·15 RELATIONSHIP BETWEEN STATISTICS AND PARAMETERS

If the characteristics of a population (its mean and standard deviation) are known, we are sometimes interested in predicting the characteristics of a sample from that population. More often in practice we are interested in the estimation of the characteristics of the population from those found for a sample. In either case, the relationship between the parameters of the population and the statistics of the sample is of importance.

Consider the raw data in Table 1·15 and the frequency histogram in Fig. 1·5 based on these data. Although the data comprise only 201 values, and therefore represent a finite population, let us suppose that they exemplify a much larger, infinite population. The mean μ of the population is 0.00; the standard deviation is 3.47. We chose two samples of $n = 42$ from this population, making the choice at random and with replacement; i.e., after each item was chosen, it was replaced in the lot and had an equal chance of being selected again. This process could have been repeated indefinitely; hence the population was by this method of sampling made infinite in size.

The raw data and the frequency histograms are given for these two samples

Table 1·15 Population Data Set

4	−6	2	1	2	2	−3	0	−1	7	−4	4	2	−1
−2	−1	10	5	3	−5	−4	1	−4	8	6	1	−1	0
−1	1	−1	−3	−4	−2	−4	−5	−2	−6	2	−10	5	−5
−1	1	6	−3	9	2	−2	6	6	0	0	−5	5	3
0	−7	−3	0	−9	3	−5	0	−2	2	6	−1	−4	−1
−2	5	4	2	1	0	0	−2	4	−1	2	0	0	3
5	−6	−3	4	−3	3	−3	3	−1	−2	−1	3	3	0
0	−6	−2	1	4	−3	−1	−2	1	1	1	3	4	1
−2	2	7	1	−2	3	2	0	−4	1	−4	−1	−2	−1
1	0	−5	5	−3	2	3	3	4	−3	4	−3	2	−3
−3	2	0	−3	0	2	1	2	0	3	−2	1	−1	1
4	−3	4	0	0	−1	7	−7	−1	2	1	−1	−2	−5
1	0	1	1	−7	−2	−5	−6	−1	−8	−4	1	−4	5
2	−3	−2	5	−1	−1	0	−2	2	2	0	−4	3	4
3	3	−2	−2	−4									

Size	Frequency	f
10	◆	1
9	◆	1
8	◆	1
7	◆◆◆	3
6	◆◆◆◆◆	5
5	◆◆◆◆◆◆◆◆	8
4	◆◆◆◆◆◆◆◆◆◆◆◆	12
3	◆◆◆◆◆◆◆◆◆◆◆◆◆◆◆◆	16
2	◆◆◆◆◆◆◆◆◆◆◆◆◆◆◆◆◆◆◆◆	20
1	◆◆◆◆◆◆◆◆◆◆◆◆◆◆◆◆◆◆◆◆◆◆	22
0	◆◆◆◆◆◆◆◆◆◆◆◆◆◆◆◆◆◆◆◆◆◆◆	23
−1	◆◆◆◆◆◆◆◆◆◆◆◆◆◆◆◆◆◆◆◆◆◆	22
−2	◆◆◆◆◆◆◆◆◆◆◆◆◆◆◆◆◆◆◆◆	20
−3	◆◆◆◆◆◆◆◆◆◆◆◆◆◆◆◆	16
−4	◆◆◆◆◆◆◆◆◆◆◆◆	12
−5	◆◆◆◆◆◆◆◆	8
−6	◆◆◆◆◆	5
−7	◆◆◆	3
−8	◆	1
−9	◆	1
−10	◆	1

Fig. 1·5   Frequency histogram (data in Table 1·15).

(see Table 1·16 and Fig. 1·6). For sample 1, the average was 0.02 and the standard deviation 2.9; for sample 2, the average was −0.69 and the standard deviation 3.3.

We see from these results, which are typical of the relationship between sample and population,

1 The mean and standard deviation of a sample may be close to, but not exactly the same as, the parameter of the population from which the sample was taken.
2 Samples differ in their estimates.

Table 1·16

(a) Sample 1, from Population in Table 1·15

1	0	−1	1	−5	−1	0	1	2	1	3	6	−1	−2
−1	−2	6	0	1	−1	−5	0	1	0	5	3	−4	5
1	4	−2	−5	4	−3	1	−3	−3	−1	−2	−1	0	−2

(b) Sample 2, from Population in Table 1·15

−3	−3	−3	3	−4	−5	−3	−7	2	0	2	−5	−3	−1
−2	−2	1	−2	0	−6	2	−1	0	−2	−6	2	6	1
7	−7	1	3	2	−1	−2	1	0	−2	6	−1	2	1

If we were to take all possible samples of size $n = 42$ from this finite population, we would obtain a set of means and a set of standard deviations, each of which could be plotted so as to display the frequency distribution of the sample means and of the sample standard deviations.

Size	Frequency		Size	Frequency	
7		0	7	◆	1
6	◆◆	2	6	◆◆	2
5	◆◆	2	5		0
4	◆◆	2	4		0
3	◆◆	2	3	◆◆	2
2	◆	1	2	◆◆◆◆◆◆	6
1	◆◆◆◆◆◆◆◆	8	1	◆◆◆◆◆	5
0	◆◆◆◆◆◆	6	0	◆◆◆◆	4
−1	◆◆◆◆◆◆◆	7	−1	◆◆◆◆	4
−2	◆◆◆◆◆	5	−2	◆◆◆◆◆◆	6
−3	◆◆◆	3	−3	◆◆◆◆◆	5
−4	◆	1	−4	◆	1
−5	◆◆◆	3	−5	◆◆	2
−6		0	−6	◆◆	2
			−7	◆◆	2

Fig. 1·6   Frequency distributions of sample 1 and sample 2 from Table 1·16.

Now, for *all* possible samples from an *infinite* population the following rules hold:

1 The average of the sample averages is equal to the population mean.
2 The standard deviation of the sample averages is equal to the standard deviation of the population divided by the square root of the sample size *n*.

These statements are a form of the Central Limit theorem: *If a population has a finite variance and a finite mean, as the sample size increases, the distribution for the sample mean approaches a normal distribution with mean μ and variance σ^2/n.*

These rules are illustrated by the plots in Fig. 1·7, where we show first the distribution of individuals in the population, next the distribution of averages of samples from the population, and finally the distribution of standard deviations of the samples.

We intend, in addition, to indicate that a frequency distribution, and a frequency plot, are applicable to any statistic, as well as to individual observations. We can calculate, for any set of statistics, a mean value and a standard deviation. The standard deviation of a statistic has a special significance: we call it the *standard error* of the statistic. We may find a standard

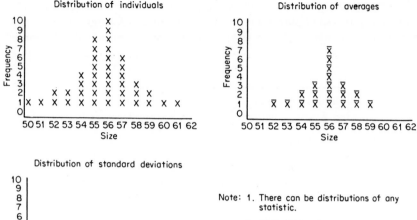

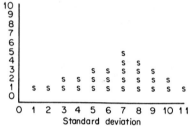

Note: 1. There can be distributions of any statistic.

2. Each distribution has a mean value.

3. Each distribution has a standard error.

$\bar{X}, \; \bar{\bar{X}}, \; s_X, \; s_{\bar{X}}, \; s_S, \; \bar{X}_S$

Fig. 1·7   Distribution of individuals, averages, and sample standard deviations from the same population.

deviation of a set of sample means; this value is the standard error of the mean. We may calculate a standard deviation of a set of sample standard deviations; this is the standard error of the standard deviations.

It is essential to remember that all statistics measured from samples are just estimates of the "true" population value, and as such they too have frequency distributions associated with them.

PROBLEMS

1·43   Consider the population consisting of the three values 2, 4, and 6.

(a) List the averages of the 9 possible samples with replacement of size $n = 2$, taken from the population.

(b) Calculate the average of the 9 sample averages and compare this value with the average of the population.

(c) Construct a frequency distribution of the sample averages.

(d) Calculate the standard deviation of the distribution of the sample averages, and compare this value with the value of the population standard deviation.

1·44   Using the population which consists of the values 8, 10, and 12:

(a) Calculate the values of the 27 possible samples of size $n = 3$ that can be taken from the population, and compute their means and standard deviations.

(b) Calculate the population mean and population standard deviation.

(c) Compute the average of the 27 sample averages, and construct the frequency distribution of the sample averages.

(d) What is the value of the standard deviation of the distribution of sample means?

(e) Construct a frequency distribution of the sample standard deviations and calculate the average value.

(f) What is the value of the standard error of the sample standard deviations?

QUESTIONS

1·1   In the study of variation of a repetitive process, how does the frequency distribution of individual samples from the population help to gain insight into the problem?

1·2   What are the parameters that are used to describe a population or universe?

1·3   What are the statistics that are used to describe the distribution of a sample?

1·4   What is the relationship between sample statistics and population parameters?

1·5   What is the difference between continuous data and discrete data? List examples of each type of data.

1·6   Why is it sometimes necessary to use cells for the tabulation of frequency of occurrence of observations?

1·7   What is the major difference between the frequency histogram and the frequency polygon?

1·8   When is it necessary to use grouped data in making a frequency distribution?

1·9   How are the cell limits arrived at when data are grouped, when making a frequency distribution?

1·10   In a grouped frequency histogram:

(a) Why should the cell limits always be one digit more precise than the raw data?

(b) When values are placed in the same cell interval, what unit of measure are they all considered as having?

(c) How many cell intervals should there be to give a good picture of the variation of the data, but still not require a great amount of work?

1·11   Why does one consider making a frequency distribution in connection with a repetitive process?

1·12   What are the measures of central tendency? Why might one be used and not the others?

1·13   How might the frequency distribution be used to compare two or more sources of the same type of data?

1·14   What is a measure of dispersion, and what does it measure?

1·15   In the calculation of the standard deviation from grouped data, why does not the cell midpoint or cell limit enter into the calculations?

1·16   When working with grouped data:

(a) Why should the larger values appear at the top of the table?

(b) In the calculation of the median, what use is made of the cumulative frequency?

(c) Why doesn't it matter which cell is selected as the assumed mean?

1·17   In the coding of raw data, to calculate the mean, what must be done to uncode the data after the calculation has taken place?

1·18   In the coding of raw data, to calculate the standard deviation, what must be done to uncode the data after the calculation is completed?

1·19   How can we use the cumulative relative frequency of grouped data to gain insight into the population from which the data came?

1·20   What is skewness, and how might we test for it?

REFERENCES

16, 22, 26, 28, 31, 33, 35, 53, 62, 67, 73, 82, 109, 120, 123, 125, 126

2 The Normal Curve

More than two centuries ago it was observed that errors of measurement seemed to fall into a definite pattern. When the length of an object was repeatedly measured, the results for many observations were found to be similar, even though any single measurement appeared to be haphazard. When recorded in large numbers, the same type of frequency distribution appeared for many kinds of measuring devices. In a different context, it was observed that the random variations in the number of heads appearing on many throws of n coins corresponded to the terms of the binomial expansion (see Chap. 20) of the expression $(\frac{1}{2} + \frac{1}{2})^n$. More important, as n increased the distribution of number of heads was seen to approach the same form as that of a series of repeated measurements.

A Belgian statistician, Quetelet (1797–1874), is thought to be the first person to have referred to these distributions as the *normal* curve of error. The term *normal* is used to identify the distribution, not to describe it. We find normal distributions in a great variety of situations, among them the following: intelligence (as measured by "intelligence tests"); dimensions of parts produced by automatic machines; the results of strength-of-materials tests on a large number of steel samples of the same batch. In fact, the failure of a statistician to find a normal distribution in a study of a process is often a clue suggesting that some factor is exerting an unusual effect on the process. This discovery may provide a reason for a further study of the process.

2·1 CHARACTERISTICS OF THE NORMAL DISTRIBUTION

The normal distribution is represented by a smooth, symmetrical, bell-shaped curve, similar to that shown in Fig. 2·1. The tails of the curve extend

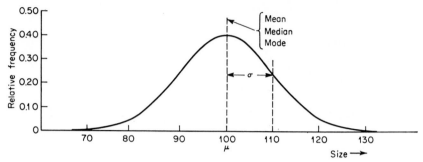

Fig. 2·1   A normal distribution, with mean = 100 and standard deviation = 10.

indefinitely in both directions from the center; they approach, but never reach, the horizontal axis. The mean, the median, and the mode (defined in Chap. 1) coincide.

The normal curve is a mathematical abstraction, a good example of a mathematical model. Distributions which are (or appear to be) approximately normal are, as we said above, commonly encountered in experimental and industrial work. We never, however, find real data which correspond exactly to the mathematical model. Fortunately, moderate departures from normality do not seriously affect the reliability of many of the decisions which may be based on the assumption of normality. It is important, however, to appreciate that the normal distribution is a limiting one—one which is approached only as the sample size, or number of trials, becomes very large.

The normal curve is of particular importance in statistics for the following reasons, among others:

1   As we pointed out above, normal (or near normal) distributions are frequently encountered.
2   The mathematical equation for the normal curve is known. The knowledge of this equation (which we discuss shortly) has led to the development of an extensive group of simple aids to calculation.

2·2 INDICATIONS OF LACK OF NORMALITY

For normal distributions, the form of the frequency histogram or polygon approaches (if the sample size is large) that shown in Fig. 2·1. Other patterns suggest that the underlying distribution is not normal. Lack of normality is suggested by multimodal patterns and by rectangular distributions, among others. Asymmetrical plots are sometimes seen in which one tail of the distribution extends farther along the X axis than does the other tail. Such nonnormal distributions are called *skewed*. The skewness is said to be to the right or the left, according to the position of the extended tail. In Fig. 2·2,

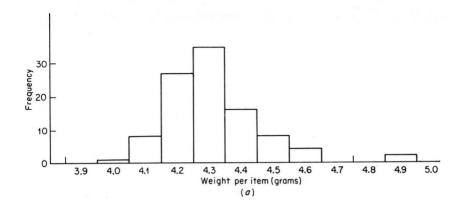

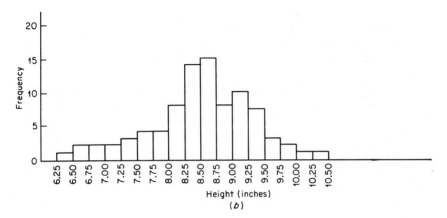

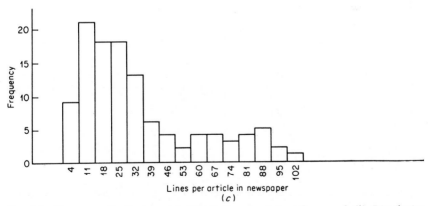

Fig. 2·2 Frequency distributions. (a) Distribution approximately normal. (b) Distribution slightly skewed to the left, perhaps normal. (c) Distribution skewed to the right, probably not normal.

graph (*a*) appears to indicate an approximately normal distribution; (*b*) is slightly skewed to the left, and (*c*) is strongly skewed to the right.

Cumulative-frequency graphs of normal distributions, when they are plotted on conventional graph paper using rectangular coordinates, give S-shaped curves, as shown in Chap. 1. When the same data are plotted on special probability graph paper, straight lines result. A significant departure from straightness is an indication that the distribution is not normal. Probability graph paper has a conventional horizontal axis with linear spaces; on this axis we plot the *X* values. The vertical axis is nonlinear: the scale is expanded at both ends and compressed in the center. The spaces on this scale are inversely related to the percentage of observations we expect to find in any specified interval of *X* values for a normal distribution; thus the use of such a scale straightens out the plot if the distribution is normal. To make such a plot is a simple way of detecting departures from normality. It is difficult to interpret, however, a slight difference from a straight line.

There are specific statistical methods for making more precise tests to see whether or not a distribution can safely be considered normal. These will be discussed in Chap. 7.

We consider the following example: A filling machine packages a mixture of powdered chemicals; 150 packages were taken at random from the production line, and each package was analyzed for the amount of one ingredient to test for filling uniformity.

The frequency plot is shown in Fig. 2·3; superimposed on the histogram is a normal curve approximating the sample distribution. The histogram is essentially symmetrical, nearly bell-shaped, and neither multimodal nor obviously skewed. The mean $\bar{X}$ of the sample is 15.75; the mode is nearly

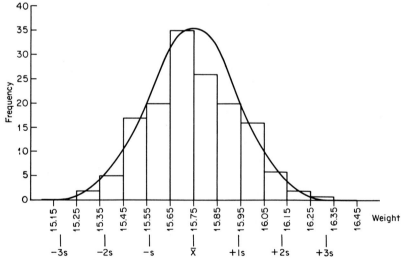

Fig. 2·3   Frequency distribution of weights of 150 samples of powder; X = weight of Na_2SO_3.

the same. In Fig. 2·4 we show the cumulative plot on probability paper. The graph is nearly straight except for the point at $X = 16.35$ indicating an abnormally large number of observations of this size. On the whole, these results indicate that the data can reasonably be treated on the assumption that they come from a normal population.

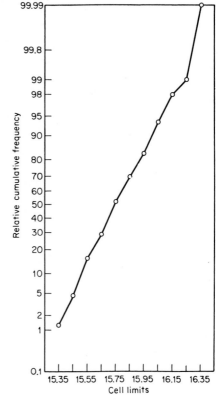

Fig. 2·4 Frequency distribution plotted on probability paper—near normal distribution.

Note that we have not "proved" that the data come from a normal population. The situation is described thus by Schlaifer:[1] "If a normal curve 'fitted' to the available historical data seems on inspection to express the decision maker's judgment concerning the probabilities about as well as any other curve he could fit, then he can rationally compute costs and make decisions on the basis of this normal distribution."

PROBLEMS

2·1   Toss 15 pennies in a random fashion by shaking them all well in your hands or a container and then allowing them to fall on a soft, flat surface such as a bed or rug.

[1] "Probability and Statistics for Business Decisions," p. 295, McGraw-Hill Book Company, New York, 1959.

(a) Toss the 15 pennies, all at the same time, a total of 25 times and record the number of heads that appear face up on the coins.

(b) Toss the pennies an additional 25 times and again record the results of the heads in a frequency distribution.

(c) Compare the results of the two frequency distributions in answer to parts (a) and (b) above, and then combine both samples into the same frequency distribution and again note the shape of the distribution.

(d) After tossing the pennies 100 times, discuss the shape the frequency distribution has assumed, and how the overall shape changed as the total number of tosses increased.

2·2　What caused variations in the number of heads that appeared in answer to Prob. 2·1?

2·3　What type of frequency distributions do the following situations present (gather the data to decide):

(a) The number of "e" letters used per line of type on a single page of this text?

(b) The weights of 50 randomly selected students, all of the same sex, correct to the nearest pound?

(c) The total values of 3 dice when thrown 100 times?

2·4　If you were to ask a large number of your friends to "guess" your age, correct to the nearest month, and you made a frequency histogram of the results, what shape do you believe the distribution would have? Why?

(a) List five populations for which the distributions of individuals taken at random would probably take on the general shape of a normal curve.

(b) List five populations for which the frequency distribution of individuals, taken at random, would not be expected to approach the normal curve, regardless of the sample size.

2·3 FORMULA FOR THE NORMAL CURVE

We said above that the mathematical equation for the normal curve is known. In one form it is

$$y = \frac{1}{\sigma\sqrt{2\pi}}\, e^{-\frac{1}{2}[(X-\mu)/\sigma]^2} \tag{2·1}$$

The symbols used in this equation are familiar; their relationship deserves comment. The factor $1/\sigma\sqrt{2\pi}$ has the effect of causing the area under the entire normal curve to total unity. The fraction in brackets in the exponent of e (which is the base of the natural system of logarithms) has the effect of expressing the separation of each X value from the mean in terms of the standard deviation of the population. The use of this exponent reduces all normal distributions to the *standard form*. Since e and π are constants, the y

value for a particular value of X is dependent only on the value of the mean μ and the value of the standard deviation σ.

2·4 STANDARD FORM OF THE NORMAL CURVE

Equation (2·1) shows that the two parameters σ and μ define a specific normal distribution. Since these two parameters may vary independently, there exists an infinite number of different normal distributions. We can simplify matters considerably if we change Eq. (2·1) by the use of this definition:

$$t = \frac{X - \mu}{\sigma} \tag{2·2}$$

to make the equation for the normal curve now read

$$y = \frac{1}{\sqrt{2\pi}} e^{-\frac{1}{2}t^2} \tag{2·3}$$

By this change we have coded the data thus:

1 We have subtracted the mean μ from each observation, and now are concerned only with deviations from the mean.
2 When we divide each deviation by σ, we express the departures of the observations in units of standard deviation or standard *score*. We end up with a single standard normal distribution, relocated to have a center at 0, and scaled to have a standard deviation of 1. Thus all normal distributions can be reduced to the same form.

(In some references, the symbol Z or u is used rather than the symbol t which we employ. In these references, the symbol t is reserved for another distribution called Student's t distribution, which we will discuss later. Since the normal distribution can be considered a limiting case for Student's distribution, we will use an appropriate subscript for t to distinguish the normal from the Student's distribution when we reach that point.)

Consider a normally distributed set of data with a mean of 135 and a standard deviation of 5, as plotted in Fig. 2·5a. If we subtract the mean from each X value, we shift the mean to 0, as shown in Fig. 2·5c. If now we divide each difference $X - \mu$ by the standard deviation 5, we specify the deviation of the observations from the mean in t units, or in units of standard deviation. A value of 145 from this population will have a deviation of $145 - 135$, or 10 from the mean, and a t value of $10/5$ or $+2$. Thus, in the standard form, the observation 145 would be plotted at $t = 2$ units to the right of the mean, which is now at 0. In the same manner, a value of 120 from this population would plot at the position $t = -3$.

A different normal population, as shown in Fig. 2·5b, has a mean of 350 and a standard deviation of 50. When converted to the standard form this population would plot as in Fig. 2·5c, just as the previous population did.

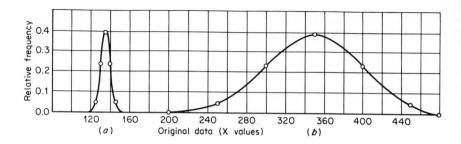

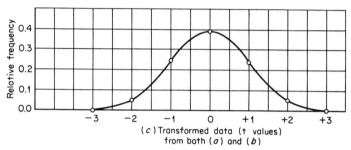

Fig. 2·5 *Normal distributions in original form and changed to standard form. (a) Original data. (b) Original data. (c) Transformed data.*

A value of 450 from the second population would plot at $t = +2$ with reference to the mean of this population.

In standard form, every normal distribution plots as a curve having these characteristics:

1 The curve is symmetrical about the ordinate having a t value of 0.
2 The curve has its peak (mean, median, and mode) at the point $(0, 1/\sqrt{2\pi})$, approximately at $(0, 0.40)$.
3 The curve has points of inflection at $t = -1$ and $t = +1$. The curve is concave downward between $t = -1$ and $t = +1$, and is everywhere else concave upward.
4 The area under the entire curve is unity; thus the whole population is represented by 1 or by 100 percent.

PROBLEMS

2·5 Find the standard score (t value) for each of the following:
(a) Mean of 50, standard deviation of 10, score of 70
(b) Mean of 50, standard deviation of 5, score of 70
(c) Mean of 50, standard deviation of 5, score of 45
(d) Mean of 75, standard deviation of 5, score of 70
(e) Mean of 75, standard deviation of 5, score of 85

2·6 The same test in statistics was given to two different groups of students. The mean score in group 1 was 80, with a standard deviation of 5; the mean score in group 2 was 75 with a standard deviation of 10.
  (a) Should a score of 85 in both groups be given the same grade?
  (b) If the top score in both groups was 95, which student with the 95 had the relatively higher standing?
  (c) Convert a score of 65 from both groups into standard scores and indicate which of the two scores is the poorer of the two.

2·7 If all scores of an examination were turned into standard scores, and then plotted against frequency, what advantages would there be in the assigning of the letter grades?

2·5 AREA UNDER THE NORMAL CURVE

Most calculations involving normal distributions are performed with the aid of tables which enable us to find what fraction of the population may be expected to fall between any two *t* values we desire. Such calculations are useful in solving problems like these:

1 What percentage of items from a production line can be expected to meet certain customer specifications?
2 How often will an adjustment need to be made in a process?
3 How large a sample must be taken to detect a specified change in a process?

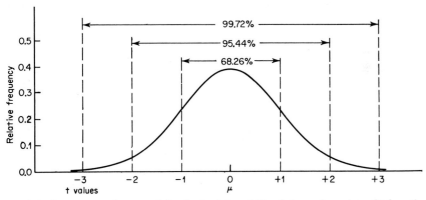

Fig. 2·6 Percentages of a normal distribution lying within whole-number values of t from the mean.

We show first in Fig. 2·6 the percentages of a normal distribution falling within whole-number values of *t* about the mean:

1 Because of the clustering of *X* values near the mean, we find that about 68 percent of all the members of the population are included within the interval bounded by $t = -1$ and $t = +1$ (that is, between $\pm 1\sigma$).

THE NORMAL CURVE   45

2 About 95 percent of the members of the population will lie within the interval bounded by $t = -2$ and $t = +2$ ($\pm 2\sigma$).

3 About 99.7 percent of the members of the population will lie within the interval bounded by $t = -3$ and $t = +3$ ($\pm 3\sigma$).

4 Only about 0.3 percent of the members of the population lie in the two tails beyond $t = -3$ to the left and $t = +3$ to the right.

The areas under the normal curve corresponding to integral values of t are very convenient to work with; they are frequently used, especially in connection with control charts.

The percentages given above for the specified intervals of t are relative frequencies. For individual observations drawn from a normal population, these relative frequencies are also probabilities of occurrence; herein lies an additional use of the percentages. If 68 percent of the members of a normal population lie within ± 1 sigma, the probability that a single observation (drawn from that population) will lie within this interval is 0.68. Because the distribution is symmetrical, the percentage of the population lying between the mean and t is about 34 percent; therefore, the probability of a single observation from a normal population lying within the interval from μ to ($\mu + \sigma$) is about 0.34. Similar probabilities apply to any region of the curve. Note especially that the probability of observing (from a normal population) a value outside the $\pm$ three-sigma interval is only 0.003; furthermore, the probability of observing a value lying this far from the mean *and* in the positive direction from the mean is only half this small probability.

A certain type of photographic lamp has an average life to failure of 30 hr with a standard deviation of 2 hr, a large number of tests indicating a normal distribution. The probability that a single such lamp will last between 28 and 32 hr before failure is 0.68, since these two values are respectively at $t = -1$ and $t = +1$ from the mean.

The probability that one such lamp will last over 30 hr is 0.50, since half the population will lie to the right of the mean. The probability that one such lamp will last over 34 hr is found as follows: The observation 34 has a t value of $+2$; the whole two-sigma interval is about 0.95 in area; half of this (0.475) lies on the high side of the mean; beyond this lies an area of $0.50 - 0.475$, or about 0.025. Thus, approximately 0.025 of the population lies beyond the specified value, and the probability of such an observation is also 0.0227.

PROBLEMS

2·8 The mean of a normally distributed set of observations is 100 with a standard deviation of 10.

(a) What percentage of the values is larger than 110?

(b) What percentage is larger than 120?

(c) What percentage is less than 90?

(d) What percentage is less than 80?

2·9 The average I.Q. of an American youth is said to be 100 with a standard deviation of 10 units.

(a) What is the probability that a single person, selected at random, has an I.Q. of over 110? over 120?

(b) What values of I.Q. would be considered as being average, if the middle 68 percent of the values are said to be normal?

2·10 The mean weight of 500 male students at a certain college was found to be 150 lb with a standard deviation of 15 lb.

(a) Assuming that the weights are normally distributed, how many students weigh more than 165 lb?

(b) How many students weigh less than 120 lb?

(c) If a student is selected at random from the 500, what is the probability that he will weigh more than 150 lb? more than 165 lb?

2·6 TABLES OF NORMAL PROBABILITIES

So far we have dealt with the area under the normal curve only for whole-number values of t. Tables, however, are available which supply similar data for fractional values of t. Table A·1 is such a table. For illustrative purposes we present in Table 2·1 values for the areas under the curve corresponding to t values in tenths. Note particularly:

1 The data in the table represent areas extending from $-\infty$ to the t value given. Thus the data are really derived from the cumulative graph of the normal curve.

2 Only positive values of t are specified. The symmetry of the distribution permits us to calculate areas for negative t values, as we shall illustrate.

The shaded area in Fig. 2·7 shows the area under the curve represented by the table value for $t = 0.5$.

Other sources of data for areas under the normal curve may present the information in a different manner. Sometimes tables give the areas from the mean of the distribution to the desired t value; such table areas are 0.5 less than those in Table 2·1. In other sources, the boundaries are from the given table value to $+\infty$; such values are equal to 1 minus those in Table 2·1. The same results are obtained by the use of any such table.

A water heater requires on the average 30.0 min to heat 40 gal of water to a specified temperature. If the heating times are normally distributed with a standard deviation of 0.5 min, what percentage of the heating times are more than 31.0 min?

We sketch in Fig. 2·8 a normal curve of distribution, with the mean and the point in question (31.0 min) indicated. Since we are interested in the percentage of times *more than* 31.0 min, we would like to know the shaded

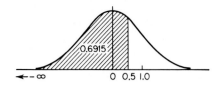

Fig. 2·7 Area under the normal curve for t = 0.5.

Fig. 2·8 Normal curve (mean = 30.0, σ = 0.5) showing area beyond X = 31.0.

area under the curve. We first calculate from Eq. (2·2) the t value for the desired point:

$$t = \frac{X - \mu}{\sigma} = \frac{31.0 - 30.0}{0.5} = \frac{1.0}{0.5} = +2.0$$

Referring to Table 2·1, we find opposite the t value of 2.0 the area 0.9772. We recall that these table values give areas under the curve from $-\infty$ to the specified t value; we, however, are interested in the remaining area under the curve. We therefore subtract from unity the table value; the shaded area under the curve is then 1.0000–0.9772, or 0.0228. Expressed as a percentage, we now know that about 2.28 percent of the heating times are over 31.0 min.

Table 2·1   Area under the Normal Curve from $-\infty$ to t

t	Area from $-\infty$ to t	t	Area from $-\infty$ to t
0.0	0.5000	2.0	0.9772
0.1	0.5398	2.1	0.9821
0.2	0.5793	2.2	0.9861
0.3	0.6179	2.3	0.9893
0.4	0.6554	2.4	0.9918
0.5	0.6915	2.5	0.9938
0.6	0.7257	2.6	0.9953
0.7	0.7580	2.7	0.9965
0.8	0.7881	2.8	0.9974
0.9	0.8159	2.9	0.9981
1.0	0.8413	3.0	0.9987
1.1	0.8643	3.1	0.9990
1.2	0.8849	3.2	0.9993
1.3	0.9032	3.3	0.9995
1.4	0.9192	3.4	0.9997
1.5	0.9332	3.5	0.9998
1.6	0.9452	3.6	0.9998
1.7	0.9554	3.7	0.9999
1.8	0.9641	3.8	0.9999
1.9	0.9713	3.9	1.0000
		4.0	1.0000

Had we been interested in knowing the percentage of the heating times which were *less than* 31.0 min, we would have used the table value itself; the required area would then have been the shaded area indicated in Fig. 2·9.

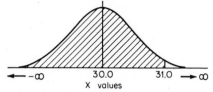

Fig. 2·9   Normal curve (mean = 30.0, σ = 0.5) showing area below X = 31.0.

In solving problems such as these, making a sketch of the curve with the desired area marked will aid in understanding the nature of the situation and the required solution.

A chemical process gives, on the average, a percent yield of 84.0; the standard deviation is 1.6. If the yield measurements indicate a normal distribution, above what value will the yield fall 10 percent of the time, that is, 10 times out of 100? A sketch of the situation (Fig. 2·10a) shows the

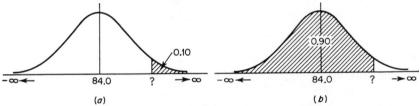

Fig. 2·10   Normal curve (mean = 84.0, σ = 1.6).   (a) Showing X value above which lies 10 percent of the area. (b) Showing that 90 percent of the area lies below the X value.

nature of the problem: we are interested in that value of X which will place 0.10 of the area under the curve into the right-hand tail of the distribution. By subtraction from unity, we know that the area under the curve from $-\infty$ to the desired X value must contain an area of 0.90. Figure 2·10b now shows as the shaded area the region which must contain 0.90 of the whole distribution.

Again we use Eq. (2·2), but with a slightly different substitution: We first find the required t value by locating in the body of Table 2·1 the value nearest 0.90; opposite this we find a t value of approximately 1.3. We now substitute known values in the equation, thus:

$$t = \frac{X - \mu}{\sigma} \qquad 1.3 = \frac{X - 84.0}{1.6}$$

Solving for X gives approximately 86.1; thus we know that 90 percent of the time the yield will fall below the value 86.1, and 10 percent of the time the yield will fall above 86.1, which is the solution to the problem.

2·11 If a process is known to have a mean of 80.0 and a standard deviation of 5.0, and the distribution of the individuals from the process is normal:
- (a) What is the probability that an individual taken at random from the process will measure more than 87.0?
- (b) What percentage of the individuals made by the process will be less than 73.0?

2·12 Consider the 500 male students whose weights are normally distributed with a mean weight of 150 lb and a standard deviation of 15 lb.
- (a) What percentage of the students weigh more than 175 lb?
- (b) What is the probability that a randomly selected male student weighs less than 145 lb?
- (c) Thirty-five percent of the male students weigh more than what value?

2·13 The mean weight of large grapefruit is 14.0 oz, with a standard deviation of 1.5 oz, as reported by a local merchant. Assume that they are normally distributed.
- (a) What percentage of the grapefruit weigh more than 16 oz?
- (b) What percentage of the grapefruit are less than 10 oz?
- (c) Above what weight does the largest 10 percent of the grapefruit fall?

2·7 TWO-TAILED AREAS

The preceding examples concerned only one tail of the normal distribution; we therefore used only one boundary to separate the shaded area of the curve from the remainder. Some problems, however, require the use of two boundaries, and then two t values are involved. The procedure for such problems is illustrated by the following example.

A car was repeatedly driven over the same route; the average gas mileage was found to be 14.5 miles per gallon. The set of data indicated a normal distribution with a standard deviation of 0.8 mile per gallon. We wish to know between what two values of mileage the middle 60 percent of the data will fall.

The problem is sketched in Fig. 2·11. We require two values of X, one being labeled X_1 and the other X_2. The shaded area is to include 0.60 of the total population; since this occupies the middle of the distribution, 0.30 must lie on either side of the mean. This leaves 0.20 for each of the tails of the distribution. Thus, for X_1 the t value is that giving an area of 0.20 (for the left-hand tail of the distribution); for X_2 the t value is that giving an area

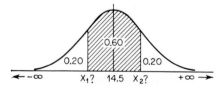

Fig. 2·11 Normal curve (mean = 14.5, $\sigma = 0.8$) showing the two X values between which 60 percent of the area lies.

of 0.80 (the whole population minus the right-hand tail area). We find these t values in Table A·1: First, for X_2 directly, the value is 0.84; then by symmetry the t value for X_1 must be -0.84. We substitute each of these values in Eq. (2·2):

$$-0.84 = \frac{X_1 - 14.5}{0.8} \qquad +0.84 = \frac{X_2 - 14.5}{0.8}$$

From these equations we find that X_1 is approximately 13.8 and that X_2 is approximately 15.2. Therefore, 60 percent of the time we would expect the gas mileages to lie between 13.8 and 15.2 miles per gallon.

The preceding example illustrates a technique that is often used in industrial operations when it is necessary to find what percentage of a product conforms to specifications as given by an engineering department or by a customer. For this method to apply to a process, data must have been collected for a long enough period to show that the observations on the process form essentially a normal population, and the mean and the standard deviation of the data must be found. (Should the distribution not be normal, other techniques would have to be used.) By calculating the plus and minus three-sigma (or plus and minus two-sigma) interval of the data, we can compare this with the specification limits. The interval between plus and minus three sigma, for example, will contain about 99.7 percent of the items made by the process. If the three-sigma interval is included within the specification limits, we know that practically all the units made by the process will conform to the requirements. Also, by comparing the sample mean with the target value we can see whether or not the units are on the average conforming to what is desired.

If, on the other hand, one (or both) of the three-sigma limits falls outside the specification limits, we know that some of the product will not conform to the prescribed limits. By calculating this percentage, we can discover how much scrap the process is making, and how troublesome inspection problems will be.

A single automatic machine produces a part for which the outside-diameter specification is 3.58 in. $\pm$ 0.05 in. A frequency distribution of a large number of sample items appears to be normal, with a mean of 3.59 and a standard deviation of 0.02. We wish to know what percentage of the parts made by this machine do *not* conform to specifications.

We note that some improvement could be made in this process if the process average could be changed to conform more closely to the target mean of 3.58; the sample study indicates that the process averages a little high. We will assume that the average is as close as we can presently set it, and continue our analysis of the percent nonconformance.

In Fig. 2·12 we sketch the distribution, indicating the upper and lower specification limits. The percentage of the parts which fail to meet specifications is shown by the shaded parts of the curve. Observe that the *process*

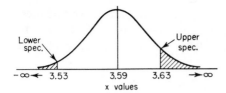

Fig. 2·12 Normal curve (mean = 3.59, σ = 0.02) showing area beyond specification limits of 3.53 and 3.63.

average is used for the mean of this curve, not the desired average (which is not being met). We now find the t values for the two specification limits, X_1 being 3.53 and X_2 being 3.63.

$$t_1 = \frac{X_1 - \mu}{\sigma} \qquad t_2 = \frac{X_2 - \mu}{\sigma}$$

$$= \frac{3.53 - 3.59}{0.02} \qquad = \frac{3.63 - 3.59}{0.02}$$

$$= -3.0 \qquad\qquad = +2.0$$

The corresponding areas under the curve are 0.0013 and 0.9772. The first area represents the left-hand shaded tail; the right-hand tail can be found by subtracting 0.9772 from 1.0000, giving 0.0228; this, added to the 0.0013 for the left-hand tail, gives 0.0241. Therefore, about 2.4 percent of the product falls outside specifications. This out-of-specification fraction of the product is small in spite of the high average of the machine because the standard deviation is small compared with the specification limit range.

PROBLEMS

2·14 Specifications for a mixed powder call for a concentration of one of the ingredients to be 15.80 grams ± 2 percent. If the mixing process gives a normal distribution with mean of 15.75 and a standard deviation of 0.19 gram, what percentage of the product fails to meet the specifications?

2·15 A paper manufacturer makes 50-lb paper. The process average, however, gives a normal distribution with a sample mean of 51.0 lb and a standard deviation of 1.0 lb.

(a) How much of the product exceeds 51.5 lb?
(b) How much of the product falls below 47.5 lb?
(c) Between what values does 65 percent of the product lie?
(d) What two values include 90 percent of the product?

2·16 The length of wire-cut nails of a certain type is normally distributed. The mean of the distribution is 2.500 in. and the standard deviation is 0.006 in.

(a) Between what two values does the middle 50 percent of the nails fall?
(b) Determine the probability that the length of a nail should equal or exceed 2.509 in.

(c) Determine the probability that a nail should measure between 2.499 and 2.510 in.

2·17 The breaking strength of a certain type of plastic varies from sample to sample in a pattern consistent with the normal curve. The mean breaking strength is 2,200 lb, and the standard deviation is 150 lb.

(a) At how many pounds of pressure, or less, can 90 percent of the samples be expected to break?

(b) What is the maximum amount of pressure the plastic should be subjected to if we expect no more than 5 percent of the samples to break?

(c) If the plastic is to have a break resistance of 2,150 lb, with a ± 10 percent allowable error, what percent of the plastic produced by the process will meet the specifications?

2·8 SPECIFICATIONS AND TOLERANCES

An important part of a contract for the production of an item is the set of *specifications* which describe in detail the characteristics of the item which is to be produced. The specifications will include, depending on the situation, dimensions, performance characteristics, color, etc. Included in the specifications will necessarily be a statement of the permitted variation in the qualities of the item to be produced; these are the *tolerances* which are to be allowed in acceptable items.

Suppose that the drawings for a manufactured part specify a tolerance for one dimension of ± 0.003 in. To interpret this specification properly, it is necessary to know what the engineer had in mind when he set this tolerance. The question is: "How many of the parts are supposed to come within this tolerance?" These possibilities exist, among others:

1 The engineer may expect *all* the parts to fall within the specified limits. The only way we can make *nearly* all (but not all) the parts fall within the tolerance is to have a manufacturing process with a very small standard deviation indeed. If $\sigma = 0.00075$ in., only about 1 part in 10,000 will not satisfy the requirement. Note, however, that the process will then necessarily make about 68 percent of the product within one-fourth of the tolerance; that is, within ± 0.00075 in. To do this would require an exceptionally well run, and no doubt expensive, process.

2 The engineer may permit as little as ¼ percent of the product to fall outside the tolerances. Now, 99¾ percent will fall within the limits; this percentage implies a three-sigma interval. Now sigma need be only 0.001 in., and about 68 percent of the parts must fall within ± 0.001 in. of the aim. To run a manufacturing process of this quality is perhaps possible.

3 If the tolerances need be met only 95 percent of the time, we are now working within a two-sigma interval, and a sigma of 0.0015 in. would satisfy this, easing the manufacturing situation considerably.

It is essential that tolerances be established in the light of a statistical analysis of the manufacturing problem. A tolerance set in terms of a plus-minus statement, without a clear understanding of what is required and what is possible, is of no value whatever to the production staff. Furthermore:

1   Specifications must be realistic; that is, they must be of significance to the consumer, and related to his true needs. In paper manufacture, to make a paper having far more than the required bursting strength is expensive and may adversely affect other important characteristics of the paper. To specify, therefore, a higher-than-necessary bursting strength may defeat the very purpose specifications are intended to meet.

2   The tolerances must be consistent with process capabilities. If impossibly tight tolerances are set, production workers will realize that they cannot be met, and may ignore them completely. Elaborate inspection methods are then required to give a product that is satisfactory to the consumer.

3   Tolerances must be set so that the product can be manufactured at a competitive cost. The desire to make an excellent product must be balanced against the cost of doing so, which may be prohibitively expensive. If the market can successfully use a product manufactured inexpensively with wide tolerance, it may be unnecessary to work with costly tighter limits.

4   Tolerances are not sacred or unchangeable. Changes may well be demanded by changing conditions of manufacture or consumption. It is important, however, that changes in tolerances and in a process be made by someone familiar both with the process capability and with the end use of the product.

5   Complete specifications should state (a) the sample size to be chosen for tests; (b) the method of selecting the sample; (c) the test methods to be applied to the sample; (d) the calculations to be performed on the data; (e) the tolerances and the fraction of conformance desired; and (f) the action to be taken on lots which fail to meet the requirements.

These recommendations are in the interest of minimizing conflicts between supplier and customer. Clearly understood, agreed-upon specifications serve to help accomplish this end.

Once specifications and tolerances have been established, the application to the inspection situation must be understood. In this connection, the discussion of significant figures in Chap. 1 is to the point. Measurement of the item should be to at least as many decimal places as in the specification. Comparison of the measurement with the specification should be made only after the measured value has been rounded to the same number of decimal places as in the tolerance. For example, if a specification is given as 0.165 ($+0.000$, -0.004) and a measured item is found to be 0.16550, we would round the measured value to 0.166, and find it to lie outside the tolerance.

PROBLEMS

2·18   If the specifications for a part are 0.15(+0.05, −0.00), would each of the following be accepted or rejected?

(a) 0.205
(b) 0.146
(c) 0.21
(d) 0.204
(e) 0.145

2·19   A normally distributed process is centered at 20.6 cc and has been maintaining a constant standard deviation of 0.5 cc. What percent of the processed items from this production line would conform to the following specification limits:

(a) 20.6 ± 1.0 cc
(b) 21.0 ± 1.0 cc
(c) 20.5 ± 0.5 cc

2·20   On a given part, an internal diameter is specified as 1.008 in. ± 0.0035 in. From a large number of mass-produced parts, the standard deviation of the measurement was estimated as being 0.0011 in. Find the relationship between the natural tolerances of the process and the specification tolerances.

2·9 DISTRIBUTIONS OF AVERAGES

To this point we have discussed the normal distribution as made of observation of individual pieces of data. In many industrial applications, samples are often taken in lots of four, or five, or more, items. Averages of these *subgroups* are used to make frequency distributions. Subgroups are used when the short-time variability of the process is believed to be relatively large, and when it is desirable to average out this variability.

More important, the frequency distribution of subgroup averages has one great advantage over distributions of individuals. Subgroup averages tend to give a normal distribution, even when the individuals come from a population that is far from normal.

This may be illustrated by using a bridge deck of cards. Here we expect a rectangular distribution, since there are four of each kind of card (ace through king). If we assign numbers 1 through 13 to each of the kinds, select at random four cards and find their average, and repeat this process, we can construct a distribution of the averages of four cards. This distribution will be found to approach normality as the sample size increases; this would not occur with individual cards. A repetition of the experiment with larger subgroups (say six or eight cards) would lead to the same result, except that the standard deviation of the distribution of subgroups would be *smaller* with a *larger* subgroup.

Similar trials with dice lead to a similar result; the frequency distribution for a single die is rectangular; for two dice, triangular. With more than two dice, the approximation to normality increases as the number of dice is increased. The principles are these:

1 Frequency distributions of subgroup averages tend toward normality, even if the parent population of individuals is not normal.
2 The standard deviation of frequency distributions of subgroup averages is smaller as the subgroup sample size increases. In fact, the standard deviation of averages varies inversely as the subgroup sample size, or

$$\sigma_{\bar{x}} = \frac{\sigma_X}{\sqrt{n}} \qquad (2\cdot4)$$

Thus we expect that the standard deviation of a set of subgroup averages $\sigma_{\bar{x}}$ will be equal to the standard deviation of the individuals σ_X divided by the square root of the subgroup sample size n. For example, we would expect that the standard deviation of averages of subgroups of four would be one-half the standard deviation of individuals from the same population.

$\sigma_{\bar{x}}$ is called the *standard error of the mean*. It is a useful number in that its use improves our ability to make inferences from samples. Because individual observations have relatively large variability σ associated with them, it is sometimes hard to tell whether an observation which is different from the mean represents a change in process level or just random variation. Since averages have smaller variability $\sigma_{\bar{x}}$, it becomes easier to distinguish between a real change and random variation.

From a process which has been producing satisfactory units of production, it was decided to take random samples of four at specified intervals of time to keep a check on the process. If the process is centered at 35.00 with a standard deviation of 0.04, what percentage of the time can we expect the subgroup *averages* to fall between the values 34.95 and 35.05 if the process remains stable?

We modify Eq. (2·2) to read

$$t = \frac{\bar{X} - \mu}{\sigma_{\bar{x}}} \qquad (2\cdot5)$$

We sketch the distribution of subgroup averages as a normal one, as in Fig. 2.13, indicate where the two values in question will fall, and then shade in the area we desire to find.

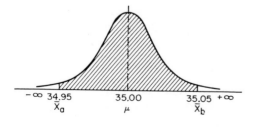

Fig. 2·13 Distribution of means of samples of n = 4, process centered at 35.00.

The solution to the problem requires finding the area of the curve between the two specified values. We apply Eq. (2·5):

$$t_a = \frac{34.95 - 35.00}{0.04/\sqrt{4}} \qquad t_b = \frac{35.05 - 35.00}{0.04/\sqrt{4}}$$

$$= \frac{-0.05}{0.02} \qquad\qquad = \frac{0.05}{0.02}$$

$$= -2.5 \qquad\qquad = +2.5$$

From the table for values of areas under the normal curve we find the area from $-\infty$ to a t value of $+2.5$ is 0.9938; from $-\infty$ to a t value of -2.5 is 0.0062. The required area is represented by the difference between these two areas, or 0.9876.

We therefore expect that about 98.8 percent of all the subgroup averages will fall between the values 34.95 and 35.05. If we were to find that an average of $n = 4$ fell outside these limits, we would have good reason to assume that the process had changed. Our reason would be that fewer than 1.2 percent of the sample averages would lie outside the limits by chance alone; thus the odds would clearly favor the belief that the process had shifted. It is upon this principle that control charts are based; these will be discussed in Chaps. 21 to 23. (Note that had individual observations, rather than averages of four, been used the limits would have been widened to 34.90 and 35.10 to keep the same percentage within limits, since $\sigma_X = 2\sigma_{\bar{x}}$ when $n = 4$.)

The preceding discussion has assumed that we really know the mean and the standard deviation of the population of interest. When, as in much research work, the mean and standard deviation are estimated from relatively small samples, we do not know μ and σ; we must then use $\bar{X}$ and s as estimates. Then we will need to use a distribution (Student's t) which approximates the normal distribution. There is a family of Student's t distributions, each one depending on the sample size for its degree of approximation to the normal curve. The larger the value of n the better the approximation to the normal distribution. When the sample size has reached about 30, the values of Student's t distribution will be close enough to those of the normal distribution so that the latter can be used in most such cases.

In general, when the standard deviation is known, the normal distribution is to be used for making statements involving averages from a population. When σ is unknown, we must use the Student's t distribution appropriate to the sample size for which s is computed. Much more will be said on this question in Chap. 3.

PROBLEMS

2·21 The standard deviation of a process is estimated to be 4.3. If the output is normally distributed and centered at 76.0, what is the probability that a sample mean of $n = 9$ will fall within the limits 74.0 to 78.0?

2·22 A process is estimated to have an average of 9.334 and a standard deviation of 2.050. Samples of 5 are taken from the process at stated intervals. Find an upper limit which will be exceeded by the sample averages not more than once in 100 times on the average, assuming that the process operates in a random manner.

2·23 From the information in Prob. 2·22, find a lower limit below which the sample averages will fall not more than 2 times in 100 on the average, provided that the process operates in a random manner.

2·24 A process is known to have an average value of 76.4 and a standard deviation of 7.6, and is operating in a random manner. Samples of $n = 8$ are taken from the process. What is the probability that a sample mean will fall:

(a) outside the limits 72.0 and 80.0

(b) within the limits 72.4 and 81.0

2·10 MEANS AND STANDARD DEVIATIONS OF SUMS AND DIFFERENCES

In many manufacturing operations, the dimension of an assembled item is the sum of the dimensions of several parts. If the parts are normally distributed and assembled in a random manner, the following statements can be made:

$$\mu_y = \mu_1 + \mu_2 + \cdots + \mu_n \tag{2·6}$$

$$\sigma_y^2 = \sigma_1^2 + \sigma_2^2 + \cdots + \sigma_n^2 \tag{2·7}$$

These equations indicate that the mean of the assembly will equal the sum of the means of the separate parts, and that the variance of the assembly will equal the sum of the variances of the separate parts. From Eq. (2·7) it can be inferred that the standard deviation of the assembly will equal the square root of the sum of the squares of standard deviations of the separate parts.

If we are given the distribution of a variable X and the distribution of another variable Y, there are several relationships we know about the sum or difference of the two distributions:

1 The mean value of the distribution of the variable $X \pm Y$ equals the mean value of X plus or minus the mean value of Y.

2 The variance of the variable $X \pm Y$ equals the variance of X plus the variance of Y, provided X and Y are independent.

3 If the distributions of X and Y are both normal, the distribution of the sum or difference, $X \pm Y$, is also normal.

Class S_1 weights for analytical balances have, according to the National Bureau of Standards, tolerances as follows (all in grams):

Value	100	50	30	20	10	5
Tolerance	±0.002	±0.0012	±0.0009	±0.0007	±0.0005	±0.00036

We wish to know what variability we may expect in combinations of these weights, such as two 100-gram weights used together. The significance of the term "tolerance" in this context is not clear; we will assume that it represents three-sigma variation. On the basis of this assumption, the standard deviation of a large set of 100-gram weights would be one-third of the tolerance, or approximately 0.00067. The standard deviation of a large number of pairs of such weights would be by Eq. (2·7)

$$\sqrt{(0.00067)^2 + (0.00067)^2}$$

or approximately 0.00094. Observe from this that the standard deviation of a set of a number of similar items is not proportionately related to the number of combined items, since the standard deviation of a set of pairs is not twice the standard deviation of a set of individuals. The reason is that it is unlikely that the largest item in set A will be combined with the largest in set B; therefore, the standard deviation of a sum will be less than the sum of the standard deviations of the individual sets as long as the selections are independent.

Similarly, in a large number of pairs of 100- and 50-gram weights the standard deviation would be $\sqrt{(0.00067)^2 + (0.0004)^2}$, or about 0.0023. From this example, it appears that the variability in a set of weights is closely related to the variability in the largest of the weights forming the combination. In general, the total variability is mainly determined by the largest variability in the component items, and is little influenced by the minor variability.

PROBLEMS

2·25 Using the data given above, would it be better on the average to use a single 100-gram weight from a large number of such weights, or to make up 100 grams by the use of ten 10-gram weights from a large number of such weights? Defend your answer.

2·26 A three-gear assembly is put together with spacers between the gears. The mean thickness of the gears is 2.055 in. with a standard deviation of 0.004 in. The mean thickness of the spacers is 0.050 in. with a standard deviation of 0.001 in.

(a) What is the mean thickness of a set of assembled parts, using 3 randomly selected gears and 2 randomly selected spacers?

(b) What will be the standard deviation of thickness of the assembled units?

2·27 If, in Prob. 2·26, the thicknesses of the parts are normally distributed, what is the probability that the assembled unit will have a total thickness in excess of 6.175 in.?

2·11 DISTRIBUTIONS OTHER THAN NORMAL

Although normal distributions are common, there are situations in which other patterns are detected. We describe some of these briefly here.

Multimodal distributions usually indicate the presence of more than one population. The frequency polygon of the heights of students in a co-educational university would probably show two peaks, inasmuch as men and women have different average heights. Examination of the situation usually indicates the cause of such a pattern.

If we were to record the results from many rolls of a single die, we would expect that each face would appear about the same percentage of the time. A frequency plot of the results would be rectangular. Other rectangular distributions are found in many games of chance; they are rare in engineering applications.

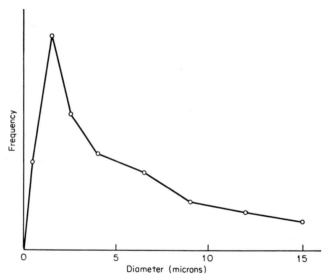

Fig. 2·14   Frequency distribution of dust particles.

Counts of the number of atmospheric dust particles of various sizes give frequency plots as in Fig. 2·14. This is a highly skewed distribution. The presence of skewness in a frequency distribution often leads to the discovery of a physical or instrumental limitation imposed on a process. There may be an upper or lower limit beyond which the process cannot possibly operate. When we attempt to count dust particles, the limitation probably is present because the measuring instrument is incapable of detecting exceptionally small particles.

Since it is often troublesome to work with nonnormal distributions, it is sometimes worth the effort to transform the data so that the plot approximates normality. Strongly skewed distributions, for example, can often be transformed to near-normal distributions by using log X rather than X for the horizontal axis of the plot. This mathematical treatment is useful for studies of particle sizes in atmospheric contamination.

Alternative methods of dealing with nonnormal distributions will be discussed in Chap. 18. These methods are useful because they require no assumptions about the normality of the data.

QUESTIONS

2·1   Describe what is meant by a "normal distribution."

2·2   How many different normal curves are there?

2·3   How many parameters are required to determine a normal curve?

2·4   Why cannot a sample be exactly normally distributed?

2·5   How can we determine if a population is normally distributed?

2·6   Are all bell-shaped distribution curves normal curves? Why?

2·7   Why are normal curves often changed into the standard form?

2·8   How is the area under the normal curve related to probability values?

2·9   How is it possible to use the same table of probability values for all normal curves, regardless of the value of the mean or standard deviation?

2·10   What is a "one-tailed" area? a "two-tailed" area?

2·11   It is necessary to have a table of normal probability values that run only from the mean to negative infinity. Why?

2·12   Discuss the difference between "natural" tolerances of a process and specification tolerances.

2·13   When can we expect a distribution to be approximately normal?

2·14   What is the relationship between the standard deviation of the distribution of individuals, and the standard deviation of the distribution of sample means?

2·15   What are the advantages of working with averages, over working with individuals?

2·16   What is the standard error of a statistic?

2·17   What type of distribution is the Student t distribution?

2·18   When two normally distributed variables are added or subtracted, and the variables are independent of one another, what are the relationships involved?

REFERENCES

16, 27, 28, 45, 46, 62, 67, 69, 73, 82, 96, 109, 115, 120, 123, 125, 126

3 Tests of Hypothesis of the Mean

We perform experiments in order to make some decision about a process. These decisions must be made on the basis of the data we collect from the experiment. Before we perform the experiment, we must know what kind of question we want to answer. Based on experience, on scientific theory, or on intuition, we set up a *hypothesis* which we will proceed to test.

The tested hypothesis may be complex, or as simple as this one: The average length of the items produced by this machine is 2.000 in. An experiment intended to test this hypothesis would involve choosing samples from the process, measuring them, and from the sample data making a decision about the population, i.e., to reject or accept the hypothesis we made at the start.

A fundamental problem exists, associated with the variability of the data obtained by experiment. We understand that a sample cannot always be expected to reflect exactly that nature of the population from which the sample was taken. Therefore, when we make judgments about the population from an examination of a sample, we cannot be wholly confident that our judgments are correct. What we would like to do is to limit our risks of error, and to understand what these risks are in any given problem. Statistical tests of significance are systematic procedures which assist us in making decisions. In this chapter we shall discuss tests of significance for the *mean*. We use the following symbolism:

1   μ_0 represents the hypothesized mean of the process level. If we attempt to set an automatic machine so that it will make parts which average

2.000 in. in length, we say that $\mu_0 = 2.000$ in. Alternatively, in a scientific experiment, μ_0 represents the hypothesized value of the speed of light or the hypothesized value of the heat of fusion of ice.

2   μ represents the mean of the population really being made or of that being measured. This mean is actually unknown; we can, however, estimate it by $\bar{X}$, the mean of the sample data.

3·1 STATEMENT OF THE HYPOTHESIS

The relationship which concerns us here is that between μ and μ_0. The question is whether or not the mean of the population of parts being made by the machine μ should be considered the same as the aim-point mean μ_0. We now must state the hypothesis which we wish to test; either of two forms might be used:

1   The set of sample data is *not* representative of the desired population; i.e., *there is a difference* between the population being made by the machine and the desired population.
2   The set of sample data *is* representative of the desired population; i.e., *there is no difference* between the population being made by the machine and the desired population.

It may seem that these two possible forms of the test hypothesis are merely conversely related, and in a sense this is true. In the practical experimental situation, however, there is a real difference in our attitude toward these two different statements. If the hypothesis is stated in the first form, and we fail to find a difference on the basis of a test, we still have not proved or disproved the hypothesis. Further tests, if carried out, might show a difference which our test has failed to discover. Even if a large number of tests were performed, and no difference were found, we might fear that still further experiment would reveal a difference. No finite number of tests would suffice to prove that no difference exists.

If, however, the hypothesis is stated in the second form—that there is no difference in the two means—even a single test, *if a difference is found*, is sufficient to disprove the hypothesis. In this case, if no difference is found, we in effect reserve judgment.

There is another reason for using the second form of the hypothesis: It is that in most situations we fear (or hope) that a difference may occur, and therefore we seek to know when such a difference exists. If a difference is found, we intend to take some kind of action; if no difference is found, we intend to continue as before. Since we will modify our procedure only when a difference is detected, it is appropriate to state our hypothesis in a form which leads to action.

It is practically and theoretically proper, therefore, to state the hypothesis being tested in this form: *There is no difference* between the mean estimated from the sample data and the assumed mean. Such a hypothesis is called a *null* hypothesis; it is invariably the form used for the purpose of statistical testing. In the example being considered, the null hypothesis is this: There is no significant difference between the mean of parts being produced by the machine and the desired mean. In symbolic form the null hypothesis is $H_0: \mu = \mu_0$; that is to say, the null hypothesis H_0 states that the mean of the population parts from the machine μ is the same as that of the desired population μ_0. To say that the two means are equal is to say that there is no significant difference between them. In the statistical test we are about to describe, if we find a genuine difference, we will reject the null hypothesis. If we fail to find a genuine difference, we will accept (or better, not reject) the null hypothesis.

Note that the word "significant" in statistical parlance means "real." Thus, an effect that shows up as a significant difference implies only that a measurable effect has been found. It does not necessarily imply that the effect is important enough to warrant corrective action in a process, since a significant effect may be unimportant for practical considerations.

The alternative hypothesis is stated thus: $H_1: \mu \neq \mu_0$. This is to say, the mean of the population of parts produced by the machine is *not* the same as the desired mean. This is the hypothesis that we actually test. If we cannot accept the alternative, then we act as if the null hypothesis were true. We take action only when we can accept the alternative hypothesis.

We state the null and the alternative hypothesis in terms of two population means. We repeat, however, that since we can rarely examine an entire population, we cannot find directly the value μ. We necessarily use the sample average $\bar{X}$ as our best available estimate of the population average. Although the alternative hypothesis H_1 reads $\mu \neq \mu_0$, it will not be μ that will appear in the calculations, but rather $\bar{X}$.

3·2 TYPES OF ERROR

The necessity for using $\bar{X}$ as an estimate of μ leads to two possible errors of judgment, similar to those discussed in Chap. 2:

1 Error of type I is that of deciding from the sample data that the machine is not producing parts of the right size, when in fact the machine is working properly. This kind of error is that of *rejecting* the null hypothesis when we should *accept* it. Associated with this kind of error is a probability value—the *alpha* (α) risk—the risk that a type I error can occur. We can, and should, set in advance of the experiment the α risk we are willing to take, and determine the statistical test accordingly. Methods for doing this will be described in what follows.

2 Error of type II is that of deciding that the sample data indicate that the machine is working properly, when in fact the machine is on the average producing parts of the wrong size. This kind of error is that of *accepting*

the null hypothesis when we should *reject* it. Associated with this kind of error is a probability value—the *beta* (β) risk—the risk that a type II error can occur. We cannot fix the β risk without being quite specific about how large a deviation from μ we want to detect. If we do not define this deviation, we must act in general terms. In practice, we usually reduce the β risk by increasing the sample size we use for the test. An increased sample size gives us a more precise measure of the population mean and therefore reduces the likelihood of type II error.

Note especially that we are never free of the risk of both types of error, though we can reduce one risk at the expense of increasing the other. For example, if we want to minimize an error of type I, we want to be quite sure, on the basis of the sample data, that the machine is performing badly. Recall, in this connection, that the ordinate of the normal curve never falls to zero; thus there is some small probability that the machine will produce parts deviating considerably from 2.000 in., even when on the average it is producing parts that are of the correct size. If, therefore, we accept the null hypothesis except when parts of extreme size appear in the sample, we reduce the probability of type I error, but at the same time we run a large risk of failing to detect a real change in machine performance.

In general, for a given sample size, decreasing the likelihood of one type of error necessarily involves an increase in the likelihood of the other type of error. The only way we can reduce both types of error is by increasing the sample size. In practice, one type of error may be more serious than the other; the specific problem will govern which needs tighter control.

3·3 RISKS OF ERROR AND THE NORMAL CURVE

To design a test of significance, we first decide on the risk of a type I error we are willing to take. The probability of making this type of error is called the *level of significance* of the test. We should specify the α risk before taking samples so that the data do not influence our judgment.

If we are dealing with a normal population we can represent the α risk as in Fig. 3·1. The curve represents the assumed normal distribution; the α risk is shown by the areas under the shaded parts of the curve. These areas are called the *critical regions*. Whenever we find the average from a sample to fall in these areas, we agree to act on the basis that the data come from a

Fig. 3·1 Critical regions for the normal curve, based on the α risk.

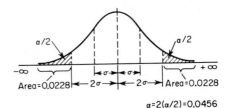

population differing from the assumed population. If an average falls in the critical regions, we reject the null hypothesis. Here we show the α risk divided into two equal parts, one for each tail of the curve; sometimes we put the entire α risk into one tail of the distribution. The distinction between two-tailed and one-tailed tests will be discussed later in this chapter.

The nature of type II error and the associated β risk is indicated in Fig. 3·2. Curve A represents the normal distribution of the hypothesized process;

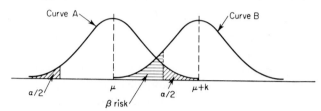

Fig. 3·2   β risk represented by area common to two normal curves.

curve B represents a different population with mean k units more than the first. The overlapping area marked "β risk" represents data properly belonging to curve B which are indistinguishable from some of the data in curve A. Even though the process has truly changed from a mean of μ to a mean of $\mu + k$, some of the data from the changed process could have come from the original process. The probability of this situation's occurring represents the β risk.

3·4   INFORMATION NECESSARY FOR A TEST OF HYPOTHESIS

Assume that we wish to test the hypothesis that the mean of the population of parts produced by a machine is 2.000 in. The null hypothesis is H_0: $\mu = \mu_0 = 2.000$ in. Assume further that we select one item from the process and find it to be 2.005 in. In the absence of other information, we can make no inference about the mean of the process from this single sample. Specifically, we have no data about the variability of the process. Therefore, we cannot judge whether a part of this size lies close to or far from the average product. What we require in addition to the sample data is knowledge of (or an estimate of) the standard deviation of the process.

Suppose now that we know that the standard deviation of the population of parts normally produced by the machine is 0.002 in. We can calculate the t value of the measurement 2.005 in. From Eq. (2·2) we find

$$t = \frac{2.005 - 2.000}{0.002} = +2.5$$

Now we can ask: How likely will it be that we find a t value as far out as either $+2.5$ or -2.5 if the process mean is really at 2.000 in.? From Table A·1 we find that the sum of the two areas under the normal curve farther from the mean than $t = 2.5$ is about 0.0124. We interpret this number as meaning this: If the process were truly giving a population with a mean of 2.000 in. and a standard deviation of 0.002, only about 1¼ percent of the product would plot as far from the mean as the piece we measured.

From this statistical test, we have two alternatives:

1 We can suppose that finding this item was a rare event and that the machine is truly operating at an average level of 2.000 in.
2 We can suppose that the average of the machine production is different from 2.000 in.

The first alternative requires that we accept the null hypothesis, the second that we reject it. If we reject the null hypothesis we are taking a risk of 1.24 percent of being wrong.

We can specify in advance of the experiment the α risk we are willing to take. For example, to set the α risk at 5 percent means that we are willing to reject the null hypothesis on the average 5 times in 100 trials when it is in fact true. (To avoid taking any α risk whatever would require never rejecting the null hypothesis, no matter how far the process deviated from 2.000 in.) Five percent of the area under the normal curve (see Table A·1) lies outside of $t = \pm1.96$. Thus we can set a decision rule: Based on an α of 0.05, if the t value obtained for any observation is larger than 1.96, reject the null hypothesis; otherwise, accept it.

3·5 THE TEST STATISTIC

The preceding section contains the essentials of every test of hypothesis. The key is the calculation of a value which we can relate to an assumed pattern of distribution which is usually the normal curve. The value is here a t number, the general formula for which is

$$t = \frac{\text{statistic} - \text{parameter}}{\text{standard error of the statistic}} \tag{3·1}$$

In the example of this chapter, the statistic is the sample value; the parameter is the assumed mean of the population; the standard error of the statistic is the standard deviation of the assumed population. By finding the t value we have coded the data so that we can make use of the relative frequencies associated with the normal curve.

The general formula for the test statistic applies in many situations. Differences in detail are found in specific cases, depending on the parameter we wish to test and on the method of finding the variability of the process measured by the standard error of the statistic.

3·6 STEPS IN HYPOTHESIS TESTING

1 State the null hypothesis and the alternate hypothesis.
2 State the assumptions involved in the test. These assumptions involve

the pattern of distribution supposed to represent the situation, e.g., normal. It is usually assumed that the sample is randomly selected; efforts must be taken to ensure the validity of this assumption.

3 Specify the risks that are to be run.

4 Determine the critical regions, which are usually derived from Table A·1 or a modification of this table.

5 Calculate the test statistic using the sample data.

6 Compare the value obtained in step 5 with the critical values found in step 4 to decide whether to accept or reject the null hypothesis.

We have so far in this chapter discussed these steps for a test of significance based on a single sample ($n = 1$). Although we prefer to obtain more data about the process, we may use a single sample in the test under the following circumstances: (1) we have evidence from past information that the process gives a normal distribution; (2) we know the standard deviation of the process (we must assume that the standard deviation is unchanged at the time the sample is taken); (3) we are interested in being relatively sure of detecting only large deviations from the hypothesized value.

PROBLEMS

3·1 Using Table A·1 find the value of t which is to be used for the following two-tailed α risks:
(a) α of 0.05
(b) α of 0.10
(c) α of 0.01

3·2 If the mean of a normal population is 34.5, with a standard deviation of 1.5:
(a) Between what two values should we expect the next random sample from the population to lie, if we are using a two-tailed α risk of 0.05?
(b) Between what two values should the next randomly selected sample from the population fall if we choose to use a two-tailed α of 0.10?

3·3 If the population mean of a normal distribution is known to be 150 and the standard deviation is known to be 10, what is the probability that a single observation from the population, selected in a random fashion, will be contained between the following values?
(a) 140 and 160
(b) 130 and 170
(c) 120 and 180
(d) 150 and 160
(e) 130 and 160

3·4 A production worker has been told that the population average of a load of springs is 14.0 grams with a standard deviation of 0.8 gram. If he selects, at random, one spring from the entire load of springs:

(a) What is the probability that the spring he selects will be less than 14.0 grams?

(b) What is the probability that the spring will weigh between 13.5 and 14.5 grams?

(c) Between what two values should he predict the weight of the spring to lie if he wishes to be correct, in the long run, 90 percent of the time?

3·5   A certain type of plastic is known to have a break resistance of 3,200 pounds per square inch (psi), with a standard deviation of 50 lb. If a single piece of the plastic is selected at random from the process, and tested for break resistance:

(a) How often should we expect it to have a break resistance of more than 3,200 psi?

(b) How often should we expect it to have a break resistance between 3,000 and 3,500 psi?

(c) How often should it have a break resistance between 3,100 and 3,300 psi?

(d) Between what two values can we expect it to be 95 percent of the time?

3·6   Rubber bands of a given type and size were tested to see how far they could be stretched prior to breaking. The elastic limit of the rubber bands turned out to give a normal distribution with a mean of 8.2 in., and a standard deviation of 0.7 in.

(a) Between what two dimensions did the middle 50 percent of the bands break?

(b) What is the likelihood that a single rubber band selected at random will break between 8.0 and 8.4 in.?

(c) How often should a rubber band break between the values 8.1 and 8.3?

3·7   An automatic bottle-filling machine is set to fill 32.0 oz with a standard deviation of 0.1 oz. The machine was found to meet these requirements. After the process was stopped, cleaned, overhauled, and restarted, it was desired to see if the machine was still set at the same level. Past history of the process indicates that the standard deviation will not change because of this stoppage.

(a) State the null hypothesis to be tested.

(b) State the alternative hypothesis.

3·7   TEST OF THE MEAN (σ KNOWN)—SAMPLE VERSUS STANDARD

If we can obtain several observations from a process, our test of significance will be based on Eq. (3·1), with these changes from the single-sample test: The statistic will be the mean of the sample data $\bar{X}$; the standard error of the statistic will be $\sigma_{\bar{x}}$. If we know the population standard deviation σ we can find $\sigma_{\bar{x}}$ from

$$\sigma_{\bar{x}} = \frac{\sigma}{\sqrt{n}}$$

Assume that we have a random sample of four items from a process which is known to have a standard deviation of 0.002 in. and that the four observations are 2.004, 2.006, 1.998, and 2.004 in. We wish to test the hypothesis that the mean of the production is 2.000 in.; i.e., H_0: $\mu = \mu_0 = 2.000$ in.; H_1: $\mu \neq \mu_0$. The assumptions we are making are normal distribution; standard deviation 0.002 in. unchanged with process level; random selection of the sample. We will specify an α risk of 0.05.

From Table A·1 for $\alpha = 0.05$ the critical regions lie beyond $t = \pm 1.96$. Thus we will reject the null hypothesis if the calculated t value exceeds this value. We calculate the test statistic:

$$t = \frac{\bar{X} - \mu}{\sigma_{\bar{x}}} = \frac{2.003 - 2.000}{0.002/\sqrt{4}} = +3.0$$

Since the calculated t value is more than the critical t value, we conclude that the process is *not* operating at an average level of 2.000 in. We are saying that it is extremely unlikely to get a t value of 3 unless a shift in level has occurred.

As compared with a test based on a single observation, the use of a sample of four items has increased the sensitivity of the experiment. Since the standard error of the mean of a set of four items is only half the standard deviation of the parent population, we will be able with a sample of four to detect a smaller deviation from the mean. Thus the use of a larger sample permits us to find smaller differences in a process, though at the cost of greater inspection effort.

3·8 ONE-TAILED VERSUS TWO-TAILED TESTS

We have so far considered tests of hypothesis for cases in which we were testing only for a change in a process, without regard to the direction of the change. When we set the α risk, we have thought of it as being divided into two parts, one in each tail of the assumed distribution, as in Fig. 3·3. Here there are two critical regions.

In some situations, however, we are interested in testing for a change in only one direction. It may, for example, be impossible for a process average to increase because of a limitation inherent in the process. In the manufacture of paper, "whiteness" estimated by reflectivity may be important. Reflectivity of light can never exceed 100 percent, and in any case high reflectivity is desirable. A test of significance in this application would be concerned only with a *decrease* in reflectivity.

If we are interested in a possible change in a process in one direction only, we say that we make a *one-tailed* test of hypothesis, and we associate the entire α risk with one critical region, as in Fig. 3·4. For one-tailed tests,

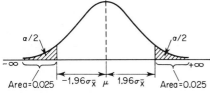

Fig. 3·3 *Critical regions for a two-tailed test of significance.*

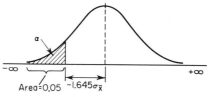

Fig. 3·4 *Critical region for a one-tailed test of significance.*

we write the hypothesis differently from those for two-tailed tests. If we are interested in only a possible *decrease* in the mean of the population, $H_0: \mu = \mu_0$ and $H_1: \mu < \mu_0$. (There is another possible region of H_1, $\mu > \mu_0$, but by the definition of the problem we are not interested in whether or not we would accept this hypothesis.) The null hypothesis is that there is no difference between the population mean μ and the assumed mean μ_0. The alternative hypothesis is that the population mean is less than the assumed mean. Similarly, if we were interested only in a possible increase in the mean of the population, $H_0: \mu = \mu_0$ and $H_2: \mu > \mu_0$ (and we ignore $\mu < \mu_0$).

Calculations for one-tailed tests are identical with those for two-tailed tests, with these exceptions: The algebraic sign of the deviation from the mean is important; we test against a one-tailed test value of α.

A machine fills opaque bottles with a liquid. We want the filling process to give a mean volume of 16.0 oz. A history of the process indicates a normal distribution of volumes, with standard deviation of 0.1 oz. We wish to know whether or not the machine is working correctly. If the level were to increase, the machine operator could detect the change almost immediately, since the bottles would overflow. We wish primarily, therefore, to be sure that the average level of the process does not *decrease*. Here we would use a one-tailed test and place the entire α risk in the *low* side of the assumed distribution.

The null hypothesis is $H_0: \mu = \mu_0 = 16.0$. The alternative hypothesis is $H_1: \mu < \mu_0$. The statement of the null hypothesis implies no difference between current process level and the target value. We will accept the alternative hypothesis only if underfilling is indicated, since that is the only situation for which we want to take action.

Figure 3·4 sketches the situation. The entire critical region is in the left-hand tail; if we choose an α risk of 0.05, the shaded area will include 5 percent of the total area under the curve. The distance from the mean of the distribution in Fig. 3·4 to the right-hand boundary of the critical region will be given by the t value of that boundary, in a negative direction from the mean. If, when we run the test, we find a positive deviation from the mean, we will act as though we accept the null hypothesis regardless of the size of the deviation from the mean since we are not interested in taking any action in this situation anyway.

A sample of 12 bottles from the process gives volumes as follows: 15.8, 16.0, 15.6, 15.7, 15.8, 15.8, 15.9, 15.9, 15.9, 15.9, 16.1, 16.0. We compute $\bar{X}$ to be 15.87 and a standard deviation that does not indicate a change from the past value of $\sigma = 0.1$.

Again we use Eq. (3·1):

$$t = \frac{\bar{X} - \mu_0}{\sigma_{\bar{x}}} = \frac{15.87 - 16.0}{0.1/\sqrt{12}} = -4.49$$

From Table A·1 we find the critical value for a one-tailed alpha of 0.05 to be -1.645. Since the calculated value lies in the critical region, as shown in Fig. 3·5, we reject the null hypothesis and accept the alternative. Thus we

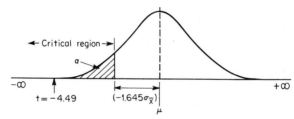

Fig. 3·5   Critical region for a one-tailed test, $\alpha = 0.05$.

have evidence that the machine is underfilling, on the basis of the sample we measured.

If the standard deviation of the process had been larger than 0.1, the sample average would have been located fewer standard deviations below the target value of 16.0—the t value of the deviation would have been smaller. For instance, if the value of sigma had been 1.0 instead of 0.1, the t value would have been found to be 0.449, and we would have accepted the null hypothesis. The inference is that if the repeatability of the process is poor so that the standard deviation is large, it takes a larger deviation from the standard to be significant.

Should the test statistic indicate that the null hypothesis is to be accepted, we must not then conclude that we have proved there is no difference between the population mean and the standard value. Rather, we have only failed to find evidence that there is a difference.

The risk we take of being wrong when we accept the null hypothesis is the β risk; we can determine the value of this risk for whatever deviation we want to detect based on the sample size that was used. In our example, we said we were intending to use $n = 12$ and set only the α risk. It is possible to find the value of the β risk which was set when a sample size of 12 was selected, but the usual method is to set the sample size as a result of pre-setting both the α risk and β risk, as we will show in Chap. 7.

PROBLEMS

3·8   Toss a coin 64 times and record the number of times that a head comes up. Then calculate the mean and the standard deviation for the distribution of heads. $\mu = Np$ and $\sigma = \sqrt{Npq}$ when $N =$ number of tosses in the population, $p =$ probability of a head, or 0.50, and $q =$ probability of a tail, or 0.50.

(a) Using an α of 0.05, test to see if the coin you used was an honest coin.
(b) Using an α of 0.01, test the hypothesis that the coin is honest.

3·9   Using the data gathered in Prob. 3·8, test the hypothesis that the number of heads that turn up in 64 tosses will be at least 30, using an α of 0.10.

3·10   A process has been operating over a long period of time and the parameters $\mu = 45.5$ and $\sigma = 2.0$ have been established from past data. The machine is moved to a new location in the plant and a sample of $n = 16$ is taken from the machine after it is restarted. If the standard deviation has not changed:

(a) Test the hypothesis that the process average has not changed, if for $n = 16$, $\bar{X} = 43.5$, using an α of 0.05.
(b) If $\bar{X}$ turned out to be 44.4, test the hypothesis that the process is at least as high as it was prior to moving the machine. Use an α of 0.05.
(c) Test the hypothesis that the process average of the machine after moving is at the most 45.5 if the $\bar{X}$ turned out to be 47.0.

3·11   The mean life of a sample of 25 light bulbs produced by a company is found to be 1,535 hr with a known sigma of 100 hr. The average of this type of light bulb in the past has been $\mu = 1,500$ hr. If the standard deviation has not changed, test the hypothesis that the new light bulbs last longer than the older ones.

(a) Use an α of 0.05.
(b) Use an α of 0.10.
(c) Use an α of 0.01.
(d) At an α of 0.05, test the hypothesis that the lifetime of the new bulbs is different from that of the older lamps.

3·9   HYPOTHESIS TEST FOR TWO SAMPLE AVERAGES, σ KNOWN

A question that arises in many kinds of experimentation is this: do two sets of sample data represent the same, or different, populations? Several different kinds of answers are possible, depending in part on the nature of the available data and on the kind of information that is sought. We here consider one situation.

The question to be answered is: Are the *means* of the two populations to be considered the same or not? We assume that we know the standard deviations of the two populations from which the samples were drawn; furthermore, we assume that these standard deviations are equal. The distribution in this instance is a plot, not of individuals, but of the differences in the means of all possible pairs of samples of size n. If we assume that the distribution of these differences is normal, we can use a test of hypothesis based on the normal curve of distribution.

A physics experiment requires measurements to be made over a period of 2 days. The measuring equipment is subject to change, and we therefore check the instrument on each of the 2 days by making a series of measurements on a standard. We are interested in knowing whether or not the measurements made on day 2 are similar to those made on day 1. If our decision is that there is no difference, we will then suppose that the instrument is performing similarly on both days and that we can combine the measurements safely.

We make four measurements on the first day, and obtain an average value of 25.0; five measurements on the second day average 26.0. Experience with the apparatus indicates that the standard deviation is 1.0. In making the test we assume normal distribution; standard deviation unchanged from day to day. The hypotheses are $H_0: \mu_1 = \mu_2$; $H_1: \mu_1 \neq \mu_2$.

To calculate the test statistic, we use Eq. (3·1) with slight modifications. As the numerator of the fraction we use the difference between the sample averages $\bar{X}_1 - \bar{X}_2$. This value is our best available estimate of the difference in the means of the two hypothetical populations. For the denominator of the fraction we need now $\sigma_{\bar{X}_1 - \bar{X}_2}$, that is, the standard error of the difference in the two means. Thus we have a test of hypothesis based on

$$t = \frac{(\bar{X}_1 - \bar{X}_2) - (\mu_1 - \mu_2)}{\sigma_{\bar{X}_1 - \bar{X}_2}}$$

in which we hypothesize that $\mu_1 - \mu_2 = 0$.

Recall from Chap. 2 that the variances of differences are additive and that the variance is reduced in accordance with sample size. Thus

$$\sigma^2_{\bar{X}_1 - \bar{X}_2} = \sigma_{\bar{X}_1}{}^2 + \sigma_{\bar{X}_2}{}^2 = \frac{\sigma^2}{n_1} + \frac{\sigma^2}{n_2} = \sigma^2\left(\frac{1}{n_1} + \frac{1}{n_2}\right)$$

and

$$\sigma_{\bar{X}_1 - \bar{X}_2} = \sigma\sqrt{\frac{1}{n_1} + \frac{1}{n_2}}$$

This last is the value of the standard error of the difference in the means of two samples of sizes n_1 and n_2.

The test statistic becomes

$$t = \frac{\bar{X}_1 - \bar{X}_2}{\sigma\sqrt{1/n_1 + 1/n_2}} \qquad (3\cdot2)$$

for the usual case in which H_0: $\mu_1 = \mu_2$ [in some problems, we prefer to hypothesize a specific difference in means, and must keep the $(\mu_1 - \mu_2)$ term in the previous equation].

If we set the α risk at 0.05, we find the critical values of t to be ± 1.96. This is a two-tailed test, since we are interested in finding a significant difference in the means in either direction.

From Eq. (3·2)

$$t = \frac{25.0 - 26.0}{1.0\sqrt{\tfrac{1}{4} + \tfrac{1}{5}}} = -1.49$$

Since the calculated t value lies between the critical t values ± 1.96, we conclude that we should not reject the null hypothesis. We have not shown that a real difference exists between the population averages estimated from the data obtained on each of the 2 days.

PROBLEMS

3·12 Identical examinations were given to 20 students in each of two colleges. The average for college A was 63; for college B 72. It is known that the standard deviation for persons taking the examination is 10. Test the hypothesis that there is no difference between the means using $\alpha = 0.05$.

3·13 Two sections of a college class in statistics each took the same examination. Section A containing 20 students obtained a mean score of 72, while section B containing 30 students had a mean score of 75. Experience with tests of this type gives us a standard deviation of 8. Test the hypothesis that there is no difference between the two sections as to their understanding of the subject matter. Use (a) a significance level of 0.05 and (b) a significance level of 0.10.

3·14 In a driver-education course it was found that the average length of time it took 16 boys to learn to drive a car was 24 days, while the average length of time for 12 girls was 30 days. Assuming the standard deviation for both boys and girls was the same, and was known to be 2 days, test the hypothesis that the boys were faster learners than were the girls of this class.

3·15 Two machines were known to have a variance of 2.5 in pounds of pressure. Machine A was modified to ensure more positive closing of the casting device, but the modification did not affect the variance. To see if the modification of the machine had influenced the level of pounds pressure, the following samples were taken from the two machines (pounds pressure):

Machine A......25.6	26.3	27.4	28.5	28.8	27.8	
Machine B......23.9	24.5	25.3	24.4	25.6	23.6	

(a) Using a significance level of 0.01, test the hypothesis that machine A now has more pressure than machine B.

(b) Using an α of 0.01, test the hypothesis that the two machines are still using the same pressure.

3·16　The bonding strength of type A glue was measured at 14,000 psi, as a result of a sample of 16 tests. Type B glue (modified version of A) was found to have a mean bonding strength of 14,500 psi as determined from a sample of 24 test specimens. The variance in both cases was known to be 90,000. Test the hypothesis that there is no difference in the bonding strengths of the two glues, using a level of significance of 0.01.

3·17　Two brands of batteries for transistor radios were tested to see if the average lifetimes were different. From a sample of 25 of brand A, $\bar{X}_a$ was found to be 32.4 hr, while a sample of 16 from brand B gave an average life of 30.8 hr. The standard deviation of brand A batteries was $\sigma_a = 2.2$ hr and for brand B was $\sigma_b = 2.2$.

(a) Test the hypothesis that there is no difference between the two population means, at a 0.05 level of significance.

(b) Test the hypothesis that brand A has a longer life, on the average, than does brand B (use an α of 0.05).

3·18　The specifications for a certain type of soap were 82.7 percent pure with a standard deviation of 4.2 percent. Two formulations were tried in an attempt to conform to the specifications. The first formulation, using a sample of 10 bars, gave $\bar{X} = 80.0$ and the desired standard deviation. The second formulation, using a sample of 12 bars, produced an $\bar{X} = 85.0$ and also met specifications with respect to standard deviation.

(a) Test the hypothesis that there is no difference between the means of the two formulations, at $\alpha = 0.05$; at $\alpha = 0.01$.

(b) Test the hypothesis that the second formulation is better than the first one in that it produces a higher percent purity, at $\alpha = 0.05$.

3·19　In a 52-card deck, such as the one used in playing bridge, the mean of the population of cards is 7.0. Select at random a sample of 9 cards and calculate the sample mean and sample standard deviation. Then select another random sample, after replacement of each card, of 16 cards and compute the average and the standard deviation of the second sample.

(a) Using the sample standard deviation as a value equal to sigma, determine if the mean of your first sample is significantly different from the population mean of 7.0.

(b) Test the hypothesis that the mean for the second sample does not differ from the population mean of 7.0, if its sample estimate of the standard deviation is assumed to be the value of sigma.

(c) Test the hypothesis that there is no difference between the population means as estimated by the two samples. (Use a pooled estimate of

the standard deviation as if it were the population parameter sigma.)
(d) Why should you expect the null hypothesis to be accepted in answer to part (c)?
(e) If, in part (c), the null hypothesis were to be rejected, what reasons might you give for this result?

3·10 TESTS OF PAIRED DATA, σ KNOWN

In the preceding test of hypothesis for two sample means, we have assumed that the samples were selected at random from the two populations of interest, and that there was no connection between them. Furthermore, we have assumed that there was no systematic factor affecting the level of observations. There are, however, situations in which there may be a need to compare two different procedures, and in which we suspect a change in the level of the response variable which we cannot control, or which we do not wish to control. Since the level of measurement may vary from sample to sample, the two observations may not be independent, but related.

Suppose that we have two instruments for measuring the hardness of a metal; we wish to know whether or not the results from the two instruments are equivalent. This is a common type of problem which is especially acute in interlaboratory comparison of data and in relating pilot-plant results to production results. Our data come from measurements with each instrument on a series of samples that have different levels of hardness. We expect that the level of the observation will change with the sample. Thus the samples are not independent.

In this example, the test for average differences discussed above, and based on Eq. (3·2), would be useless, since the standard deviation for each instrument would be greatly affected by the "true" hardness of the sample which is of no immediate concern. We are here interested in the standard deviation of the *instruments*, not of the *samples*. With a large sample standard deviation, we would have difficulty in detecting instrumental differences.

In this kind of situation, instead of testing the instrument means, we test the average differences between the instruments for each sample in pairs. In this test, we assume that any extraneous factors affect each member of the pair in the same manner, and that this effect is essentially an increase or decrease by some constant.

The null hypothesis is that of no difference between pairs. The test of hypothesis will be based on the test statistic

$$t = \frac{\bar{d} - \delta}{\sigma_{\bar{d}}} \tag{3·3}$$

where $\bar{d}$ = average difference between sample pairs

δ = hypothesized pair difference (here, and usually, equal to zero)

$\sigma_{\bar{d}}$ = standard error of pair differences, which we will base on a known standard deviation of differences

Table 3·1 Paired Comparison of Automobile Gasoline Mileages

Sample	Regular, X_1	High test, X_2	$d = X_2 - X_1$
1	20	25	5
2	19	26	7
3	22	23	1
4	15	17	2
5	21	21	0
6	18	23	5
			$\bar{d} = 3.3$

Six automobiles were used to test two types of gasoline. The response variable was miles per gallon. The data were arranged in pairs, one pair representing the two gasolines tested in each car, because it was feared that the cars were considerably different in performance. The null hypothesis is that no difference exists between gasolines, or that no difference will be found in paired comparisons. Thus H_0: $\delta = 0$. The alternative hypothesis is H_1: $\delta \neq 0$. We intend to test at $\alpha = 0.05$, for which the critical value of t is ± 1.96.

The observations as collected are shown in Table 3·1. Past history indicates that $\sigma_d = 1.5$; this is the standard deviation of the differences in data pairs. For samples of 6,

$$\sigma_{\bar{d}} = \frac{\sigma_d}{\sqrt{n}} = \frac{1.5}{\sqrt{6}} = 0.6$$

Now we apply Eq. (3·3): $t = 3.3/0.6 = 5.5$. The calculated value exceeds the critical value for the chosen α risk; hence we have evidence that the high-test gasoline gives a greater number of miles per gallon as tested in these six automobiles.

Table 3·2 Yields in Pounds from 10 Paired Plots

Plot number	Without fertilizer	With fertilizer	Difference
1	140.4	170.5	30.1
2	174.7	207.4	32.7
3	170.2	215.9	45.7
4	174.6	209.0	34.4
5	154.5	171.6	17.1
6	185.0	201.2	16.2
7	118.9	209.9	91.0
8	169.8	213.3	43.5
9	174.7	184.1	9.4
10	176.7	220.4	43.7
Total	1,639.5	2,003.3	363.8

3·20 An experiment was carried out to determine whether or not there was an effect due to fertilizer on the yield of a farm crop. A farm was divided into 10 plots of equal size; each plot was subdivided into 2 equal subplots, one of which was treated with fertilizer and one not. In Table 3·2 are given the results, the response variable being yield in pounds.

If the standard deviation of the differences was known and equal to 24.5 lb, test the hypothesis that the use of the fertilizer had no effect, using a level of significance of 0.10.

Table 3·3 Thickness Measurements with Two Machines

Sample number	Machine A	Machine B
1	0.0053	0.0060
2	0.0094	0.0092
3	0.0155	0.0163
4	0.0460	0.0465
5	0.0885	0.0890

3·21 Two machines were used for measuring the thickness of two types of paper. The data are as shown in Table 3·3. If the standard deviation of the differences is known to be 0.0003 in., test the hypothesis that the two machines measure similarly over all thicknesses tested, using an α of 0.05.

3·11 TESTS OF HYPOTHESIS, σ UNKNOWN

In discussing tests of significance so far, we have been bound by the assumption that sigma was known. This assumption is not justified in many cases. For us to know the standard deviation requires a long history of data and also requires that we know that the process has had an essentially unchanged variability throughout this history. In particular, we must assume that the variability of the process has not changed at the time we selected the sample.

We now discuss some statistical tests of significance which are more frequently used than the preceding ones because they are more generally applicable in that they do not presuppose that the process standard deviation is known.

The difference between the methods about to be described and the preceding ones is this: We will estimate the variability of the parent populations from the *sample* observations themselves. We will now suppose that we do not know σ, and we will use s found from the sample (even a small sample) as the estimate of the process standard deviation. We first develop a concept which is needed for the understanding of these methods.

3·12 DEGREES OF FREEDOM

Let us assume that we are able to secure only two pieces of data about a process, and that the observations are 2 and 4. From these two observations alone we can calculate two statistics: We can estimate the mean, since $\bar{X} = 3$; we can also estimate the standard deviation by calculating the s value. For the latter calculation, we use Eq. (1·6).

$$s = \sqrt{\frac{\sum (X_i - \bar{X})^2}{n - 1}} = \sqrt{\frac{1 + 1}{1}} = 1.4$$

Note that we started with two (assumed independent) pieces of information; from these we estimated an average, and this average was used in turn to calculate the s value. The total number of *degrees of freedom* we have is equal to the number of independent pieces of data we have secured, that is, n. We say that one degree of freedom is used in estimating the mean, and we then have left $n - 1$ degrees of freedom for the calculation of the estimate of the standard deviation. This is the significance of the $n - 1$ term in the formula for calculating s.

The concept of degrees of freedom may be restated thus: If we have two pieces of information, and we further know the mean, we can freely choose one of the two observations; the other, however, is fixed by the value of the mean. If the average of two numbers is known to be 10, we can freely choose any number we please as one of the two numbers: let it be 15; now the other number must be 5 if the mean is to be 10. One degree of freedom has been, as it were, consumed in the determination of the mean.

Similarly, if there are three pieces of data, and the mean is determined, we may choose two of the numbers freely, but the third is fixed by the value of the mean. In general, then, the number of degrees of freedom (symbolized by ν, Greek lowercase letter "nu") left from a sample of size n after the estimate of the mean is $n - 1$; the 1 that is subtracted represents the single degree of freedom that is "used" in the calculation of the estimate of the mean.

In a somewhat different context, we may wish to draw a straight-line graph from measured data; to do this we require at least two plotted points, or two observations. If we have *only* two, once we draw a line between the points, we have no way of determining whether or not the line is correct; i.e., we have no remaining observations to permit us to estimate the error that was contained in the two observations, and therefore no way of estimating the correctness of the line that we have drawn. The two observations involved only two degrees of freedom and we exhausted both of these in drawing the line. If we were to secure another observation we would then have another degree of freedom which we could use for the estimation of error. That we "lost" two degrees of freedom in this example is associated with the equation for a straight line. The equation contains two constants— slope and intercept—both of which must be determined when we draw the line. Thus in this case, the degrees of freedom remaining for error = $n - 2$.

To summarize, the total number of degrees of freedom from an experiment is equal to the number of independent observations we have made. Every value estimated from these data requires the use of one degree of freedom. Hence, we often determine the number of degrees of freedom remaining by subtracting from the total those which have been consumed in the required estimations.

The importance of the concept of degrees of freedom is this: The value of ν determines in a general way the reliability of any estimated number. If, for example, only one degree of freedom is associated with a calculated s value, we can place little confidence in that s value, in the same sense that the standard deviation of individual observations σ is greater than that for averages $\sigma_{\bar{x}}$. We expect that we can rely more on the data as the sample size increases; so also, as ν increases, we can be more confident of the computation we have made. In the tests of significance about to be discussed, we will see that we make repeated references to the value of ν. In particular, different table values of the statistics to be used are given which are determined by the available value of ν used to estimate the standard deviation.

3·13 STUDENT'S t DISTRIBUTION

Let us suppose that we take at random a large number of samples of size n from a normal population. For each sample we find the statistic t_ν:

$$t_\nu = \frac{\bar{X} - \mu}{s_{\bar{x}}} \tag{3·4}$$

If we now plot the resulting distribution of t_ν values, we obtain curves (one for each value of ν) which have these characteristics (see Fig. 3·6):

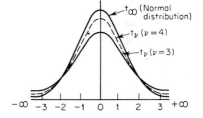

Fig. 3·6 Curves for Student's t distributions, for $\nu = \infty, 4, 3$.

1 The curves resemble in general shape the normal distribution in standard form.
2 The curves have greater standard deviations than the standard normal distribution (which has a standard deviation of 1).
3 The exact curve shape varies with ν. The curve of the t distribution is considerably different from the normal distribution if ν is small, and becomes closer to the normal distribution as ν increases. In fact, the t values for the normal distribution can be considered to be the limiting case for the t_ν distributions; that is, $t = t_\infty$ ($\nu = \infty$).

3·14 TEST OF SAMPLE AGAINST STANDARD, σ UNKNOWN

We use the Student's t distribution as the basis for tests of significance in a way similar to that in which we used the normal distribution, with these differences: We estimate the standard deviation from the sample (instead of using a known σ) and we necessarily use the particular t_v distribution appropriate to each different sample size.

In Table A·2 we find a set of t_v distribution values; these are arranged according to the value of v, i.e., $(n - 1)$ and of α. Values of v are found in the left-hand column. The desired α risk is found in the uppermost row for two-tailed tests and in the bottom row for one-tailed tests. We illustrate the use of this table in the following.

For a bottle-filling machine, the long-time process average is 50.0 ml. We wish to know if the present machine operation agrees with the past record. We take six samples at random from the process and find 53, 48, 53, 55, 57, and 49 ml. The sample average $\bar{X}$ is thus 52.5. We wish to test this sample mean against the known population mean (50.0). Not knowing the standard deviation of the process, we estimate it by calculating the sample s value; it is 3.5. We follow this sequence of steps for the test of hypothesis:

1 The tested variable is the volume of liquid per bottle.
2 Assumptions: normal distribution of population of sample averages; random sampling.
3 Hypotheses:

$$H_0: \mu = \mu_0 (50.0) \qquad H_1: \mu \neq \mu_1$$

NOTE: As before, the hypotheses are stated in terms of the parent population mean μ and the standard μ_0; the test statistic, however, is stated in terms of $\bar{X}$, since this value is what we actually have available.

4 Test statistic:

$$t_v = \frac{\bar{X} - \mu}{s_{\bar{X}}}$$

Since the sample size is 6, we have $v = n - 1$ or $v = 5$ for the t_v test. The value of v for t_v is the same as that for s.
5 The significance level: we set the α risk at 0.05.
6 This will be a two-tailed test, since we want (as the hypotheses are stated) merely to detect whether or not there is a difference between the present process average and the long-time process average; we are not interested in specifying a particular direction of the possible difference.
7 We find in Table A·2 the required t_v value, thus: We enter the table in the row for the appropriate value of v (here $v = 5$, one fewer than the sample size of 6); the t value is given in this row for various α risks; under the number 0.05 (two-tail values) we read 2.5706. If the calculated t_v value lies beyond the critical value in either the negative or positive

direction, we will reject the null hypothesis. Thus the critical regions for this test lie beyond $t_\nu = -2.5706$ and $t_\nu = +2.5706$.

8 We compute t_ν by

$$t_\nu = \frac{\bar{X} - \mu}{s/\sqrt{n}} = \frac{52.5 - 50.0}{3.4/\sqrt{6}} = 1.7$$

9 Conclusion: Since the calculated t_ν value is smaller than the critical value, we accept the null hypothesis. We have not shown from these data that the current process average is significantly different from the long-run average.

Had we wished from the same data to answer the question: is the present process consistent with a specification of a filling volume *not greater than* 50.0 ml, we would have then been involved with a one-tailed test. The hypotheses would then have been written

$$H_0: \mu = \mu_0 = 50.0 \qquad H_1: \mu > \mu_0$$

The critical region would then have been one-sided, and we would have entered the table on the same row but would have found the desired α risk on the bottom horizontal line. The one-tailed critical value for $\nu = 5$ and an α risk of 0.05 is read as 2.0150. The observations in this example would have caused us to accept the null hypothesis for the one-tailed test also. Had we been interested in a one-tailed test involving whether or not the volumes were consistent with a specification of *not less than* 50.0 ml, we would have prefixed a minus sign to the table t_ν value.

The preceding examples involve a test of a sample from a process against a known average process level; exactly the same method is used for testing a sample against a specification. Here the long-run behavior of the process is not known, but a standard can be established. A t_ν test can be used, therefore, to decide (with any desired risk of type I error) whether a measuring device is performing correctly, or whether or not a process is conforming to a desired mean.

Observe from Table A·2 how the critical values vary with sample size for a given α risk. At the 0.05 level, for example, with $\nu = 1(n = 2)$ we require a value of more than 12 in order to permit us to reject the null hypothesis. With such a small sample, the deviation of the sample average from the mean must be more than 12 times the standard deviation before we can safely say that the sample is significantly different from the mean. If ν is increased only to 2 ($n = 3$), the required t_ν value falls to only a little more than 4. The limit, for $\nu = \infty$, is slightly less than 2 (1.9600, which is the t value for sigma known).

From such observations, these conclusions may be reached: We cannot detect a small difference between a sample mean and a standard if we are using a small sample size. The improvement associated with an increase in n is most noticeable below $n = 5$. Beyond this sample size, the critical values of t_v fall only slowly; once the sample size is as large as 10, a great increase in n is needed to change the critical t_v values. Beyond $n = 30$, the Student's t test is practically equivalent to the normal distribution test; that is to say, the Student's t distribution beyond $n = 30$ is practically equivalent to a normal distribution in standard form.

Observe, also from Table A·2, that as the specified α risk is reduced, the critical t_v value increases, more noticeably for small sample sizes. For example, for $v = 1$ ($n = 2$), at an α risk of 0.05 the required t_v value is nearly 13, but for an α risk of 0.005, the table t_v value is nearly 130. If we wish to run a small risk of error in rejecting the null hypothesis, we must find a large difference between the sample mean and the standard mean. With larger sample sizes, this change in critical t_v value is greatly reduced; thus we can operate successfully with a small risk of type I error only with large sample sizes. Here again, however, the advantage of a large sample size decreases, once the sample size is as large as about 10, and beyond $n = 30$ relatively little improvement in the sensitivity of the t test occurs.

PROBLEMS

3·22   The target value for a given process is 16.0 (μ_0). From the process these random samples were taken: 15.8, 15.6, 16.1, 15.6, 16.2, 15.9, 16.0.
  (a) Using an α of 0.10, test the hypothesis that the process average is equal to 16.0.
  (b) Using an α of 0.05, is the process average at most 16.0?

3·23   An egg timer was sold as a 3-minute timer. To test the hypothesis that the process which made the timers was really producing 3-minute timers, the following measurements were obtained from 10 examples selected at random from the shipping department (times given in minutes): 3.5, 3.4, 3.5, 3.3, 3.3, 3.4, 3.1, 2.9, 3.1, 3.0.
  (a) Using an α of 0.10, test the hypothesis that the timers are 3.0-minute timers.
  (b) Using an α of 0.05, are they 3.0-minute timers?
  (c) Test the hypothesis that the timers are at least 3.0-minute timers. (Use an α of 0.05.)

3·24   Using a watch with a second hand, try to estimate a 1-min time interval. Start by looking at the watch; then look away until you believe that 1 min has gone by; then look at the watch again and record the actual elapsed time (in seconds).
  (a) Make 10 such observations, and then test the hypothesis that the elapsed time is 60 sec, using an α of 0.05.
  (b) If your sample average is higher than 60 sec, test the hypothesis that the elapsed time is at least 60 sec. If the sample average is less than 60, test the hypothesis that the elapsed time is at the most 60 sec. Use $\alpha = 0.05$.

3·15 HYPOTHESIS TEST FOR TWO SAMPLE AVERAGES, σ UNKNOWN

In most practical situations in which we wish to test for a difference in the means of two processes, we will not be confident that the standard deviations are really known. We base our test of hypothesis on an equation like Eq. (3·2)

$$t_{(v_1+v_2)} = \frac{(\bar{X}_1 - \bar{X}_2) - (\mu_1 - \mu_2)}{s_{(\bar{X}_1 - \bar{X}_2)}} \tag{3·5}$$

in which we usually hypothesize that $(\mu_1 - \mu_2)$ is zero, and in which we have substituted an estimate of the sample standard deviation s for the population standard deviation σ. We will estimate the standard deviation from the sample data.

Inasmuch as we have two samples, we have two estimates of the standard deviation. We will assume here that these two estimates are, in fact, both estimates of a single population σ, and we will pool them. Again, we make use of the rule that variances are additive. Here we will weight each variance according to the number of degrees of freedom $(n - 1)$ associated with it; thus the pooled variance will reflect (through the value of v) the degree of reliability we place in it.

The pooled estimate of the variance we find by

$$s_p{}^2 = \frac{v_1 s_1{}^2 + v_2 s_2{}^2}{v_1 + v_2} \tag{3·6}$$

The pooled variance $s_p{}^2$ is the sum of the product of each sample variance multiplied by its degrees of freedom v, divided by the total degrees of freedom $(v_1 + v_2)$ for the two samples. The square root of this pooled variance is the pooled standard deviation s_p. It is this value for the standard deviation which we use in calculating the statistic on which the test of hypothesis is based.

$$t_{(v_1+v_2)} = \frac{\bar{X}_1 - \bar{X}_2}{s_p\sqrt{1/n_1 + 1/n_2}} \tag{3·7}$$

If the two populations have the same mean, the t_v value calculated from Eq. (3·7) follows the t_v distribution with degrees of freedom equal to $v_1 + v_2$. Thus we can compare the calculated t_v with the critical value to determine whether or not the sample means are significantly different.

Two samples of paper, produced by two machines, were tested for breaking strength. The data were

Machine A....	61	52	45	46		
Machine B....	48	71	69	68	58	70	64

We wish to know whether or not these data indicate a difference in the average breaking strength of the paper produced by the two machines.

From the sample data we compute the following:

$$\text{Machine A} \ldots \quad \bar{X}_A = 51.0 \quad s_A{}^2 = 54.0 \quad \nu_A = 3$$
$$\text{Machine B} \ldots \quad \bar{X}_B = 64.0 \quad s_B{}^2 = 69.7 \quad \nu_B = 6$$

We assume that the sample variances are estimates of the same quantity, and therefore pool them by Eq. (3·6):

$$s_p{}^2 = \frac{(3 \times 54.0) + (6 \times 69.7)}{3 + 6} = 64.44$$

From this, $s_p = 8.03$.

The test of hypothesis on the sample means now follows:

1   Random variables: $\bar{X}_A$ and $\bar{X}_B$, the mean breaking strength of each type of paper.
2   Assumptions: normal distribution of populations; independence of the two sets of observations; equal population variances.
3   Hypotheses:

$$H_0: \mu_A = \mu_B \qquad H_1: \mu_A \neq \mu_B$$

4   Test statistic: $t_{(\nu_1 + \nu_2)}$ as calculated by Eq. (3·7).
5   Level of significance: $\alpha = 0.05$, with a total of 9 degrees of freedom.
6   Critical regions found in Table A·2: $t_{9,0.05} = \pm 2.26$. If the calculated t_ν value lies beyond the critical value (in either the plus or minus direction) we will reject the null hypothesis. This is therefore a two-tailed test.
7   Calculation of t_ν:

$$t_{(\nu_1 + \nu_2)} = \frac{\bar{X}_1 - \bar{X}_2}{s_p \sqrt{1/n_1 + 1/n_2}} = \frac{51.0 - 64.0}{8.03 \sqrt{1/4 + 1/7}} = -2.58$$

8   Conclusion: Since the calculated t_ν value lies beyond the critical value (in the minus direction) we reject the null hypothesis, and conclude that the average breaking strengths of the two types of paper are not the same.

PROBLEMS

3·25   Two types of seat belts were tested to test the hypothesis of no difference between types, with the following results in pounds tensile strength:

$$\text{Type A} \ldots \quad 3{,}200 \quad 3{,}300 \quad 3{,}400 \quad 3{,}200$$
$$\text{Type B} \ldots \quad 3{,}000 \quad 3{,}200 \quad 3{,}000 \quad 3{,}100 \quad 3{,}200$$

Using a significance level of 0.10, are the seat belts different with respect to strength?

3·26   Two bowlers each bowled 20 games on the same two alleys within the same period of time. Bowler A had a mean score of 167.3 with a standard deviation of 5.5, while bowler B had a mean score of 170.1 with a standard deviation of 5.3.

(a) Test the hypothesis that bowler B is a better bowler than A, using an α risk of 0.05.

(b) Test the hypothesis that the two bowlers are of equal skill at a significance level of 0.10.

3·27 Random samples of 20 bolts manufactured by machine A and 10 bolts manufactured by machine B showed a mean length of 3.345 in. and 3.346 in., respectively. The standard deviations as calculated from the samples showed for machine A 0.001, and for machine B 0.002. Test the hypothesis that the machines are both making bolts of equal size, using an α of 0.05.

3·28 Random samples were taken from a production line where the total weight per package of chemicals was the measured response variable. Thirty random samples from the day shift had a mean weight of 18.2 oz with a standard deviation of 0.63 oz, while 30 samples taken from the night shift had a mean weight of 17.8 oz with a standard deviation of 0.54 oz. Using a significance level of (a) 0.05 and (b) 0.01, test the hypothesis that the day shift is putting more chemicals into the packages.

3·29 Two groups of persons with the same type of illness were treated in the same way with the exception that group A, with 100 persons, was given a new type of drug, while the 80 persons in group B did not get the drug. The average length of time of illness for group A was 13.2 days with a standard deviation of 1.5 days. The average length of time of the illness of the B group was 13.8 days with a standard deviation of 1.7 days.

(a) Test the hypothesis that the new drug is effective in shortening the length of the illness, using a level of significance of 0.05.

(b) Test the hypothesis that there is no difference in the length of illness in the two groups, using an α of 0.05.

3·16 TESTS OF PAIRED DATA, σ UNKNOWN

The test of hypothesis for paired data from populations with unknown standard deviation follows the method based on Eq. (3·3) with the substitution of the sample estimate of the standard deviation for the population standard deviation:

$$t_v = \frac{d - \delta}{s_{\bar{d}}}$$

where $s_{\bar{d}}$ is $s_d/\sqrt{n}$, s_d being the estimate, from the sample data, of the standard deviation of differences, and n is the number of *pairs* of data. The number of degrees of freedom for this test is $n - 1$.

Two densitometers are to be compared. These instruments measure the silver deposit in a photographic negative. Readings are made with each of the instruments on a set of different silver deposits; the entire set is measured once on each instrument. Since the level of measurement will vary, we treat the data in pairs; the *difference* in readings is the random variable. The data are presented in Table 3·4.

The mathematical sign of the differences must be retained in finding the sum and the average of the differences. From the data we find $\bar{d}$, the average

Table 3·4

Densitometer	Sample number						
	1	2	3	4	5	6	7
A	0.20	0.50	0.90	1.10	1.50	2.00	2.10
B	0.10	0.55	0.80	1.00	1.30	1.80	2.00
Difference, d_i	+0.10	−0.05	+0.10	+0.10	+0.20	+0.20	+0.10

difference in the readings which is here 0.011; and s_d, the standard deviation of the set of differences which is here 0.084. The test procedure is

1 The random variable: d, the difference between the paired readings of the two instruments.
2 Assumptions: normal distribution of differences; difference independent of the level of measurement.
3 Hypotheses:
$$H_0: \delta = \delta_0 = 0 \qquad H_1: \delta \neq 0$$
where δ is the mean of the population of differences.
4 The test statistic: t_v from
$$t_v = \frac{\bar{d}}{s_d/\sqrt{n}}$$

If there is no difference between the instruments, the t_v values follow the Student's t distribution for $v = n - 1$, where n is the number of *pairs* of data. The numerator of this fraction contains only one term, because we are hypothesizing that $\delta = 0$. By using this test statistic, we are again removing the influence of within-instrument variation in order to keep the standard deviation limited to random error.

5 For $v = 6$ we find the two-tailed critical value of t_v at the 0.05 level in Table A·2. It is 2.45. The calculated t_v value must exceed this number (either positively or negatively) in order for us to reject the null hypothesis. Here we use a two-tailed test because as the hypotheses are stated we are interested in a significant difference in either direction.
6 Calculation:
$$t_2 = \frac{0.11}{0.084/\sqrt{7}} = 3.45$$

Since this value exceeds the critical value, we reject the null hypothesis and conclude that we have demonstrated, from these data, that the two instruments differ.

PROBLEMS

3·30 An analyst wishes to determine whether potentiometry and coulometry are equally good methods for determining the concentration of lead iodide in solution. Five samples of varying concentration are prepared, and a portion of each sample is analyzed by both methods. Use the data to decide whether or not the methods are equivalent at a confidence level of 0.05, and at a confidence level of 0.01.

Sample	Potentiometry	Coulometry
1	0.0015	0.0026
2	0.0003	0.0005
3	0.0029	0.0022
4	0.0018	0.0016
5	0.0046	0.0051

3·31 Two different calorimeters were used to measure the heat required to melt 1 gram of ice. Eight independent determinations were made with each instrument. Do the instruments differ at the 0.05 level of confidence?

Calorimeter A	Calorimeter B
79.0	78.0
78.5	78.0
78.0	77.0
78.5	78.0
76.5	76.0
77.7	77.0
78.5	76.0
78.7	77.0

3·32 Two analysts, A and B, determined the percentage of hydroquinone in a photographic developing solution for each of 7 different samples of developer. Do the two analysts agree?

Sample	1	2	3	4	5	6	7
Analyst A....	10.5	10.7	9.6	10.8	7.6	15.2	8.7
Analyst B....	10.9	10.5	9.8	10.8	7.5	15.3	8.8

3·33 Does the pressure of a reaction tank affect the yield of product based on the following data representing 8 *independent* runs at each pressure?

Pressure A....	37	31	34	26	29	28	30	32
Pressure B....	32	30	33	31	29	32	30	30

3·34 The yield of a production process was measured both by an assay method and by chemical analysis on each of 10 successive batches. At the 0.05 level, do the methods differ (a) by a hypothesis test on the means; (b) by a paired-comparison method?

Assay	81	78	91	82	73	66	82	88	96	81
Chemical....	82	70	80	75	75	69	80	85	80	80

3·35   A micrometer was tested against a series of standard blocks. Is the instrument in error?

Standard	Reading
0.0003	0.0002
0.1000	0.1000
0.2000	0.1997
0.3000	0.2998
0.4000	0.4003
0.5000	0.4999
0.6000	0.6000
0.7000	0.6998
0.8000	0.7999
0.9000	0.8998
1.0000	1.0004

3·36   Two suppliers each submitted a series of samples of their product. Is the average product different?

Supplier A....	94	97	97	95	97	94	96	95	92
Supplier B....	96	99	97	99	97	90	95	89	

QUESTIONS

3·1   What is a test of hypothesis, and when should it be used?

3·2   When making a test of hypothesis, how do we test the null hypothesis?

3·3   Why is it necessary to state the alternative hypothesis when setting up the null hypothesis for test?

3·4   What are the two types of errors one might make as a result of running a statistical test of hypothesis?

3·5   What are the risks which are associated with the two types of errors we might make in hypothesis testing?

3·6   If the sample size remains constant, for a hypothesis test, what effect will changing the α risk have on the β risk?

3·7   How is it possible to reduce both the α risk and the β risk at the same time in a test of hypothesis?

3·8   What is level of significance?

3·9   If the null hypothesis is accepted, what are we really saying about the hypothesis?

3·10   What do we mean when we say we have rejected the null hypothesis?

3·11   What is the general formula for the test statistic?

3·12   If we set the α risk, what effect does increasing the sample size have on the hypothesis test for the means, if sigma is known and remains constant?

3·13   In the test of hypothesis of the means, what probability scale is being used when:
(a) sigma is known
(b) sigma is unknown

3·14   What is the difference between a one-tailed and a two-tailed test of hypothesis of the means?

3·15   When should a one-tailed test of hypothesis for the mean be used?

3·16   When sigma is known, how does one look up the critical regions of a test of hypothesis of the mean when:

(a) a one-tailed test is used

(b) a two-tailed test is used

3·17   In hypothesis tests of the mean, what is the difference between testing sample mean versus a standard, and testing sample mean versus sample mean?

3·18   What is a pooled variance, and when should we pool variances?

3·19   What is the difference between a t value and a t_v value?

3·20   When should the data be paired, and the differences of the paired data be used as the test of the hypothesis?

3·21   If a process is said to have a known sigma, what is implied about the process?

3·22   What are the differences between hypotheses for the mean, when testing sample mean versus sample mean, first with the sigmas known, and then with the sigmas unknown?

3·23   What steps are involved in the running of a test of hypothesis?

3·24   What role does degrees of freedom play in looking up the critical regions when:

(a) sigma is known

(b) sigma is not known

3·25   What is the general shape of the Student's t distribution?

(a) When does it approach the normal-curve distribution?

(b) When does it become the normal curve?

3·26   List the six tests of hypotheses discussed in this chapter, and tell when each one should be used.

3·27   In a test of hypothesis of the mean, why should a sample of more than one observation be used to estimate the population mean?

3·28   When working with large sample size, say $n = 100$ or more, why is it possible in hypothesis tests of means to use the normal curve in place of the Student's t distribution to find the critical regions, even if sigma is not known?

3·29   Why is it seldom the case that the standard deviation of the differences of paired data is known?

3·30   Why should the α risk be set prior to running a test of hypothesis?

REFERENCES

16, 28, 31, 33, 37, 46, 47, 62, 67, 73, 81, 82, 115, 121, 123, 125

4 Interval Estimates for the Mean

Suppose that we are interested in *knowing* the mean of a population. This is a problem which is different from that of deciding whether or not this mean deviates from some hypothesized value. We sample the population and compute an estimate $\bar{X}$, the sample mean. Now we want to know how good this estimate is. If we were to take another sample, we almost surely would obtain a different $\bar{X}$ as another estimate of the population mean μ. Thus a single value of a sample mean is insufficient to enable us to specify the value of the population mean. It is more sensible to find a region of reasonable values within which the true population mean actually lies.

As we have said, this problem differs in approach from the hypothesis testing described in the previous chapter. Perhaps we are interested in getting an estimate of the number of good parts an operator can make in 1 hr. A sample might be 60 pieces. The question now is, at what level is this operator working? This is not the same question as, is this operator matching the general average of 65 pieces as other operators have done in the past? We may at times be interested in answering the latter question—if we want to decide whether or not this operator needs more training or deserves a raise or a reprimand. But at other times we are not interested in reaching decisions of this type. All we want to know presently is at what level is this specific operator performing. In fact, in many cases we have no basis for even setting up a hypothesis to test. The machines or method or material the operator is using may be new, so that there is no standard of comparison.

Even when we have such a basis, we may want to estimate the level of operation if we should find his performance significantly different from the value of the null hypothesis.

4·1 POINT ESTIMATES OF PARAMETERS

We speak of a sample mean as a *point estimate* of the population mean. In general, a point estimate is a *single* statistic which we use as the best single indication of a population parameter. In order to find the average level at which a process is operating, we take a sample of the process output, find the mean of this sample, and then consider the mean of the sample as an estimate of the mean of the entire product. Similarly, a sample standard deviation is a point estimate of a population standard deviation.

A satisfactory point estimate has these characteristics, among others:

1 It is *unbiased*, meaning that the average value for a large number of similar samples approaches the parameter value. The sample mean is an unbiased statistic: If we were to take *all* possible samples of size n from a given population, we would find that the average $\bar{X}$ of all the sample averages would be the same as the population average μ. Thus the mean of the sample averages would approach the value of μ as the number of samples were increased.

On the other hand, the sample standard deviation is *not* unbiased. A value of $\bar{s}$, even if found for many samples of the same size from a fixed population, does not properly estimate σ. In fact, the value of $\bar{s}$ is consistently smaller than the value of σ which we would like the sample standard deviations to estimate.

Although s is a biased estimate of σ, the sample variance s^2 is an unbiased estimate of the population variance σ^2. We can thus average a set of sample variances for samples taken from a fixed population with the belief that the average thus found does not consistently falsely estimate the population variance. This advantage of the variance, as contrasted with the standard deviation, is related to the difference between averaging a set of numbers and averaging their squares. For example, the mean of the set of values 3, 4, and 5 is 4. If we square these values we obtain 9, 16, and 25. The mean of the squares is approximately 16.7; the square root of this number is 4.1. Thus we see that the square root of the average of the squares is larger than the average of the original numbers. Similarly, the square root of the average variance will be larger than the average standard deviation, thus correcting the bias inherent in the use of the standard deviation itself.

2 It is *consistent*, meaning that the value of the statistic approaches the value of the parameter as the sample size increases. This is true of the sample mean: as we increase the value of n, the value of $\bar{X}$ comes closer and closer to the value of μ. Similarly, the sample variance is a consistent estimator of the population variance.

Even though we have, in the sample mean and variance, unbiased and consistent estimators of the population parameters, we still have, in a point estimate, only a single value of the statistic as a basis for an inference about the value of μ or σ. From a single fixed population, we can obtain many different values of $\bar{X}$; hence we can obtain many different point estimates of the value of μ. The likelihood that any specific $\bar{X}$ will equal μ is insignificantly small. Thus a point estimate alone gives us only little evidence about the true value of the population mean. The preceding remarks apply equally to the estimation of the population variance.

Interval estimates are better than point estimates because they enable us to say, as point estimates do not, where the population parameter probably lies.

4·2 INTERVAL ESTIMATES, σ KNOWN (SINGLE MEAN)

Beginning with a point estimate $\bar{X}$ of a population mean μ, we can increase the usefulness of the estimation by calculating an interval estimate of the parameter value. We desire a technique of estimation that will include the parameter in the interval a specific fraction of the time. Once we have found the interval, we will be able to say that the mean of the population lies within a specific region, and we will be able to say what the probability is that such a statement is correct.

Let us specify that we want a probability of 95 percent that we will be correct when we say the population mean lies within the desired interval. We base the calculation of the interval on the standard normal form

$$t = \frac{\bar{X} - \mu}{\sigma_{\bar{X}}}$$

From Table A·1 we find the *two*-tailed value of $t_{0.05}$ to be 1.96. The inference is that since for a mean μ, a sample $\bar{X}$ will lie within $\pm 1.960\sigma_{\bar{X}}$ 95 percent of the time, then 95 percent of the time that we place an interval of $\pm 1.96\sigma_{\bar{X}}$ around $\bar{X}$ it should include μ. That is to say, 95 percent of the time the following statements will be true:

$$-t_{\alpha/2} < t < +t_{\alpha/2}$$

$$-1.96 < \frac{\bar{X} - \mu}{\sigma/\sqrt{n}} < +1.96$$

$$\bar{X} - \frac{1.96\sigma}{\sqrt{n}} < \mu < \bar{X} + \frac{1.96\sigma}{\sqrt{n}}$$

or, in general,

$$\bar{X} - \frac{t_{\alpha/2}\sigma}{\sqrt{n}} < \mu < \bar{X} + \frac{t_{\alpha/2}\sigma}{\sqrt{n}} \tag{4·1}$$

The number 0.95 is called the *confidence coefficient*; this value depends on the percentage of times we expect the true value of the population mean to lie within the calculated interval. The end points of the interval are known as *confidence limits*, and the interval itself is the *confidence interval*.

If we use 0.95 as the confidence coefficient, then $t = \pm 1.96$. We are saying that we expect to be correct 95 percent of the time when we state that the population mean falls within the confidence interval about the sample mean. Thus we have in this case 95 percent confidence of making a correct statement. If we use a confidence coefficient of 0.90, we find $t = \pm 1.645$, and expect to be correct 90 percent of the time when we estimate the population mean to fall within the interval $\bar{X} \pm 1.645\sigma/\sqrt{n}$. We have narrower (more precise) limits when we use a confidence coefficient of 90 percent rather than 95 percent, but less assurance of being correct. Unless we have a strong reason to choose another level, it is customary to use a 95 percent confidence coefficient.

In Eq. (4·1) we have a statement of the expected position of the population parameter μ. In this statement, these factors are involved: the value of a sample statistic $\bar{X}$, the confidence coefficient we wish to use (which fixes t), the population standard deviation, and the sample size. For a given α, σ, and n, the *width* of the confidence interval is determined, but the *actual values* of the limits will vary with the observed value of $\bar{X}$. If we calculate a long series of such intervals, most of them will contain the population mean, but some of them will not. If we set a 95 percent confidence coefficient, in the long run 5 percent of the intervals will not contain the mean.

The *width* of the confidence interval depends on the values of α, σ, and n. The confidence limits are close together if α is large (and t is therefore small), but this means that we have less confidence in our ability to estimate the parameter. The confidence interval can be narrowed if the population σ can be reduced, but this often requires an improvement in instrumentation we can hardly afford. The most practical way to reduce the size of the confidence interval is to increase the sample size. By so doing, we use a distribution of averages that is tighter. Thus we narrow the confidence interval at the cost of increased effort and diminished returns as n is made larger. In Chap. 6 we discuss methods of determining the sample size required to attain any desired confidence level.

We wish to know the average percent of impurities in a large container of benzene. We know from many analyses that the standard deviation of individual tests is 0.05 percent. We will find a confidence interval for the mean at the 95 percent level. We take 10 samples, analyze them, and find the sample mean to be 2.48 percent. The t value for 95 percent confidence is, as before, found to be ± 1.96. We substitute the known values in Eq. (4·1):

$$2.48 - \frac{1.96 \times 0.050}{\sqrt{10}} < \mu < 2.48 + \frac{1.96 \times 0.050}{\sqrt{10}}$$

$$2 \cdot 449 < \mu < 2.511$$

Thus there is a probability of 95 percent that we are right when we say the true mean of the population lies between the values of 2.449 and 2.511 percent.

PROBLEMS

4·1   A random sample of $n = 16$ was taken from a process, and $\bar{X}$ was calculated as being 2.5. If the standard deviation of the process is known to be 0.4:

(a) Place 90 percent confidence limits on the population mean.
(b) Place 95 percent confidence limits on the population mean.

4·2   From $n = 25$, $\bar{X} = 365$, and $\sigma = 30$:
(a) Calculate $\sigma_{\bar{X}}$.
(b) Place 90 percent confidence limits on the population mean.
(c) Place 95 percent confidence limits on the mean.

4·3   Two populations were tested to see if they could be thought of as being the same.
From population A, $\bar{X}_A = 85.0$, $n_A = 16$, $\sigma_{\bar{X}} = 2.0$.
From population B, $\bar{X}_B = 84.0$, $n_B = 9$, $\sigma_{\bar{X}} = 1.5$.
(a) Place 95 percent confidence limits on the mean of population A.
(b) Place 95 percent confidence limits on the mean of population B.
(c) By looking at the answers to parts (a) and (b), would you say the population means are, or could be, equal?

4·4   Twenty persons with normal hearing were tested for reaction time to an auditory stimulus and $\bar{X}$ was found to be 0.145 sec. If the standard deviation of reaction time is known to be 0.025 sec place confidence limits on the mean reaction time for normal-hearing people to this stimulus.
(a) Find 95 percent confidence limits.
(b) Find 90 percent confidence limits.

4·5   The standard deviation of a population is known to be 4.0. Samples of different sizes were taken from the population, and in every case the value of $\bar{X}$ was 100.0. Place 95 percent confidence limits on the mean for (a) $n = 4$; (b) $n = 9$; (c) $n = 100$. What relationship exists between the width of the confidence interval and the sample size, other factors being constant?

4·6   How does the width of the confidence interval change when the selected α risk changes, other factors being held constant?

4·3   CONFIDENCE LIMITS FOR μ, σ UNKNOWN

When we wish to place confidence limits on a population parameter and we do not know the value of the population standard deviation, we work with the value of s found from the sample we have taken. We base the confidence-limit calculations now on Student's t distribution, and thus on the value of $t_{v, \alpha/2}$ where the subscript signifies the number of degrees of freedom $(n-1)$ and the subscript $\alpha/2$ is 0.025 if we wish to use a 95 percent confidence level, and 0.05 if we are satisfied with 90 percent confidence. Thus $t_{9,0.025}$ is the table value for a sample size of 10 and a confidence level of 95 percent.

The confidence limits are found from

$$\left(\bar{X} - t_{\nu,\alpha/2}\frac{s}{\sqrt{n}}\right) < \mu < \left(\bar{X} + t_{\nu,\alpha/2}\frac{s}{\sqrt{n}}\right) \tag{4.2}$$

Equation (4·2) differs from Eq. (4·1) not only in the method of finding the t values for the confidence coefficients, and in the substitution of s for σ but also in that it implies a collection of confidence intervals. Each different sample we take will give us values of $\bar{X}$ and s that will probably differ from those we would obtain from another sample taken from the same population. Thus the values of the confidence limits will change with the value of $\bar{X}$ and the width of the confidence interval will change with the value of s, even for constant sample size. Furthermore, for small sample sizes, the value of t_ν is large, and the confidence interval will be large. We pay for our ignorance (in this case of σ) by having a wider, less informative, confidence interval. As n increases, the confidence coefficient t_ν approaches the value of t used when sigma is known.

In a situation involving the chemical analysis of copper in an alloy, we wish to place 95 percent confidence limits on the mean percentage of copper. We have no information about the standard deviation of such analyses. For randomly selected samples from a single lot, we find these observations: 26.72, 26.65, 26.74, 26.65, 26.70, 26.68. From these data we find $\bar{X} = 26.69$ and $s = 0.037$. We note that $n = 6$ and therefore $\nu = 5$.

The table value for $t_{5,0.025}$ is 2.5706. Substitution in Eq. (4·2) gives

$$[26.69 - (2.5706 \times 0.037)] < \mu < [26.69 + (2.5706 \times 0.037)]$$

$$\frac{\sqrt{6}}{26.59 < \mu < 26.79}$$ $$\sqrt{6}$$

$$26.65 < \mu < 26.73$$

Thus we can say with 95 percent confidence that the true population mean lies between the two limits we have found.

PROBLEMS

4·7 From a sample of $n = 14$, the sample average $\bar{X}$ was found to be 35.0, and the sample variance was computed as 16.0. Place the following confidence limits on the population mean:

(a) 95 percent confidence limits
(b) 90 percent confidence limits
(c) 50 percent confidence limits

4·8 Place 95 percent confidence limits on the population mean if the sample average is 35.0, and the sample variance is 16.0, using the following sample sizes: (a) $n = 5$; (b) $n = 10$; (c) $n = 15$; (d) $n = 30$.

4·9 Place 95 percent confidence limits on the population average if the sample average of 35.0, as determined from $n = 15$, has a sample variance of (a) 20; (b) 10; (c) 4.

4·10 An incoming lot of solvent is tested for pH (a measure of acidity). Five randomly selected samples gave measurements of 5.6, 5.7, 5.6, 5.6, and 5.1.

(a) Determine the mean and the standard deviation of the sample.
(b) Place 90 percent confidence limits on the mean pH value for the solvent.
(c) Place 95 percent confidence limits on the mean pH value of the solvent.

4·11 A dairy refuses to accept milk which gives, by a standard test, more than 5,000 colonies of bacteria per milliliter. A shipment of 100 cans of milk is tested, using a random sample of ten 1-ml samples. The following data are obtained: 5,370, 4,890, 5,100, 4,500, 5,260, 5,150, 4,900, 4,760, 4,700, 4,870.

(a) Place 95 percent confidence limits on the bacteria count.
(b) Place 90 percent confidence limits on the bacteria count.
(c) Test the hypothesis that this lot meets the required specification of not more than 5,000 colonies.

4·12 Select a population—any one you please.
(a) Identify the population and specify the response variable.
(b) Take from the population a sample of $n = 16$.
(c) Compute the sample average and the sample standard deviation.
(d) Place 95 percent confidence limits on the population mean.

4·4 CONFIDENCE LIMITS FOR THE DIFFERENCE OF TWO POPULATION MEANS, σ KNOWN

We calculate confidence limits for the difference between two population means when we want to estimate the size of the difference between them. (We would use a test of hypothesis if we wanted to know whether or not the two population means should be considered as the same.) Again we resort to the basic concept of the t distribution:

$$t = \frac{\text{statistic} - \text{parameter}}{\text{standard error of the statistic}}$$

Here we use a modification of Eq. (4·1):

$$[(\bar{X}_1 - \bar{X}_2) - t_{\alpha/2}\sigma_{\bar{X}_1 - \bar{X}_2}] < (\mu_1 - \mu_2) < [(\bar{X}_1 - \bar{X}_2) + t_{\alpha/2}\sigma_{\bar{X}_1 - \bar{X}_2}] \quad (4·3)$$

where $\bar{X}_1$ and $\bar{X}_2$ are the two sample means, and μ_1 and μ_2 are the two population means.

The application of this rule requires that we have knowledge of the standard error of the difference in sample averages $\sigma_{\bar{X}_1 - \bar{X}_2}$, and we must specify as before the confidence level at which we wish to work.

We can find the standard deviation of the difference between two sample averages if we know the standard deviation of each population from which the samples were taken. Assuming that the observations are independent, we can use the formula by which the variances of the two populations are added:

$$\sigma_{\bar{x}_1 - \bar{x}_2} = \sqrt{\frac{\sigma_1^2}{n_1} + \frac{\sigma_2^2}{n_2}} \tag{4·4}$$

Equation (4·4) can be simplified if the population standard deviations are equal:

$$\sigma_{\bar{x}_1 - \bar{x}_2} = \sigma \sqrt{\frac{n_1 + n_2}{n_1 n_2}} \tag{4·5}$$

and, if we can arrange the data collection so that the sample sizes are equal, making $n_1 = n_2 = n$,

$$\sigma_{\bar{x}_1 - \bar{x}_2} = \sigma \sqrt{\frac{2}{n}} \tag{4·6}$$

The point of such a calculation as we have just described is this: Having reason to believe that there is a difference between two population means, we have taken samples from the two populations, and we have estimated the size of the difference. Since we have used only sample means, and not the population means, we have only a point estimate of the difference between the two values of μ. This estimate will be reasonably reliable if the sample sizes are large; we would, however, rather have an interval estimate of the difference. To make an interval estimate permits us to assign some degree of confidence to the interval within which we believe that the true but unknown estimate lies.

It is often unreasonable to try to set two machines or two processes to exactly the same mean level. We usually will be satisfied if the difference is small enough to cause both lots of product to be acceptable to the buyer or other user. If we have the variance of the two machines under control, we will be concerned only when the means of the two processes have drifted a specific distance apart. As long as the mean values of the two populations remain sufficiently close together, we can avoid the costly procedure of trying to find causes for a small difference. If the calculated values for the confidence limits include the value zero, we often act, as we do in hypothesis testing, as if no difference exists. If we are necessarily using a small sample size, the confidence interval will be large. To accept the supposition that (if zero lies within the interval) no difference exists is similar to running a test of significance with a small sample. In this circumstance, that we have not been able to accept an alternative hypothesis does not mean that the hypothesis is not true.

The use of equal sample sizes has two results: It simplifies the calculation of the confidence limits; it minimizes the variance of the difference in the sample means and thus reduces the interval between the confidence limits. Since in most cases we can select the sample size within practical limits, it is well to see to it that both samples are of the same size.

To set confidence limits on the difference between two population means, we now require a specified α risk, from which we find the table value of t; the population standard deviations; the sample means computed for each of the two samples, preferably of equal values of n.

We make a study of two machines for the automatic production of tungsten lamps. The response variable is the life of the lamps to failure. A sample of 150 items from machine A had an average life of 1,400 hr; a similar sample of 100 items from machine B had a mean life of 1,200 hr. Past records indicate that the standard deviation of the items produced by machine A is 120 hr, and by machine B 80 hr. We wish to place 95 percent confidence limits on the difference in the average lifetimes of the populations of the items produced by the two machines. We first apply Eq. (4·4):

$$\sigma_{\bar{X}_A - \bar{X}_B} = \sqrt{\frac{120^2}{150} + \frac{80^2}{100}} = 12.6$$

and now Eq. (4·3):

$$[(1,400 - 1,200) - (1.96 \times 12.6)] < (\mu_A - \mu_B) < [(1,400 - 1,200) + (1.96 \times 12.6)]$$

$$175 < (\mu_A - \mu_B) < 225$$

Thus we can be 95 percent confident that the difference of the two population means lies between 175 and 225 hr.

PROBLEMS

4·13   Two machines, A and B, both make the same product. We believe that there is a difference in the two population means because of different setup conditions. From machine A, we take a random sample of 200, and find $\bar{X} = 123.4$ cc, with a standard deviation of 3.6 cc. A sample of 200 from machine B gave $\bar{X} = 132.6$ cc and a standard deviation of 3.7 cc. Because the sample standard deviations are similar, and because of the large sample sizes, we will suppose that we know the population standard deviation. Place confidence limits on the difference of the population means, (a) using 90 percent confidence; (b) using 95 percent confidence.

4·14   Because of a change in raw materials, a change in the level of operation of a process is suspected, though it is believed that the variance is stable at 0.36. The former raw materials, for a sample of 9, gave a mean value of 23.7 grams; the new materials, for a sample of 15, gave a mean of 24.2 grams. Place confidence limits on the difference of the two means, at (a) 90 percent confidence; (b) 95 percent confidence.

4·15 From a calculation of 95 percent confidence limits on the difference of the two means for which data are given below, decide whether or not the two means are the same:

$n_1 = 14; \bar{X}_1 = 0.557; \qquad n_2 = 12; \bar{X}_2 = 0.538$

Standard deviation is known to be stable and to be 0.002.

4·16 Place 95 percent confidence limits on the difference of two means for each of the following, using $\sigma = 4.0$ for all; in each case, $X_1 = 40.0$ and $X_2 = 45.0$:

    (a) $n_1 = 16, n_2 = 16$
    (b) $n_1 = 30, n_2 = 2$
    (c) $n_1 = 2, n_2 = 30$
    (d) $n_1 = 14, n_2 = 18$

4·17 From your answers to Prob. 4.16, what conclusions do you draw about the effect of changing the sample sizes, when the total sample size is constant?

4·5 DIFFERENCE OF TWO POPULATION MEANS, σ UNKNOWN

A method similar to that of the preceding section may be applied to situations in which we wish to place confidence limits on the difference between two population means $\mu_1 - \mu_2$, without prior knowledge of the population standard deviations. The method is based on Student's t_v distribution, here

$$t_v = \frac{(\bar{X}_1 - \bar{X}_2) - (\mu_1 - \mu_2)}{s_{\bar{X}_1 - \bar{X}_2}} \tag{4·7}$$

We will estimate the variability of the data from two samples, and we will find confidence limits by a rule similar to Eq. (4·3), with the substitution of $s_{\bar{X}_1 - \bar{X}_2}$ for $\sigma_{\bar{X}_1 - \bar{X}_2}$:

$$[(\bar{X}_1 - \bar{X}_2) - t_{v,\alpha/2}s_{\bar{X}_1 - \bar{X}_2}] < (\mu_1 - \mu_2) < [(\bar{X}_1 - \bar{X}_2) + t_{v,\alpha/2}s_{\bar{X}_1 - \bar{X}_2}] \tag{4·8}$$

We assume that both populations have the same variability—a common value of σ. We find a pooled estimate of this standard deviation from the two values s_1 and s_2 by a method which weights more heavily the sample standard deviation with the larger number of degrees of freedoms, since we attach more validity to the statistic from the larger sample:

$$s_{\bar{X}_1 - \bar{X}_2} = \sqrt{\frac{s_p^2}{n_1} + \frac{s_p^2}{n_2}}$$

where

$$s_p^2 = \frac{v_1 s_1^2 + v_2 s_2^2}{v_1 + v_2}$$

and therefore

$$s_{\bar{X}_1 - \bar{X}_2} = \sqrt{\frac{v_1 s_1^2 + v_2 s_2^2}{v_1 + v_2}\left(\frac{1}{n_1} + \frac{1}{n_2}\right)} \tag{4·9}$$

The value of v we will assign to $s_{\bar{X}_1 - \bar{X}_2}$ and to the value of t_v in Eq. (4·7) is the sum of the values of v for the two samples.

A random sample of 10 items from filling machine A had a mean weight of 350 grams, with a sample standard deviation of 10 grams. A similarly chosen sample of 14 items from filling machine B had a mean weight of 325 grams with $s = 8$ grams. We wish to place 95 percent confidence limits on the difference of the mean weights for the two machines as they are now operating.

From Eq. (4·9) we find

$$s_{\bar{X}_1 - \bar{X}_2} = \sqrt{\frac{9(10)^2 + 13(8)^2}{9 + 13} \left(\frac{1}{10} + \frac{1}{14} \right)} = 3.67$$

The table value for $t_{22,0.025}$ is 2.07. We now use Eq. (4·8):

$$[(350 - 325) - (2.07 \times 3.67)] < (\mu_1 - \mu_2)$$
$$< [(350 - 325) + (2.07 \times 3.67)]$$
$$17.4 < (\mu_1 - \mu_2) < 32.6$$

We have 95 percent confidence that the two machines are giving average filling weights that agree to between 17 and 33 grams, machine A being higher. We can compare our specification with this and have a basis for decision whether or not to adjust the machines for closer agreement.

Very often we are concerned with the comparison of two populations which we believe to be the same but which we would like to test to see if they are. In experimental work, or research and development work, the question of equality of results of means from two sources of supply or materials is often encountered. In this situation, the question must be answered by taking samples from the two independent populations and then running a hypothesis test of the means. The standard deviations of the two populations are not known and must therefore be estimated from the samples themselves. If the hypothesis of differences of the means indicates there is a difference, then the next question to be asked is, "How much of a difference is there?"

In placing confidence limits on the difference of two means, we usually are dealing with unknown standard deviations, which we must estimate from samples. The width of the confidence interval will as before depend on sample sizes, the confidence coefficient, and the standard error of the means. There are two populations and perhaps two different sample sizes. If each of the two populations has the same variance, or if they can be assumed to be essentially equal in variance, the standard error of the difference of the two means is given by Eq. (4·8). If, on the other hand, the two population variances do differ, a more complex (Fisher-Behrens) formulation must be used (see Chap. 5).

PROBLEMS

4·18 Consider the following: $n_a = 9$, $n_b = 14$, $\bar{X}_a = 50$, $\bar{X}_b = 55$, $s_a = 2$, and $s_b = 3$.

(a) Place 90 percent confidence limits on the difference of the two population means.

(b) Place 95 percent confidence limits on the difference of the two population means.

4·19 Using the same information as in Prob. 4·18, place confidence limits on the difference of the two means if:

(a) $n_a = 20$, $n_b = 20$

(b) $n_a = 100$, $n_b = 100$

4·20 A certain machine can bind books at the rate of 660 per hour with a standard deviation of 36, as calculated from 9 hr running time. A new operator running the same machine found that over an 8-hr run he was binding 600 books per hour with a standard deviation of 50 books per hour. Place 95 percent confidence limits on the difference of the two population means.

4·21 A difference in mean values of a certain impurity was suspected between two types of coated paper used in a graphic arts process. Five samples from each type of paper were taken and the results were as follows:

Brand A: $\bar{X}_a = 0.100$, $s_a = 0.023$
Brand B: $\bar{X}_b = 0.076$, $s_a = 0.020$

(a) Place 99 percent confidence limits on the difference of the mean values of impurity.

(b) Place 90 percent confidence limits on the difference of the mean values of impurity.

(c) Explain why the confidence intervals are so large.

4·6 CONFIDENCE LIMITS FOR PAIRED DATA, σ UNKNOWN

A drug is to be tested for its effect on blood pressure. We believe that the likely effect is that of a pressure decrease, and we want to estimate this decrease and place 95 percent confidence limits on the amount of change. A sample of 10 persons was used for the experiment, with results as follows (the response variable is blood pressure, coded):

	1	2	3	4	5	6	7	8	9	10	$\bar{X}$
Before taking drug	14	15	12	9	14	12	10	9	13	12	12.0
After taking drug	10	12	12	7	15	10	7	8	11	11	10.3

We have no a priori reason to suppose that the level of blood pressure is similar in the 10 different persons used as subjects in the experiment. The data indicate, in fact, that they differed considerably in this respect. If we apply Eq. (4·8) to these data and place confidence limits on the difference between the two mean values, we will include in the term $s_{\bar{x}_1-\bar{x}_2}$ between-person variability as well as the change (if any) induced by the drug. The inflation of this estimate of the sample standard deviation will have the effect of widening the confidence limits on the difference in the two means. From Eq. (4·9) we find $s_{\bar{x}_1-\bar{x}_2}$ to be 2.3, and the confidence limits on the difference of the two means from Eq. (4·8) to be $-3.1 < (\mu_1 - \mu_2) < 6.5$, from which result we might well conclude that no effect of the drug had been demonstrated by the experiment.

For data such as these, which can logically be arranged in pairs, and in which we fear the presence of large pair-to-pair variability, a better technique is available. We will ignore the values of X, and use instead the *differences* in the paired values as the statistic. By this method, we will eliminate between-person variability and thus reduce the width of the confidence interval. For each pair of data we find the difference d, from these the average difference $\bar{d}$, and also from the set of differences the standard deviation s_d. We let δ signify the difference in the two populations; it is now on δ that we wish to place confidence limits. The rule we use is analogous to those we have used before:

$$\left(\bar{d} - t_{v,\alpha/2} \frac{s_d}{\sqrt{n}}\right) < \delta < \left(\bar{d} + t_{v,\alpha/2} \frac{s_d}{\sqrt{n}}\right) \qquad (4\cdot10)$$

The calculation of s_d is carried out as shown in Table 3·1. Now application of Eq. (4·10) gives

$$\left(1.7 - t_{9,0.025} \frac{1.5}{\sqrt{10}}\right) < \delta < \left(1.7 + t_{9,0.025} \frac{1.5}{\sqrt{10}}\right)$$

$$\left(1.7 - 2.26 \frac{1.5}{\sqrt{10}}\right) < \delta < \left(1.7 + 2.26 \frac{1.5}{\sqrt{10}}\right)$$

$$(1.7 - 1.1) < \delta < (1.7 + 1.1)$$

$$0.6 < \delta < 2.8$$

This result enables us to say, with 95 percent confidence, that a real decrease in blood pressure is associated with the use of the drug, and that this decrease is between 0.6 and 2.8. Note that we have paid a price for the use of Eq. (4·10): we had a total of 18 degrees of freedom for the method of Eq. (4·8); when we use the sample differences, we have only 9 degrees of freedom for the calculation of the confidence limits. We are willing, however, to sacrifice the 9 degrees of freedom for the elimination of the between-pair variability.

Assumptions	Population standard deviation	Confidence limits placed on:	Formula
Single population, normal	Known	Population mean	(4·1)
Single population, normal	Unknown	Population mean	(4·2)
Two populations, normal, independent	Known	Difference in means	(4·3)
Two populations, normal, independent	Unknown	Difference in means	(4·8)
Two populations, normal, dependent	Unknown	Average difference in pairs	(4·10)

PROBLEMS

4·22   Two sensitometers were used to expose adjacent areas of several different strips of photographic film. The optical densities of the processed silver images were as given in Table 4·2.

Table 4·2   Optical Density of Silver Images

Film strip	Machine A	Machine B
1	1.47	1.49
2	1.42	1.41
3	1.54	1.57
4	1.38	1.41
5	1.41	1.45
6	1.47	1.46
7	1.56	1.60
8	2.01	2.10
9	2.00	2.05
10	2.45	2.55

(a) Place 95 percent confidence limits on the difference of the two means by using the paired-data technique.
(b) Why was it necessary to use the paired-data technique?

4·23   To determine the difference in measuring the concentration of lead iodide in solution by two methods, 5 samples of varying concentrations were prepared and submitted to each of two methods of analysis. The results were as listed in Table 4·3. Place 95 percent confidence limits on the difference of the means of the two types of analysis.

Table 4·3   Determination of Lead Iodide

Solution	Analysis 1	Analysis 2
1	0.0015	0.0026
2	0.0003	0.0005
3	0.0029	0.0022
4	0.0046	0.0051
5	0.0018	0.0016

4·24   Two types of cooking oil were used to test their influence on unpopped kernels of popcorn. Samples of 7 brands of popcorn were split into equal amounts; one-half of each brand was tested using each brand of cooking oil. The results were as listed in Table 4·4, where the response variable is volume of unpopped kernels.

Table 4·4   Cooking Oils for Popping Corn

Brand	Oil A	Oil B
1	13 cc	11 cc
2	6 cc	5 cc
3	37 cc	29 cc
4	15 cc	10 cc
5	22 cc	21 cc
6	49 cc	37 cc
7	9 cc	9 cc

(a) Place 90 percent confidence limits on the difference of the mean amounts of unpopped kernels from the two types of oil.
(b) Which of the two types of cooking oil would you recommend to ensure best results for popping corn? Why?

QUESTIONS

4·1   What is the purpose of running a test of hypothesis on two population means?

4·2   What is the reason for placing confidence limits on an estimate of the population mean?

4·3   For a point estimate to be realistic, what characteristics must it have?

4·4   What is meant by an interval estimate?

4·5   Why is an interval estimate a probability statement?

4·6   What role does the confidence coefficient play in the size of the confidence interval?

4·7   What is the probable error of the mean?

4·8   When the standard deviation of the population is known, what factors control the width of the confidence interval?

4·9   In the calculations for the 95 percent confidence interval for the mean with the standard deviation of the population known, what causes the interval to shift on the response scale when several different samples are taken?

4·10   In computing confidence limits on the mean, what is the difference between knowing sigma and not knowing sigma?

4·11   Under what conditions is it permissible to use the normal curve in the calculation of confidence limits for the mean when the standard deviation is not known?

4·12   When comparing two populations, if the standard deviations are known and equal, but the sample sizes are not equal, how is the pooled estimate of the standard error of the means arrived at?

4·13   If two populations are thought to differ in their mean values, and their standard deviations are not known, how can confidence limits on the difference of the means be computed?

4·14   In confidence-limit problems, why do we not become involved with risks of making a type I or type II error?

4·15   When making an interval estimate of the difference between two population means, why is it best to have the sample sizes the same?

4·16   State reasons for an interest in knowing the size of the difference between two means.

4·17   Problems involving the placing of confidence limits on the means of paired data are often referred to as the difference of two means. How do the two types of problems differ? How are they alike?

4·18   If we were to use paired-data techniques in a situation where they were not required, what would be the effects on the confidence intervals?

4·19   In general, how can a confidence interval be made smaller in width?

4·20   In general, when do we use the Student t_v distribution when placing confidence intervals on the means?

REFERENCES
2, 5, 16, 28, 29, 33, 37, 38, 43, 62, 67, 79, 81, 82, 118, 121, 125, 126

5 Statistical Inferences about Variances

In Chap. 3 we described statistical methods for making inferences about population means. By techniques that are similar in principle, we can make inferences about population *variability*, i.e., about population variances.

We make hypothesis tests of variances when we want to answer questions such as these: Is the variability of a current process consistent with the variability of the process as it has been known to operate over long periods of time? Is the variability of a current process less than that of the former process? Of two processes, which has the lesser variability?

Suppose that a packaging process has a long record of filling boxes to an average weight of 1.0 lb with a standard deviation of 0.05 lb, and thus a variance of 0.0025. An adjustment is made to the filling machine. We now wish to know whether or not the adjustment has affected the process variability. We take a sample of 5 packages, and find the sample standard deviation to be 0.10 lb and thus the sample variance to be 0.0100.

That the sample variance of 0.0100 is four times the former population variance of 0.0025 should not necessarily be taken as meaning that the new process variability is abnormally large. There is some probability that this sample, with this variance, could have been obtained from a process with variance no larger than that of the original process. We need to explore the relationship between the variance of a population and the variance we can expect to find for a sample of $n = 5$, or for any other sample size.

5·1 THE CHI-SQUARE DISTRIBUTION

In order to test a sample mean against a hypothesized population mean, we made use of a t distribution and a t statistic. From the t distribution we obtained a table of t values with which we could compare the calculated t statistics. Similarly, for testing a sample variance against a hypothesized population variance, we make use of a χ^2 (read "chi-square") distribution.

We assume that we are concerned with a normal distribution of observations which are independent of each other. If these assumptions are valid, a distribution of a set of sample variances (the samples having been taken from a fixed population) will follow, when plotted in cumulative form, a specific curve which varies with sample size. We can use such a plot, and table values associated with the plot, as the basis for a test of significance of sample versus population variance.

The distribution we use is the χ^2 distribution; it is related to the t distribution. For a population of known mean and standard deviation, the χ^2 distribution is in fact the sum of the t values squared; that is,

$$\chi_v^2 = \sum_1^n t^2 = \sum_1^n \left(\frac{X - \mu}{\sigma}\right)^2$$

The subscript v used with χ^2 identifies the number of degrees of freedom that defines the specific χ^2 distribution. If both μ and σ are known, $v = n$, the number of observations taken and summed as $\Sigma\,t^2$. If the observations X_i are normally distributed with known σ, but with unknown mean, we estimate the population μ by the value of $\bar{X}$. In this case, we lose a degree of freedom because we have computed the value of $\bar{X}$ from the data; this was discussed in Chap. 3. The value of χ^2 is now found from

$$\chi_v^2 = \sum_1^n \left(\frac{X - \bar{X}}{\sigma}\right)^2 = \sum_1^n \frac{(X - \bar{X})^2}{\sigma^2} \tag{5·1}$$

and the value of v here is equal to $n - 1$.

We can rearrange the numerator of the last fraction in Eq. (5·1) on this basis: The sample variance is defined as

$$s^2 = \frac{\Sigma\,(X - \bar{X})^2}{n - 1}$$

Therefore, the numerator $\Sigma\,(X - \bar{X})^2 = (n - 1)s^2$ and

$$\chi_v^2 = \frac{(n - 1)s^2}{\sigma^2} = \frac{vs^2}{\sigma^2} \tag{5·2}$$

We use the term vs^2 as the numerator of a test statistic to test a sample variance against a population variance, or against a standard variance, which appears in the denominator. Thus the test statistic is the fraction in Eq. (5·2), which is to say that the value of χ^2, for samples of a given size

STATISTICAL INFERENCES ABOUT VARIANCES   109

taken from a given population, is a measure of ν times the size of the expected ratio of the sample variance to the population variance.

There are as many χ^2 distributions as there are sample sizes. The frequency plot for $n = 5$ is shown in Fig. 5·1. The distribution is far from normal,

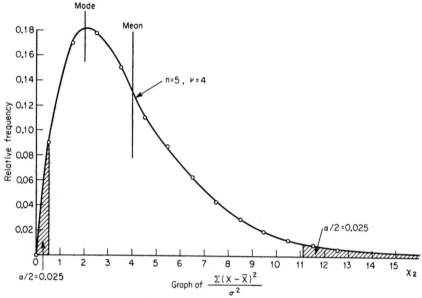

Fig. 5·1 χ^2 distributions for sample size of five.

unlike the sampling distribution of means. For small values of n the distribution is strongly skewed to the right, since variances less than zero cannot occur. The skewness is less as the sample size increases. The mean of the distribution is equal to ν, as Eq. (5·2) requires; the mode occurs at $\chi^2 = \nu - 2$.

We will use the χ^2 distribution as the basis for a test of significance of variances, in a manner similar to our use of the normal distribution as a basis for tests of the mean. What we will do is to test whether or not a sample variance fits the distribution expected for the statistic in cases where we will have hypothesized a specific value of σ^2. If the sample variance does not fit into the expected distribution, we will have reason to believe that the sample has come from a population with variance other than that which we have hypothesized. The method requires that we calculate a statistic according to Eq. (5·2) and compare this statistic with a table value. We will assign $\nu = n - 1$ degrees of freedom to the statistic.

5·2 α RISKS AND THE χ^2 DISTRIBUTION

Since the χ^2 distribution is skewed, not symmetrical, a two-tailed test of hypothesis that equally divides the α risk requires that the critical regions for the test be unequally spaced from the mean of the distribution. Consider Fig. 5·2, where we show the distribution of χ^2 for $n = 6$, or $\nu = 5$. We intend, as in tests of hypothesis for the mean, to specify two critical regions; if the calculated value of the test statistic lies in either of these regions, we intend to reject the hypothesis of no difference between the variances we are comparing.

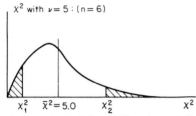

Fig. 5·2   Two-tailed critical regions, χ^2 distribution.

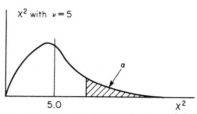

Fig. 5·3   One-tailed critical region, χ^2 distribution.

The mean of this distribution is at $\nu = n - 1 = 5$. In Fig. 5·2 the shaded areas represent the critical regions. The critical value in the left-hand tail will lie at a value less than 5, and that for the right-hand tail at a value greater than 5. The two values will not lie at equal distances from the mean because of the asymmetry of the distribution. We are testing to see if the χ^2 value $\nu s^2/\sigma^2$ we obtain from our sample is one that would be reasonable to find if σ^2, which we only estimate by s^2, actually equals σ_0^2. Thus we expect χ^2 to approximate the value 5. We know that sampling variability will make it unlikely for us to obtain exactly the value 5. The critical regions define how far from 5 we can expect the computed value of χ^2 to fall.

If we wish to choose a two-tailed α risk of 0.05, the two shaded areas in Fig. 5·2 will each contain an area of 0.025. In Table A·3 the areas of the critical regions are given always in terms of the right-hand areas. Thus the position of the point χ_1^2 marks off an area of 0.975 to the right; the position of the point χ_2^2 marks off an area of 0.025 to the right.

In most situations we fear, and would like to be warned of, only an *increase* in the variability of the process. Hence we often perform one-tailed tests of hypothesis for the variance, to protect ourselves against an increase in process nonuniformity. For a one-tailed test of hypothesis of variance, we consider the entire α risk to be represented by the area under the right-hand tail of the curve, as in Fig. 5·3. In such a situation, we would find only a single table value of χ^2.

5·3 HYPOTHESIS TESTS OF VARIANCE, POPULATION VERSUS STANDARD

If we wish to test the hypothesis that the population variance is equal to some given value, or to some standard, the null hypothesis is $H_0: \sigma^2 = \sigma_0^2$.

σ^2 is the population variance, and σ_0^2 is the hypothesized variance. If this hypothesis is correct, for all possible samples of size n, the statistic

$$\sum (X - \bar{X})^2/\sigma^2,$$

or the equivalent vs^2/σ^2, will follow the χ^2 distribution with $n - 1$ degrees of freedom.

To test the null hypothesis, we calculate the χ^2 value for the sample of size n which we have taken from the population under test, and compare the calculated value with the table values according to v and the α risks we have set. If the value of the calculated statistic falls in the critical regions (the shaded areas in Fig. 5·2) we take this as evidence that the null hypothesis was incorrect. If the value of the statistic does not lie in the critical regions but instead lies in the region of acceptance, we fail to find evidence sufficient to cause us to reject the null hypothesis.

This test of hypothesis is based on the ratio of the statistic s^2 (a point estimate of the population variance) to the true variance σ^2. If the ratio is unreasonably large, the indication is that the true variance is greater than the hypothesized variance. If the ratio is unreasonably small, there is indication that the true variance is less than the hypothesized variance.

A foundry has over a long period of time produced castings with a variance in weight of 4.0. A new process has been developed which is more economical and which gives the same average as the former process. The new process will be adopted if the variance of the new method is no more than the past variance. The question is: Is the new process variance consistent with a population variance of 4.0, or is it representative of a higher variance?

We take a random sample of six castings; the weights in pounds are 49, 55, 53, 57, 48, 53. From these data, the sample variance s^2 is 11.9. The steps in the test of significance are

1 Assumptions: normal distribution of the variable; randomly selected sample of independent observations.
2 Hypotheses: H_0: $\sigma^2 = \sigma_0^2$; H_1: $\sigma^2 > \sigma_0^2$. In these statements of the hypotheses, σ^2 signifies the variance of the new process population; σ_0^2 signifies the variance of the known (former) population.
3 The test statistic:

$$\chi^2 = (n - 1)\frac{s^2}{\sigma_0^2}$$

NOTE: We state the hypotheses in terms of σ^2, but we necessarily estimate σ^2 by using s^2, calculated from the sample data.
4 Level of significance: $\alpha = 0.05$; $v = 5$.
5 According to the statements of hypothesis, this is a one-tailed test; we are interested only in knowing whether or not the sample variance *exceeds* what would be expected from an unchanged population. According to Table A·3 for one-tailed tests we enter the table at the specified α

risk. Thus, to find the critical value of χ^2, we enter the table in the column headed 0.05; opposite $v = 5$ we read 11.07. Unless the table value is exceeded by the calculated value, we will accept the null hypothesis and reject the alternate that the variance has increased.

6   Substituting in Eq. (10·3), we obtain

$$\chi^2 = \frac{5 \times 11.9}{4.0} = 14.9$$

7   Conclusion: Since the computed value (14.9) is greater than the critical value (11.07) we reject the null hypothesis and conclude that the sample variance is not representative of a population with a variance of 4.0 lb. On the basis of the sample we have secured, we therefore may state (with a risk of error implied by the selected α) that the new process is making a product that is more variable than the established process.

PROBLEMS

5·1   The specified standard deviation of a certain process is equal to 0.010 in. Test the hypothesis that the population standard deviation is equal to the standard of 0.010, at the 5 percent level, with the intention to detect the possibility that the standard deviation may become too large. A sample of 10 parts is measured and recorded as follows:

10.011, 10.031, 10.025, 10.021, 10.000, 9.975, 9.980, 9.995, 9.970, 9.998

5·2   The standard deviation of a given process which has been operating "in a state of statistical control" is known to be 2.55 grams. After a small change has been made on the process the sample standard deviation ($n = 10$) was found to be 3.30 grams.

(a) Using an α of 0.05, test the hypothesis that the variance of the process after modification has not increased.

(b) Test the same hypothesis using an α of 0.10.

5·3   If the sample size used in Prob. 5·2 had been 50, and the sample standard deviation was the same value, test the hypothesis that the variance has not increased, using an α of 0.05.

5·4   The shelf life of a photographic chemical is advertised as being 120 days with a standard deviation of 10 days. Tests were run on 16 samples of the chemical for useful shelf life and the results showed an average shelf life of 118 days with a standard deviation of 15 days. At an α of 0.05, was there reason to doubt the claim being made by the producer of the chemical?

5·4 HYPOTHESIS TESTS OF VARIANCE, POPULATION VERSUS POPULATION

Although tests of population variance against a standard variance are useful, we are often concerned with situations in which we wish to compare the variability of one population with that of another. We may want to know whether or not two methods of measurement (two different techniques of chemical analysis, two different measuring instruments) are similar in variability. Similarly, we may wish to know whether or not two processes are delivering goods that vary in the same manner.

We will here be concerned with two samples, each from a population having possibly different variance. We will estimate the population variance from the sample variance. Just as we examined the expected relationship between sample variance and population variance by means of the χ^2 distribution, we now examine the expected relationship between sample variances.

5·5 THE F DISTRIBUTION

We imagine a single hypothetical normal population with variance σ^2. From this population we take all possible samples of size n_1, and for each sample we find the variance s_1^2. From this same population we take all possible samples of size n_2, and find the set of sample variances s_2^2. Now, we find all the possible ratios of the paired sample variances s_1^2/s_2^2. From these ratios we construct a frequency distribution; this is called an F distribution.

An F distribution displays the expected pattern of sample variance ratios for a *common* σ^2. By the use of this distribution, we can test whether or not a specific observation of a sample variance ratio indicates that the two samples really came either from a single population or from two populations having equal variances. As in other tests of hypothesis, we specify an α risk and state the conclusion of the test only to some degree of confidence.

An F test helps us decide whether or not two processes have similar variability. Thus an F test is often a useful preliminary when we intend to test a sample mean against another sample mean. In Chap. 3 we assumed no significant difference in the variability of two populations whose means we were testing; an F test now enables us to test this assumption. Furthermore, F tests are used in the analysis of data from an experiment in which we try to discover which of several factors affect a process. This type of data analysis is called *analysis of variance*; it is the subject of Chaps. 8 to 11.

Since in performing F tests we are concerned with two samples, we will have two different values of ν if the samples are of different size. Thus there are many different F distributions, forming a family of frequency curves. A change in ν for either sample changes the distribution. Some representative curves are shown in Fig. 5·4. The curves are similar to those for the χ^2 distribution because the F statistic is a ratio of χ^2 values. The mean of each curve lies at 1, since the distribution is based on the assumption of a common population variance. If two samples are in truth drawn from the same population, and the sample variances happen to be equal (an unlikely event), the ratio of the variances will be 1. If they differ, the ratio will be other than 1. Chance variations cause such ratios to be rarely exactly 1. The skewness

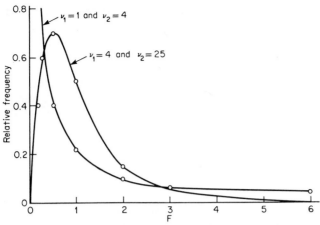

Fig. 5·4   F distributions for various sample sizes.

of the curves is caused by the impossibility of finding negative ratios, though the ratios can be large.

As in other tests of significance, we set the desired α risk, and on the basis of this (and sample sizes) fix critical regions; the values of the critical regions tell us when the ratio deviates so far from 1 that we reject the null hypothesis.

In Table A·4 are found some useful values of various F distributions. On facing pages are to be found critical values of the F distributions for a given desired α risk. Across the top of the pages are given some possible values of ν_1 (degrees of freedom, $n - 1$) for the first sample; in the left-hand column are found values of ν_2 for the second sample. Note that all entries in the body of the table are larger than 1. Because of this, these tables are used in slightly different ways, depending on whether we are running a one-tailed or a two-tailed test. If we are running only a one-tailed test, so that the hypotheses are $H_0: \sigma_1 = \sigma_2$; $H_1: \sigma_1 < \sigma_2$, we always form the test-statistic ratio $s_2{}^2/s_1{}^2$, that is, with the numerator containing the sample variance that, if it is abnormally large, will cause us to reject the null hypothesis. If we are running a two-tailed test, we always place the larger sample variance over the smaller one. By so doing, we will reject the null hypothesis if the two variances are sufficiently different, regardless of which is the larger. Here our hypotheses are $H_0: \sigma_1 = \sigma_2$; $H_1: \sigma_1 \neq \sigma_2$ and we will reject the null hypothesis if either variance is significantly larger than the other.

We wish to test to see whether or not the variability of two similar machines should be considered the same. We take two sets of random samples; the coded data are

Machine A....	35	38	36	40	38	35	39	
Machine B....	40	35	40	38	35	38	35	40

1   Assumptions: normality of each of the two populations; random selection of data; independence of the two sets of data. (The last assumption means

that the selection of items from the first machine was unaffected by the items selected from the second.)

2 Hypotheses:

$$H_0: \sigma_1{}^2 = \sigma_2{}^2 \qquad H_1: \sigma_1{}^2 \neq \sigma_2{}^2$$

The null hypothesis is that there is no significant difference in the variances of the two populations from which these samples were taken; the alternate hypothesis is that a significant difference exists. Thus this is a two-tailed test.

3 The test statistic: $F = s_1{}^2/s_2{}^2$, where $s_1 > s_2$.

4 Calculations give for machine A a variance of 38.9 and for machine B a variance of 40.4; we therefore designate the data from machine B as sample 1 and the other data as sample 2.

5 Level of significance: 0.05. Since we are looking only for a difference, without specification of the direction of the difference, we are running a two-tailed test. We therefore divide the α risk by 2, and enter the table on the pages which give F ratios for an α risk of 0.025. The number of degrees of freedom for machine B (which we have agreed to designate sample 1) is 7; for machine A, ν is 6. The table F ratio is found to be 5.70. Unless the calculated F ratio exceeds this value, we will accept the null hypothesis.

6 The calculated F ratio is 40.4/38.9, or 1.04.

7 Conclusion: We have not demonstrated by these data that there is a difference in the variances of the populations of the product of the two machines. We therefore accept the hypothesis that the variability of the two processes is the same.

Note that the asymmetry of the F distribution is shown by the difference between the table value for $\nu_1 = 7$ and $\nu_2 = 6$, which was found to be 5.70, and the table value for the reverse order of the two ν values, which is 5.12. Although in this case the table F ratio is far greater than the calculated ratio, it is always essential to distinguish between sample 1 and sample 2, and to associate the proper number of degrees of freedom with each sample. This is especially true where there is a small value of ν for one sample. (We write the generalized designation of the table F ratio as $F\nu_1, \nu_2, \alpha$.) Observe that $F_{3,1,0.025}$ is 864.16, but $F_{1,3,0.025}$ is only 17.443.

PROBLEMS

5·5 Samples of two different types of plastic were tested for breaking strength. The following data were obtained:

Type A: 60, 53, 45, 47 lb
Type B: 47, 72, 69, 67, 59, 71, 63 lb

Test to see whether or not the two types of plastic have the same variability in breaking strength.

5·6   Two machines are set up to do the same type of filling operation. From machine A, a sample of 20 filled bags (response variable—ounces) produced a sample variance of 0.14. From machine B a sample of 15 filled bags produced a sample variance of 0.24. Test the hypothesis that the two machines have the same variance in the filling operation, using the following levels of significance: (a) $\alpha = 0.05$; (b) $\alpha = 0.10$.

5·7   Twenty random samples of coal were taken from the same lot. Ten of these were analyzed for percentage carbon content by one laboratory, the remaining 10 by another laboratory. The sample variance as determined by laboratory A was 18.5 while the sample variance as determined by laboratory B was 7.8. Test the hypothesis of equal variances of the carbon content of the coal as measured by the two laboratories.

5·8   Two types of drying conditions were used to remove moisture from a wood product. The weights prior to drying and after drying were recorded and the percent loss of weight was used to measure the efficiency of the types of drying conditions. From type A drying conditions, a sample of 20 items produced a sample variance of 10 and a sample of 25 items from condition B produced a sample variance of 30.

(a) Test the hypothesis of equal variances in the two types of drying conditions, using an α of 0.05.

(b) Test the hypothesis that type A conditions are better than type B conditions, using $\alpha = 0.05$. ("Better" means the variance is smaller.)

5·9   From population A, sample variance is computed as 14, and from population B, the sample variance is found to be 20. Test the hypothesis of equal variances of the two populations if the sample sizes were as follows (use a constant α):

(a) $n_a = 4, n_b = 4$
(b) $n_a = 10, n_b = 4$
(c) $n_a = 4, n_b = 10$

5·6   HYPOTHESIS TESTS FOR MEANS FOLLOWING TEST FOR VARIANCES

In Chap. 3 we performed hypothesis tests for means always on the assumption that the population variances were equal. In the χ^2 and F tests we now have one way to test the validity of this assumption. Before running a hypothesis test of means, we should always check the variances to see whether or not they are equal.

If we run an F test on sample variances, and accept the null hypothesis, we have the right to pool the variances and then to proceed with the t_v test of the means. This procedure is based on our belief, confirmed by the F test, that the two sample variances s_1^2 and s_2^2 are estimates of the same population variance σ^2. In the pooling process, we use a weighted average of the sample variances, to give greater weight to that variance from the larger sample size.

If, on the basis of the F test, we reject the null hypothesis, we then conclude that the sample variances indicate that we are really dealing with two populations of different variability. The question now arises: Have we the right to perform a test of the means in situations in which the variances are unequal? Strictly, the answer to this question must be "no" for this reason: The t distribution is based on the pattern of sample means obtained from a single population. If we have, in reality, two populations, with variances σ_1^2 and σ_2^2, there is no single t distribution that fits. Intuitively, it seems reasonable, furthermore, that if we are concerned with two populations of considerably different variance, we must find a very large difference in sample means before we can attribute significance to this difference. In Fig. 5·5,

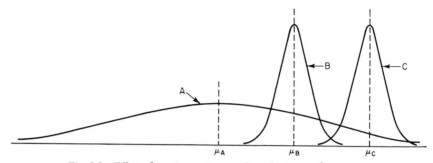

Fig. 5·5   Effect of varying variance on hypothesis tests for means.

the populations represented by curves A and B have different means. Because population A has a large variance, the distributions overlap despite the difference in the means. Thus samples from population B could not be distinguished from those from population A (although some of A could be distinguished from those of B). For samples to permit us to say with ease that the means differ, the situation would need to involve a comparison of the populations represented by curves such as A and C, for which we have a much larger difference in the means. For the populations represented by curves B and C, on the other hand, because they are equal in variance, we could expect to detect the difference in the means with samples of reasonable size. Thus it is worth the effort to make the population variances similar, by process adjustment or otherwise.

It is true that once a test of significance indicates that the variances differ, we know the populations are not the same. This knowledge may be sufficient without worrying about the means. In fact, in this case the question of whether or not the population means differ often becomes academic and can be ignored.

In spite of what we have just said, we will sometimes (because the practical situation seems to demand it) resort to approximate tests of means even though these tests are not theoretically entirely justifiable. Even though we will have rejected the null hypothesis for the variances, we may continue with a test of the means. We recognize that the second test is not based on as firm principles as we would like, but we wish to extract as much information as possible from data that are not all we would really like to have.

We make no *serious* error in testing means using a pooled variance from sample variances that have been shown to be significantly different if

1   The practical situation indicates that the variances ought to be equal, even though a significance test shows that they are different. We may have reason, for example, to be firm in our belief that two sets of measurements with the same micrometer ought to have about the same variability. We would, on the other hand, by no means hold to this belief in comparing micrometer measurements with yardstick measurements.

2   The variances are not far different, say not exceeding the 0.01 level of the F ratio.

If neither of the circumstances above is valid, but necessity nevertheless requires a test of the means, the following approximation to the Fisher-Behrens test may be used, with the understanding that this technique gives only a rough approximation to the desired answer. If the differences are large, the approximation becomes very poor.

We have two samples of sizes n_1 and n_2, with significantly different variances s_1^2 and s_2^2. We wish to test the hypothesis of no difference in the means $\bar{X}_1^2$ and $\bar{X}_2$. The test statistic, as before, is

$$t_v = \frac{\bar{X}_1 - \bar{X}_2}{\sqrt{s_1^2/n_1 + s_2^2/n_2}} \tag{5·3}$$

We must have a single value of v in order to enter the t_v table. We find this value by

$$v = \frac{(s_1^2/n_1 + s_2^2/n_2)^2}{(s_1^2/n_1)^2/v_1 + (s_2^2/n_2)^2/v_2} \tag{5·4}$$

We round off to the nearest integer the value found from Eq. (5·4).

PROBLEMS

5·10   A sample of 50 items supplied by company A yielded a mean value of 1,752, with a sample standard deviation of 102. A sample of 40 items of similar items supplied by company B had a mean value of 1,713 with a sample standard deviation of 90.

(a) Test the hypothesis of no difference in the two company variances, using a level of significance of 0.05.

(b) Run the test of hypothesis of no difference in the two population means, using $\alpha = 0.05$.

5·11   During a given week a process turned out 220 items of product, the mean weight of which was 2.46 lb, with a standard deviation of 0.57 lb.

During the next week when a different raw material was used, the mean weight of 205 items turned out was 2.55 lb with a standard deviation of 0.48 lb.

(a) Test the hypothesis of no difference in the variances of the production of the two different weeks using $\alpha = 0.10$.

(b) Test the hypothesis of no difference in the mean of the population with old materials and that of the population using the new materials, using $\alpha = 0.05$.

5·12   Two students set out to cover the same 10-mile distance by walking at constant speeds. The time it takes to cover each mile is measured and the results are as follows: Student A walks at an average rate of 4.6 miles per hour, with a standard deviation of 0.4 mile per hour. Student B has an average of 4.9 miles per hour, with a standard deviation of 0.8 mile per hour.

(a) Test the hypothesis that the two students are walking at the same degree of consistency, using an α of 0.10.

(b) Test the hypothesis that the two students have the same average miles per hour walking speed, using $\alpha = 0.10$.

5·13   Two populations with the same average have the following standard deviations as indicated by sample sizes of $n = 12$. Test the hypothesis of equal variances in all cases, using $\alpha = 0.05$.

(a) $s_a = 4.0, s_b = 4.5$
(b) $s_a = 4.0, s_b = 8.0$
(c) $s_a = 4.0, s_b = 16.0$

5·7   CONFIDENCE LIMITS FOR VARIANCE

By an argument similar to that used in Chap. 4, we can base interval estimations for variance on the appropriate sampling distribution, either χ^2 or F. We use the χ^2 distribution when we want to place confidence limits on a single population variance, and the F distribution when we want to place confidence limits on the ratio of two population variances.

If we wish to place 90 percent confidence limits on the variance based on $n = 10$, we have $\nu = 9$. We conceive of the distribution of χ^2 to be marked off in such a way that 5 percent lies in either tail, as in Fig. 5·2, so that 90 percent of the time the value of $\nu s^2/\sigma^2$ will lie within the central region. We find the table values of χ^2 for the positions 0.95 and 0.05, $\nu = 9$, to be 3.33 and 16.92. Now, with 90 percent confidence,

$$3.33 < \frac{\nu s^2}{\sigma^2} < 16.92$$

which can be modified to read

$$\frac{\nu s^2}{16.92} < \sigma^2 < \frac{\nu s^2}{3.33}$$

or, in general,

$$\frac{\nu s^2}{\chi^2_{\nu, \alpha/2}} < \sigma^2 < \frac{\nu s^2}{\chi^2_{\nu, 1-\alpha/2}} \qquad (5\cdot5)$$

For a sample of $n = 10$, we find the sample variance to be 20. We wish to place 95 percent confidence limits on the population variance. We find $\chi^2_{9, 0.025}$ to be 19.03, and $\chi^2_{9, 0.975}$ to be 2.70. We apply Eq. (5·5):

$$\frac{9 \times 20}{19.03} < \sigma^2 < \frac{9 \times 20}{2.70}$$

$$9.4 < \sigma^2 < 66.6$$

Thus, with 95 percent confidence, we can state that the population variance lies between the values 9.4 and 66.6. These limits are wide because the value of n was as small as 10. A good estimate of a population variance will be obtained only when the sample size is large.

Confidence limits for the population standard deviation are based on Eq. (5·5) by taking the square root of the values found for the variance. This method is satisfactory for large sample sizes only, since a bias will be present in the limits for standard deviation if the sample size is small.

We can find interval estimates for the ratio of two population variances by the use of the F distribution. This leads to a rule similar to Eq. (5·5):

$$\frac{s_1^2/s_2^2}{F_{\nu_1, \nu_2, \alpha/2}} < \frac{\sigma_1^2}{\sigma_2^2} < \frac{s_1^2/s_2^2}{F_{\nu_1, \nu_2, 1-\alpha/2}}$$

The value for $F_{\nu_1, \nu_2, \alpha/2}$ can be found in the F tables. The value of $F_{\nu_1, \nu_2, 1-\alpha/2}$ cannot be found in the F tables since such tables are one-tailed. We can, however, find the value of $F_{\nu_1, \nu_2, 1-\alpha/2}$ by looking up $F_{\nu_2, \nu_1, \alpha/2}$ (note the reversed order of the degrees of freedom) and taking the reciprocal of the table values. Thus we arrive at

$$\frac{s_1^2/s_2^2}{F_{\nu_1, \nu_2, \alpha/2}} < \frac{\sigma_1^2}{\sigma_2^2} < \frac{s_1^2}{s_2^2} \times F_{\nu_2, \nu_1, \alpha/2} \qquad (5\cdot6)$$

in which the order of the values of ν should be observed.

We wish to place confidence limits on the ratio of two population variances. A random sample of $n_1 = 10$ gave a value for s_1^2 of 38; another random sample of $n_2 = 8$ gave a value for s_2^2 of 42. We apply Eq. (5·6), noting that $\nu_1 = 9$, $\nu_2 = 7$, and that at an α risk of 0.05 we have $F_{9,7,0.025} = 4.82$ and $F_{7,9,0.025} = 4.20$. Thus we calculate

$$\frac{38/42}{4.82} < \frac{\sigma_1^2}{\sigma_2^2} < (38/42)4.20$$

$$0.19 < \frac{\sigma_1^2}{\sigma_2^2} < 3.8$$

Since the confidence limits for the ratio of the two variances include the value 1, we could conclude that the two sample variances are estimates of the same population variance. The values for the boundaries of the interval estimate indicate that either of the two sample variances could be the larger in another pair of samples similarly chosen.

PROBLEMS

5·14 Place 95 percent confidence limits on the population standard deviation, if the value of $s = 0.03$, as determined from a sample size of (a) $n = 5$; (b) $n = 10$; (c) $n = 20$.

5·15 Place 95 percent confidence limits on the population variance if the value of $s = 0.03$ as determined from a sample size of (a) $n = 5$; (b) $n = 10$; (c) $n = 20$.

5·16 A sample of 100 measurements of the lifetime of a certain type of television picture tube had a standard deviation of 6.0 hr.
(a) Place 95 percent confidence limits on the population variance.
(b) Place 90 percent confidence limits on the population variance.
(c) Place 90 percent confidence limits on the population standard deviation.

5·17 Population A had a variance of 3.4 as determined from a sample of 20 items, while population B was estimated to have a variance of 5.2 as determined from a sample of 16 items.
(a) Place 95 percent confidence limits on the population variance of population A.
(b) Place 95 percent confidence limits on the population variance of population B.
(c) Place 95 percent confidence limits on the ratio of the two population variances.
(d) Place 90 percent confidence limits on the ratio of the two population variances.

5·18 The yield of a production process was studied by both an assay and a chemical method on each of 10 successive batches. The sample variance of the assay method was 0.35 while the sample variance of the chemical method was 0.28.
(a) Place 90 percent confidence limits on the ratio of the two variances as determined by the two methods.
(b) Place 95 percent confidence limits on the ratio of the two variances.

QUESTIONS

5·1 Why is it necessary to test to see if the population variance has changed, if an improvement is made in the average setting of the process?

5·2   Can a process have its average value changed, while the standard deviation of the process remains unchanged? Explain.

5·3   When comparing two populations to see if they can be thought of as being equal, why is testing the population means only one-half the comparison procedure?

5·4   When comparing a population variance with a given standard, what probability distribution is used?

5·5   When comparing two sample variances to see if they came from the same population or equal populations, what is the probability distribution that is used?

5·6   How is the χ^2 distribution related to the t distribution?

5·7   What is the general shape of the χ^2 distribution for small sample sizes?

5·8   Where does the mode of the χ^2 distribution fall?

5·9   Why is it not correct to use equal distances to the critical regions when using the χ^2 distributions?

5·10   When looking up values in the χ^2 table,
(a) what information must be known
(b) how does a sketch of the distribution help

5·11   Under what conditions would a one-tailed area be used in the χ^2 distribution?

5·12   When would it be wise to run a hypothesis test of the population variance?

5·13   When is the null hypothesis rejected, when we are testing the population variance against a standard value?

5·14   What does it mean when we have accepted the null hypothesis of population variance equal to a standard?

5·15   What is the general shape of the F distribution for small sample sizes?

5·16   Where does the mode occur for the F distributions?

5·17   When should we consider running an F test?

5·18   What does it mean to reject the null hypothesis of equal variances?

5·19   Why should the larger of the two sample variances be placed over the smaller variance when running an F test?

5·20   If there were three populations to be tested for equal variance, how might the problem be approached?

5·21   When entering the F distribution table, what information is required?

5·22   When testing the means of two populations, what are the reasons for running an F test prior to running the proper t test?

5·23   What troubles are encountered when testing for the equality of two means, when the hypothesis of equal variances is rejected?

5·24   Very often, two variances are assumed equal, and the proper hypothesis test for the means is run. What conditions permit variances to be assumed equal?

5·25 What is the relationship between the formulas for the confidence limits for the population standard deviation and for the population variance, if the confidence level is the same in both cases?

5·26 Why are confidence limits placed on population variances?

5·27 What is the difference between $\alpha/2$ and $1 - \alpha/2$?

5·28 What, if any, is the relationship between the F distribution, and the χ^2 distribution?

5·29 How can the width of the confidence intervals for the variance be decreased?

5·30 Once the width of the confidence interval for the variance has been set, explain how it might be possible to arrive at the proper sample size to be used.

REFERENCES

16, 28, 33, 34, 35, 37, 47, 62, 67, 73, 81, 82, 123, 125, 126

6 Sample Size for Variables

A question often asked of statistical personnel is this: How large a sample size is needed in order to detect a specified effect? Such a question arises, or should arise, in situations like this: We have a currently operating process, and we wish to investigate whether or not a change in procedure has an effect on the nature of the product. We may be concerned with a manufacturing operation in which humidity is possibly a factor; we change the humidity and compare the product made under the new condition with the former product. It is an economic, as well as a statistical, problem to decide how much of the product (made under the new condition) we need to examine before we can decide whether or not the change has had an effect.

6·1 PROCESS CHANGE AND SAMPLE SIZE

We picture the situation in Fig. 6·1: The initial process, before the change is made, is described by a curve representing a normal population with a mean of μ and a standard deviation of σ. We assume that the record of the process is sufficiently complete for us to know both these parameters.

When the change is made in the operating condition, either of two types of new populations may result. The new population may have a mean of μ_1 or μ_2, depending on whether the average process level falls or rises. We assume that we do not know which of these hypothetical populations is more likely. We further assume that the standard deviation of the process is unchanged even though the process level may change. This assumption

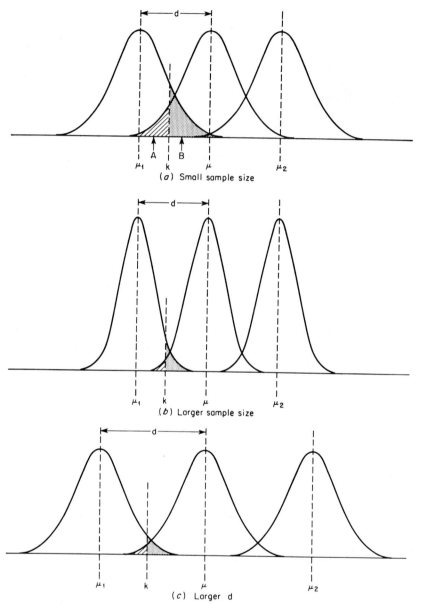

Fig. 6·1   Factors affecting the distinguishability of populations of different means. (a) Small sample size. (b) Large sample size. (c) Large difference in means.

is justified in many, if not most, industrial operations. The operator has considerable control over the process level, but the variability of the process is mainly determined by random causes and is not greatly modified by process adjustments.

If the frequency distributions are based on samples of relatively small size, our estimate of the process standard error of the mean is necessarily large, since $\sigma_{\bar{x}} = \sigma/\sqrt{n}$. If, on the other hand, the distributions are based on large sample sizes, our estimates of the standard error of the mean are smaller. The difference between these two situations is suggested by Fig. 6·1. In Fig. 6·1b, as compared with Fig. 6·1a, the smaller standard errors permit us more easily to distinguish from each other populations of small difference in the means. That the estimate of the standard error is reduced for larger sample sizes was discussed in Chap. 2. The extent of the reduction is defined by Eq. (2·4).

In a situation in which we wish to detect a change in a process, we hypothesize an original distribution and possible new distributions. Since the tails of a normal distribution never fall to a frequency of zero, regardless of how far we move from the population mean, there is always some overlap of the original distribution and the hypothesized new distributions. The preceding statement implies that there is always some probability that any sample we collect may have come from (1) the original population, so that any change we have introduced in process conditions had no real effect; or (2) a really changed distribution and a new population. Thus we cannot have *absolute* assurance that the decision we reach based on any sample is correct. If, however, the original and the hypothesized new distributions are far apart, we can reach a conclusion with great confidence. Conversely, the degree of confidence we desire helps us decide how narrow we want the distributions to be (to cut down on the overlap), and this in turn dictates how large a sample we must take.

6·2 SAMPLE SIZE AND RISKS

The overlap of the populations, and consequently the uncertainty, can be made small under any combination of the following circumstances:

1 The standard deviation of the process is small. Usually we have little control over this factor; sometimes we must try to reduce the standard deviation before we can successfully find the effect of a change in the process.

2 The distance d between the mean of the original population μ and that of the possible new populations μ_1 or μ_2 is great, as in Fig. 6·1c. This is to say that, if we are interested in detecting only a large change in a process, we can do so in spite of the presence of great variability in the process. Conversely, we have great difficulty in detecting a small process change if the variability of the process is large.

3 We use a larger sample size, as in Fig. 6·1b.

The inferences are these: If we wish to find a large process change, which implies a wide separation of the hypothetical populations, we can do so with a small sample; even with a small sample the overlap between widely separated populations will not be troublesome. If, however, we must detect a small process change to a high degree of probability, we will be forced to use a large sample size in order to reduce the sample standard deviation. Since the selection of the sample size is usually under our control, the specification of sample size is essential to the success of industrial experimentation.

Note the position of the arrow A in Fig. 6·1a. Let us suppose that a sample from the (possibly changed) process were to plot here. What judgment should be made? We could suppose from this sample that the population μ has changed, and that the sample properly belongs to a new population whose mean μ_1 is smaller than μ. (Note that we cannot specify what this mean μ_1 is in actuality, but only that it is smaller than the original population mean.)

We run a risk of error in making this judgment, since sample A might well, with measurable probability, truly belong to the left-hand tail of the original population. The risk of this kind of error is the *alpha* (α) risk. The size of this risk is (for the specified cutoff point shown at k) indicated by the crosshatched area in the figure.

Note now the position of the arrow B in Fig. 6·1a indicating the position at which another sample plots. We could suppose from this sample that the population has not changed and that the sample properly belongs to the original population of mean μ. If we make this judgment, we now run a risk of a different kind of error: this sample might well, with measurable probability, truly belong to the right-hand tail of a different population of mean μ_1. The risk of this kind of error is the *beta* (β) risk; the size of this risk is (for the same position of the cutoff point k) represented by the stippled area in the figure.

Observe that the relative sizes of the α and β risks are determined by the position of the line placed at k. This position in effect specifies a decision line; when a sample average lies beyond this point, it indicates that the process has changed. For a given value of n, this supposition involves specific α and β risks which determine the position k. In Fig. 6·2a we show two possible positions of k. If we set the limit at k_1, we have established an α risk represented by the left-hand tail of the population of mean μ; the β risk is represented by the right-hand tail of the population of mean μ_1. If we shift the decision line to the position k_2, we have now a larger α risk than for position k_1 but a smaller β risk. We show in Fig. 6·2b that, by increasing the sample size and thereby reducing the standard deviation of sample averages, we can reduce both the α and β risks. Observe here that the tail areas beyond the position k are reduced in both populations.

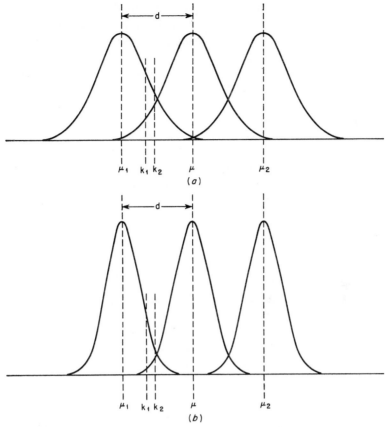

Fig. 6·2 *Change in α and β risks with position of confidence limit k.* (a) *Small sample size.* (b) *Large sample size.*

If we are trying to detect merely a change in process level, we will count it a success if the change is in either direction—above *or* below the original process level. Thus the α risk is *two-tailed*; we split the desired α risk into two equal parts and assign one part to each of the tails of the original distribution. The β risk is equivalent to a *one-tailed* α risk, but it involves a hypothetical new population. It is one-tailed, since only one side of either of the possible new populations can be involved.

6·3 FACTORS AFFECTING SAMPLE SIZE (σ KNOWN)

It should now be seen that the problem of determining the correct sample size for variables is answerable if we know these factors:

1 The standard deviation of the process.
2 The change in the process mean which we wish to detect.
3 The α and β risks we wish to assume.

The problem of determining sample size should be stated in this way: Given a process standard deviation, what value of n will enable us to find a specified difference in the means of two populations with specified α and β risks?

How large a sample size must we use to discover a difference of 0.04 in. between a standard of 1.00 in. and a comparison (secondary) standard, when the standard deviation of the measurement process is known to be 0.07 in.? We will assume an α risk of 0.05 and a β risk of 0.10. The risks we have chosen are commonly used in industrial applications; the choice of these risks is based on a compromise between the desire to obtain reliable results and the necessity for restricting sample size to a reasonable level.

We sketch the situation in Fig. 6·3. The known values are $d = 0.04$ in.; $\mu = 1.00$ in.; $\alpha/2 = 0.025$; $\beta = 0.10$. We do not know the position of k

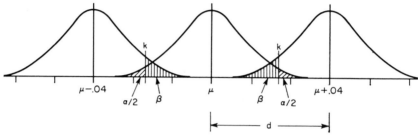

Fig. 6·3   Relationships of factors determining sample size.

which gives the desired areas under the tails of the normal distributions, nor do we know the value of n which determines the spread of the distributions of sample averages. The solution to the problem requires the following steps:

1   We write an expression for the t value of the position k with reference to the population of mean μ:

$$t = \frac{k - \mu}{\sigma/\sqrt{n}} \tag{6·1}$$

2   From Table A·1 we find the t value for the α risk (0.025) to be 1.96. We substitute this and the other known values in Eq. (6·1) and get

$$1.96 = \frac{k - 1.00}{0.07/\sqrt{n}}$$

When we clear of fractions and transpose, we have

$$k - 1.00 = 1.96 \times \frac{0.07}{\sqrt{n}} \tag{6·2}$$

3 Another relationship involving the desired quantities is obtained from knowledge that position k must produce an area of 0.10 under the left-hand tail of the hypothetical distribution having a mean of $1.00 + 0.04$, since 0.04 is the difference we wish to detect. The value 0.10 fixes a t value which is found, also in Table A·1, to be 1.28. We prefix a negative sign to this t value, since it lies in the left-hand tail of the distribution we are now considering. We substitute this and the other known values in Eq. (6·1):

$$-1.28 = \frac{k - (1.00 + 0.04)}{0.07/\sqrt{n}}$$

Rearranging this expression as before, we have

$$k - (1.00 + 0.04) = -1.28 \times \frac{0.07}{\sqrt{n}} \tag{6·3}$$

4 We now have in Eq. (6·2) and (6·3) two equations in two unknowns. We can eliminate the terms k and 1.00 from these equations by subtracting the second from the first. Thus

$$0.04 = 3.24 \times \frac{0.07}{\sqrt{n}}$$

$$\sqrt{n} = \frac{3.24 \times 0.07}{0.04}$$

$$\sqrt{n} = 5.67$$

$$n = 32.2$$

Thus we require a sample size of 33.

5 Now we can compute k by substitution of n in either Eq. (6·2) or Eq. (6·3).

In step 4 of the solution to this problem, we eliminated both k and the mean of the original population. We arrived at the final answer using only the following values: the necessary difference d, the population standard deviation σ, and the t values corresponding to the α and β risks we wished to assume. Furthermore, the signs of the two t values will always be opposite, since they involve the right-hand tail of one distribution and the left-hand tail of another distribution. Therefore, in the subtraction process in step 4, we really use the arithmetic sum of the t values. Thus we write

$$n = \left[(t_{\alpha/2} + t_\beta) \frac{\sigma}{d} \right]^2 \tag{6·4}$$

The required sample size depends only on the t values, determined by the risks we wish to assume, and on the ratio of the process standard deviation to the difference we wish to detect.

Equation (6·4) involves a two-tailed α risk because the statement of the problem required only that a difference be found. If the problem had instead called for finding an increase in the variable (or a decrease) we would have solved the problem in the same manner, except that we would have found t_α instead of $t_{\alpha/2}$.

The application of Eq. (6·4) to find the sample size for variables requires that we know or assume that

1 The populations of interest are normally distributed.
2 The standard deviation of the population is known and is unchanged when the mean of the population changes.
3 We can specify the difference in means we wish to detect.
4 We can specify the risks of error we can tolerate.

In practical industrial and experimental situations, the first two of the requirements above are usually not difficult to meet. For averages, the distribution is normal regardless of the nature of the parent population. If we do not know σ, the techniques of the next section of this text will provide a solution, with a slightly larger sample size as the penalty for our ignorance of the population standard deviation.

It is the fourth requirement in the list that is often troublesome. The specification of the risks of error is usually an engineering rather than a statistical problem. No estimate of the necessary sample size is possible, however, unless the α and β risks are fixed. Furthermore, to make judgments on the basis of test data without first determining the required extent of the testing procedure is indefensible.

PROBLEMS

6·1 How many tests need we run to establish a difference of 1.00 between the mean and a theoretical value, if the standard deviation is 0.50 and acceptable risks are $\alpha = 0.05$ and $\beta = 0.05$?

6·2 Find the sample size for the same data as in Prob. 6·1 except that the β risk is raised to 0.10.

6·3 Using the same information as in Prob. 6·1, find the required sample size if
(a) $\alpha = 0.10$, $\beta = 0.05$
(b) $\alpha = 0.10$, $\beta = 0.10$

6·4 What generalizations can be made on the basis of the answers to Probs. 6·1, 6·2, and 6·3?

6·5 How many tests need we run if the process standard deviation is 1.0 and the risks are constant at $\alpha = 0.5$, $\beta = 0.10$, to establish
(a) a difference of 0.5   _0.05_
(b) a difference of 1.0
(c) a difference of 2.0

6·6 How many tests need we run to establish a difference of 0.5 between the mean and a theoretical value, for constant risks of $\alpha = 0.05$, $\beta = 0.10$, when the standard deviation of the process is (a) 0.5; (b) 1.0; (c) 2.0.

6·4 SAMPLE SIZE FOR t TEST FROM TABLES

If we do not know the population standard deviation, and if we do not wish merely to define the minimum detectable difference in terms of an unknown value of σ, two alternatives are available to us. We may estimate the value of the standard deviation on the basis of a preliminary experiment, in which we use a small sample. From this sample we find s and substitute it for σ in Eq. (6·4). A better procedure is to estimate the sample standard deviation and to use Table A·11 to find the sample size.

If we want to detect a shift in the process average of 0.4 or greater, what is the minimum sample size if we have an estimate of sample standard deviation of 0.5 and we assume $\alpha = 0.05$ and $\beta = 0.01$? To use Table A·11, we first find

$$D = \frac{\delta}{s} = \frac{0.4}{0.5} = 0.8$$

We enter Table A·11 at $D = 0.8$; in the third major column, headed 0.05, and under the value of 0.01 for β, we find the number 31. This is the required sample size to permit us to find the difference in the means with the assumed α risks.

If we apply Eq. (6·4) with the same data, assuming the sample standard deviation to be an approximation of σ, we find the required sample size to be 30, a little smaller than that from the table. This observation illustrates these principles:

1. The sample size is larger when we do not know σ. The larger sample size is the price we pay for less knowledge of the process.
2. The increase in sample size is small for larger sample sizes. Above $n = 30$ the difference is inconsiderable.

We can, in the same manner, use Table A·11 to find the required sample size with no knowledge whatever of σ, and without even estimating σ. What we need to do, in this situation, is to specify only the value of k in Fig. 6·3. Thus we say that we want to detect a difference of 1 or 2 or 3 standard deviations, and use 1 or 2 or 3 as the value of D to enter Table A·11.

PROBLEMS

6·7 What sample size is required to detect a shift in the process average of the following sizes, if we agree to an α risk of 0.05 and a β risk of 0.10:
   (a) Shift of 0.5 standard deviations
   (b) Shift of 1.0 standard deviations
   (c) Shift of 1.5 standard deviations
(NOTE: We wish to detect a shift of the process average in either direction, and we do not know sigma but have an estimate of it.)

6·8 If we assume risks of $\alpha = 0.05$ and $\beta = 0.05$, and wish to detect a shift of the process average of one-half standard deviation in either direction, how large a sample size should we use to ensure our chances of detecting such a shift when it occurs?

6·9 Find the sample size required to detect a shift of the process average of one standard deviation in either direction, under the following sets of risks (sigma unknown):

(a) $\alpha = 0.05, \beta = 0.01$
(b) $\alpha = 0.05, \beta = 0.05$
(c) $\alpha = 0.05, \beta = 0.10$
(d) $\alpha = 0.05, \beta = 0.50$
(e) $\alpha = 0.01, \beta = 0.10$
(f) $\alpha = 0.10, \beta = 0.10$

6·5 SAMPLE SIZE FOR THE DIFFERENCE IN TWO MEANS

By an extension of the methods of the preceding section, we can, if we know σ, determine the sample size required to detect any specified difference between two population means. We can use

$$n = 2\left[(t_\alpha + t_\beta)\frac{\sigma}{d}\right]^2 \tag{6·5}$$

We must assume that the standard deviation of the process is unchanged with a hypothetical change in process average.

If we wish to detect a difference in the process mean of 0.15 or larger, when the process standard deviation is known to be 0.10, and we assume α and β risks each of 0.05, substitution in Eq. (6·5) gives

$$n = 2\left[(1.96 + 1.65)\frac{0.10}{0.15}\right]^2 = 2(3.61 \times 0.67)^2$$
$$= 11.5$$

Thus we would use a total sample size of 24, 12 from the original population and 12 from the hypothetical new population.

If we do not know σ but have a value for a sample standard deviation from a few preliminary runs, we can apply Table A·12. Using the same values for α, β, and δ as immediately above, but assuming now that we are estimating s to be 0.10, the table requires us to use a total of 13, rather than the 12 we found when we assumed that the population standard deviation was known. The use of this table involves a "guesstimation" of the standard deviation, and an assumption that the population standard deviation is unchanged with a change in the process level.

6·6 SAMPLE SIZE FOR CONFIDENCE LIMITS

Questions about sample size are sometimes posed in relationship to confidence intervals, or can be so interpreted. How many chemical determinations are needed in order to have 95 percent confidence that the true mean lies not farther from the sample average than 0.1, if σ is known to be 0.3? What we are here trying to do is to bound μ and to determine the sample size required to limit the confidence interval to ± 0.1.

The confidence interval when σ is known is

$$\left(\bar{X} - t_{\alpha/2}\frac{\sigma}{\sqrt{n}}\right) < \mu < \left(\bar{X} + t_{\alpha/2}\frac{\sigma}{\sqrt{n}}\right)$$

The difference d that the statement of the problem permits is expressed by the term $t_{\alpha/2}\sigma/\sqrt{n}$ in the formula for confidence limits. Thus

$$d = t_{\alpha/2}\frac{\sigma}{\sqrt{n}}$$

and

$$n = \left(t_{\alpha/2}\frac{\sigma}{d}\right)^2 \tag{6·6}$$

Substitution in Eq. (6·6) of the data above gives

$$n = \left(1.96 \times \frac{0.3}{0.1}\right)^2 = 34.6$$

Thus we would use a sample size of 35.

Note that no question of the β risk arises in connection with sample size and confidence limits. We cannot say how far the sample size may lie from the true value but can only define a confidence interval within which (to some probability) the true mean lies.

If we wish to determine a sample size needed for a specified confidence-interval determination for a mean without knowledge of the population standard deviation, we again resort to Table A·11. We wish to obtain an interval estimate of μ and want to know how large a sample size we must use to limit the confidence interval to $\bar{X} \pm 0.3$ with 95 percent confidence. We do not know sigma, but we have an estimate of the standard deviation of 0.4 on the basis of a small experiment. Here $\delta = 0.3$, $s = 0.4$; thus $D = 0.75$. We enter Table A·11 under $\alpha = 0.05$ and under $\beta = 0.50$ (since β is unspecified). We find in the body of the table that $n = 9$.

PROBLEMS

6·10 From only a few preliminary tests, we estimate the standard deviation of a process to be about 3.37. Find the sample size to give 95 percent confidence that the true mean of the process does not deviate from the sample mean by more than 1.50.

6·11 We wish to estimate the average shelf life of a large batch of chemicals. How large a sample must be tested to be 95 percent confident that the true mean will not differ from the sample average by more than 2.5 days? We believe from previous data that the standard deviation of shelf life is about 5.0 days.

6·12 Use the information in Prob. 6·11 to answer the question of sample size, with these changes:

(a) The standard deviation is estimated at 3.0 days.
(b) The standard deviation is estimated at 7.0 days.
(c) The difference we wish to find is 7.0 days and the standard deviation is estimated at 5.0 days.

6·13 How large a sample would be required in Prob. 6·11 if:

(a) the desired confidence level were 98 percent
(b) the desired α risk were 0.10

6·7 SAMPLE SIZE FOR TESTS OF VARIANCES

From tables similar to those used for finding required sample sizes for tests of the mean, we can find approximate sample sizes needed to detect specified changes in variance from a population variance, or between two sample variances. If we are studying a population of known variance σ_0^2, we define the variance change in terms of the ratio to the hypothesized variance.

Limiting values of the detectable ratio are given in the body of Table A·13, with corresponding values of α and β at the top of the table and the necessary number of degrees of freedom (and therefore sample size) at the left. Assume that we wish to detect a variance that is 3 times the population variance, with α and β risks both equal to 0.05. We find the value 3 (the ratio) to lie in the column for $\alpha = \beta = 0.05$ between the rows with $v = 15$ and 20. Linear interpolation indicates that the value of v should be 19, and the sample size therefore should be 20.

If we want to detect a *decrease* in variance, we use the same table, but with reversed values of α and β if the values of the assigned risks are different, and at the value of the reciprocal of the ratio we wish to discover. For example, if we wish to detect a variance $\frac{1}{5}$ of the hypothesized variance, at $\alpha = 0.05$ and $\beta = 0.01$, we find $1/\frac{1}{5} = 5$ in the table for $\alpha = 0.01$ and $\beta = 0.05$ to lie at the value of $v = 12$. What we have done here is conceptually to interchange the roles of original population and hypothetical new population so as to permit the use of a single table for either an increase or a decrease of variance.

Table A·14 is used in a manner similar to that for the preceding and permits us to determine sample size required for an F test when we can specify the limiting ratio of sample variances and the risks of the two kinds of error. If the desired ratio is 3, and $\alpha = 0.05$ and $\beta = 0.10$, we find that the value of v (for both samples) is approximately 30.

6·14   In a given experiment we wish to detect an increase in the variance if it exceeds twice the variance as indicated by past runs of similar tests. How large a sample should be used in the experiment if
(a) $\alpha = 0.05$, $\beta = 0.05$
(b) $\alpha = 0.05$, $\beta = 0.10$
(c) $\alpha = 0.10$, $\beta = 0.05$
(d) $\alpha = 0.10$, $\beta = 0.10$

6·15   A production line has remained stable over a period of time with respect to the average output, but a decrease in the variance of output is suspected. How large a sample should be taken from the production line if we wish to detect a decrease of 0.5 of the past value of output variance, if
(a) $\alpha = 0.10$, $\beta = 0.05$
(b) $\alpha = 0.01$, $\beta = 0.10$

6·16   We wish to test two population variances to see if there is a difference between them, and we desire to detect a difference whereby one variance exceeds the other by a multiple of 4. How many samples of each population should be taken if
(a) $\alpha = 0.05$, $\beta = 0.05$
(b) $\alpha = 0.01$, $\beta = 0.01$
(c) $\alpha = 0.05$, $\beta = 0.01$

6·17   How many samples from each of two populations should be taken to ensure, at an α of 0.05 and a β of 0.10, the detection of a ratio of variances of the magnitude of
(a) 1.5 times
(b) 2.0 times
(c) 2.5 times
(d) 3.0 times

QUESTIONS

6·1   When the question is asked, "How large a sample size is required?" what information must be available before the question can be answered?

6·2   If the calculated sample size is too large from an economic point of view, what can be done to reduce the sample size?

6·3   Even when the standard deviation is known, how is it possible to make two types of errors when judging the process level on the basis of a sample statistic?

6.4   Other conditions being fixed, show that a position of k which gives a very small α risk necessarily gives a large β risk.

6·5   What distinguishes between two-tailed and one-tailed situations in assigning the α risk?

6·6 Why is the β risk always taken as one-tailed in determining sample size?

6·7 For a problem in which the standard deviation is known, how does the required degree of confidence influence the sample size?

6·8 For fixed confidence levels and standard deviation, how does the size of the difference sought influence the sample size?

6·9 For agreed-upon confidence levels and desired difference, how does the estimate of the standard deviation influence the required sample size?

REFERENCES

16, 28, 30, 31, 33, 35, 37, 59, 62, 67, 73, 76, 78, 81

7
Additional
Uses of
the Chi Square
Distribution

In Chap. 5 we discussed the use of the χ^2 distribution as the basis of tests of significance for the variance. This same distribution can be used to see whether or not an observed set of data agrees with an expected set. In a situation in which we can state (in quantitative terms) the results to be expected, we can use the χ^2 distribution to compare observations with expected results.

7·1 OBSERVED VERSUS EXPECTED RESULTS

Consider a trivial example: we wish to test whether or not a coin is symmetrical and honestly tossed. We make a series of trials, say 100. We find that heads appears 60 times and tails 40 times. Our belief is that with a sample size as large as 100 the number of heads will be close to 50. It is tempting to infer from this set of trials that some nonchance factor is operating. There is, however, even for an honest coin a strong probability that the result we have obtained (or even a worse imbalance) will occur. We should not, therefore, make a judgment about the coin or the tossing without some statistical basis for this judgment.

As for other tests of significance, we can set a desired α risk and compare a calculated statistic with a table value. The table we will use is that of the χ^2 distribution. The test statistic is

$$\chi^2 = \sum \frac{(O - E)^2}{E} \tag{7·1}$$

where O is the observed frequency of occurrence of a specified event; E is the expected frequency. The application of Eq. (7·1) requires that we do the following:

1 Find the difference between each observed frequency and the corresponding expected frequency.
2 Square these differences.
3 Divide each squared difference by the corresponding expected frequency.
4 Sum the resulting quotients.

The sum of these quotients approximately follows the χ^2 distribution if there is *no difference* between the observed and the expected frequencies. Thus at a chosen level of significance (α risk) we can test a set of data against the table value of χ^2. What we are doing is testing whether the results we observe are consistent with what we might expect to find in tossing an "honest" coin. That is, we compare the observed data with the theoretical frequencies.

We apply this method to the problem of the coin:

1 For ease in organizing the data, and especially in the more complicated situations to be discussed later, we arrange the observations as in Table 7·1. Such an arrangement is a *contingency table*. The observed frequency

Table 7·1   1 × 2 Contingency Table

Heads	Tails	Total
60	40	100
(50)	(50)	(100)

appears in the center of each cell of the table; the expected frequency is placed in the same cell, enclosed in parentheses. The contingency table used here is a 1×2 table: that is, the table contains *one* row of data and *two* columns of data. The observed entries in the table are designated by the general symbol X_{ij}; the subscript i indicates the row of the table in which the entry appears; the subscript j indicates the column of the table. Thus the symbol X_{12} designates the observation in the first row and in the second column of the table. We read this symbol "X sub one, two."
2 We choose a level of significance: Let it be 0.05. Thus the α risk is 0.05.
3 The number of degrees of freedom we will have for this test is equal to the number of *unrestricted* choices we have in filling out the contingency table. Here ν is only 1: Since we know that the total number of tosses was 100, once we specify the number of heads, the number of tails is fixed. We have therefore only *one unrestricted choice*—that of the entry under "heads"—and therefore only one degree of freedom.

4 The hypothesis we are testing is that the frequency of occurrence of heads and tails is *not different* from that which could reasonably have been obtained with an unbiased coin. Thus

$$H_0: (O - E)^2 = 0 \qquad H_1: (O - E)^2 > 0$$

5 This statement of the hypotheses implies a one-tailed test, since we will reject the hypothesis of no difference if the number of heads is either too large or too small. We therefore find the table value for χ^2 with $\nu = 1$ and for an α risk of 0.05. The table χ^2 value is 3.84. Only if the value calculated from Eq. (7·1) is greater than the table value will we reject the null hypothesis.

6 Calculation, by substitution in Eq. (7·1):

$$\chi^2 = \frac{(60 - 50)^2}{50} + \frac{(40 - 50)^2}{50} = \frac{100}{50} + \frac{100}{50} = 4.0$$

7 Conclusion: Since the calculated value does exceed the table value, our data do not indicate that the coin is false. Hence we reject the hypothesis that the coin is an honest one, honestly tossed. Note that we might have desired to run a 1.0 percent risk of error of type I (i.e., calling the coin tosser a cheat when he is really honest). In the outcome of the test of hypothesis, we would have found the table value of $\chi^2_{1,0.01}$ to be 6.63. We then with the same data would have rejected the null hypothesis and would have considered the data to show that the coin was false. It is therefore necessary to establish *in advance* a level of significance and to adhere to it; only by such a procedure can we say definitely what is the conclusion of the test.

A hand-fed printing press was subject to what seemed to be an unreasonable number of chokes (or stoppages) caused by jamming of the sheets fed into the press. A test was run to see whether or not different operators had different degrees of difficulty with the press. Each operator fed the same number of sheets into the press, and the number of chokes for each operator was counted:

Operator	A	B	C	D	Total
Chokes	5	8	9	10	32

If there were no difference between operators, the expected number of chokes for each of the four would be 8—one-fourth the total number of 32. Thus, in the 1×4 contingency table (Table 7·2) 8 appears as the expected value in each cell of the table.

Table 7·2   1 × 4 Contingency Table

Operator	A	B	C	D	Total
Number of chokes	5 (8)	8 (8)	9 (8)	10 (8)	32 (32)

The number of degrees of freedom is here 3: There are three unrestricted observed values in the contingency table; the fourth value is fixed by the other three and by the total number of chokes. Using again a 5 percent α risk, we find the table value for $\chi^2_{3,0.05}$. It is 7.81 for the one-tailed test.

The calculated value for χ^2 is

$$\chi^2 = \Sigma \frac{(O - E)^2}{E} = \frac{(5 - 8)^2}{8} + \frac{(8 - 8)^2}{8} + \frac{(9 - 8)^2}{8} + \frac{(10 - 8)^2}{8}$$

$$= 9/8 + 0/8 + 1/8 + 4/8 = 1.75$$

Since the calculated value is less than the table value, we accept (on the basis of these data) the hypothesis that there is no demonstrated difference between operators. The inference is that the difficulty lies elsewhere than with the pressmen. This conclusion is perhaps unexpected, since the original data show that operator D had twice as many chokes as operator A. Without running the test of hypothesis we might well have been tempted to blame operator D for his seemingly bad performance. Indeed, he still may have been a source of trouble; our data have not proved that he was free of blame. The conclusion that we have found no difference in operators might be caused by the small number of sheets the operators ran. Had the operators run six times as many sheets and had exactly six times the number of chokes each, E would have been 48, and the number of chokes per operator would have been 30, 48, 54, and 60. Now χ^2 would be found to be 10.5, and we would reject the null hypothesis.

In this example, had the operators not run the same number of sheets through the press, we would have found it necessary to find the expected number of chokes by using the ratio of the total number of sheets to the total number of chokes. We illustrate this method in the following example.

Six drugs were tested for their efficacy on a group of patients on the basis of whether or not an examining physician found that the patient improved over the course of treatment. The original data are shown in Table 7·3 and the corresponding contingency table in Table 7·4. The expected number of improved cases is found thus: If there were no difference in the effect of the six treatments, we would expect the fraction of patients showing improvement to be in the same ratio as the totals in the last column of

Table 7·3   Results of Tests of Six Treatments (Drugs)

Treatment	A	B	C	D	E	F	Total
Number of patients	54	47	52	53	49	52	307
Patients improved	16	13	15	8	15	2	69

Table 7·4   1 × 6   Contingency   Table   Based   on   Data   in Table 7·3

Treatment	A	B	C	D	E	F
Patients improved	16 (12.1)	13 (10.6)	15 (11.7)	8 (11.9)	15 (11.0)	2 (11.7)

Table 7·3. This ratio is 69/307. If we multiply by this ratio the number of patients treated with each drug, we find the expected number of improved patients if there were no difference in the efficiency of each drug. Thus, for drug A, we multiply 69/307 × 54 to get 12.1. Similarly, we find the expected values for each of the other treatments.

From the contingency table, we find the calculated value of χ^2 as before:

$$\chi^2 = \frac{(16 - 12.1)^2}{12.1} + \frac{(13 - 10.6)^2}{10.6} + \frac{(15 - 11.7)^2}{11.7}$$
$$+ \frac{(8 - 11.9)^2}{11.9} + \frac{(15 - 11.0)^2}{11.0} + \frac{(2 - 11.7)^2}{11.7}$$

$$\chi^2 = 12.9$$

The table value of $\chi^2_{5,0.05} = 11.07$. Since the calculated value exceeds the table value, we conclude that the data indicate a difference from the expected values with an α risk of 0.05. It should be noted that the largest contributor to the calculated value of χ^2 is the last term. The original data show that only two patients were improved when treated with drug F. Additional tests, to be discussed in Chap. 11, permit us to decide whether we are right in suspecting that only the results with drug F are different from chance expectations.

PROBLEMS

7·1   Toss a coin 50 times in random fashion; record the number of heads. Test the hypothesis that the coin was symmetrical at a significance level of 0.05.

7·2   Roll a single die a large number of times. Based on your observations, test the hypothesis that all sides of the die are equally likely to face up.

7·3   In a true-false test, a student answered 60 correctly out of 100 questions.

(a) If you wish to test whether the student really knew something about the material on which the test was based, state the null and the alternate hypotheses.

(b) At an α risk of 0.10, did the student simply guess?

7·4   At an α risk of 0.05, how many questions must a student answer correctly on a 100-question true-false test in order to show that he knows something about the test content?

7·5 Shuffle a deck of playing cards; select a card at random and record its suit; replace the card, reshuffle, and draw again. Repeat a large number of times.

(a) Test the hypothesis at a significance level of 0.10 that the selection was random.

(b) Test the hypothesis that red cards were selected as often as black cards.

7·2 2 × 2 CONTINGENCY TABLES

A different problem involving coin tossing takes this form: A believer in extrasensory perception claims to have the gift of controlling the fall of a coin. In a series of 200 tosses, he tries in each toss to cause the coin to fall "heads." Since the coin may be nonsymmetrical, we compare his results with those of a nongifted person who tosses the same coin 100 times.

The hypothesis we wish to test here is different from the preceding ones. Before we tested the agreement of a set of data with a hypothetical probability. Now we will test the consistency of two sets of data, without any assumption about the hypothetical probability.

The results of the test are presented in Table 7·5. This is a 2 × 2 contingency table. The rows display the results for the two persons and the columns the two different outcomes.

Table 7·5   *Data from a Coin-tossing Experiment*

	Heads	Tails	Total
Not gifted	60 (64)	40 (36)	100
Gifted	132 (128)	68 (72)	200
Total	192	108	300

We calculate the expected values from the data themselves, as follows: The "gifted" person obtained 132 heads; he made 200 out of the total of 300 trials and would be expected (if there were no difference between persons) therefore to obtain $\frac{2}{3}$ of the total number of heads. Thus the expected value of heads for the "gifted" person is $\frac{2}{3} \times 192$ or 128. The other expected values are obtained similarly.

Now we can find a value of χ^2 by Eq. (7·1):

$$\sum \frac{(O - E)^2}{E} = \frac{(60 - 64)^2}{64} + \frac{(40 - 36)^2}{36} + \frac{(68 - 72)^2}{72} + \frac{(132 - 128)^2}{128}$$

$$= 1.04$$

It might be thought that we would now have available more degrees of freedom for the χ^2 test than in the previous examples of this chapter. Instead, we still have only one degree of freedom. This is the reason: Since we know the totals for each row and column, placing a single value in any one of the four cells of the table will determine what the remaining cells must contain. We have thus only one unrestricted choice of the table values, and therefore only one degree of freedom. If we choose an α risk of 0.05, the table value for $\chi^2_{1,.05} = 3.84$. Since the calculated value of χ^2 is less than this table value, we conclude that the data show no inconsistency, and we cannot suppose that the "gifted" person from this experiment has shown an ability to control the fall of a coin, as compared with the "nongifted" person.

If we wish to test whether two operators perform similarly in an operation, we may determine for each operator the number of defective items in a series of production pieces. For operator A we find 100 defectives in a total of 1,000 items; for operator B we find 60 defectives in a total of 500 items. We prepare Table 7·6, including the totals for each row and column and the grand total.

Table 7·6   Results of Inspection of Two Processes

Operator	Defective	Nondefective	Total
A	100 (X_{11})	900 (X_{12})	1,000 ($T_{1.}$)
B	60 (X_{21})	440 (X_{22})	500 ($T_{2.}$)
Total	160 ($T_{.1}$)	1,340 ($T_{.2}$)	1,500 ($T_{..}$)

Our test is merely to see whether or not the two operators are to be considered as similar. Thus we find the expected values for each cell as in the previous example: For operator A, who made 1,000 of the total of 1,500 items, he would be expected to produce 1,000/1,500 of the total of 160 defective items. Thus $1,000/1,500 \times 160 = 106.7$, which appears in parentheses as the expected value in Table 7·7. Similar calculations give the expected values for the other cells in the contingency table.

Table 7·7   2 × 2 Contingency Table (Data in Table 7·6)

Operator	Defective	Nondefective
A	100 (106.7)	900 (893.3)
B	60 (53.3)	440 (446.7)

From these data we find a calculated χ^2 value as before; it is 1.4. If we test again at the 0.05 level of α, with $\nu = 1$, we conclude by comparison with the table value of about 3.84 that the data do not show a significant difference between the number of defectives obtained by the two tested operators.

The method we have described can be expanded to include any number of samples and any number of processes or levels of operation. It is necessary to increase correspondingly the number of rows and columns of the contingency table. The value of ν is equal to the product of the number of rows and columns of the contingency table, each reduced by 1; that is,

$$\nu = (r - 1)(c - 1)$$

where r and c are the numbers of rows and columns, respectively.

The use of the χ^2 distribution is not economical when r and c are large; other methods to be described in Chaps. 8 to 11 are more efficient. Furthermore, the use of a χ^2 test merely permits us to conclude that the observed results are probably not due to chance alone; we cannot solely from this test determine which of the results are responsible for the nonchance effect.

PROBLEMS

7·6   Three bowlers bowled 3 games each on the same set of bowling alleys. Test the hypothesis that the bowlers are equal in ability if the scores were bowler A—120, 130, 105; bowler B—145, 130, 120; bowler C—185, 180, 175. (Set the significance level prior to running the test.)

7·7   In each of three shifts at a factory, all workers performed the same jobs; the shifts were of the same size. Accidents requiring the attentions of the nurse for a 4-week period were shift A—34, 42, 39, 17; shift B—17, 25, 22, 19; shift C—11, 11, 13, 15. Test the hypothesis that the shifts are equally accident-prone at a significance level of 0.05.

7·8   Three methods of packing dishes for shipping were tested over a 4-week period; a count was made of the number of dishes per 1,000 that were chipped or broken. The results were method A—1, 3, 3, 4; method B—4, 4, 6, 5; method C—12, 10, 11, 15.

(a) Merely by looking at the data, what do you believe should be concluded from the experiment?

(b) At a significance level of 0.05, test the hypothesis that all three methods of packing dishes are equally good.

7·9   Test the hypothesis that two students did equally well in all subjects if their final grades in 10 courses were as follows:

Course	1	2	3	4	5	6	7	8	9	10
Student A....	80	60	85	90	85	75	80	65	70	80
Student B....	75	85	70	85	95	60	60	90	70	75

Compare the averages of the two students.

7·3 YATES CORRECTION FACTOR

When the sample size is small, a better fit of data to the theoretical χ^2 distribution is obtained by applying a small constant to the observed values. This is essentially a correction for continuity. The χ^2 distribution is a continuous function but our data are discrete. The constant is $\frac{1}{2}$; $\frac{1}{2}$ is added to the observed value if it is smaller than the expected value for a cell; $\frac{1}{2}$ is subtracted from the observed value if it is larger than the expected value for a cell.

The application of this factor to the first coin-tossing example would give a calculated value for χ^2 of 3.61 instead of the 4.0 we found before. For large numbers, as in this example, the use of the Yates correction factor has little effect. The factor should, however, be used for small sample sizes, where the values both of O and of E are small.

7·4 TEST FOR HOMOGENEITY OF VARIANCES

It is often useful to compare more than two variances to discover whether or not there is a significant difference among them. One method requires as many F tests as there are pairs of variances to be compared. This may be very laborious: To test each possible pair of variances in a set of 6 would necessitate the use of 15 F tests.

Cochran's test for homogeneity of variances is based on the ratio of the largest variance to the sum of all the variances; this test determines whether or not the largest variance is to be considered an extreme case.

Because it is easier to compute the range than the variance, a control chart for ranges is often used to test the homogeneity of the dispersions of a set of samples, since the range is itself a measure of the lack of repeatability (as is the variance).

Bartlett makes use of the χ^2 distribution to test the homogeneity of several variances. For this test, χ^2 is calculated as the difference between (1) the total number of degrees of freedom times the natural logarithm of the pooled estimate of variance and (2) the sum of the products of the individual degrees of freedom and the natural logarithms of the individual estimates of variance:

$$\chi^2 = [(\ln \bar{s}^2) \sum (n_i - 1)] - \sum [(n_i - 1) \ln s_i^2]$$

where $\bar{s}^2 = $ pooled estimate of variance
$n_i = $ number of terms used in the calculation of s_i^2
$s_i^2 = $ set of individual variances

In common logarithms,

$$\chi^2 = 2.303\{(\log \bar{s}^2) \sum (n_i - 1) - \sum [(n_i - 1) \log s_i^2]\}$$

If all the variances are estimated from the same sample size, all values of n_i

are equal. The relationship then becomes

$$\chi^2 = 2.303(n - 1)(k \log \bar{s}^2 - \sum \log s_i^2) \tag{7·2}$$

where k = number of variances being compared

Bartlett found that the value of χ^2 from Eq. (7·2) was slightly biased on the high side and should be corrected by a factor which varies according to the values of v:

$$C = 1 + \frac{1}{3(k - 1)}\left[\sum \frac{1}{n_i - 1} - \frac{1}{\Sigma(n_i - 1)}\right]$$

If all values of n_i are the same, the correction factor reduces to

$$C = 1 + \frac{k + 1}{3k(n - 1)} \tag{7·3}$$

The corrected χ^2 value is symbolized by χ_C^2; it is equal to χ^2 from Eq. (12·2) divided by C from Eq. (7·3). Note that the value of C is always greater than 1; hence the value of χ_C^2 will always be less than the uncorrected value of χ^2. If the initial calculation of χ^2 does not cause the null hypothesis of $s_1^2 = s_2^2 = \cdots = s_k^2$ to be rejected, there is little need to correct for the possible bias, and C need not be found.

Ten machines produce the same kind of items; we wish to test whether or not the machines have the same variability. We wish, therefore, to test the homogeneity of 10 variances. We select at random a sample of 6 items from each machine, and calculate the sample variances. In this example, $n = 6$ and $k = 10$. The variances were found to be 9, 3, 4, 3, 5, 4, 2, 4, 5, 3.

We now proceed to test the hypothesis that all variances are the same. H_0: $\sigma_1^2 = \sigma_2^2 = \cdots = \sigma_{10}^2$. H_1: All σ_i^2 are not equal. We intend to test the hypothesis at a significance level of 0.05.

The test statistic is given by Eq. (7·2) and corrected by Eq. (7·3) if required:

$$\begin{aligned}
\chi^2 &= 2.303(n - 1)(k \log \bar{s}^2 - \sum \log s_i^2) \\
&= 2.303(6 - 1)(10 \log 4.2 - 5.89) \\
&= 3.91
\end{aligned}$$

The table value of $\chi^2_{9,0.05} = 16.919$. Since the uncorrected value for Bartlett's test is less than this, we can accept the null hypothesis without correcting for bias. Thus we can conclude from the evidence of the samples we selected that we have failed to find reason to believe that the machines have different variances.

Had the calculated value exceeded the table value, we would have found the value of C from Eq. (7·3):

$$C = 1 + \frac{k - 1}{3k(n - 1)} = 1 + \frac{11}{30 \times 5} = 1.07$$

We would then have divided the value from Eq. (7·2) by this factor and would have again compared the result with the table value.

PROBLEMS

7·10 The data in Table 7·8 were computed from a set of equal sample sizes, $n = 5$. Test the data for homogeneity of variance.

Table 7·8   Data for Tests of Homogeneity of Variance

Sample number	s_i^2	$\log s_i^2$	$v\, s_i^2$
1	0.675	9.82930–10	2.7000
2	2.050	0.31175	8.2000
3	1.000	0.00000	4.0000
4	0.375	9.57403–10	1.5000
5	4.750	0.67669	19.0000
6	5.075	0.70544	20.3000
7	1.000	0.00000	4.0000
8	5.325	0.72632	21.3000
$k = 8$		$\Sigma = 1.82353$	$\Sigma = 81.0000$

7·11 At an 0.05 level of significance, test the hypothesis of homogeneity of the following sample variance ($n = 9$ for all samples): 0.0045, 0.0047, 0.0049, 0.0039, 0.0036, 0.0040.

7·12 In a study of the effects of drying temperatures on the dimensional stability of photographic film base, it was believed that the variances at 4 different temperatures were the same. After 36 pieces of film were tested at each of the 4 temperatures, the resulting variances were 0.455, 0.523, 0.538, 0.856.

(a) Test the assumption of equal variances for the 4 temperatures at a significance level of 0.10.
(b) Test the assumption at a significance level of 0.01.

7·5 TEST FOR GOODNESS OF FIT

A χ^2 test may be usefully applied in situations in which we wish to decide whether a set of observations fits a theoretical or previously determined relationship. The theoretical distribution may be normal, binomial, or hypothetical.

For a pair of dice, the binomial distribution is the expected pattern. On the average, this distribution predicts that 7 should appear about 1/6 of the time and that 11 should appear about 1/18 of the time. If we were to toss a pair of dice 360 times, 7 should appear nearly 60 times and 11 should appear nearly 20 times. If in a given experiment we observe 7 a total of 74 times and 11 a total of 24 times, we may prepare the contingency table as in Table 7·9.

Table 7·9   Data for Goodness-of-fit
Test

	Sum of dice		
	7	11	Total
Observed frequency	74	24	98
Expected frequency	60	20	80

Because there are only two cells in the table, there will be one degree of freedom for the calculation of χ^2.

$$\chi^2 = \sum \frac{(O - E)^2}{E} = \frac{(74 - 60)^2}{60} + \frac{(24 - 20)^2}{20}$$
$$= 4.07$$

The table value for $\chi^2_{1,0.05}$ is 3.84. Thus we reject the null hypothesis and conclude that we have found evidence that the dice were not honest.

We can apply the principles involved in goodness-of-fit tests to many situations. We said in Chap. 2 that one test for normality of observed data is whether or not the points appear to give a straight line when they are plotted on probability graph paper. We also said that the results of such a test are difficult to interpret. Equation (7·1) now offers an alternative method which lets us decide the question of normality to any degree of confidence we desire.

The method of applying the goodness-of-fit test for normality to frequency distributions is as follows (data from Fig. 2·4, calculations in Table 7·10):

Table 7·10   Calculations for Goodness-of-fit Test for Normality
$\bar{X} = 15.75$; $s = 0.20$; $n = 150$

UCL	Observed frequency, O	$UCL - \bar{X}$	t value $\dfrac{UCL - \bar{X}}{s}$	Table value	Difference	E	$\dfrac{(O - E)^2}{E}$
16.35	1	0.60	3.0	0.9987	0.0049	0.74	0.0914
16.25	2	0.50	2.5	0.9938	0.0166	2.50	0.1000
16.15	6	0.40	2.0	0.9772	0.0440	6.60	0.0545
16.05	16	0.30	1.5	0.9332	0.0919	13.79	0.3542
15.95	20	0.20	1.0	0.8413	0.1498	22.47	0.2715
15.85	26	0.10	0.5	0.6915	0.1915	28.73	0.2594
15.75	35	0.00	0.0	0.5000	0.1915	28.73	1.3684
15.65	20	−0.10	−0.5	0.3085	0.1498	22.47	0.2715
15.55	17	−0.20	−1.0	0.1587	0.0919	13.79	0.7472
15.45	5	−0.30	−1.5	0.0668	0.0440	6.60	0.3879
15.35	2	−0.40	−2.0	0.0228	0.0166	2.50	0.1000
15.25	0	−0.50	−2.5	0.0062			
						Total	4.0060

1 Using suitable cell intervals (as described in Chap. 1) classify the data in a frequency distribution. The numbers in each cell are the O values which will be used in the χ^2 calculation, Eq. (7·1).

2 Find the mean and standard deviation of the sample data.

3 Find the t value for the upper cell limit (UCL) of each cell. This is done by subtracting the calculated mean from the UCL, and then dividing this difference by the value of s.

4 From Table A·1 find the area under the curve corresponding to the t value assigned to each UCL.

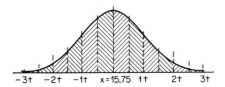

Fig. 7·1   Assumed normal distribution, with areas corresponding to frequency-distribution cell intervals.

5 Find the differences in the successive pairs of areas. Refer to Fig. 7·1, which shows that what we have done so far is to reduce the data to the standard form. The last calculation gives us the expected areas under each section of the normal curve corresponding to the cells used in making the frequency distribution.

6 Use these differences as expected fractions of the population for each cell: multiply each of them in turn by the sample number (here 150). These products are the E values for each of the cells.

7 From the O values (step 1 above) and the E values just found, apply Eq. (7·1) as before. Here we have 3.7521.

For this test, the number of degrees of freedom is the number of cells k reduced by 3; that is, $\nu = k - 3$. Three degrees of freedom have been lost because we have fixed three characteristics: the sum of the frequencies, the mean, and the standard deviation. If we intend, therefore, to test at a significance level of 0.05, we find the table value of $\chi^2_{9,0.05}$; it is 16.9. Since the calculated value is less than this, we conclude that the data do not indicate other than a normal distribution.

PROBLEMS

7·13   Mendelian principles of inheritance predict the occurrence of types of peas with specific parent plants in the ratio $9:3:3:1$ for the kinds smooth and yellow, smooth and green, wrinkled and yellow, wrinkled and green. From an experiment, a count gave, in order, frequencies of 315, 108, 101, and 32. At a significance level of 0.05, do the data agree with theory?

7·14 When 10 unbiased coins are tossed, the expected percentage of heads is as follows:

Number of heads	10	9	8	7	6	5	4	3	2	1	0
Percentage ...	0.10	0.98	4.39	11.72	20.51	24.61	20.51	11.72	4.39	0.98	0.10

(a) Toss 10 coins 100 times and record the number of heads appearing on each toss.

(b) Test the hypothesis that the coins are unbiased at the 0.05 significance level.

(c) How is the expected frequency of heads determined?

7·15 In a final examination in statistics, the following grades were received by the students. Test to see whether or not the grades were distributed normally.

Data: 80, 70, 90, 75, 55, 80, 80, 65, 100, 75, 60, 60, 75, 95, 80, 80, 90, 85, 70, 95, 75, 70, 85, 80, 80, 65, 65, 50, 75, 75, 85, 85, 90, 70

QUESTIONS

7·1 When a calculated χ^2 value is too large, we reject the null hypothesis. Is it possible for the calculated value to be too small? What would you infer if the calculated value were too small?

7·2 What is meant by "the additivity of χ^2"?

7·3 Under what conditions does the χ^2 distribution approach a normal distribution? Can this answer be inferred from the values in the table of the χ^2 distribution?

7·4 Draw a 3 × 4 contingency table. Fill in the symbols that represent the entries in the table and the totals.

7·5 How many degrees of freedom are available for a χ^2 test based on a 3 × 4 contingency table?

7·6 In a test for homogeneity of variances,

(a) What is meant by the rejection of the null hypothesis?

(b) What is meant by the acceptance of the null hypothesis?

(c) If the null hypothesis is rejected, what further question must be answered?

7·7   What can be inferred if a goodness-of-fit test leads to a rejection of the hypothesis of conformance?

7·8   Suggest relationships which might be tested by a goodness-of-fit test.

REFERENCES
16, 45, 46, 73, 74, 115, 121, 123, 125

8 Analysis of Variance

In Chap. 4 we described tests of hypothesis for the means of two populations. We are now about to discuss situations in which we hypothesize more than two possible populations, and in which we wish to test the hypothesis of no difference among them. For example, we may be interested in the performance of several operators, or several machines, or several processes. To deal with such a plurality of possible populations, we need a method different from that which we described before.

If we hypothesize more than two populations, we could run a series of t tests, by which we would test each mean against each of the other means. We would require three such tests for three means, six such tests for four means, and for many more means an awkwardly large number of t tests. Furthermore, each individual t test has associated with it an α risk of making a type I error. If we were to make many such tests, we would be almost certain of drawing a wrong conclusion in at least one test.

The method of data analysis we are about to describe is an extension of t tests of hypothesis for the mean. The method permits us in a *single* test and with a *single* α risk to answer questions of this kind: Do the data indicate that the members of a set of hypothesized population means differ among themselves? Are these differences significantly different from a chance result? The method, called *analysis of variance*, is especially useful when it is applied to complex situations. We deal first with a simple case.

8·1 SINGLE-FACTOR CASE

We wish to test the hypothesis of no difference among three different brands of automobile tires. We choose a *response variable*; that is, we decide on a measure of performance suitable to the situation. Here we choose as the response variable the number of miles (in thousands) at which the tread marks disappear. We test randomly selected samples of each of the three brands. We fear that there will be variations among the tires of each brand, and we therefore test three of each. The repetition of a test is called *replication*; here we have three replicates of each test.

A tabulation of the nine values of the response variable would, most probably, include nine different observations. The differences among the nine values we may ascribe to two different sources: (1) a possible real difference among the brands; (2) all other sources of variation, including untested factors and chance causes, all of which we here lump together and call *experimental error*. In a well-designed experiment, we plan to reduce the causes of experimental error to those sources of variability which cannot be eliminated.

Since we have tested, in this experiment, three brands of tires, we say that we have tested three *levels* of the factor under study. If we average the values of the replicates for each level, we obtain three means, and thus three values which we designate as $\bar{X}_i$, where the subscript i is an index standing for 1 or 2 or 3, the specific factor level being averaged.

We want now to test the data to distinguish among several possible cases illustrated in Fig. 8·1. In Fig. 8·1a we show a single normal curve, representing a single population. We indicate three values of $\bar{X}_i$; each of these is a different point estimate of a single common population mean μ. That these three values of $\bar{X}_i$ are different is attributable only to the expected differences among samples taken from the same population. The situation shown in Fig. 8·1b is quite different. Here we show three different populations; for each of these we have a single point estimate of three different means μ_1, μ_2, and μ_3. That these three sample means are different is attributable to real differences among the population means. Still a third possibility exists, as shown in Fig. 8·1c: Two of the sample means are point estimates of a single population mean, but the third is an estimate of a different population mean.

We want our test to distinguish between the first of the cases in Fig. 8·1 and the other two. The null hypothesis we are to test is $H_0: \mu_1 = \mu_2 = \mu_3$; that is, the true population mean is the same for each level of the factor we sample and compute an $\bar{X}$ for, and we really have only one population. We can extend this kind of null hypothesis to include any number of hypothesized means, and thus to any number of levels of a single factor. The alternative hypothesis is this: Not all the means are equal. Thus we will reject the null hypothesis if even *one* of the population means (or factor levels) differs from the others.

The test procedure we will use is based on an *analysis of variance* (abbreviated ANOVA) which involves this concept: We estimate the variance of all

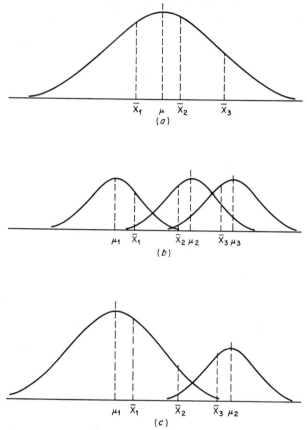

Fig. 8·1 Possible relationships between a set of samples and the populations from which the samples were taken.

the data without regard to cause; we partition this total variance between (1) the tested factor and (2) experimental error; we compare these two variances by means of an *F* test.

8·2 ASSUMPTIONS IN SINGLE-FACTOR ANOVA

Before we proceed with the calculations that are preliminary to the test of significance, we state the assumptions under which the method is valid.

1   We are interested in this first case in only one of the possible influencing factors on the response variable. This factor may be the temperature at which the process is run; the levels of the factor would be any desired values of the temperature. The response variable may be yield. If we

make such an experiment, we are assuming that we are not interested in any other factor (pressure, ingredients, method of combination) that might affect yield. If other factors are believed to have an effect, we must either fix them at some level (perhaps arbitrary) or try to keep these effects from biasing the responses for the factor under study.

2 We must assume that the effects of chance and of nontested factors are equal and normally distributed at all levels of the tested factor. This is equivalent to the assumption that the variance within a level is constant at all tested levels. If only temperature is being tested, we believe that the effects of variations in pressure, ingredients, etc., are unchanged with temperature level.

3 We assume that the experiments made at each level of temperature are independent and that the results obtained at one level are not affected in any systematic fashion by any other source of variability that can be carried over to the results obtained at another level. If we believe that independence cannot be safely assumed, we must take different steps, some of which will be described later.

The assumptions we have stated are implied in the test procedure: We will test an estimate of variance *among* the levels of the tested factor against a pooled estimate of variance *within* the levels of the factor. Unless the experiments are independent, no clear interpretation attaches to a variance among the levels. Unless the within-level variances are similar, we have no right to pool them.

8·3 A LOOK AT VARIATION

Before we describe the details of the calculations of the ANOVA needed for a single-factor experiment, we consider three idealized examples. In Table

Table 8·1 Single-factor Experiment with Only Within-level Variation

Level	Replicate			
	1	2	3	$\bar{X}$
1	5	6	7	6.0
2	7	5	6	6.0
3	6	7	5	6.0
				$\bar{\bar{X}} = 6.0$

8·1 we show the nine values (coded) which might have been obtained from a tire experiment.

We have arranged the data in this way: In the three rows we group the values of the response variable according to level; in the columns we place the values from the replicates at each of the levels. In this table the values of $\bar{X}_i$ (the row averages) are identical. Thus we attribute to experimental error the differences within rows (among replicates). We need for these data make no further computations to see that there is no basis to suspect a difference among levels, since the means are the same. Such data represent a situation in which we would conclude that the only source of differences in the values of the response variable is other than the factor we have tested; we would for these data accept the null hypothesis. (These data are, it should be admitted, not realistic, since if experimental error exists we would expect that the sample $\bar{X}_i$ values would vary as well as the within-level observations. In fact, if σ^2 were the population variance within groups, we would expect on the average that the variance of averages of three would be $\sigma^2/3$. We are in these examples using idealized data to make a point.)

Table 8·2   Single-factor Experiment with Only Between-level Variation

Level	Replicate			
	1	2	3	$\bar{X}$
1	6	6	6	6.0
2	5	5	5	5.0
3	7	7	7	7.0
				$\bar{\bar{X}} = 6.0$

In Table 8·2 we show a quite different kind of result: Here there are no differences whatever within rows; thus the value of the variance for experimental error is zero. The differences in the data are wholly attributable to the differences among brands of tires. Without calculations of the variances, we conclude that the experiment has resulted in data that require us to reject the null hypothesis. (A real experiment rarely produces data which give a zero value for experimental error. The occurrence of such a result should cause us to suspect that the tests were crudely performed.)

In Table 8·3 we show data more nearly representative of a real situation. Differences appear among rows and within rows. Now the question is this: Is the variance *among* row averages large enough, in comparison with the variance *within* rows, to permit us to say that a real difference probably exists among the tire brands? To answer this question, we must now compute the two variances and compare them by an F test. For us to decide that the among-row variance is real, it must be the larger of the two variances by a factor greater than that attributable to chance.

Table 8·3   Single-factor Experiment with Both Within-level and Between-level Variation

Level	Replicate			
	1	2	3	$\bar{X}$
1	5	6	7	6.0
2	6	4	5	5.0
3	7	6	8	7.0
				$\bar{\bar{X}} = 6.0$

8·4 THE MATHEMATICAL MODEL

A useful way of specifying, in compact form, our hypothesis about the situation is by means of a statement called a *mathematical model*. Such a statement defines our belief about the possible sources of variation in the data in such a way as to clarify the relationship that we expect to test.

For the single-factor case, the mathematical model is

$$X_{ij} = \mu + R_i + \epsilon_{j(i)}$$

The statement is interpreted in this way: Each piece of data (X_{ij} value) in the array is what it is because of three components: the general level of the data μ; a possible effect associated with different levels of the factor contained within rows R (where R_i is the deviation of the ith row from the overall mean μ); the effect of error ϵ. The subscript for X is ij, specifying a value in one of the i rows and in one of the j columns where in this case we place the replicates. The subscript for R is i, specifying one of the i levels of the factor being tested. The subscript for the error term is written $j(i)$, a special way of indicating that the replicates are contained within the rows. We speak of such containment of data as "nested"; more will be said on this point later.

In the ANOVA to be described, we will estimate the variance of the row effects R_i by the amount that the variance *among* the row averages exceeds the random variability of averages were there no row effects. We will estimate the variance of $\epsilon_{j(i)}$ by the pooled variance of the *within*-row data. We will compare, by an F test, the *among*-row variance with the *within*-row variance. If we find the row effect to be too small to be significant, we will then interpret this result as giving us no reason to reject a different and simpler mathematical model:

$$X_{ij} = \mu + \epsilon_{j(i)}$$

That is, we accept the null hypothesis that the row deviations from the general average are zero: $H_0: R_1 = R_2 = R_3 = 0$. If we accept the null hypothesis, we do so because we have no reason to believe that there is any contribution associated with the difference among rows; our samples reflect only the grand average plus random error. If we accept the null hypothesis, we consider that the test gives us no reason to doubt that the pattern is as shown in Fig. 8·1a, and as represented by the data in Table 8·1.

The basis for ANOVA, then, from this mathematical model, is this: A calculated variance of means (levels of a tested factor) includes two possible parts—random error, and a real difference among levels. We estimate random error by a pooled *within-row* variance; we estimate the real difference among levels by subtraction of this random error divided by c (number of columns) from the variance of the means computed from the grand average of all the data.

The model for the F test is

$$F = \frac{\text{variance of means}}{\text{variance for error}} = \frac{\sigma^2/c + \sigma_r^2}{\sigma^2/c}$$

where σ^2/c is the variance for error (we divide by c since we have c values in each row) and σ_r^2 is the variance associated with a real difference among levels.

If σ_r^2 is zero, meaning that no difference among levels exists, the population F ratio is 1. (Because of sampling, we would not expect to get exactly 1. The F distribution indicates how large a number we can expect when the population value really is 1.) Thus the null hypothesis for this test is H_0: $\sigma_r^2 = 0$. The implication is that the F ratio is 1. If there is a real row effect, $\sigma_r^2 > 0$, and the F ratio is significantly greater than 1.

8·5 ANOVA FROM VARIANCE FORMULA

The two variances we need for the F test of hypothesis can be found from the fundamental variance formula

$$s^2 = \sum \frac{(X - \bar{X})^2}{n - 1} \qquad (8·1)$$

Equation (8·1) gives the *total* variance if we make these substitutions in it:

1 For X in the formula, we substitute X_{ij}, each value of the response variable in the table of data.
2 For $\bar{X}$ in the formula, we substitute the average of all the data $\bar{\bar{X}}$.
3 For n in the formula, we substitute the entire sample size.

The total variance as found by this formula is in this one-factor case to be analyzed into two parts—one associated with a possible between-level effect, the other associated with random error. [Note that the denominator of the fraction in Eq. (8·1) is the total number of degrees of freedom available to us. Some of these will be associated with the among-row effect, some with error.]

Equation (8·1) gives the variance *among* the factor levels if we change it to read

$$s_{\bar{X}}^2 = c \sum \frac{(\bar{X}_i - \bar{\bar{X}})^2}{r - 1} \qquad (8·2)$$

where c = number of columns as we have ordered the observations
$\bar{X}_i$ = any one of the row averages
r = number of rows

The numerator of the fraction in Eq. (8·2) is the sum of the squares of the deviations of the averages at each level from the grand average; the denominator is the number of degrees of freedom for the sample averages. We use c as a multiplier for the numerator in order to weight the computation in accordance with the number of observations that enter into each mean.

Similarly, we find the variance attributable to error by

$$s_e^2 = \sum \frac{(X_{ij} - \bar{X}_i)^2}{r(c - 1)} \qquad (8·3)$$

The numerator of the fraction in Eq. (8·3) specifies the sum of the squares of the differences between every observation and the mean of the level in which this observation appears. The divisor is the number of degrees of freedom for error. We subtract 1 from the number of replicates c because our calculation of the mean at each level consumes one degree of freedom for every such calculation.

Note that the sum of the degrees of freedom in Eq. (8·2) which is $(r - 1)$ and that in Eq. (8·3) which is $r(c - 1)$ is the total value of ν for the experiment, which is $n - 1$, inasmuch as $rc = n$.

Using the data as contained in Table 8·3, from Eq. (8·2) we have

$$s_{\bar{X}}^2 = \frac{3[(6.0 - 6.0)^2 + (5.0 - 6.0)^2 + (7.0 - 6.0)^2]}{3 - 1}$$

$$= \frac{3(0 + 1 + 1)}{2} = 3.0$$

From Eq. (8·3) we have

$$s_e^2 = \frac{\begin{array}{c}(5 - 6.0)^2 + (6 - 6.0)^2 + (7 - 6.0)^2 \\ + (6 - 5.0)^2 + (4 - 5.0)^2 + (5 - 5.0)^2 \\ + (7 - 7.0)^2 + (6 - 7.0)^2 + (8 - 7.0)^2\end{array}}{3(3 - 1)}$$

$$= \frac{(1 + 0 + 1) + (1 + 1 + 0) + (0 + 1 + 1)}{6}$$

$$= \frac{6.0}{6} = 1.0$$

Thus the estimate of the variance due to levels (brands of tires) plus chance and nontested factors is 3.0, and the estimate of the variance within levels (due only to chance and nontested factors) is 1.0. The ratio of these two values 3.0 we compare with the table of F values for the values of v, which are 2 and 6. We find the table value for $F_{2,6,0.05}$ to be 5.14. We conclude, at this level of significance, that the variance of the sample means is not large enough to indicate a significant difference, and we accept the null hypothesis.

8·6 ANOVA BY SUMS OF SQUARES

A systematic method of carrying out the calculations needed for analysis of variance is based on a modification of Eq. (8·1) to read

$$s^2 = \frac{\sum X^2 - (\sum X)^2/n}{n-1} \tag{8·4}$$

It is convenient to use this relationship for finding variances because it obviates the necessity of finding means and deviations from the means. We can use instead sums and squares of the observations, hence the expression "sums of squares."

We will use the notation in Table 7·6 to specify the sums: T specifies a total; a subscript dot specifies the composition of the sum. Thus $T_i.$ signifies the total of the ith row, the dot indicating that the total includes all column values in that row. $T_2.$ specifies the total of all the values in the second row. $T..$ specifies the grand total of all the observations, summing over all rows and columns. We use this notation in Table 8·4.

Table 8·4   Data for Calculating Sums of Squares

Level	Replicate			$T_i.$
	1	2	3	
1	5	6	7	18
2	6	4	5	15
3	7	6	8	21
				$T.. = 54$

The total sum of squares (abbreviated SST) is the numerator of the fraction in Eq. (8·4). Thus

$$\text{SST} = \sum X_{ij}{}^2 - \frac{\left(\sum X_{ij}\right)^2}{n}$$

or, using the T notation,

$$\text{SST} = \sum X_{ij}{}^2 - \frac{(T..)^2}{n} \tag{8·5}$$

Applying Eq. (8·5) to the data in Table 8·4, we obtain

$$\text{SST} = (5^2 + 6^2 + 7^2 + \cdots + 8^2) - \frac{54^2}{9} = 12$$

The value of 12 for SST is the numerator of Eq. (8.4). It is this total which we will partition.

Similarly, we use the row totals $T_i.$ to calculate a sum of squares for rows SSR:

$$\text{SSR} = \sum \frac{(T_i.)^2}{j} - \frac{(T..)^2}{n}$$

We divide the sum of the squares of the row totals by the number of values making up each row, and subtract the same term as before. Thus

$$\text{SSR} = \tfrac{1}{3}(18^2 + 15^2 + 21^2) - \frac{54^2}{9} = \tfrac{1}{3} \times 990 - 324 = 6.0$$

To find the sum of squares for error SSE we must go back to the original table of data (Table 8·3) and calculate the sum of the squares of the deviations of all the data from the corresponding row average. Thus

$$\text{SSE} = \sum (X_{ij} - \bar{X}_i)^2$$

or, in the T notation we are using,

$$\text{SSE} = \sum X_{ij}^2 - \frac{\sum T_i.^2}{j} = (5^2 + 6^2 + \cdots + 8^2) - \tfrac{1}{3}(18^2 + 15^2 + 21^2)$$
$$\text{SSE} = 6.0$$

An important characteristic of the sums of squares we have calculated is that they are additive. In particular, the total sum of squares SST is the sum of the term we have called SSR plus that assigned to error SSE. We could in this case, therefore, find the value of SSE by subtraction:

$$\text{SSE} = \text{SST} - \text{SSR}$$

The application of this rule would give 6.0 for SSE, as we found by direct computation.

If we now divide the sums of squares SSR and SSE by the value of v appropriate to each, we will have the variance terms we need for the F test. The value of v for the whole experiment is $n - 1 = 9 - 1 = 8$. The value of v for rows (the among-level effect) is one fewer than the number of rows = $r - 1 = 2$.

The value of v for error is $r(c - 1)$ following this reasoning: Each row has c values (here 3); each of these rows contributes $c - 1$ degrees of freedom to the error term; since we have r such rows, the total value for v is $3 \times 2 = 6$.

Just as the sums of squares for the variances total to give SST, so the values for ν for the among-level effect and error total to give the number of degrees of freedom for the whole experiment.

The quotient obtained when we divide each SS term by its value of ν is called the <u>mean square value</u> for each source of variation; the data are conventionally assembled in an ANOVA table, as in Table 8·5. Such a table

Table 8·5　Summary ANOVA (Data from Table 8·4)

Source	Sum of squares	ν	Mean square
Tires	6.0	2	3.0
Error	6.0	6	1.0
Total	12.0	8	

(handwritten annotations:) Sum of Squares ✓ ; Mean square →

$$F = \frac{MST}{MSE} = \frac{3}{1}$$

systematizes the calculations required for the F test which follows.

Three micrometers are used in an inspection procedure; we wish to know whether or not we can use them interchangeably. The null hypothesis is that there is no difference in the sets of readings obtained from the different micrometers A, B, and C, which are the three levels of this test. Thus $H_0: \mu_A = \mu_B = \mu_C$; that is to say, the variance among these means is zero. The alternative hypothesis is that at least one of the means differs from the others, and that the variance among the means is thus other than zero. We will test at the 0.05 level of significance.

The response variable is the difference in thousandths from the nominal value for a standard block. We make four replicate measurements of the block with each micrometer in random sequence; the data are in Table 8·6.

Table 8·6　Micrometer Readings

Micrometer	Replicates				
	1	2	3	4	$T_{i\cdot}$
A	1	6	0	0	7
B	2	6	3	6	17
C	3	9	8	7	27
					$T_{\cdot\cdot} = 51$

Here $n = 12$, and the total for ν is 11. Since $r = 3$, the value of ν for rows is $r - 1 = 2$. For error, the value of $\nu = r(c - 1) = 3 \times 3 = 9$. We note that the sum of the values of ν for rows and for error equals the total for the experiment.

The calculations follow:

$$SST = \sum X_{ij}^2 - \frac{T..^2}{n}$$

$$= (1^2 + 6^2 + \cdots + 7^2) - \frac{51^2}{12}$$

$$= 325.00 - 216.75 = 108.25$$

$$SSR = \frac{(7^2 + 17^2 + 27^2)}{4} - \frac{51^2}{12}$$

$$= 266.75 - 216.75 = 50.00$$

$$SSE = SST - SSR$$

$$= 108.25 - 50.00 = 58.25$$

The summary ANOVA is given in Table 8·7. The calculated F ratio is 25.00/6.47 or 3.86. The table F ratio at the 0.05 level of significance is 4.26.

Table 8·7   Summary ANOVA (Data from Table 8·6)

Source	Sum of squares	ν	Mean square
Micrometers	50.00	2	25.00
Error	58.25	9	6.47
Total	108.25	11	

Even if the true ratio were only 1, we could expect to get a ratio as large as 4.26 for a sample as often as once in twenty times. We therefore accept the null hypothesis, and conclude that there is insufficient evidence from this test to suppose that the micrometers cannot be used interchangeably.

PROBLEMS

8·1   Apples can be stored in various ways. Four different kinds of storage conditions were tested, and the time before spoilage (in months) was recorded as in Table 8·8.

Table 8·8   Storage Life of Apples in Months

Storage condition	Replicate		
	1	2	3
A	7	8	7
B	3	3	5
C	3	4	3
D	9	7	10

(a) By looking at the data, make a judgment about the effect of storage conditions on the response variable.

(b) Perform an analysis of variance, and test at the 0.05 level for a significant difference associated with storage conditions.

8·2   Three types of float cutoffs were tested in a bottle-filling machine. The response variable is level of liquid in the hopper. Using an α risk of 0.05, test for a significant difference associated with type of float cutoff on the basis of the data in Table 8·9.

Table 8·9   Liquid Level with Three Types of Float Cutoff

Float	Replicate			
	1	2	3	4
1	11.3	11.7	10.5	11.8
2	10.5	11.2	11.0	10.8
3	11.0	10.4	10.9	11.3

8·3   An operator, equally efficient on each of three desk calculators, computed the standard deviations of 5 sets of sample data on each machine. The times required for the computations were recorded as in Table 8·10. Is there evidence of a difference in the calculators?

Table 8·10   Computation Times for Three Desk Calculators

Calculator	Replicate				
	1	2	3	4	5
A	32.5	32.5	33.0	32.5	33.0
B	35.0	34.0	34.5	34.0	33.5
C	30.5	31.5	31.0	32.0	31.5

8·7   SUMMARY

We have so far dealt with the analysis of variance for a set of data obtained because we were interested in the possible effect of only a single factor— e.g., kind of tire or kind of micrometer. ANOVA is applicable to situations in which we wish to test the effect of more than one factor; it is in such situations, in fact, that ANOVA is most useful. Before we discuss more involved cases, we summarize the principles which we have illustrated by the single-factor case.

1   ANOVA is a method of partitioning the total variability of a set of observations among the possible sources of variability.

2   ANOVA incorporates F tests to evaluate the mean squares.

3 The conclusion, after the completion of the F tests, is a statement of this nature: The variance associated with a tested factor is greater than (or not greater than) that associated with chance and untested factors, at a specified level of significance. We cannot by ANOVA *explain* the variability in the observations in a cause-and-effect sense, since there may exist associations among variables due to a multiplicity of causes.

4 ANOVA is valid only if the data have been appropriately obtained. The hypotheses to be tested and the level of significance should be stated in advance of the data collection. Since ANOVA assumes that sources of variation other than tested factors are attributable to error, it is necessary that the observations be made so that nontested sources of variation are random in their effect.

8·8 TWO-FACTOR ANOVA

We return to the data from the test of micrometers in Table 8·6, and reconsider the observations on the assumption that each of the four readings was made by a different operator. We now have the possibility that either or both of *two* main factors contribute to the differences in the data. The effect of the main factor "micrometers" will be shown by differences among the *row* totals; the effect of the main factor "operators" will be shown by differences among the column totals. The data and the row and column totals are given in Table 8·11. In this table, we have added to the T values

Table 8·11   Two-factor Nonreplicated Experiment

Micrometer	Operator				
	O_1	O_2	O_3	O_4	$T_{i\cdot}$
M_1	1	6	0	0	7
M_2	2	6	3	6	17
M_3	3	9	8	7	27
$T_{\cdot j}$	6	21	11	13	$T_{\cdot\cdot} = 51$

used before the symbol $T_{\cdot j}$, which represents the total summed over all rows, for the column specified by the subscript j.

We have only a single observation for each combination of the two factors "micrometer" and "operator." Thus the data were collected *without replication*. We will be unable to find a value for experimental error directly, based on the differences between replicates at the same combination of factor levels. What we will do is to find a total sum of squares, a sum of squares for rows, a sum of squares for columns, and a remainder when we subtract the last two from the total. This remainder, called a residual, we will attribute to chance causes, and we will use this as a substitute for the error term.

The use of a residual sum of squares for error assumes that there is no joint effect between operators and micrometers. That is, we assume that one operator is no more likely to make an error with one micrometer than with another, and that operator variability is of the same size no matter what micrometer is used. When the opposite is true, that is, when the effect of one factor varies with the level of the other factor, we say that *interaction* occurs. In a nonreplicated two-factor experiment, we must assume that no interaction occurs since we have no estimate of error against which we can test the possible interaction.

The mathematical model for this two-factor nonreplicated experiment is

$$X_{ij} = \mu + R_i + C_j + \epsilon_{ij}$$

that is to say, each observation X_{ij} is assumed to be determined by four possible effects: the general mean μ of the data, a possible row effect (here micrometers), a possible column effect (here operators), and error.

We may now test two hypotheses: (1) there is no significant effect associated with micrometers; (2) there is no significant effect associated with operators. Our test will be based on F ratios, comparing mean squares for rows and columns against the residual mean square.

In constructing the ANOVA table, the calculations for SST and SSR will be exactly the same as for the earlier experiment with micrometers; also, the total value of v and the value of v for rows will be unchanged. We now calculate a sum of squares for columns (SSC) in a manner similar to that for finding SSR, as follows:

$$SSC = \sum \frac{(T_{\cdot j})^2}{r} - \frac{(T_{\cdot \cdot})^2}{n}$$

The uncorrected sum of squares for columns will be the first of these two terms, or

$$\frac{6^2}{3} + \frac{21^2}{3} + \frac{11^2}{3} + \frac{13^2}{3} = 255.67$$

Observe that each $(T_{\cdot j})^2$ is divided by 3, the number of items making up the column totals. There will be $c - 1$ degrees of freedom for columns, where c is the number of columns. The ANOVA table is given in Table 8·12.

Table 8·12   Summary ANOVA (Data from Table 8·11)

Source	Sum of squares	v	Mean square
Micrometers	50.00	2	25.00
Operators	38.82	3	12.94
Residual	19.43	6	3.24
Total	108.25	11	

Note that the residual value of v is found by subtracting from the total number of degrees of freedom the value of v for rows and columns. The value of v for the residual also equals the product of the values of v for rows and columns, or $(r - 1)(c - 1)$. The residual sum of squares = SST — (SSR + SSC).

We now test each of the mean squares for rows and columns against the residual mean square by calculating the F ratio. For rows, the ratio is 25.00/3.24 or 7.71. The table value for $F_{2,6,0.05}$ is 5.14. We therefore reject the null hypothesis and find the row factor (micrometers) significant. The F ratio for columns is 12.94/3.24 = 3.99. The table value for $F_{3,6,0.05}$ is 4.76. We therefore accept the null hypothesis here, and find that the column factor (operators) is not significant.

If the results of this ANOVA are compared with those from the earlier illustration involving micrometers, these differences appear:

1 The two-factor experiment requires the same number of observations (12) as did the single-factor experiment. The two-factor design is therefore more economical, since it enables us to test two factors with the same cost in time and labor as we used before for a single factor.

2 The sum of squares for the residual (assumed to be caused by error) is in this analysis smaller than before. The two-factor analysis in effect assigns to a second factor a portion of what was before considered to be error.

3 The mean square for the residual is considerably smaller; therefore, the F ratio is larger. The result is a more sensitive F test. Even though the number of degrees of freedom for the residual is reduced, we can often detect, by a two-factor analysis, significant variance differences which the single-factor test cannot find. Note that by the single-factor test we concluded that there was no evidence of a significant effect due to micrometers; the two-factor analysis was sensitive enough so that a significant difference (at the 0.05 level) was discovered. The result was due to the removal of the assignable cause (operators) from the error term.

8·9 SECURING DATA FOR A TWO-FACTOR ANALYSIS

For the type of ANOVA just illustrated, data must be collected for each level of each factor at each level of the other factor. Here, for example, it was necessary for *each* operator to work with *each* micrometer. If the situation were to involve the effect on the quality of a product where two chemical ingredients were important, it would be necessary to prepare sufficient batches of the product so that each level of each component would be coupled with each level of the other component. Such a set of experiments is termed *crossed*.

In order to avoid systematic errors due to nontested factors, randomization of the sequence of preparation of the batches would be necessary, as well as attention to possible sources of error in preparation or measurement of the quality of the product.

PROBLEMS

8·4 pH measurements were taken from the top, middle, and bottom of 6 samples of different soils. The data are given in Table 8·13.

Table 8·13 pH Readings of Soil Core Samples

Soil sample	Top	Middle	Bottom
1	7.5	7.6	7.2
2	7.2	7.1	6.7
3	7.3	7.2	7.0
4	7.5	7.4	7.0
5	7.7	7.7	7.0
6	7.6	7.7	6.9

(a) Write the mathematical model for the experiment.
(b) State the assumptions involved in the experiment.
(c) Write the null and alternative hypotheses under test.
(d) Is there a real difference among samples?
(e) Is there a real difference associated with position within the soil sample?

8·5 Three analysts made determinations of the percentage of total solids in wet brewers' yeast, with results as in Table 8·14.

Table 8·14 Percentage of Solids in Brewers' Yeast

Analyst	Sample				
	1	2	3	4	5
A	10.1	4.7	3.1	3.0	7.8
B	10.0	4.9	3.1	3.2	7.8
C	10.2	4.8	3.0	3.1	7.8

(a) What is the model for the experiment?
(b) Is there a sample-to-sample difference?
(c) Is there a real difference among analysts?

8·6 Four kinds of gasoline were used in three makes of automobile; the response variable was miles per gallon, as in Table 8·15.
(a) In securing the data, how might the effects of nontested factors be minimized?

Table 8·15 Data from Test of Four Gasolines
with Three Automobiles

Gasoline	Automobile		
	A	B	C
1	16.4	15.9	17.0
2	15.6	14.9	16.8
3	14.8	15.0	15.3
4	18.4	17.9	17.6

(b) Do the data indicate a difference among gasolines?
(c) Does the make of automobile make a difference?

8·7   Consider the data in Table 8·16.

Table 8·16   Data from a Two-factor Experiment

Factor II levels	Factor I levels			
	A	B	C	Total
a	1	2	3	6
b	4	5	6	15
c	7	8	9	24
d	10	11	12	33
Total	22	26	30	78

(a) Looking at the data, make a judgment whether or not either of the factors has a significant effect.
(b) Estimate the size of the error term.
(c) How could such a set of data be obtained?
(d) Is it necessary to calculate the data for ANOVA? Explain.

8·8   Run an analysis of variance for the data in Table 8·17.

Table 8·17   Data from a Two-factor Experiment, One Unusual Observation

Factor II levels	Factor I levels			
	A	B	C	Total
a	1	2	3	6
b	4	5	6	15
c	7	8	9	24
d	100	11	12	123
Total	112	26	30	168

(a) How does one unusual piece of data affect the row variance as compared with the column variance?

(b) From the data alone, should we suspect that the odd value is false? How could we find out?

8·10 TWO-FACTOR ANALYSIS CONTINUED

There are many situations in which the preceding method of analysis (which ignores the possibility of interaction) is inappropriate. Such situations involve two or more factors whose influence may well not be independent; i.e., the influence of one factor may depend on the level of the other factor. Examples are

1 Different catalysts and different temperatures may be used in the "cracking" of crude petroleum. The effect of temperature depends on the catalyst used; some catalysts are sensitive to temperature changes, others are less so.

2 Flavors in food products are dependent in part on the amounts of sugar and salt used. The influence of the amount of sugar depends on the amount of salt. If there is little salt, the amount of sugar is important; if much salt is used, the effect of the amount of sugar is less.

3 In finishing machine parts, the quality of the surface is determined in part by a plating operation and a buffing operation. The results produced with different buffing methods vary with the plating method.

4 In medical prescriptions, synergistic action is encountered. A drug has a different potency, depending on whether it is used alone or with another drug.

In these and similar cases, an *interaction* is said to occur. In our treatment of ANOVA up to now we have assumed no significant interaction. If we suspect the possibility of interaction, we must obtain data in such a way as to permit us to test for the suspected interaction effect. The requirement for a two-factor case is that we must replicate or have some previous estimate of random error.

Replication means the repetition of test procedures. This repetition allows us to obtain an estimate of error from the differences in the results of the repetitions. Replication must be distinguished from *duplication*, which ordinarily means merely making two or more measurements on a single sample which has come from one treatment, or one experimental procedure. To replicate requires the preparation of two or more *independent* samples treated as nearly alike as possible. Any differences we observe among samples we attribute to chance factors.

The mathematical model for a replicated two-factor experiment is

$$X_{ijk} = \mu + R_i + C_j + (R \times C)_{ij} + \epsilon_{k(ij)}$$

This model differs from the previous one in this respect: the addition of the term $(R \times C)_{ij}$. This term represents the joint influence of the row and column factors, and thus the interaction.

The mathematical model thus indicates that each single observation (for any column, row, and replication) depends on the mean of the population from which the observation is drawn, an influence due to the row factor, an influence due to the column factor, an interaction, and error. Now, the total sum of squares in the ANOVA will be made up of four parts, thus, SST = SSR + SSC + SSI + SSE, where the new term SSI represents sum of squares for interaction.

We take as our example this problem: In the manufacture of a photographic film emulsion, it is customary to mix the ingredients and to allow the mixture to stand for a certain time at a certain temperature. During this time, the characteristics of the mixture are thought to change; one of the most important changes is that of the speed of the film that is finally made from the emulsion. We are interested in discovering whether, for a particular set of ingredients, the time and the temperature are really important. We believe that there may be an interaction, and we desire to know also whether the interaction is significant here.

Since this problem involves two factors (time and temperature) plus a possible interaction, we wish to obtain a separate estimate of error. We therefore plan to twice replicate; that is, we will prepare eight batches of the mixture, two for each combination of time and temperature. The data tabulation will contain two estimates of speed for each of the time-temperature combinations, each estimate having been obtained from a separate batch. (Had we merely prepared one batch of ingredients for each time and temperature, and made two measurements on each batch, we would have duplicated, not replicated.)

In order to reduce the systematic effects of nontested factors, we randomize the sequence of batch preparation and testing. We hope by this method that any effects due to time, to operator, to machine changes, etc., will not seriously affect the results.

From the data secured by such an experiment, we can test three null hypotheses. They are (1) that changes in the time of holding the mixtures produce no effect; (2) that changes in temperature produce no effect; (3) that there is no interaction of time and temperature. We plan to test these hypotheses by means of three F ratios, using the estimate of error as the basis for comparison, and to test at the 0.05 level of significance.

Table 8·18 displays the results of the set of tests, the numbers in the body of the table being percentage change from a standard formulation. The negative numbers in the table indicate a reduction in speed. The negative signs must be taken into account in the sums, but not in the squares, since the square of a negative number is positive.

Three different kinds of totals appear in the tabulation. Thus we need three subscripts for identifying the totals: the subscript i indicates that the

Table 8·18 Two-factor Replicated Experiment

Time	Temperature		
	50°	70°	$T_{i..}$
1 hr	10 30 $T_{1j.}$ $\overline{40}$	-20 -40 $\overline{-60}$	-20
2 hr	0 40 $T_{2j.}$ $\overline{40}$	50 40 $\overline{90}$	130
$T_{.j.}$	80	30	110 $T_{...}$

total applies to a row; the subscript j indicates that the total is for a column, and the subscript k indicates that the total is for the replicates. A subscript (.) in the place of one of the letter subscripts indicates that the total includes the row, column, or replicates so indicated. For example, $T_{i..}$ indicates the total of a particular row for all columns and replicates in that row; $T_{.j.}$ indicates a column total for all row and replicate numbers, etc. $T_{...}$ indicates the grand total. $T_{ij.}$ specifies the total of replicates in a particular row and column: thus $T_{12.}$ is here -60.

In this notation, the sum of squares calculations are carried out as follows:

$$\text{SST} = \sum (X_{ijk})^2 - \frac{(T_{...})^2}{n} \text{ where } n \text{ is } 8$$

$$= 10^2 + 30^2 + (-20)^2 + \cdots + 50^2 + 40^2 - \frac{110^2}{8}, \text{ or } 7{,}187.5$$

$$\text{SSR} = \frac{1}{ck} \sum (T_{i..})^2 - \frac{(T_{...})^2}{n}$$

where ck is the product of the number of columns and replicates; it is the number of *individual* observations making up each row total. Thus

$$\text{SSR} = \frac{(-20)^2}{4} + \frac{130^2}{4} - \frac{110^2}{8}, \text{ or } 2{,}812.5$$

Similarly,

$$\text{SSC} = \frac{1}{rk} (T_{.j.})^2 - \frac{(T_{...})^2}{n} = \frac{80^2}{4} + \frac{30^2}{4} - \frac{110^2}{8}, \text{ or } 312.5$$

To find the sum of squares for interaction, we use the totals of the replicates (symbolized by $T_{ij.}$), square these, and divide by the number of items making up these totals, that is, the number of replicates which is here 2. This sum of squares, however, includes the differences in rows and columns which we have already calculated. We therefore subtract these, and then apply the correction factor as before. Thus

$$\text{SSI (sum of squares for interaction)} = \frac{1}{k} \sum (T_{ij.})^2 - \text{SSR} - \text{SSC}$$

$$- \frac{(T_{...})^2}{n}$$

$$\text{SSI} = \frac{40^2}{2} + \frac{(-60)^2}{2} + \frac{40^2}{2} + \frac{90^2}{2} - 2,812.5 - 312.5 - \frac{110^2}{8}, \text{ or } 2,812.50$$

The degrees of freedom in this example will be
  Total, $n - 1$, or 7
  For rows, $r - 1$, or 1
  For columns, $c - 1$, or 1
  For interaction, $(r - 1)(c - 1)$, or 1
  For error, the remaining degrees of freedom are 4.
The data are summarized in Table 8·19. The calculated F ratios are in each case the quotient of the mean square for time, temperature, and interaction

Table 8·19   Summary ANOVA (Data from Table 8·18)

Source	Sum of squares	v	Mean square	F ratio
Time	2,812.5	1	2,812.5	9
Temperature	312.5	1	312.5	2
Interaction	2,812.5	1	2,812.5	9
Error (by subtraction)	1,250.0	4	312.50	
Total	7,187.5	7		

and the error mean square. The table value for $F_{1,4,0.05}$ is 7.71. Our conclusions are: We reject the null hypothesis for time and for interaction; we accept the null hypothesis for temperature. Thus the experiment has not shown an effect associated with a change in temperature but has shown a change associated with a change in time. That the factor interaction has a significant effect means that a change in the level of temperature changes the direction in which a change is induced by time. Thus even though the average effect of temperature is small, we cannot ignore it as a factor since it does affect the potency of the time factor.

8·11   GRAPHING INTERACTION
A useful way of showing interaction is by a plot of the $T_{ij.}$ values against treatment time, for each temperature used in the emulsion experiment. Such a graph appears in Fig. 8·2. The presence of an interaction is indicated by the nonparallelism of the two lines. The plots show that at the lower

temperature both treatment times produce the same speed value; a temperature rise is associated with a speed *increase* for one treatment time and a speed *decrease* for the other treatment time. Such a relationship between

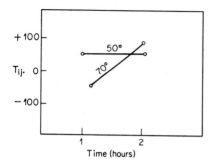

Fig. 8·2 Graph of interaction.

effects is what is meant by interaction. If in such plots as these the lines intersect (or would intersect if extended) an interaction effect is suggested by the data.

PROBLEMS

8·9 A twice-replicated two-factor experiment was run to test the effect of catalyst and temperature on the yield of a process, with results as given in Table 8·20.

Table 8·20 Effect of Temperature and Catalyst upon Yield

Temperature	Catalyst		
	A	B	C
160°	1 3	−2 −4	3 2
180°	0 4	5 4	0 −3

(a) Write the mathematical model for this experiment.
(b) Is there a difference among catalysts at the 0.05 level of significance?
(c) Is there a difference associated with temperature at the 0.05 level?
(d) Plot the temperature-catalyst interaction, and test to determine whether or not it is significant.

8·10 In Table 8·21 are given the results of a test of the effect of temperature on the breaking strength of a fabric. Analyze the data for a significant effect associated with (a) temperature; (b) fabric; (c) interaction.

Table 8·21   Effect of Temperature on the Breaking Strength of Fabric

Fabric	Temperature			
	210°	215°	220°	225°
A	1.8	2.0	4.6	7.5
	2.1	2.1	5.0	7.9
B	2.2	4.2	5.4	9.8
	2.4	4.0	5.6	9.2
C	2.8	4.4	8.7	13.2
	3.2	4.8	8.4	13.0

8·11   Four types of developers and four methods of agitation were tested for an effect on the resolving power of an aerial film, with results as given in Table 8·22.
(a) Does either of the factors have an effect?
(b) Is there a significant interaction?

Table 8·22   Resolving Power of an Aerial Film

Agitation method	Developer			
	A	B	C	D
I	98	114	102	103
	109	101	96	111
	100	113	110	106
II	118	102	113	103
	106	106	117	119
	116	110	116	116
III	107	106	111	101
	112	98	100	105
	105	93	109	94
IV	128	113	119	101
	119	119	106	108
	126	111	107	107

QUESTIONS

8·1   If we are interested in a comparison of three population means, could we arrive at some decision about the equality of the means using t tests alone? Explain.

8·2   When we use the analysis of variance techniques, are we testing hypotheses of the means, or of the variances? Explain.

8·3   When running an analysis of variance of a single factor, do we need to have the same number of replications for each level of the factor? Explain.

8·4   In the single-factor case, if we do not have an external estimate of the error, where do we obtain such an estimate in ANOVA?

8·5   What are the assumptions in single-factor ANOVA?

8·6   Why should we write our mathematical model to a problem before we consider the ANOVA table?

8·7   What information is contained in the mathematical model?

8·8   Do all experiments contain experimental error? Explain.

8·9   In the ANOVA table, why do we run F tests when the hypotheses are about means?

8·10   What part of the formula for variance does the sum of squares come from?

8·11   Why is the sum of squares divided by the degrees of freedom in the ANOVA table?

8·12   If all the data in an experiment were made into a frequency distribution, and the standard deviation of the distribution was computed, what value in the ANOVA table would it be, if any? Explain.

8·13   When using the T notation, what do the following mean:

(a) $T_{.i}$

(b) $T_{...}$

(c) $T_{.j}^2$

8·14   Is it always possible to obtain experimental error from direct calculation, rather than from subtraction of the terms in ANOVA from total sum of squares? Explain.

8·15   In a two-factor case, with replication, how many hypotheses are there under test?

8·16   What is meant by a "partitioning of sum of squares"?

8·17   In the two-factor case without replication, why should we have an external estimate of error if we wish to test for an interaction between the two factors?

8·18   Can you tell by looking at the mathematical model if there has been replication of the experimental runs? Explain.

8·19   If there is a factor which affects the response variable and we do not test for this factor or allow for its being held constant, what effect will the factor have on the values in the ANOVA table?

8·20   Can a sum of squares value ever turn out to be a negative value? Explain.

8·21   How does graphing the two-factor examples aid us to understand interaction if it exists?

REFERENCES

16, 19, 25, 28, 31, 33, 37, 39, 46, 47, 59, 60, 68, 70, 73, 82, 123

9 Components of Variance

An analysis of variance, as we have seen in Chap. 8, is a technique by which we can assign to each of the factors tested in an experiment some portion of the total variability. After we have carried out this analysis, we can by tests of significance determine whether or not the variance assigned to a specific factor is larger than that which we plausibly assign to chance.

We have so far restricted our discussion to situations in which

1 We always test the factor effect against an estimate of error.
2 We are interested only in knowing whether or not a factor has an effect.

We now point out that there are other situations in which it is incorrect to use the error-variance estimate as the denominator of the F ratio, and other situations in which we wish to estimate the *magnitude* of the variability, rather than merely to state whether or not the effect is present.

The subject of this chapter—components of variance—includes the analysis of the specific portions of the total variance assigned to the various factors. For each mean square term we will now identify the *expected mean squares* which are mathematical models associated with the individual sources of variation. We will use the expected mean squares to help us decide which comparisons are valid. The principle is this: An F test in an ANOVA can be correctly interpreted when only a *single* cause contributes to any possible difference between the two variances comprising the ratio used in the F computation.

Thus we will add to the summary ANOVA table an additional column headed "Expected mean squares." The items in this column will direct us to the appropriate F tests. The values in the mean square column are sample estimates of the expected mean squares, in the same sense that s^2 is an estimate of σ^2. This has meaning only when the levels of the factors involved have been randomly selected. We are now led to a discussion of the distinction between random and fixed levels of a factor.

9·1 FIXED VERSUS RANDOM FACTORS

Suppose that we use four operators as a factor in an experiment. We consider the factor "operators" as *fixed* if we are interested in the performance of these operators and no others. We consider the factor "operators" as random if we intend from the experiment to make inferences about a large group of operators from which we have taken only a random sample for the experiment.

If a factor has fixed levels, every one of these levels is included in the experiment, and we make inferences only about those we have tested. If a factor has random levels, we are really interested in making inferences about a large population but are in the position of having to choose from this population only a sample. We intend, however, on the basis of the results of the experiment, to make inferences about the entire population.

We may be dealing with a process in which a hundred machine operators participate; we wish to know whether or not the factor "operator" affects the process significantly. Not wishing to test all the operators, we select at random four of them for the experiment. In this case, the factor "operator" is a *random* one.

If we were dealing with a process in which four operators and no others were concerned and we wished to know whether these operators differed in their effect on the process, we would in the experiment test all of them. In this case, the factor "operator" would be a *fixed* one.

9·2 ANALYSIS OF VARIANCE MODELS

The distinction between fixed and random factors is, we shall see, important because it relates to the nature of the hypotheses we will test and to the method of data analysis we will use.

For a single-factor experiment, we write the mathematical model thus:

$$X_{ij} = \mu + A_i + \epsilon_{j(i)}$$

The model attributes the value of any specific observation X_{ij} to three possible causes: the mean of the population μ, the effect of the tested factor A_i, and the error effect $\epsilon_{j(i)}$. This mathematical model applies to every single-factor experiment.

There are, nevertheless, differences among single-factor experiments. In a fixed-model experiment, the levels being tested (levels such as different machines, different processes, different lots of raw material) are a set of fixed but unknown *constants*. In a random-model experiment, the levels chosen are supposedly representative of *variable* quantities that form a (presumably) normal distribution of their own.

The null hypotheses we test differ, depending on whether we have a fixed or a random model. For the fixed model, $H_0: T_i = 0$; that is to say, the hypothesis is that the effect of the tested factor T is zero at any level T_i of the factor. To test this hypothesis, we perform an F test of significance against *error*, and reject the null hypothesis if we find that the calculated F ratio departs significantly from a chance result. For the random model, on the other hand, the null hypothesis is $H_0: \sigma_T^2 = 0$; that is to say, the variance among *all possible levels* of the factor, of which we have sampled only a few, is zero. For the random model, we shall see that we often do not test the factor effect against error mean square. Furthermore, if we are using a random model (and only then) we can estimate the extent to which different factors contribute to the total variance.

PROBLEMS

9·1 We wish to test for an effect of the location of food items in large supermarkets on the total sales. We choose one food item, and display the item in each of four different store locations. The response variable is sales in dollars.

(a) Write the mathematical model for the experiment.
(b) Is "location" a fixed or random factor? Explain.
(c) Is "food item" a factor in this experiment? Could it be a factor? Explain.
(d) What are the hypotheses under test in this experiment?

9·2 A class in statistics believed that statistical problems could be solved more quickly by the use of desk calculators. Six students of about equal ability were selected; two used desk calculators, two used mathematical tables, and two used paper-and-pencil methods. The response variable was time to complete a problem set.

(a) Write the mathematical model for this experiment.
(b) Is "student" a factor? Explain.
(c) Is "method" a fixed or random factor? Explain.
(d) State the hypotheses under test.

9·3 Describe briefly four experiments testing fixed factors. Explain why the factors are fixed.

9·4 Describe briefly four experiments using random factors. Identify the populations from which the factors are taken.

9·3 EXPECTED MEAN SQUARES

We now extend the summary ANOVA table to include one more column. Here we will place mathematical terms which will express our belief about the sources of variation which contribute to the mean square terms of the table. Each entry in the expected mean squares (EMS) column will specify the nature and the source of the variation which is included in the value of the mean square term in the same row. We need the EMS terms because they direct us to the correct construction of the F ratios for tests of significance, and because they permit us (for random models) to calculate components of variance.

We use as an example this two-factor experiment: We measured the effective speed of three different brands of photographic film as exposed in four different printers. Two replicates were made of each test. The films and the printers were the only ones of interest; thus both these factors were fixed. The replicates were, however, only some of many replications that we might have made; thus the factor "replications" was random. From the result of this experiment, we can make inferences about only those films and printers we tested. The estimate of error is, however, here and always random. Our estimate of the error mean square is (by the very nature of error) only a sample of the many somewhat different error estimates that repetitions of the test might give us.

The mathematical model for this experiment is

$$X_{ijk} = \mu + F_i + P_j + (F \times P)_{ij} + \epsilon_{k(ij)}$$

To the subscripts are assigned these meanings: i, any of the $f = 3$ films; j, any of the $p = 4$ printers; k, either of the $r = 2$ replications. The statement of the mathematical model is an expression of our belief that each observation X_{ijk} has a value that is determined by (1) the general level of the X values; (2) a possible effect associated with films; (3) a possible effect associated with printers; (4) a possible interaction effect; (5) an effect associated with error.

The null hypotheses are $H_0: F_i = 0$; $H_0: P_j = 0$; $H_0: (F \times P)_{ij} = 0$. Each of these null hypotheses states that the effect of the factor or interaction is zero at any level. The alternative hypothesis in each case is that the effect is other than zero at one level at least.

9·4 ANOVA WITH EMS

The summary ANOVA in Table 9·1 has been calculated from the observations by the methods of Chap. 8. In this table appears a new column headed EMS, standing for "expected mean squares." The entry opposite F_i reads $8 \Sigma F_i^2/(f - 1) + \sigma_e^2$. This entry states that the variance associated with the column effect in the original data is composed of eight times a term analogous to the variance associated with films (but not quite a variance since we deal with the whole population of films we are interested in) plus the variance associated with error. The factor 8 arises from our having used

4 printers and 2 replications in the experiment. The divisor $f - 1$ is the value of ν for this factor. Similarly, the entry in the second row of the table gives the expected mean squares for the factor P_j as $6 \Sigma P_j^2/(p - 1) + \sigma_e^2$.

Table 9·1 Summary ANOVA Table, Two Fixed Factors

Source	Sum of squares (SS)	Degrees of freedom	Mean square (MS)	Expected mean square (EMS)
F_i	6.0	2	3.0	$\dfrac{8 \sum F_i^2}{f - 1} + \sigma_e^2$
P_j	24.0	3	8.0	$\dfrac{6 \sum P_j^2}{p - 1} + \sigma_e^2$
I_{ij}	12.0	6	2.0	$\dfrac{2 \sum I_{ij}^2}{(f - 1)(p - 1)} + \sigma_e^2$
Error	6.0	12	0.5	σ_e^2
Total	48.0	23		

When we examine the EMS column of the summary ANOVA table, we see that the EMS terms for both factors and for interaction contain only *one* source of variation in addition to error. It is because of this composition of the variances that it is right for a *fixed-model* experiment to test each variance against error. Since each variance includes one source of variation other than error, a significant departure of the F ratio from unity will cause us to reject the null hypothesis. That is, if there is indeed no film effect or no printer effect, the numerators in each case will simplify to σ_e^2. Therefore, we would expect σ_e^2/σ_e^2 to equal 1, subject only to random sampling errors.

We therefore perform the F tests using the estimate of error variance as the denominator of the F ratio. The complete table is given in Table 9·2; the results cause us to reject the null hypothesis for the interaction term and also for the printers. Using the same level of significance, 0.05, the null hypothesis for the films is also rejected.

9·5 DETERMINATION OF EMS

The entries in the EMS column are found by an algorithm—a stepwise procedure that leads to the desired results. We first prepare a two-way table as shown in Table 9·3. The row headings are the terms in the mathematical model, exclusive of the general level μ. The column headings are the individual identifying subscripts i, j, and k together with their uppermost values. Over each subscript we place F or R according to whether the factor identified by the subscript below is fixed or random.

Table 9·2 ANOVA Summary and F Ratios, Two Fixed Factors

Source	SS	ν	MS	EMS	Calculated F ratio	Table F ratio
F_i	6.0	2	3.0	$\dfrac{8\sum F_i^2}{2} + \sigma_e^2$	6.0	$F_{2,12,0.05} = 3.88$
P_j	24.0	3	8.0	$\dfrac{6\sum P_j^2}{3} + \sigma_e^2$	16.0	$F_{3,12,0.05} = 3.49$
I_{ij}	12.0	6	2.0	$\dfrac{2\sum I_{ij}^2}{6} + \sigma_e^2$	4.0	$F_{6,12,0.05} = 2.99$
Error	6.0	12	0.5	σ_e^2		
Total	48.0	23				

Table 9·3 Two Stages in the Preparation of an EMS Coefficients Table

(a)

	F i; 3	F j; 4	R k; 2
F_i		4	2
P_j	3		2
I_{ij}			2
$e_{k(ij)}$	1	1	

(b)

	F i; 3	F j; 4	R k; 2
F_i	0	4	2
P_j	3	0	2
I_{ij}	0	0	2
$e_{k(ij)}$	1	1	1

We now consider each row in turn. For every row, we enter in the corresponding space the value of each column subscript that is *missing* in the *row* identifying symbol. Thus the first row is identified as F_i. Since the first column is marked i, we make for the moment no entry in this first cell. We do make an entry of 4 (there are 4 printers) in the next cell of this row, since this cell is in the column marked j, and j does *not* appear in the row identification. Similarly, we enter 2 (there are 2 replicates) in the next cell.

For the second row, we enter 3 (there are 3 films) in the first cell, nothing in the next cell, and 2 in the last cell. For the third row, our only entry is 2 in the last cell, since only in the last column do we find a subscript that is missing from the row identifying symbol.

The last row requires special treatment, since it involves what is called a "nested" factor, i.e., replications. The pair of replicates is compared within each cell, since experimental conditions for the pair were supposedly identical (except for random causes). The variance estimates from the within-cell data are pooled (averaged together) to give an overall estimate of within-cell variability—random error. (We will discuss nested factors in more detail in Chap. 10.) In this last row, we place 1 (regardless of the number of replicates) under the corresponding column heading for each subscript in parentheses. Thus far we have completed Table 9·3a.

Now we consider each column in turn. For each column involving a fixed factor (headed with F) we place 0 in each empty cell of the column. For each column involving a random factor (headed with R) we place 1 in each empty cell in the column. The two-way table is now completely filled in; we show it as Table 9·3b.

This table is used to find the coefficients of the terms in the EMS column of the ANOVA summary table by the following steps:

1   We begin at the top of the table and with the first factor.
2   We ignore the column or columns that contain nonnested (nonparenthetical) subscripts appearing in the factor identification.
3   We find the products of the remaining numbers in each of the rows that contain the subscript identification of the factor. If an entry stands alone, we take that entry as the product.
4   We continue in the same way for the remaining factors.

When we apply this method for the first factor F_i, we ignore the first column since it is headed with i. The remaining numbers in the first row (headed by i) give a product of 8. This is the value of the coefficient to be used with the term involving i, and we therefore write $8 \Sigma F_i^2/(f-1)$. We go on to the next row. It is headed only by subscript j; since no subscript i appears in the row heading, we skip this row. The third row is headed ij; we here find the subscript i in which we are presently interested. The product of the remaining numbers in the third row is, however, zero; therefore, the variance σ_{ij}^2 is *not* a component of the variance of the factor we are considering. Finally, the last row contains in the heading the subscript i (among others); we find the product of the remaining values to be 1, and we therefore include $1\sigma_e^2$ with the components of variance for F_i. The complete expression for the EMS for F_i is thus $8 \Sigma F_i^2/(f-1) + \sigma_e^2$, as shown initially in Table 9·1.

Similarly, for the second factor we ignore the column headed j, and find the products of the remaining values in each row that contains j in the row heading. In this way all the EMS entries in Table 9·1 are found for each of the factors, for interaction, and for error.

The composition of the EMS terms shows us that it is correct to test each of the factors and interaction for significance against the MS for error. The results of this decision lead to the F ratios shown in Table 9·2. Our conclusions are that all the tested sources are significant at the 0.05 level of significance.

PROBLEMS

9·5   A study is made of the variation in weights of filled packages from four filling machines, the only ones in use in a specific plant. Eight randomly selected packages are taken from each line and the weight of each package is recorded.
(a) Write the mathematical model for this experiment.
(b) State the test hypotheses.
(c) Develop the EMS terms for the ANOVA.

9·6   An argument among students about whether four professors are equally late to classes is to be settled by a record of minutes late during two semesters.
(a) What are the factors under test?
(b) Are the factors fixed or random? Explain.
(c) Write the mathematical model for this experiment.
(d) State the test hypotheses.
(e) Work out the EMS terms for the ANOVA if each professor meets 30 classes per semester.

9·7   Consider the following mathematical model:

$$X_{ijk} = \mu + A_i + B_j + AB_{ij} + \epsilon_{k(ij)}$$

where $a = 6$
        $b = 4$
        $r = 2$
(a) Work out the EMS terms for ANOVA if factors A and B are fixed.
(b) Complete the general form of the ANOVA summary table.
(c) State the hypotheses under test.

9·6   DETERMINATION OF EMS FOR RANDOM FACTORS

We now reconsider the example we have been using on this supposition: We have selected for testing by a random method three films out of many possible films, and four printers out of many possible printers. We wish now to make our conclusions apply to the entire set of films and to the entire set of printers. Now the factors F_i and P_j are random, not fixed as they were before.

The mathematical model is unchanged. We state, however, the null hypotheses differently from those we used for the fixed model. For the present random model they are H_0: $\sigma_F^2 = 0$; $\sigma_P^2 = 0$; $\sigma_{(F \times P)}^2 = 0$. The alternative hypotheses are H_1: $\sigma_F^2 \neq 0$; $\sigma_P^2 \neq 0$; $\sigma_{(F \times P)}^2 \neq 0$.

The analysis of variance procedure will be unchanged except for the calculation of the EMS terms; we will see that this difference requires a different ratio for the F tests for the main effects. Note that in Table 9·4 there are these differences from Table 9·3b:

Table 9·4 Table for Calculation of EMS Coefficients, Both Factors Random

	R i; 3	R j; 4	R k; 2
F_i	1	4	2
P_j	3	1	2
I_{ij}	1	1	2
$e_{k(ij)}$	1	1	1

1   R appears as part of the heading for the first and second columns because now the factors associated with the subscripts i and j are *random*.
2   As a consequence, we have entered 1 in the table in the cells where we entered 0 before.

When we now calculate the EMS terms, the entries for F_i and P_j will change. To find the EMS term for F_i, we suppress as before the first column, and find again for the first row the product $8\sigma_i^2$. The second row we ignore as before. For the third row, however, the presence of 1 in the second cell gives product of 2 for the remaining numbers instead of the 0 we found previously. Therefore, we annex the term $2\sigma_{ij}^2$ to the EMS term for F_i. Similarly, we find it necessary to include the expected variance for interaction in the EMS for P_j.

We present in Table 9·5 the summary ANOVA containing the same data as in Table 9·1, but with the EMS column appropriate to all factors random, rather than fixed. We see from the EMS terms that it is proper to test the interaction effect against error, since the EMS for interaction includes only a *single* possible effect in addition to error. We must, however, test the effect of F_i and P_j *not against error* but against the interaction mean square, since we have found the effect of interaction to be significant. Had we found the interaction effect to be not significant, we might have pooled the interaction MS term with the error MS term on the basis of the belief that (since inter-action was not found to be significant) the interaction term was an additional error estimate.

Here, with a significant interaction effect, to test the main effects against error would be incorrect because the EMS terms for these effects contain *two* sources of variation besides error. To test the main effects against

Table 9·5 Summary ANOVA Table; Two Random Factors

Source	SS	ν	MS	EMS	Calculated F ratio	Table value F ratio
F_i	6.0	2	3.0	$8\sigma_i^2 + 2\sigma_{ij}^2 + \sigma_e^2$	1.5	$F_{2,6,0.05} = 5.14$
P_j	24.0	3	8.0	$6\sigma_j^2 + 2\sigma_{ij}^2 + \sigma_e^2$	4.0	$F_{3,6,0.05} = 4.76$
$(F \times P)_{ij}$	12.0	6	2.0	$2\sigma_{ij}^2 + \sigma_e^2$	4.0	$F_{6,12,0.05} = 2.99$
Error	6.0	12	0.5	σ_e^2		
Total	48.0	23				

interaction is correct because the EMS terms for F_i and P_j differ from that for interaction only by the presence of the term $8\sigma_i^2$ for F_i or $6\sigma_j^2$ for P_j. If we find a significant difference by the F test, we can by this method attribute that difference to a single effect. If we were to test the main effects against error, we would not be sure if we found a significant difference to which effect we should assign the result.

The F ratio found for F_i is therefore $3.0/2.0 = 1.5$, and that for P_j is $8.0/2.0 = 4.0$. The book F ratios are included in Table 9·5. Comparison of the computed and book values for the F ratios leads to the following conclusions:

1 The interaction term is significant at the 0.05 level.
2 The effects of the factors F_i and P_j are not significant at the 0.05 level when compared with the interaction effect.

Once we find that the interaction is significant, we cannot (despite conclusion 2 above) safely ignore either factor although the results of the experiment indicate that the interaction effect dominates the average effect of the main factors. The effect of factor F can be described only when we know the level of factor P. The effect of changing brands of film is opposite for the different printers.

PROBLEMS

9·8 A manufacturer is concerned with the burst strength of plastic bags. Eight bags are randomly chosen from the production line and tested for burst strength at each of four different places within each bag.
(a) Set up the ANOVA summary table, including the EMS terms.
(b) What are the hypotheses under test?
(c) Write the mathematical model.

9·9 We wish to investigate the differences in bowling scores for different bowlers and different bowling halls. From a list of weekly bowlers, 4 are selected at random; each bowls 10 games in each of 3 bowling halls selected at random from a total of 10 establishments.

(a) Write the mathematical model for the experiment.

(b) List the factors under test, and identify each as being fixed or random.

(c) Write the hypotheses being tested.

(d) Determine the EMS terms for the ANOVA.

9·10 Consider the mathematical model

$$X_{ijk} = \mu + A_i + B_j + AB_{ij} + \epsilon_{j(ij)}$$

where $a = 5$
$\quad\quad b = 4$
$\quad\quad r = 7$

(a) Find the EMS terms if all factors are random.

(b) Find the EMS terms if all factors are fixed.

(c) List the hypotheses under test if all factors are random.

In Table 9·6 we show the generalized method for finding sums of squares, degrees of freedom, mean squares, and expected mean squares for any

Table 9·6   ANOVA for Two-factor Experiments, Random or Fixed Factors

Source	Sum of squares	v	Expected mean squares
Factor A (rows)	$\dfrac{1}{cn} \Sigma A_i{}^2 - \dfrac{1}{rcn} T\ldots^2$	$r - 1$	$\sigma_e{}^2 + n\left(1 - \dfrac{c}{C}\right)\sigma_{AB}{}^2 + cn\sigma_A{}^2$
Factor B (columns)	$\dfrac{1}{rn} \Sigma B_j{}^2 - \dfrac{1}{rcn} T\ldots^2$	$c - 1$	$\sigma_e{}^2 + n\left(1 - \dfrac{r}{R}\right)\sigma_{AB}{}^2 + rn\sigma_B{}^2$
$A \times B$ (interaction)	$\dfrac{1}{n} \Sigma (AB)_{ij}{}^2 - \dfrac{1}{rcn} T\ldots^2$ $- SS_A - SS_B$	$(r - 1)(c - 1)$	$\sigma_e{}^2 + n\sigma_{AB}{}^2$
Replications (error)	$\Sigma X_{ijk}{}^2 - \dfrac{1}{n}\Sigma(AB)_{ij}{}^2$	$rc(n - 1)$	$\sigma_e{}^2$
Total	$\Sigma X_{ijk}{}^2 - \dfrac{1}{rcn} T\ldots^2$	$rcn - 1$	

Factor A at r out of R possible levels.
Factor B at c levels out of C possible levels.
Replicates per cell at n replicated out of N possible replicates.
For fixed model: $r = R$; $c = C$.
For random model: $R \to \infty$; $C \to \infty$.
For mixed model: If factor A is fixed, $r = R$; if factor B is fixed, $c = C$.
$\quad\quad\quad\quad$ If factor A is random, $R \to \infty$; if factor B is random, $C \to \infty$.
For all models: $N \to \infty$.
Example: A fixed ($r = R$); B random ($C \to \infty$).
Expected mean squares:
Factor A, $\sigma_e{}^2 + n(1 - 0)\sigma_{AB}{}^2 + cn\sigma_A{}^2$ or $\sigma_e{}^2 + n\sigma_{AB}{}^2 + cn\sigma_A{}^2$.
Factor B, $\sigma_e{}^2 + n(1 - 1)\sigma_{AB}{}^2 + rn\sigma_B{}^2$ or $\sigma_e{}^2 + rn\sigma_B{}^2$.
Interaction $\sigma_e{}^2 + n\sigma_{AB}{}^2$.
Error $\sigma_e{}^2$

two-factor experiment, with fixed or random factors. In this table, A and B specify the factors arranged in the rows and columns of the ANOVA table. The subscript i specifies any one of the levels in the row; i.e., $i = 1, 2, 3, \ldots, r$, where r is the number of items in the row. Similarly, $j = 1, 2, 3, \ldots, c$, where c is the number of items in the columns, and $k = 1, 2, 3, \ldots, n$, where n is the number of replicates in each cell of the table.

Table 9·6 indicates that all the calculations for a two-factor ANOVA are the same for fixed and random models, with the essential exception of the EMS terms. For a fixed model, the interaction is associated with the specific levels being tested and is thus not (as in the random model) a population variance. For this reason, the interaction term drops out of the EMS terms for the main effects in the fixed model.

9·7 EMS FOR MIXED MODELS

Consider a manufacturing process in which only two different techniques may be employed; there may, on the other hand, be many operators involved in the process. If we perform an experiment to discover whether or not the factors "technique" and "operator" have an effect on the quality of the product, we would necessarily test both techniques and expect the conclusion of the experiment to apply only to these two. Thus the factor "technique" is fixed. We would, however, select for testing only some of the many operators; thus the factor "operators" would be random.

We have in this example a *mixed* model, inasmuch as one of the factors is fixed and the other random. The null hypothesis for the factor "techniques" is $H_0: T_i = 0$. The null hypothesis for the factor "operators" is $H_0: \sigma_0^2 = 0$.

To find the EMS terms, we apply the method indicated by Table 9·7, on the assumptions that we test four randomly selected operators ($o = 4$) and make three replications of each run ($r = 3$). The EMS terms are

1 For T_i: $12\sigma_i^2 + 3\sigma_{ij}^2 + \sigma_e^2$
2 For O_j: $6\sigma_j^2 + \sigma_e^2$
3 For I_{ij}: $3\sigma_{ij}^2 + \sigma_e^2$
4 For $\epsilon_{k(ij)}$: σ_e^2

Table 9·7 Expected Mean Squares for Mixed Model

	F $i; 2$	R $j; 4$	R $k; 3$	EMS
T_i	0	4	3	$12\sigma_i^2 + 3\sigma_{ij}^2 + \sigma_e^2$
O_j	2	1	3	$6\sigma_j^2 + \sigma_e^2$
I_{ij}	0	1	3	$3\sigma_{ij}^2 + \sigma_e^2$
$e_{k(ij)}$	1	1	1	σ_e^2

This result indicates that, for this mixed model, we should test the interaction effect against error, the operator effect against error, and the technique effect against interaction. These conclusions from the EMS calculations are perhaps unexpected; they illustrate the necessity, for every experiment of this type, of finding the EMS terms in order to know how the F tests should be run.

PROBLEMS

9·11   To study the variability in results of 4 laboratories, we select 3 solutions from a list of frequently used solutions and send one-fourth of each solution to each laboratory. The laboratories are asked to make 2 analyses for a given chemical using each solution.
(a) Write the mathematical model for the experiment.
(b) What are the hypotheses under test?
(c) Work out the EMS terms for the ANOVA table.

9·12   Consider the following:

$$X_{ijk} = \mu + A_i + B_j + AB_{ij} + \epsilon_{k(ij)}$$

where $a = 6$
     $b = 4$
     $r = 3$
(a) Work out the EMS terms if factor A is fixed and the rest are random.
(b) Work out the EMS terms if factor B is fixed and the rest are random.
(c) What would the EMS terms be if all factors are fixed? If all are random?

9·13   To investigate the variability in yield of a given process, we wish to consider 3 possible factors and their possible interactions. Factor A is chosen to have 4 levels, random, factor B to have 5 levels, also random, and factor C to have 3 levels, fixed. Two replicates per treatment combination will be used.
(a) Write the mathematical model for the experiment.
(b) Compute the EMS terms for the ANOVA table.
(c) State the hypotheses under test.

9·14   In each of the problems above, state how the F tests should be run if:
(a) all interactions are found to be significant
(b) all interactions are found not to be significant
(c) only two-factor interactions are found to be significant

9·8   COMPONENTS OF VARIANCE ANALYSIS

We indicated in the introduction to this chapter that we can, in the appropriate circumstances, estimate the *size* of the effect attributable to the tested sources of variation in a process. For this purpose, we require an ANOVA and the EMS terms for each factor. The method can be used only for a *random* model. As with any other process of estimation, we should remember to use confidence statements.

A machine produces sheet plastic. The foreman is required to keep the thickness of the sheet to nearly a standard value but is asked to increase the running speed of the machine. He thinks that an increase in speed will change the thickness of the sheet and also that the diameter of a specific machine roller has an important effect. An experiment is carried out to estimate the influence of these factors on sheet thickness. (Note that the experiment is not intended to discover whether or not an effect exists but estimate the magnitude of the variability due to these sources.)

Six randomly chosen speeds and four randomly chosen roller diameters were used; replicate runs were made for each possible combination of the levels of the two factors. The mathematical model for the experiment is

$$X_{ijk} = \mu + D_i + S_j + (S \times D)_{ij} + \epsilon_{k(ij)}$$

The raw data (differences in hundredths of an inch from the aim point) and the totals are given in Table 9·8.

Table 9·8 *Data from Experiment on Machine Production (Random Model)*

Roller diameter (sixteenths of an inch)	Running speed						Total
	14	20	25	27	28	30	
13	0	−2	1	4	0	4	
	−1	−1	−2	3	−1	1	6
16	1	2	0	1	3	1	
	−3	0	3	2	2	1	13
21	0	−1	0	3	3	6	
	3	0	−2	3	0	2	17
24	2	−1	1	−1	0	1	
	3	−1	2	3	3	2	14
Total	5	−4	3	18	10	18	50

We will use ANOVA to estimate the components of variance. This is a different problem from the earlier ones in this chapter where we dealt with tests of significance. The analogy is with the difference between tests of hypothesis for the mean and confidence intervals. Thus we omit the *F* tests for this operation.

The EMS calculations are given in Table 9·9, and the summary ANOVA in Table 9·10. The computations for the ANOVA table follow:

Table 9·9   Expected Mean Squares for Machine Experiment (Data in Table 9·8)

	R $i; 4$	R $j; 6$	R $k; 2$	EMS
D_i	1	6	2	$12\sigma_i^2 + 2\sigma_{ij}^2 + \sigma_e^2$
S_j	4	1	2	$8\sigma_j^2 + 2\sigma_{ij}^2 + \sigma_e^2$
$(D \times S)_{ij}$	1	1	2	$2\sigma_{ij}^2 + \sigma_e^2$
$e_{k(ij)}$	1	1	1	σ_e^2

Table 9·10   Summary ANOVA (Data from Tables 9·8 and 9·9)

Source	Sum of squares	ν	Mean squares	Expected mean squares
Diameter	5.42	3	1.805	$12\sigma_i^2 + 2\sigma_{ij}^2 + \sigma_e^2$
Speed	47.67	5	9.533	$8\sigma_j^2 + 2\sigma_{ij}^2 + \sigma_e^2$
$S \times D$	56.83	15	3.788	$2\sigma_{ij}^2 + \sigma_e^2$
Error	60.00	24	2.500	σ_e^2
Total	169.92	47		

Correction factor $CF = \dfrac{50^2}{48} = 52.08$

$$SSR = \tfrac{1}{12}(690) - CF = 5.42$$
$$SSC = \tfrac{1}{8}(798) - CF = 47.67$$
$$SSI = \tfrac{1}{2}(324) - (CF + 5.42 + 47.67) = 56.83$$
$$SST = 222 - CF = 153.92$$
$$SSE = 169.92 - (5.42 + 47.67 + 56.83) = 60.00$$

To find the components of variance, we work upward in the ANOVA table, as follows:

1 We have a mean square for error of 2.500; according to the EMS term for error, this is an estimate of the variance associated with error σ_e^2.
2 We have a mean square for interaction of 3.788. According to the EMS term in this row of the ANOVA table, this MS is an estimate of $2\sigma_{ij}^2 + \sigma_e^2$.

We equate these two expressions, and substitute the estimated value for σ_e^2:

$$2\sigma_{ij}^2 + \sigma_e^2 = 3.788$$
$$2\sigma_{ij}^2 + 2.50 = 3.788$$
$$\sigma_{ij}^2 = 0.644$$

3 For the factor speed, we similarly equate the EMS expression with the MS term estimated by the experiment and substitute the presently known values:

$$8\sigma_j^2 + (2\sigma_{ij}^2 + \sigma_e^2) = 9.533$$
$$8\sigma_j^2 + 3.788 = 9.533$$
$$\sigma_j^2 = 0.718$$

4 For the factor diameters, we proceed in the same fashion:

$$12\sigma_i^2 + (2\sigma_{ij}^2 + \sigma_e^2) = 1.805$$
$$12\sigma_i^2 + 3.788 = 1.805$$

We are led, if we follow the mathematics here, to an impossible result— a negative variance; hence we suppose that the experiment indicates that our best estimate of the variance associated with the factor diameters is that it is zero.

5 We find the total estimate of variance for a single observation from any random selection of a roller and speed by summing the values found in the preceding steps:

$$\sigma_t^2 = \sigma_e^2 + \sigma_{ij}^2 + \sigma_j^2 + \sigma_i^2$$
$$= 2.50 + 0.644 + 0.718 + 0.000$$
$$= 3.862$$

Thus $\sigma_e^2/\sigma_t^2 = 2.50/3.862$, or about 65 percent. Similarly, σ_{ij}^2 is about 17 percent of the total, σ_j^2 about 19 percent, and σ_i^2 is zero. (These calculations are only point estimates of the contributions of the factors to the variance, and only rough approximations indeed.) We can now say that each response value differs from the average of the data for reasons that can be attributed to error (about 65 percent), interaction (about 17 percent), and speed (about 19 percent).

9·9 CONFIDENCE LIMITS ON COMPONENTS OF VARIANCE

In the ANOVA Table 9·10, the expected mean square column shows the expected parameter values of the sample estimates in the mean square column. The MS term is a sample estimate of the corresponding EMS term, just as s^2 is an estimate of σ^2.

Now, we can place confidence intervals about the estimate of the components. We proceed to place 95 percent confidence intervals on the parameters, here the components of variance.

We use the following symbolism: $\hat{\sigma}^2$ signifies an estimate of a population value of variance; we read the symbol "sigma hat squared." Thus $\hat{\sigma}_e^2$ signifies the estimate of the error variance; it is an approximation to σ_e^2, the population error variance. Our best estimate of the value of σ_e^2 is, in this example, the error mean square, or 1.83. We treat this value of $\hat{\sigma}_e^2$ just as we would a value of s^2 and use the χ^2 distribution to place our confidence limits on the parameter. Thus (see Chap. 5)

$$\frac{\nu_e \hat{\sigma}_e^2}{\chi^2_{\nu_e, \alpha/2}} < \sigma_e^2 < \frac{\nu_e \hat{\sigma}_e^2}{\chi^2_{\nu_e, 1-\alpha/2}}$$

$$\frac{24 \times 2.50}{\chi^2_{24,0.025}} < \sigma_e^2 < \frac{24 \times 2.50}{\chi^2_{24,0.975}}$$

$$\frac{60}{39.4} < \sigma_e^2 < \frac{60}{12.4}$$

$$1.52 < \sigma_e^2 < 4.84$$

For the interaction, we have $2\hat{\sigma}_{ij}^2 + \hat{\sigma}_e^2$ as an estimate of $2\sigma_{ij}^2 + \sigma_e^2$. We also know from the mean square column that $2\hat{\sigma}_{ij}^2 + \hat{\sigma}_e^2 = 3.79$. Using the value for $\hat{\sigma}_e^2$, and solving for $\hat{\sigma}_{ij}^2$, we have

$$2\hat{\sigma}_{ij}^2 + \hat{\sigma}_e^2 = 3.79$$

$$\hat{\sigma}_e^2 = 2.50$$

$$\hat{\sigma}_{ij}^2 = \frac{3.79 - \hat{\sigma}_e^2}{2}$$

$$\hat{\sigma}_{ij}^2 = 0.64$$

Here we are involved with two estimates of variance instead of one. Thus we will deal with the ratio of variances.

$$F = \frac{\text{EMS (interaction)}}{\text{EMS (error)}} = \frac{M_{ij}'}{M_e'}$$

We have M_{ij}/M_e as an estimate of M_{ij}'/M_e', and the 95 percent confidence interval is

$$\frac{M_{ij}/M_e}{F_{\nu_{ij}, \nu_e, 0.025}} < \frac{M_{ij}'}{M_e'} < \frac{M_{ij}}{M_e} F_{\nu_e, \nu_{ij}, 0.025}$$

where

$$\frac{M_{ij}'}{M_e'} = \frac{2\sigma_{ij}^2 + \sigma_e^2}{\sigma_e^2} = \frac{2\sigma_{ij}^2}{\sigma_e^2} + 1$$

$$\frac{M_{ij}/M_e}{F_{15,24,0.025}} - 1 < \frac{2\sigma_{ij}^2}{\sigma_e^2} < \frac{M_{ij}}{M_e} F_{24,15,0.025} - 1$$

$$\frac{1}{2}\left(\frac{3.79/2.50}{2.44} - 1\right) < \frac{\sigma_{ij}^2}{\sigma_e^2} < \frac{1}{2}\left[\frac{3.79}{2.50}(2.70) - 1\right]$$

$$0 < \frac{\sigma_{ij}^2}{\sigma_e^2} < 1.55$$

Since variances cannot be negative, the ratios cannot be either; so the minimum limit is 0. We therefore have a point estimate for the interaction value 0.98, and a 0.95 confidence interval of 0 to 2.29.

For speed:
$$8\hat{\sigma}_j^2 + 2\hat{\sigma}_{ij}^2 + \hat{\sigma}_e^2 = 9.53$$
$$8\hat{\sigma}_j^2 + 3.79 = 9.53$$
$$\hat{\sigma}_j^2 = 0.72$$

Our point estimate is then 0.72, and the 0.95 confidence interval is as follows:

$$\frac{M_j/M_{ij}}{F_{v_j, v_{ij}, 0.025}} < \frac{M_j'}{M_{ij}'} < \frac{M_j}{M_{ij}} F_{v_{ij}, v_j, 0.025}$$

$$\frac{M_j'}{M_{ij}'} = \frac{M_{ij}' + 8\sigma_j^2}{M_{ij}'} = \frac{8\sigma_j^2}{M_{ij}'} + 1$$

$$\frac{M_j/M_{ij}}{F_{5,15,0.025}} - 1 < \frac{8\sigma_j^2}{M_{ij}'} < \frac{M_j}{M_{ij}} F_{15,5,0.025} - 1$$

$$\frac{1}{8}\left(\frac{9.53/3.79}{3.58} - 1\right) < \frac{\sigma_j^2}{M_{ij}'} < \frac{1}{8}\left[\frac{9.53}{3.79}(6.43) - 1\right]$$

$$0 < \frac{\sigma_j^2}{M_{ij}'} < 1.90$$

For diameters:

$$\frac{M_i/M_{ij}}{F_{v_i, v_{ij}, 0.025}} < \frac{M_i'}{M_{ij}'} < \frac{M_i}{M_{ij}} F_{v_{ij}, v_i, 0.025}$$

$$\frac{M_i'}{M_{ij}'} = \frac{12\sigma_i^2 + M_{ij}'}{M_{ij}'} = \frac{12\sigma_i^2}{M_{ij}'} + 1$$

$$\frac{M_i/M_{ij}}{F_{3,15,0.025}} - 1 < \frac{12\sigma_i^2}{M_{ij}'} < \frac{M_i}{M_{ij}} F_{15,3,0.025} - 1$$

$$\frac{1}{12}\left(\frac{1.81/3.79 - 1}{4.15}\right) < \frac{\sigma_i^2}{M_{ij}'} < \frac{1}{12}\left[\frac{1.81}{3.79}(14.25) - 1\right]$$

$$0 < \frac{\sigma_i^2}{M_{ij}'} < 0.48$$

We have now estimated the four variances σ_e^2, σ_i^2, σ_j^2, and σ_{ij}^2, associated with error, diameter variation, speed variation, and interaction. As they are only estimates they are subject to some uncertainty, and before using them as a basis for future action we must calculate this uncertainty.

A convenient method of expressing the uncertainty of an estimate of any statistic is to calculate its confidence limits. This we have done for our four variance estimates. We now know that the error variance σ_e^2 probably lies between 1.39 and 3.01 (with 95 percent confidence); the best estimate is 1.82.

The ratio of the variance due to interaction to the error variance probably lies between 0 and 2.29 (with 95 percent confidence); our best estimate of the ratio is 0.98.

The confidence limits may indicate that the variances have not been estimated with sufficient precision. If so, the remedy is to carry out further experimental work. We may find it necessary to increase n or to increase the number of levels of each factor until the required precision is obtained.

If there is some reasonable estimate of error variance prior to any testing, some guide to the amount of work necessary to detect a specified difference is possible.

PROBLEMS

9·15   A study was made of the amount of water contained in cans of food as they are sold from the leading stores in the area. Three stores were selected at random; in each store, 4 kinds of foods were selected, and 2 cans of each type were chosen for analysis. The drained-off water was measured and the results were recorded in Table 9·11.

Table 9·11   Water Content of Canned Foods

Food	Store A	Store B	Store C
Peas	3.24	3.16	2.96
	3.56	3.26	3.01
Corn	3.92	3.81	3.76
	3.86	3.80	3.75
Pears	9.13	8.86	8.70
	9.23	8.79	8.75
Cherries	8.35	8.11	7.94
	8.29	8.24	7.99

(a) Complete the analysis of variance for this set of data (factors to be considered as random).
(b) Estimate the components of variance.
(c) Place 95 percent confidence intervals on the components of variance.

9·16   Table 9·12 contains the results of a two-factor experiment with both factors random.

(a) Complete the analysis of variance table.
(b) Estimate the components of variance.
(c) Place 90 percent confidence intervals on the components of variance.

Table 9·12   Two-factor Experiment,
            Both   Factors   Random

	A_1	A_2	A_3
B_1	8.76	8.39	8.23
	8.72	8.43	8.27
B_2	8.03	7.86	7.72
	8.11	7.84	7.79
B_3	6.61	6.32	6.21
	6.77	6.23	6.13

QUESTIONS

9·1   When calculating the EMS term, all factors being fixed, what are the hypotheses that are being tested?

9·2   When all factors are random factors, what are the hypotheses that are being tested?

9·3   When all factors are fixed factors in an analysis of variance, why is it permissible to test each factor against the error term without first calculating the EMS terms?

9·4   When calculating the EMS terms, why should the model be written prior to anything else?

9·5   What role does the EMS play in the running of the F tests in analysis of variance?

9·6   What is the difference between fixed and random factors?

9·7   What would be a "mixed model"?

9·8   Would it be possible to have a situation where some of the factors were fixed factors and others were random factors? Explain.

9·9   Why is error always considered to be a random factor?

9·10   What is an algorithm?

9·11   List the three rules to be followed in filling in the table for the calculation of the EMS terms.

9·12   In making an EMS table, what constitute the row headings? What provides the column headings?

9·13   Why is it incorrect to test all random factors against the error term?

9·14   Why is it not necessary to indicate whether an interaction is fixed, random, or mixed?

9·15   If an interaction turns out to be not significant, the components of variance for it are crossed out. Why?

REFERENCES
10, 31, 35, 59, 60, 68

10 Crossed and Nested Experiments

In our discussion of analysis of variance up to now we have assumed that all tested factors were *crossed*. By the term "crossed" we mean that each level of each factor is combined in the experimental runs with each level of each other factor. If the factors to be tested are "operator" and "instrument" we may arrange the runs in the experiment so that every operator is required to use every instrument; thus the factors "operators" and "instruments" are crossed.

Other examples of crossed experiments are these: (1) The yield of a chemical process is found when the process is operated at each of three different temperatures at each of four different pressures. (2) The efficacy of a medical preparation is found for compounds of each of three concentrations of ingredient A at each of three levels of ingredient B.

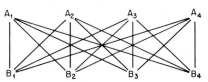

Fig. 10·1 Relationship of factors in a crossed experiment.

Figure 10·1 is intended to suggest the relationship of the experimental runs in a crossed experiment involving two factors (*A* and *B*) each at four levels, and singly replicated. One line (each representing a single run) is drawn from *each* level of *A* to *each* level of *B*. The observations for a crossed,

Table 10·1   Scheme of a
Crossed Experiment
(Two Factors, Each
at Four Levels,
Twice Replicated)

Factor A	Factor B			
	B_1	B_2	B_3	B_4
A_1	X_{000} X_{001}	X_{010} X_{011}	X_{020} X_{021}	X_{030} X_{031}
A_2	X_{100} X_{101}	X_{110} X_{111}	X_{120} X_{121}	X_{130} X_{131}
A_3	X_{200} X_{201}	X_{210} X_{211}	X_{220} X_{221}	X_{230} X_{231}
A_4	X_{300} X_{301}	X_{310} X_{311}	X_{320} X_{321}	X_{330} X_{331}

twice-replicated experiment would be recorded as in Table 10·1. The mathematical model for this experiment is

$$X_{ijk} = \mu + A_i + B_j + (AB)_{ij} + \epsilon_{k(ij)}$$

where μ = population average of all factors and levels under study, usually estimated by $\bar{\bar{X}}$, the grand average of all observations in the experiment

A_i = effect of factor placed in rows

B_j = effect of factor placed in columns

$(AB)_{ij}$ = effect of interaction of main factors

$\epsilon_{k(ij)}$ = effect of random error estimated from replicates

10·1 NESTED EXPERIMENTS

In many situations, we find it impossible or undesirable to cross the factors being tested in an experiment. A carload of drums of chemical is inspected for percentage composition. Four drums are taken at random from each car; three samples are taken at random from each drum; two analyses are performed on each sample. Here, there would be no reason to compare drum 1 from car A with drum 1 from car B, since we have no basis for supposing that any meaning attaches to the designation of a drum as 1 or 2.

In such a *nested* experiment, some of the factors are contained within other factors as subclasses; so that they cannot be compared at each level of every other factor. In the example above, the drums are nested within cars, samples are nested within drums, and analyses are nested within samples. Any difference between the first and second analysis of a given sample is meaningful only for a specific sample. We sketch the situation in Fig. 10·2.

In ANOVA the type of relationship between factors—crossed or nested—will determine in part the method of data analysis we will use. In a nested

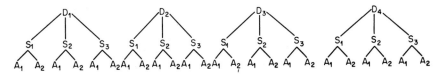

Fig. 10·2 Relationship of factors in a nested experiment. D = drums; S = samples; A = analyses.

experiment, we can order the factors, and thus designate them as main factors and subgroup factors; in a crossed experiment we cannot order the factors in this way.

PROBLEMS

10·1 For each of the following situations: (a) list the factors being tested; (b) indicate the number of levels of each factor; (c) identify each factor as crossed or nested; (d) identify each factor as being fixed or random.

(a) A company manufactures bricks to be used as furnace linings. Only 2 plants produce bricks; each plant has 3 storage locations for maintaining water content in bricks. We select 2 bricks from each location and check the moisture content of each brick.

(b) Samples of iron ore are sent to 4 different laboratories for analysis; each laboratory assigns 3 technicians to carry out the analysis. Each technician makes 2 determinations of iron content on each sample.

(c) A manufacturer has 3 methods of making cartons. A sample of 5 cartons is taken from each process; 2 tests of burst strength are made on each carton.

(d) To investigate the effect of the data of planting and type of fertilizer on the yield of soybeans, 2 planting dates were used with 4 types of fertilizer. Each type of fertilizer was applied to samples planted at each of the dates. The yield of soybeans was found at harvest time.

10·2 For each of the situations in Prob. 10·1, identify the response variable, with special attention to an unmistakably clear definition of the measurement to be used.

10·3 Each of a set of 3 standards is measured on each of 2 spectrophotometers in each of 3 different laboratories. In each laboratory, each of 2 operators measured the standards on each instrument.

(a) List the factors under test.

(b) Indicate the number of levels of each factor.

(c) Indicate whether each factor is crossed or nested.

10·2 MATHEMATICAL MODEL FOR NESTED EXPERIMENTS

Assume that we are concerned with a two-factor experiment, with factor B being nested within factor A. Each factor is at four levels, and the experimental runs are twice replicated.

The plan of the experimental runs and the observations are shown in Table 10·2. Compare this table with Table 10·1, and note that in this nested experiment each set of four levels of B is grouped (nested) within one of the levels of A. In addition each pair of the replicates is nested within one level of B. It is convenient to speak of a nested (or hierarchical) experiment as comprising factors in *tiers*: if factor B is nested within factor A, we refer to tiers A and B.

Table 10·2   Scheme of a Nested Experiment (Two Factors, Each at Four Levels, Twice Replicated)

A_1				A_2				A_3				A_4			
B_1	B_2	B_3	B_4	B_5	B_6	B_7	B_8	B_9	B_{10}	B_{11}	B_{12}	B_{13}	B_{14}	B_{15}	B_{16}
X_{111}	X_{121}	X_{131}	X_{141}	X_{251}	X_{261}	X_{271}	X_{281}	X_{391}	$X_{3,10,1}$	$X_{3,11,1}$	$X_{3,12,1}$	$X_{4,13,1}$	$X_{4,14,1}$	$X_{4,15,1}$	$X_{4,16,1}$
X_{112}	X_{122}	X_{132}	X_{142}	X_{252}	X_{262}	X_{272}	X_{282}	X_{392}	$X_{3,10,2}$	$X_{3,11,2}$	$X_{3,12,2}$	$X_{4,13,2}$	$X_{4,14,2}$	$X_{4,15,2}$	$X_{4,16,2}$

The mathematical model for this nested experiment is

$$X_{ijk} = \mu + A_i + B_{j(i)} + \epsilon_{k(ij)}$$

The model implies that each observation is what we find it to be because of the level of the population mean μ, a possible A effect, a possible B effect within tier A (we indicate nesting by the use of parentheses placed about i), and error nested within tier B.

If we compare this model with that for the crossed experiment, we see that the essential difference is in the absence of an interaction effect. For the crossed model, interaction was separable from the main effects. In the present nested model, the B effect and the AB interaction are added together to give the B effect within tier A. This statement means that, in the ANOVA for a nested experiment, the sum of squares for B within A equals the sum of squares for B plus the sum of squares for AB.

When we are concerned with a nested experiment, we therefore cannot find an interaction effect since interaction may be defined as the differential effect of one factor at different levels of a second factor. Such an effect implies a crossed classification.

The difference between crossed and nested experiments cannot always be seen by the manner in which the data are presented in tabular form. It is for this reason that we should always write the mathematical model for the experiment that we intend to carry out. It is useful to employ a conventional method of writing the subscripts in the mathematical model, and to use parentheses to indicate a nested factor.

To understand whether a two-factor experiment, for example, is crossed or nested, we must know how the problem is set up. In a biological experiment, we may be testing the effect of a spray on each of three plants; we may take for examination four leaves from each of the plants. In the absence of further information, there is no reason to associate any single leaf from one plant with any specific leaf from another plant. Here we would consider the factor "leaf" nested within the factor "plant." The subgroup factor "leaf" would thus have 12 levels. Four levels of this factor are associated with each of the three levels of the main group factor "plant."

We might, on the other hand, in a similar experiment select leaves from four prearranged locations. We might pick leaves from the top of the plant, slightly above the middle, slightly below the middle, and from the bottom. Under these circumstances, we would have designed a two-factor crossed experiment. The main factor would be "plant" and the second factor would be "location" within plant. We would be able to compare leaves taken from the top of the plants with those taken from the bottom of the plants, etc.; thus we would have a crossed classification.

10·3 ANOVA FOR NESTED EXPERIMENTS

The methods of ANOVA used for nested experiments differ in only a few respects from those discussed in Chap. 8. The differences are, however, important. We demonstrate the methods by an example.

Two tank cars of gasoline are received for inspection. Four samples are taken at random from each car. Two measurements of octane rating are made on each sample. Each set of four samples is thus nested within cars; each of the sets of two measurements is nested within samples. We note, for the purposes of finding EMS terms, that the factor "cars" is fixed, inasmuch as only two cars were available and both were tested; the factors "sample" and "analysis" are random. When we compute the components of variance, we will follow the rules of Chap. 9, but the EMS values for the nested experiment will differ from those of crossed experiments.

The data (coded by subtracting 80) are given in Table 10·3. Here we also give the column (sample) totals, the car totals, and the grand total. We will

Table 10·3   Data for a Nested Experiment

	Car 1				Car 2			
Samples	1	2	3	4	5	6	7	8
Analysis 1	7	10	6	4	8	11	9	6
Analysis 2	7	9	8	4	7	10	8	6
Total $T_{ij.}$	14	19	14	8	15	21	17	12
Total $T_{.j.}$	55				65				$T... = 120$

use these totals in finding sums of squares. Note that the row total is not shown. There is no significance to be attached to row totals, since the grouping of the samples as "1" and "2" is entirely arbitrary.

The mathematical model for this experiment is

$$X_{ijk} = \mu + \overline{C_i} + \overline{S_{j(i)}} + A_{k(ij)}$$

C (car) is the first factor, shown in Table 10·3 in the major columns; i has values of 1 or 2. S (sample) is the second factor, shown in subcolumns within the major columns; j has values 1 through 4. A (analysis) is the third factor, shown in rows within subcolumns; k has values of 1 or 2.

That we write the subscript for S as $j(i)$ is intended to show that the j levels of S are nested within the levels i. Similarly the subscript for A is $k(ij)$, the parentheses being used to show that the k levels of analysis are nested within the levels both of i (cars) and of j (samples).

Note the symbols used for the different totals. $T...$ is the grand total of all X values. It will be recalled that the dot in the subscript replaces the levels over which the total is taken. Here we have the total over all levels of all factors. $T_{ij.}$ represents the totals of the two analyses for each sample. $T_{i..}$ represents the total for each car over all four samples and both analyses of each sample. There is no row total.

In the ANOVA, we will take the pooled variability between pairs of analyses as our estimate of error. Thus we will test to see if the variability found among the samples is larger than that associated with analysis alone. Then we will test to see if the variability between cars is greater than that accounted for by variability among samples. The best way to see the reasons for this procedure is to examine the EMS terms in Table 10·4. The composition of these terms indicates that we should test factor S against factor A, and factor C against factor S.

Table 10·4   Estimated Mean Squares for Nested
Experiment (Data in Table 10·3)

	F i; 2	R j; 4	R k; 2	EMS
C_i	0	4	2	$\sigma_a^2 + 2\sigma_s^2 + 8\sigma_c^2$
$S_{j(i)}$	1	1	2	$\sigma_a^2 + 2\sigma_s^2$
$A_{k(ij)}$	1	1	1	σ_a^2

The degrees of freedom for the ANOVA table are found as follows:

1   For the cars, v is the number of cars (which is 2) less 1, or 1.
2   For samples, we have 4 measurements for each car; we reduce this by 1,

giving $v = 3$. Pooling the degrees of freedom for the 2 cars, we have 6 degrees of freedom associated with samples.

3   For analyses, we have 2 measurements for each sample; we reduce this by 1, giving $v = 1$. For the 8 samples there will be a total of 8 degrees of freedom associated with analyses.

4   The sum of the three values of v just found is $1 + 6 + 8 = 15$. This result agrees with the rule that the total value of v is one fewer than the sample size, which is here 16.

The determinations of the values of v in steps 1 to 3 above are in each case based on the loss of one degree of freedom for each total that is associated with the factor under consideration.

The sums of squares for the ANOVA table are found as follows:

1   The total sum of squares is found in the usual way: $\Sigma X^2 - (T...^2)/16 = 962.00 - 900.00 = 62.00$. We have subtracted the correction factor from the sum of the squares of the observations.

2   The sum of squares for cars SSC is also found in the same way as in Chap. 9: $\Sigma (T_{.j.})^2/8 - (T...^2)/16 = 906.25 - 900.00 = 6.25$. We here divide each $T_{.j.}^2$ term by 8 because there are 8 values that make up each of the totals for cars.

3   It is in the determination of the sum of squares for samples SSS that we use a calculation different from that used in crossed classifications. Since samples are nested within cars, we must find a correction factor for *each* car separately; this correction factor is $(T_{.j.})^2/8$, which we have already found in step 2 just above. Thus, for the first car, we find the sum of squares as

$$\frac{14^2 + 19^2 + 14^2 + 8^2}{2} - \frac{(55)^2}{8} = 30.375$$

The denominator of the first fraction is 2 because each squared term is composed of the two replicates. Similarly, for the second car, we find

$$\frac{15^2 + 21^2 + 17^2 + 12^2}{2} - \frac{(65)^2}{8} = 21.375$$

The whole sum of squares for samples is the sum of the two terms just calculated, or 51.75.

4   The sum of squares for analyses SSA is found by the sum of the squares of the observations, from which we subtract $\frac{1}{2}$ the sum of the squares of the sample totals. Thus $\Sigma X_{ijk}^2 - \frac{1}{2} \Sigma (T_{ij.}^2) = 962.00 - 958.00$, or 4.00.

The value of the mean square for each of the factors is, as before, found by dividing the sums of squares by the appropriate value of degrees of freedom.

The complete summary ANOVA table is given in Table 10·5. The composition of the EMS terms shows that we should test the MS value for samples against the MS term for analyses. Since this is a ratio of two variances, we use the F ratio and find the value to be 8.62/0.50, or 17.24. Comparing this value with $F_{6,8,0.05} = 3.58$ we find variation between samples, within cars, to be significant at the 0.05 level. In fact, it is also significant at the 0.01 level.

Table 10·5   Summary ANOVA (Data in Tables 10·3 and 10·4)

Source	Levels	D.F.	SS	MS	EMS
Between cars	$c = 2$	$2 - 1 = 1$	6.25	6.25	$\sigma_a^2 + 2\sigma_s^2 + 8\sigma_c^2$
Between samples	$s = 4$	$2(4 - 1) = 6$	51.75	8.62	$\sigma_a^2 + 2\sigma_s^2$
Between analyses	$a = 2$	$2(4)(2 - 1) = 8$	4.00	0.50	σ_a^2
Total	$csa = 16$	$16 - 1 = 15$	62.00		

To test the variation between cars, we must compare the MS term for cars with the MS term for samples and not with analyses as we would have in a crossed classification. There is no significant car-to-car variation greater than sampling variability.

PROBLEMS

10·4 Randomly chosen samples of a brand of cigarette were sent to 4 independent testing laboratories for an analysis of tar content. Each laboratory was instructed to assign 3 technicians to carry out independent tests. Each technician made 2 determinations of the tar content. The coded results were as given in Table 10·6.

Table 10·6   Measurements of Tar Content of Cigarette Samples

Technician	Laboratory			
	1	2	3	4
A	5 4	4 4	6 8	7 6
B	3 3	6 4	5 7	7 6
C	2 3	3 3	7 8	7 8

(a) Write the mathematical model for this experiment.
(b) Find the sum of squares for factor L (laboratories).
(c) Find the sum of squares for factor T (technicians) within L.
(d) Find the sum of squares for error.
(e) Complete the analysis of variance for this experiment.
(f) Now suppose that the experiment was completely crossed. Use the data in Table 10·6 and prepare an ANOVA table for a crossed classification.
(g) Compare the sum of squares of T within L [from (c) above] with the sum of squares for T and the sum of squares for $T \times L$ as found in (f) above.
(h) Work out the components of variance for both the nested and the crossed versions of the experiment. Compare your results.

10·4 COMPONENTS OF VARIANCE FOR NESTED CLASSIFICATION

When we are dealing with factors nested within other factors, the variability associated with a nested factor is included within the total variability found in a factor classified in an upper tier. Thus the determination of components of variance is especially useful for nested classifications, since this determination enables us to estimate the contribution of each factor to the total variability.

We use the methods of the preceding chapter to find the expected mean squares (the parameters to which the MS terms are approximations) and to place confidence limits on these parameters. We recall that we can find components of variance only when the levels of the factors have been randomly selected. No meaning attaches to components of variance when the levels have been deliberately chosen, as is the case when we select a low concentration as one level and a high concentration as another level.

In an orchard in an industrial valley, the trees developed a rash of specks. An investigation was carried out to find out the extent of difference among trees and on different parts of a tree. Three trees were chosen at random from the orchard. Four leaves were taken at random from each tree. Each leaf was cut in half to make two samples for each leaf. The number of specks per half leaf is the response variable in Table 10·7. The mathematical model for the experiment is

$$X_{tls} = \mu + T_t + L_{l(t)} + S_{s(tl)}$$

The factor T (trees) was used at $t = 3$ random levels; the factor L (leaves) was used at $l = 4$ random levels, this factor being nested in T; the factor S (samples) was used at $s = 2$ random levels, this factor being nested in L.

By the algorithm described in Chap. 9, we find the EMS terms as in Table 10·8. The calculating methods for the ANOVA are given in Table 10·9, and the results of the application of these methods in Table 10·10.

Table 10·7   Data for a Two-factor Nested Experiment

	Tree A				Tree B				Tree C			
Leaf	1	2	3	4	5	6	7	8	9	10	11	12
	16	10	11	9	15	9	11	12	17	10	10	8
	14	10	9	7	17	8	10	11	16	11	9	10
Leaf total	30	20	20	16	32	17	21	23	33	21	19	18
Tree total	86				93				91			

Table 10·8   Expected Mean Squares for a Two-factor Nested Experiment

	R $t; 3$	R $l; 4$	R $s; 2$	EMS
T_t	1	4	2	$8\sigma_t^2 + 2\sigma_l^2 + \sigma_s^2$
$L_{l(t)}$	1	1	2	$2\sigma_l^2 + \sigma_s^2$
$S_{s(tl)}$	1	1	1	σ_s^2

Table 10·9   Calculations for ANOVA for a Two-factor Nested Experiment

Source	SS	ν
Among trees	$\dfrac{1}{8}\sum T_{i..}^2 - \dfrac{1}{24}T_{...}^2$	$t - 1$
Among leaves within trees	$\dfrac{1}{2}\sum T_{tl.}^2 - \dfrac{1}{8}\sum T_{i..}^2$	$t(l - 1)$
Between samples within leaves	$\sum X_{tls}^2 - \dfrac{1}{2}\sum T_{tl.}$	$tl(s - 1)$
Total	$\sum X_{tls}^2 - \dfrac{1}{24}T_{...}^2$	$tls - 1$

Table 10·10   Components of Variance for a Two-factor Nested Experiment

Source	SS	ν	MS	EMS
Trees	3.25	2	1.625	$8\sigma_t^2 + 2\sigma_l^2 + \sigma_s^2$
Leaves within trees	186.25	9	20.694	$2\sigma_l^2 + \sigma_s^2$
Samples within leaves	13.00	12	1.08	σ_s^2
Total	202.50	23		

We add these notes:

1 In Table 10·9, for the sum of squares among trees, the multiplier $\frac{1}{8}$ is used since we are interested in the tree average, and 8 observations were added to give the tree total.
2 The "among leaves within trees" term is the result of pooling (that is, adding the sums of squares and the values of ν) the variances found for leaves measured within each tree.
3 Since each tree may be at a different level, the leaves from each tree are compared with their tree average, which differs for each set of leaves.
4 The "between samples within leaves" pools the variance between each of the 2 samples in each of the 12 sets of replicates.

We proceed now to find the components of variance, and to place confidence limits about the population parameters. We recall that the values in the MS column of the ANOVA Table 10·10 are estimates of the expected mean squares. Thus we have as an estimate of the parameter σ_s^2 the value 1.08. We again designate a population estimate by the use of the symbol $\wedge$; thus the symbol $\hat{\sigma}_s^2$ specifies the estimate of the population variance among samples. The value 1.08 is therefore a point estimate of the value of the population variance. To place confidence limits on this value, we treat $\hat{\sigma}_s^2$ as a sample variance s^2, and use the χ^2 distribution (as described in Chap. 5). Thus confidence limits for this variance are found from

$$\frac{\nu_s \hat{\sigma}_s^2}{\chi_{\nu_s,\alpha/2}^2} < \sigma_s^2 < \frac{\nu_s \hat{\sigma}_s^2}{\chi_{\nu_s,1-\alpha/2}^2} \tag{10·1}$$

where ν_s is the value of degrees of freedom associated with samples, and the value of χ^2 is determined by the number of degrees of freedom and the level of confidence we wish to assume. To find the 0.95 confidence interval, we substitute in Eq. (10·1):

$$\frac{12 \times 1.08}{\chi_{12,0.025}^2} < \sigma_s^2 < \frac{12 \times 1.08}{\chi_{12,0.975}^2}$$

$$\frac{12.96}{23.34} < \sigma_s^2 < \frac{12.96}{4.40}$$

$$0.55 < \sigma_s^2 < 2.95$$

Thus we have evidence of a real, if slight, difference among samples taken from the same leaf.

For leaves within trees, we move up to the next row in Table 10·10 and equate the MS and the EMS terms, substituting the estimates of the population parameters, and again using the symbol $\hat{\sigma}$ for these estimates:

$$2\hat{\sigma}_L^2 + \hat{\sigma}_s^2 = 20.694$$

We substitute our point estimate for $\hat{\sigma}_s^2$ which is 1.08, and find $\sigma_L^2 = (20.69 - 1.08)/2 = 9.80$. We can now compare this estimate of within-leaf variance with the within-sample variance by an F ratio. Refer to Chap. 9, in which we discussed confidence limits for the ratio of two variances. Here we find M_L/M_S, that is, the ratio of the mean square for leaves to the mean square for samples within leaves. We expect that this ratio will approximate that of the corresponding population values, which we designate by primes, i.e., M_L'/M_S'. Now, to find the 95 percent confidence interval, we use the following, as shown in Chap. 9:

$$\frac{M_L/M_S}{F_{v_L,v_S,0.025}} < \frac{M_L'}{M_S'} < \frac{M_L}{M_S} F_{v_S,v_L,0.025} \tag{10·2}$$

In this case,

$$\frac{M_L'}{M_S'} = \frac{2\sigma_L^2 + \sigma_S^2}{\sigma_S^2} = \frac{2\sigma_L^2}{\sigma_S^2} + 1$$

and therefore

$$\frac{1}{2}\left(\frac{M_L/M_S}{F_{v_L,v_S,0.025}} - 1\right) < \frac{\sigma_L^2}{\sigma_S^2} < \frac{1}{2}\left[\frac{M_L}{M_S}(F_{v_S,v_L,0.025}) - 1\right]$$

Substituting the known values:

$$\frac{1}{2}\left(\frac{20.69/1.08}{F_{9,12,0.025}} - 1\right) < \frac{\sigma_L^2}{\sigma_S^2} < \frac{1}{2}\left[\frac{20.69}{1.08}(F_{12,9,0.025}) - 1\right]$$

$$\frac{1}{2}\left(\frac{19.16}{3.44} - 1\right) < \frac{\sigma_L^2}{\sigma_S^2} < \tfrac{1}{2}[(19.16)(3.87) - 1]$$

$$2.28 < \frac{\sigma_L^2}{\sigma_S^2} < 36.57$$

Similarly, we have from the ANOVA table an MS value of 1.625 for trees, which we equate to the estimated values corresponding to the EMS term for trees:

$$8\hat{\sigma}_T^2 + 2\hat{\sigma}_e^2 \times \hat{\sigma}_S^2 = 1.625$$

$$8\hat{\sigma}_T^2 = 1.625 - (2\hat{\sigma}_L^2 + \hat{\sigma}_S^2)$$

$$8\hat{\sigma}_T^2 = 1.625 - 20.694$$

$$8\hat{\sigma}_T^2 = -19.069$$

which leads us to a calculated negative variance. Since a genuine negative variance is impossible, we must suppose that the variance $\hat{\sigma}_T^2$ is really zero.

We summarize the results of our components of variance analysis of the two-factor nested experiment for the data in Table 10·11:

Table 10·11   Estimates of Variance for a Two-factor Nested Experiment

Source	Point estimate	0.95 confidence interval
Samples within leaves	$\hat{\sigma}_s^2 = 1.08$	$0.55 < \hat{\sigma}_s^2 < 2.95$
Leaves within trees	$\hat{\sigma}_1 = 9.81$	$2.28 < \dfrac{\hat{\sigma}_1^2}{\hat{\sigma}_s^2} < 36.57$
Trees	$\hat{\sigma}_t = 0$	$0 < \dfrac{\hat{\sigma}_t^2}{\hat{\sigma}_s^2} < 0.26$

1 We have failed to find evidence of a difference among trees.
2 We have evidence of a slight difference among samples from the same leaf.
3 The major source of variation is among leaves within trees.

PROBLEMS

10·5 To study the ability of various products to prevent rusting in automobiles driven in areas where salt is used to remove snow from highways, products from several companies were studied. From 4 different companies, 2 grades of rust preventives were chosen, and 3 automobiles were undercoated with each product, one product per car. After one winter of driving the test cars in the same city, the amount of rust formation was measured in exposed areas. The total amount of rust formed per car is shown in Table 10·12.

Table 10·12 Amount of Rust Formed per Car

Car	Company A		Company B		Company C		Company D	
	Good grade	Poor grade	Good grade	Poor grade	Good grade	Poor grade	Good grade	Poor grade
1	35	87	20	90	44	53	38	90
2	40	93	22	85	45	59	35	85
3	38	90	20	84	46	48	30	80

(a) Complete the ANOVA table from the data.

(b) Calculate the EMS values for the experiment.

(c) Place 95 percent confidence intervals on the components of variance.

10·6 Consider an experiment involving 5 castings from each of which 2 samples are taken, and finally 2 determinations of hardness are made from each sample. The results are shown in Table 10·13.

Table 10·13 Hardness Determinations for Five Castings

	Casting 1		Casting 2		Casting 3		Casting 4		Casting 5	
Samples	A	B	A	B	A	B	A	B	A	B
Determinations	136	99	150	130	95	105	140	120	80	120
	144	105	140	135	110	100	130	110	85	115

(a) Write the mathematical model for the experiment.

(b) Complete the ANOVA table and the EMS terms for the tests.

(c) Place 90 percent confidence intervals on the components of variance.

10·5 NESTED SAMPLING STUDY

The analysis of a chemical mixture for an essential component involves a very laborious analytical procedure. If we are testing a lot of the mixture for the percentage of this component, we must make a decision on this question: To obtain a good estimate of the amount of the component present, should we take many samples and make few analyses on each sample, or should we take few samples and make many analyses on each sample? The decision is best made on the basis of a components of variance analysis. We would like the contribution of among-sample and among-analysis variances to be of the same order of magnitude. If the component of variance for analyses is small, but that for samples is large, we will do well to make few analyses per sample and take many samples.

An experiment in which we have made four analyses on each of three samples results in the data in Table 10·14. By the methods we have previously described, we prepare the ANOVA summary in Table 10·15.

Table 10·14   Data for Sampling Study

Sample	A	B	C
Analyses	5	3	7
	5	3	8
	4	2	7
	4	3	8
Total	18	11	30

Table 10·15   Summary ANOVA for Data in Table 10·12

Source	SS	ν	MS	F	EMS
Samples	46.17	2	23.08	74.4	$4\sigma_S^2 + \sigma_A^2$
Analyses	2.75	9	0.31		σ_A^2
Total	48.92	11	23.39		

From Table 10·15, our point estimates of the population variances are

for $\hat{\sigma}_A^2$: 0.31
for $\hat{\sigma}_S^2$: $\frac{1}{4}(23.08 - 0.31) = 5.69$

The subscript A designates analyses; the subscript S designates samples. The multiplier $\frac{1}{4}$ is used because four analyses are included in each sample. We use Eq. (10·1) to place confidence limits on the variance for analyses:

$$\frac{\nu_A \hat{\sigma}_A^2}{\chi^2_{\nu_A, 0.025}} < \sigma_A^2 < \frac{\nu_A \hat{\sigma}_A^2}{\chi^2_{\nu_A, 0.975}}$$

$$\frac{9 \times 0.31}{19.0} < \sigma_A^2 < \frac{9 \times 0.31}{2.70}$$

$$0.15 < \sigma_A^2 < 1.03$$

Also,

$$\frac{4\sigma_S^2 + \sigma_A^2}{\sigma_A^2} = \frac{M_S'}{\sigma_A^2}$$

By Eq. (10·2),

$$\frac{M_S/M_A}{F_{\nu_S, \nu_A, 0.025}} < \frac{M_S'}{M_A'} < \frac{M_S}{M_A} (F_{\nu_A, \nu_S, 0.025})$$

From these two relationships,

$$\frac{1}{4}\left(\frac{M_S/M_L}{F_{2,9,0.025}} - 1\right) < \frac{\sigma_S^2}{\sigma_A^2} < \frac{1}{4}\left[\frac{M_S}{M_A}(F_{9,2,0.025}) - 1\right]$$

$$\frac{1}{4}\left(\frac{23.08/0.31}{5.71} - 1\right) < \frac{\sigma_S^2}{\sigma_A^2} < \frac{1}{4}\left[\frac{23.08}{0.31}(39.39) - 1\right]$$

$$3.01 < \frac{\sigma_S^2}{\sigma_A^2} < 733$$

We have found from this analysis that the component of variance for analyses is small and that the component of variance for samples is at least several times as large. To develop an economical testing procedure, we should plan to make fewer analyses of more samples.

We change the sampling plan to make only 2 analyses of each of 7 samples, obtaining the data in Table 10·16 and the summary ANOVA in Table 10·17.

Table 10·16   Data from Additional Sampling Study

Samples	D	E	F	G	H	I	J
Analyses	5	6	8	4	6	5	7
	5	5	7	4	5	5	5
Total	10	11	15	8	11	10	12

Table 10·17   Summary ANOVA for Data in Table 10·16

Source	SS	ν	MS	F	EMS
Samples	14.0	6	2.33	4.66	$2\sigma_S^2 + \sigma_A^2$
Analyses	3.5	7	0.50		σ_A^2
Total	17.5	13			

Application of the components of variance analysis gives

$$\hat{\sigma}_A^2 = 0.50 \quad \text{and} \quad \hat{\sigma}_S^2 = 0.92$$

In addition,

$$0.22 < \sigma_A^2 < 2.07$$

and

$$0 < \frac{\sigma_S^2}{\sigma_A^2} < 12.78$$

By changing the testing procedure, we have made the two components of variance comparable in magnitude, thus making the procedure more efficient.

10·7   A study was made of the variability of amounts of paint used in "paint by numbers" art kits. Eight art kits were chosen at random from all available production in a given plànt. Four pictures can be painted from the material supplied in each kit. The amount of paint used in the 4 pictures for each of the 8 kits is shown in Table 10·18. Show the analysis of variance table for the data contained in Table 10·18.

Table 10·18   Variability of Paint Used in Paint-
by-numbers Sets

Art kit							
A	B	C	D	E	F	G	H
0.98	0.91	0.78	0.95	0.86	1.08	1.00	1.05
0.86	0.87	0.93	1.25	0.78	0.92	0.87	0.83
1.01	1.00	0.68	1.06	1.00	0.93	0.79	1.03
1.03	0.99	0.83	1.00	1.04	1.16	0.76	1.00

10·8   From the information contained in Prob. 10·7, compute the components of variance and their 95 percent confidence limits.

10·9   Comment on the efficiency of the testing procedure used in Prob. 10·7.

10·1   Define a factorial experiment. Give an example.

10·2   Define a "crossed" experiment. Give an example.

10·3   Define a "nested" experiment. Give an example.

10·4   Explain why we do not test for interaction in a nested experiment.

10·5   Is it possible for *all* factors in an experiment to be nested? Explain.

10·6   Why is error always treated as nested?

10·7   What is the difference between "residual" and "error"?

10·8   What is the importance of replications in an experiment?

10·9   Explain how subscripts are used to distinguish nested from crossed factors.

10·10   Why should the mathematical model be written before performing ANOVA?

10·11   Is it possible for an experiment to have some factors that are crossed and other factors that are nested? Explain.

10·12   In hierarchical (nested) experiments, what hypotheses can be tested? How do we determine the *F* tests to be run?

10·13   Both factorial and hierarchical experiments are cases of analysis of variance with multiple classification. How do these types of experiments differ?

10·14   What type of question is being asked when we are interested in running the F tests at the end of the analysis of variance?

10·15   What type of questions are being asked when we are interested in placing confidence intervals on the components of variance?

REFERENCES

5, 10, 28, 42, 59, 68, 73, 79, 126

11 Studying Individual Effects

The method of analysis of variance (described in Chap. 8) permits us to answer the questions such as: Does this factor have an effect which is real when compared with the effect associated with experimental error? The answer to such a question is merely "yes" or "no," to a specified probability level.

If the process of ANOVA indicates that we have found no significant effect for any of the tested factors, we then proceed to test for other possible factors which might be significant but were overlooked in our first analysis of the problem. If we do discover that a tested factor has a significant effect, it is fruitless to stop with this discovery. Merely to say, for example, that temperature affects the yield of a manufacturing process is to say very little about the process. Once ANOVA has shown that a factor has a significant effect, these further questions should arise:

1 Of the tested levels of the significant factor, which make the largest contribution to the effect? Such a question is appropriate if we have used raw materials supplied by five different sources and have found that a statistically significant difference exists among suppliers. It now becomes necessary to find out whether the discovered difference is associated with a single supplier, or whether the five suppliers form homogeneous groups between which differences exist.

2 Is there a functional relationship between the response variable and the significant factor? This kind of question is appropriate if we have tested

the effects of the concentrations of the ingredients in a chemical manufacturing operation and have found that one ingredient significantly affects the yield. We should now establish the kind of relationship that exists between the concentration of this ingredient and the process yield. Once we know the nature of the relationship, we can decide in what direction the concentration should be changed in order to improve the process.

Of these two types of questions, the first is appropriately asked about factors which are not quantitative, for example, different processes, different machines, different operators. None of these is described in numerical terms. The second type of question is significant for factors like temperature, pressure, and concentration, which can be related to a numerical scale.

If the experiment has not shown that a factor has an effect, there is no reason to examine the data from *this* experiment further, as far as this factor is concerned. We may have had reason to suppose that temperature affected a process. If we perform an experiment on the process, and conclude that we have not demonstrated an effect associated with temperature, we abandon these data even though in the future we may reexamine this factor over a wider range of levels or with more precise testing. In particular, it is pointless to plot a graph of response variable versus temperature if temperature has not been shown to have a significant effect on the process.

11·1 INSPECTION OF DATA

Often a careful inspection of the data, especially when the data are in coded form, is sufficient to suggest the sources of variation in the data. Consider Table 11·1, which displays the results of a crossed experiment involving three inspectors of raw materials from three suppliers. As the data are arranged here, it may not be obvious that there is a pattern of variation. Since the identification of inspectors and suppliers as 1, 2, 3 is arbitrary, we may rearrange the data as in Table 11·2. Now we see a set of numbers which increase both left to right and downward. Such a pattern suggests not only that both tested factors are significant but also that the response variable is consistently greater both for S_3 and for I_2.

Table 11·1 Data from Inspection by Three Inspectors of Material from Three Suppliers

Inspectors	Suppliers			
	S_1	S_2	S_3	$T_{i.}$
I_1	4	3	10	17
I_2	13	10	20	43
I_3	10	8	15	33
$T_{.j}$	27	21	45	

Table 11·2 Data from
Table 11·1
Rearranged

Inspectors	Suppliers			
	S_2	S_1	S_3	$T_i.$
I_1	3	4	10	17
I_3	8	10	15	33
I_2	10	13	20	43
$T._j$	21	27	45	

A situation like that in Table 11·3 requires careful examination. The data show that only one of the response items is much different from the other eight. ANOVA would here probably compel the rejection of the null hypothesis for the factors in both rows and columns. The rejection of the null hypothesis, however, would really be based on the presence of only one unusual piece of data. When such results appear, it is well to reevaluate the experiment and the data collection. There may have been a mistake in measurement or in recording the data, or else some major break-through may have been achieved.

Table 11·3 Data from Fully Crossed
Experiment with One
Unusual Response

	A_1	A_2	A_3
B_1	3	4	3
B_2	4	3	3
B_3	20	3	4

PROBLEMS

11·1 Using the data as presented in Table 11·1, code each value by subtracting 3 from the recorded value.
  (a) Does the relationship between the levels of the factors change?
  (b) What would happen if the grand average were subtracted from each of the observations?

11·2 Using the data as presented in Table 11·1, add the value 155 to each observation.
  (a) Does the relationship between levels of the factors appear to change?
  (b) How has the mathematical model for the new data been changed? Explain.
  (c) Has the variance of the set of observations from Table 11·1 been changed when we increase or decrease the value of the response variable?

11·3 Consider the following observations: 2,168, 2,171, 2,169, 2,173, 2,180.

(a) Do the numbers appear to vary? Calculate the range.
(b) Code the numbers by subtracting 2,168 from each value, and calculate the range.
(c) Do the coded values vary more than the original ones?

11·4 Perform the analysis of variance for the data in Table 11·3.

(a) Consider all factors fixed.
(b) Consider both factors random.
(c) How does the presence of one extreme value influence the results of ANOVA?
(d) What results would you expect to get in either fixed or random factor ANOVA if the value 20 were replaced by a 3 or a 4?

11·2 GRAPHICAL ANALYSIS

When we cannot discover a pattern from a close examination of the data, plotting the data will often be useful. Table 11·4 gives in coded form the results of an experiment in which the sharpness of photographic images was determined for four different machine processors using four different developing solutions in a crossed experiment.

Table 11·4   Data from Experiment Testing Developers and Machines; Response— Sharpness of Photographic Image

Machine	Developers			
	D_1	D_2	D_3	D_4
1	40	20	48	38
2	25	5	20	5

We have plotted the data in Fig. 11·1. That the four values of sharpness from machine 1 all lie above those from machine 2 suggests that there is a difference between these machines. If we plot the averages of the two graphs as X values, D_3 would give the highest sharpness. (Note that these comments are illustrative only. The widths of the confidence limits about these points would be large enough to make it impossible for us to be sure of the differences just mentioned.)

If we wish to investigate the methods of obtaining the highest possible degree of sharpness, the graphs indicate that machine 1 should be used with D_3 and this combination should be used as the starting point for a future test.

In such graphs, an interaction would be shown by an intersection (real or projected) of the two lines. If no interaction exists, the two lines will tend to be parallel.

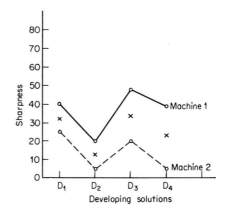

Fig. 11·1 Graphical analysis of data from Table 11·4.

Note that the plotted points on the graph in Fig. 11·1 are by no means intended to suggest a functional relationship between the response variable and the factor "developing solutions," since no significance can be attached to the positions of the points on the horizontal axis. The graphs do, however, facilitate the detection of patterns of variation that may be obscure in the tabulated data.

When replicate runs are made of each combination of levels of factors, graphs may be made of the averages of the replications, or of the totals of the replicates. In either case, the patterns, if they exist, will be the same. If the averages of the replicate values of response are plotted, and vertical lines (representing the range of the replicated values) are drawn through the averages, then the graph will also depict the estimation of variation for comparison purposes. Such a line, showing the range of the replicates, should be drawn to be symmetrical about the average. In general, the simpler the graph, the better it can be used to display patterns which might exist.

PROBLEMS

11·5   Replot the data in Table 11·4, using the horizontal axis for machines at two levels, thus showing the patterns for the four developing solutions. What inferences are suggested by this new graph?

11·6   Refer to Table 10·6.
(a) Plot the averages per cell using Fig. 11·1 as a model.
(b) Plot the totals per cell.
(c) Compare the answers to parts (a) and (b) above, commenting on the patterns which might exist.

11·7   Plot the data in Table 11·1.
(a) Place suppliers on the horizontal axis.
(b) Place inspectors on the horizontal axis.

11·8   What inferences do you make from these plots?

11·9 Plot the data in Table 11·2 and compare with the graphs made for Prob. 11·7. Comment on the effect of the rearrangement of the data.

11·3 LEAST SIGNIFICANT DIFFERENCE

Once ANOVA has led to the rejection of the null hypothesis for a tested factor, we may discover which levels of the factor contribute to the effect by a statistical test based on the concept of confidence limits. The test is based on the *least significant difference*—LSD. The value of the LSD is that difference in two means necessary to permit us to suppose that the means involve different populations.

The value of the LSD is related to the t_ν distribution by the following:

$$\text{LSD}_\alpha = t_{\nu, \alpha/2} \sqrt{\frac{2s_e^2}{n}} \tag{11·1}$$

LSD_α is the least significant difference for a specified α risk; $t_{\nu, \alpha/2}$ is the table t_ν value for $\alpha/2$ and the number of degrees of freedom associated with the error mean square in the ANOVA summary table; s_e^2 is the mean square for error in the ANOVA summary table; n is the number of observations involved in each treatment mean. Note that the term $\sqrt{s_e^2/n}$ is the standard error of the mean as described in Chap. 2. The factor 2 appears in Eq. (11·1) because in this test we are always comparing two means.

We performed an analysis of variance whereby we tested two factors—method of mixing and machine—for their effect on the yield of a manufacturing process. We found method of mixing to have no significant effect. We found, however, that the factor "machine" was significant.

For four runs, the mean yields for the three tested machines were: A, 25; B, 30; C, 50. Note that we have ordered the means in accordance with their size.

The question we wish to answer is: Are all the machines different in their effect, or is one alone probably responsible for the significance of the factor "machine"? The answer will depend upon the estimate of error in the experiment. If the estimate is very small, we may conclude that the three machines are all different; if the estimate is large, we may conclude only that machine C differs from machines A and B.

If the MS for error in the summary ANOVA table is 100, with $\nu = 6$, we now apply Eq. (11·1), assuming that we desire an α risk of 0.05:

$$\begin{aligned}
\text{LSD}_{0.05} &= t_{0.025,6} \sqrt{2 \times 100/4} \\
&= 2.447 \times \sqrt{50} \\
&= 17.3
\end{aligned}$$

The number we have just found is the required difference in two sample means to permit us to say, at the 0.05 significance level, that a real difference exists between them. Since the difference between the average yields for machines A and B is less than the calculated LSD, we have not shown that they are different. The difference between the means for machines B and C is, on the other hand, greater than the calculated LSD. Our conclusion is that the factor "machine" was found to be significant in the ANOVA because machine C produces a yield greater than the yield produced by the other two tested machines.

11·4 MULTIPLE RANGE TEST

The multiple range test, developed by Duncan in 1955, is similar to the test for least significant difference. If we rearrange the formula for LSD to read

$$LSD_{0.05} = \sqrt{2}\, t_{0.025} \sqrt{\frac{s_e^2}{n}} \qquad (11\cdot2)$$

the multiplier $\sqrt{2}\, t_{0.025}$ is called the *significant studentized range*. The significant studentized ranges for the 5 percent level are equal to $\sqrt{2}\, t_{0.025,\nu}$, and those for the 1 percent level are equal to $\sqrt{2}\, t_{0.005,\nu}$, where ν is the degrees of freedom associated with the mean square for error from the ANOVA table. The product of $\sqrt{2}$ and the appropriate t_ν value is the significant studentized range for the comparison of two means, when the means have been arranged in order of size.

The multiple range test is best used in situations where more than two means are being compared. In a two-factor experiment where factor A has four levels and factor B has five levels, factor A has been found to be significant. We now wish to compare the means of the four levels of factor A to see how they should be grouped. The means of the four levels of A are $\bar{X}_1 = 18.50$, $\bar{X}_2 = 12.50$, $\bar{X}_3 = 20.00$, and $\bar{X}_4 = 35.00$. We first arrange the means in order according to size:

$$\begin{array}{cccc} \bar{X}_2 & \bar{X}_1 & \bar{X}_3 & \bar{X}_4 \\ 12.50 & 18.50 & 20.00 & 35.00 \end{array}$$

The standard error of the mean, which is required to compute the shortest significant range, is $s_{\bar{X}} = \sqrt{s_e^2/n}$. (Assume that the ANOVA table shows an MS for error of 33.97 with 12 degrees of freedom.) n will equal 5 in this case, as there are 5 levels of B (essentially replicates) used in the calculation of each mean value. Then $s_{\bar{X}} = \sqrt{33.97/5}$, or 2.61. If we now choose the 0.05 level of significance, the significant studentized ranges can be obtained from Table A·15. Using ν_2 equal to 12 (the ν of MS for error from ANOVA), and g equal to 2, 3, and 4 (where g is the number of means in a group to be compared), we find the tabulated values to be: for $g = 2$, 3.08; for $g = 3$, 3.23; and for $g = 4$, 3.33. Each of these table values must then be multiplied by the standard-error value of 2.61, and the resulting SSR values are

$$\begin{array}{cccc} g & 2 & 3 & 4 \\ SSR & 8.04 & 8.43 & 8.69 \end{array}$$

We now find the differences between the means in the following way: $4 - 3$, $3 - 2$, $2 - 1$, $4 - 2$, $3 - 1$, and $4 - 1$. In the first three differences there were two means involved; so their differences are compared with the SSR value for $g = 2$. If the difference of the means is less than the SSR value, we draw a line under the two means to signify that they are not significantly different. If the difference exceeds the value for SSR of $g = 2$, no line is drawn. The fourth and fifth differences involve three means and are compared with the value of SSR for $g = 3$. The line is once again drawn or not drawn as the difference is found to be smaller or larger than the corresponding SSR value. The last difference is from four means and is therefore compared with SSR for $g = 4$.

The result of the comparisons is as follows:

$$\bar{X}_2 \quad \bar{X}_1 \quad \bar{X}_3 \quad \bar{X}_4$$

We see that $\bar{X}_4$ is different from the other three means and may therefore be considered as being significantly higher in value. The values of $\bar{X}_2$, $\bar{X}_1$, and $\bar{X}_3$ appear to be similar, and the factor appears to have been significant because of the lack of agreement on the part of the fourth level.

In general, if there are k levels of the significant factor, there will be $(k/2)(k - 1)$ comparisons which can be made.

PROBLEMS

11·10   A two-factor experiment was run with factor A at 4 levels and factor B at 5 levels. As no interaction was expected, replicate runs of the treatment combinations were not run. Therefore the experiment required 20 runs. Analysis of variance of the data indicated that factor A was not significant but that factor B was. The average values of the 5 levels of factor B were found to be 44.54, 35.90, 27.00, 60.64, and 44.82, each average the result of 4 observations. From the ANOVA table, the value of the mean square for error was found to be 101.7.

(a) Calculate the standard error of the mean.
(b) Using a significance level of 0.05, compute the value of LSD.
(c) Which of the 5 levels of factor B can be considered as being similar?
(d) How should the levels of factor B be grouped to indicate the number of populations they represent?

11·11   In a two-factor ANOVA, both factors were found to be significant. The averages of the levels of factor A were $A_1 = 334.4$, $A_2 = 296.5$, $A_3 = 342.4$, $A_4 = 346.6$. The averages of the levels of factor B were found to be $B_1 = 468.4$, $B_2 = 455.3$, $B_3 = 396.1$. With two replicates per treatment combination, the experimental error was found to be 236.95 with 12 degrees of freedom, providing mean square for error of 19.75.

(a) Using a level of significance of 0.05, compute the value for LSD for the two factors.

(b) Which of the levels of factor A can be considered as equal?

(c) Which of the levels of factor B can be considered as equal?

11·12   Using the data from Table 11·8, rank the means of the factor "concentrations" by the multiple range test, at the 0.05 level.

11·13   Using the data from Table 11·8, rank the means of the factor "heating times" by the multiple range test, at the 0.05 level.

11·14   Comment on what happens when the multiple range test is used to rank means of a factor which is not found to be significant. Why should the ranking of means be used only on significant factors?

11·5 ORTHOGONAL COMPARISONS

When we select the levels of a factor for testing, we can often choose them so as to answer questions other than merely "are the levels different?" Although we may want to know whether or not the levels do differ in their effects, we also often want to know if different combinations of the levels have different effects.

If we are concerned with four operators, we would run an experiment to find out whether or not operators have an effect on the yield of a given process. Perhaps two or more similar machines are also to be tested. Such an experiment would call for a two-factor arrangement, with multiple levels of each factor. The point now is that, in the selection of the four operators, we may also choose to investigate a possible difference in operators due to training or sex or some other factor.

We now decide to select the operators in the following way: (1) Two operators have had considerable experience on the machine, but for quite different lengths of time. (2) Two operators are new to the machine. Of the latter two, one will be instructed orally and the other will be given a set of written instructions. By this method of selection, we may now look for answers to these questions:

1   Is the yield affected by different operators?

2   Is there a difference between experienced and inexperienced operators?

3   Does length of experience make a difference?

4   For new operators, is there a difference in effect between oral and written instructions?

If we expect the answer to the first question to be "No," then we would not want to test the other three questions. But if we expect the answer to be "Yes," then we can gain insight into the other questions by subdividing the sum of squares of the factor "operators." We can then test the subdivided

SS values against the error mean square to find which of the groupings of the levels are statistically significant.

We may make as many comparisons among the levels as we have degrees of freedom. A necessary condition is that the comparisons must be independent; that is, we may subdivide the factor "sum of squares" (here operators) in three different ways, but the ways must be genuinely different, with no duplication of the subdivisions.

Independent comparisons of this type are called *orthogonal*. The method of ensuring that the comparisons are orthogonal to each other is by the preparation of a table of coefficients, following certain rules. We illustrate the method with data taken from Table 8·11. In the experiment which gave rise to the data, we found we were interested in evaluating four operators and three micrometers.

We have a basis for believing O_1 and O_3 might be alike as they were trained in the same laboratory, while O_2 and O_4 are new and relatively inexperienced. The three comparisons we wish to make are:

1 $(O_1 + O_3) - (O_2 + O_4)$, a comparison of the trained operators contrasted to the new ones
2 O_1 versus O_3, a comparison of the two trained operators
3 O_2 versus O_4, a comparison of the new operators

For the first comparison, we assign $+1$ values as coefficients to the two levels (O_1 and O_3) forming the first part of the pair, and -1 values to those in the second part (O_2 and O_4). (The assignment of signs to the first or second parts is arbitrary, but the same sign must be given to every term in each pair of the comparison.) Thus we have the first row of coefficients in Table 11·5. It is highly desirable, though not necessary, to have the row total sum to zero to facilitate the calculations to follow. The significance of the first row of coefficients in Table 11·5 is that we wish to compare O_1 and O_3 against O_2 and O_4. We have now set up one comparison.

The second comparison will involve only O_1 and O_3. We assign the coefficient 0 to the pair in which we are for the moment not interested. We assign $+1$ to O_1 and -1 to O_3. Now we have the second row of coefficients in Table 11·5. This row also sums to zero, the first consideration, and the two rows also obey a basic requirement for independence (orthogonality): *The sum of the products of the coefficients of the two "independent" comparisons, multiplied column by column and added, is zero.* The products are in order $+1 \times +1 = +1$; $-1 \times 0 = 0$; $+1 \times -1 = -1$; $-1 \times 0 = 0$. These products sum to zero, and thus comply with the requirement.

The same rules apply to the third row of coefficients. It is at this point that the rules will determine which comparison we are now permitted to make. We wish to compare O_2 and O_4. We assign the coefficient 0 to the other two operators, $+1$ to O_2 and -1 to O_4. With this, we have now the complete set of coefficients in Table 11·5. The last set we entered obeys

Table 11·5 Orthogonality Coefficients for Four Levels of One Factor

Level				Divisor
O_1	O_2	O_3	O_4	
+1	−1	+1	−1	$4 \times 3 = 12$
+1	0	−1	0	$2 \times 3 = 6$
0	+1	0	−1	$2 \times 3 = 6$

the first rule for orthogonality and, when paired with each of the other two rows, the second rule as well.

Observe that had we wished to compare O_1 with O_2, we would have written the sequence of coefficients $+1$, -1, 0, 0. The application of the second rule, however, would show that this comparison is not orthogonal to the first two and is therefore by the rules forbidden once the first two comparisons are selected.

In order properly to weight the new SS terms, which we will compare with the error MS, we divide each term by the sum of the squares of the coefficients, multiplied by the number of items that contributed to each column total. In this example, we use as divisor the sum of the squares of the coefficients in each row times 3. The required divisors appear in the last column of Table 11·5.

We proceed now to apply the coefficients to each total, sum the result, square it, and apply the divisor:

$$\frac{(+6 - 21 + 11 - 13)^2}{12} = 24.08$$

For the second comparison,

$$\frac{(+6 + 0 - 11 + 0)^2}{6} = 4.17$$

For the third comparison,

$$\frac{(+0 + 21 - 0 - 13)^2}{6} = 10.67$$

We have now divided the original SS term for operators into three parts, each part having one of the original three degrees of freedom. We test each of the new SS terms against error MS, which is 3.24. The calculated F ratios are 7.43, 1.29, 3.29. The table F ratio for 1 and 6 degrees of freedom is 3.78 for $\alpha = 0.10$. At the 0.10 level of significance, we reject the null hypothesis for the first and accept it for the second and third. We thus conclude that we have evidence to indicate that the operators are performing differently among themselves (we also would have found this from the regular ANOVA— see Chap. 8). Now we also believe that O_1 and O_3 differ from each other and from O_2 and O_4 as a pair. In the future use of micrometers, we would suppose that the measurements made by O_2 and O_4 are equivalent but that an effort should be made to bring the work of O_1 and O_3 into agreement with that of the other two operators.

From the same experiment, we have data on three micrometers. Suppose M_1 is our original instrument and M_2 and M_3 are recent additions. Since we have two degrees of freedom for the factor "micrometers" we may make two orthogonal comparisons for the levels of this factor and thus determine the manner in which the micrometers differ.

Here we have an odd number of levels of the tested factor, and we must weight the coefficients to balance out the effects. Thus, if we wish to compare M_1 with the M_2, M_3 pair we would want to contrast M_1, with the average of M_2 and M_3 or M_1 versus $\frac{1}{2}M_2$ and $\frac{1}{2}M_3$. Since fractions are a nuisance we write the coefficients $+2$, -1, -1. The use of the coefficient $+2$ for the single term weights it appropriately for comparison with the sum of the other two levels. One more comparison is now permitted. Application of the two rules for orthogonality will show that the second comparison may involve only M_2 and M_3 and the coefficients $+1$ and -1. The complete set of coefficients is given in Table 11·6, along with the divisor calculated as before. Note here that 4 terms make up each $T_{i.}$ value.

Table 11·6   Orthogonality Coefficients for a Three-level Experiment

	Level			Divisor
	M_1	M_2	M_3	
	$+2$	-1	-1	6×4
	0	$+1$	-1	2×4

The subdivision of the ANOVA SS term is: for the first comparison,

$$\frac{[+2 \times 7 - 17 - 27]^2}{24} = 37.5$$

and for the second comparison,

$$\frac{(0 + 17 - 27)^2}{8} = 12.5.$$

A comparison with the ANOVA in Chap. 8 shows that the SS of 50.0 given there has been broken down into two additive components. The calculated F ratios, based again on the error MS, are 11.55 and 3.86. Comparing these with the table F ratios for 1 and 6 degrees of freedom, we find both comparisons to have an effect which is significant at the 0.10 level. We therefore conclude that no two of the micrometers are performing similarly. The old and new micrometers are different, and in fact, the new ones do not agree either.

PROBLEMS

11·15   For a factor tested at each of 4 levels, make 2 different tables of orthogonal coefficients to permit the making of 3 comparisons of the data.

11·16   An experiment was conducted to test the ability of men and women, with and without experience, to perform an assembly operation. Eight replications of each treatment were used. The replication sum of squares was found to be 89.65, and the error sum of squares was 127.62. The treatment totals were

> Men with no experience 92.7
> Men with little experience 82.8
> Men with much experience 93.3
> Women with no experience 60.2
> Women with little experience......73.5

(a) Make a table of orthogonal coefficients showing the comparisons that should be made.
(b) Develop an ANOVA table showing the breakdown of the treatment sum of squares into components.
(c) Complete the ANOVA and state your conclusions.

11·17   List 5 levels of a factor and state the questions you might ask about the levels.
(a) Set up a table of orthogonal coefficients, and indicate what question each line of coefficients is related to.
(b) Set up a second table of orthogonal coefficients, different from the first, and indicate what questions are related to each line of coefficie

11·18   Comment on the possible use of orthogonal coefficients as used with:
(a) quantitative factors
(b) qualitative factors
(c) four-dimensional graphs, or n-dimensional space problems

11·19   The questions to be asked of the factors under test should be considered prior to the collection of any data. Very often this will influence the levels which are chosen for the factor.
(a) Why should the questions to be answered in the experiment be listed prior to laying out the design of the experiment?
(b) If orthogonal comparisons are made on the basis of looking at the results of the data, what false influences might affect the conclusions?

Under certain conditions, we can use the method of partitioning an SS term to discover whether or not a specified kind of functional relationship probably exists between a response variable and the levels of a factor. The conditions are these: The factor must be quantitative, and the levels of the factor must be equally spaced.

If we wish to apply this method in the study of the effect of temperature on a process, the temperatures used in the experiment may form a series such as 60°, 75°, 90°, 105°. Alternatively, we could test a series of concentrations of an ingredient and express the levels as the logarithms of the concentrations. The method about to be described can be used if the logarithms form an arithmetic series. If the concentrations are 5, 10, 20, 40 grams per liter, the log concentrations are 0.7, 1.0, 1.3, and 1.6, and the conditions are fulfilled. We could also have the levels equally spaced in the sense that the change in yield from one level to the next is expected to be equal; and the experiment could be run to check this.

For this kind of analysis we make use of coefficients like those used above for partitioning SS terms. The coefficients we will use here are fixed by the nature of the functional relationship we are testing for. The types of relationship are three:

1   A linear function, which means that a straight line would reasonably represent the data if they were plotted.
2   A quadratic function, which means that a simple curve would represent the data. The curve will usually have a single maximum or a single minimum, though the maximum or minimum may lie outside the range of the tested data.
3   A cubic function, which means that a more complex curve would be required to represent the plotted data. Cubic relationships typically have both a maximum and a minimum over a sufficient range.

Examples of these three kinds of functional relationships are shown in Fig. 11·2.

In practice, we find that linear and quadratic relationships are frequent. Cubic relationships are rare, and functions of greater complexity are encountered so rarely that we generally ignore the possibility of finding a functional relationship of the fourth degree or beyond.

The number of allowable comparisons is, as before, determined by the number of degrees of freedom for the significant factor. If we have in the experiment used only two levels of the factor, $v = 1$, and we can make no judgment about the nature of the relationship. We must assume a linear relationship since two points do not allow us to check a departure from a straight line. If we have used three levels, we can make two comparisons and can thus test to see which of two possible relationships exists. Since linear

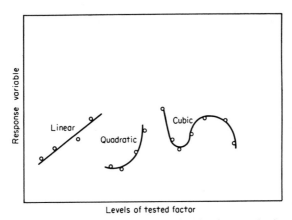

Fig. 11·2   Examples of linear, quadratic, and cubic functional relationships.

and quadratic relationships are most common, if $v = 2$ at least, we will test for those. If the number of levels is four or more, we can test for three possible relationships.

The test method requires a partitioning of the SS term for the factor, using the coefficients in Table 11·7. The coefficients follow the rules for orthogonal comparisons. Furthermore, they are selected so as to display, if they are plotted against equal horizontal intervals on a graph, the patterns typical of the kind of relationship they represent. We plot in Fig. 11·3 the coefficients for $n = 4$, that is, for four levels of the tested factor. The coefficients for L (linear) plot as a straight line. The coefficients for Q (quadratic) plot as approximately a parabola, typical of a second-degree equation. The coefficients for C (cubic) give a plot similar to that for an equation of the third degree.

Table 11·7   Coefficients for Testing for Functional Relationships

$n = 3$		$n = 4$			$n = 5$			$n = 6$		
L	Q	L	Q	C	L	Q	C	L	Q	C
-1	$+1$	-3	$+1$	-1	-2	$+2$	-1	-5	$+5$	-5
0	-2	-1	-1	$+3$	-1	-1	$+2$	-3	-1	$+7$
$+1$	$+1$	$+1$	-1	-3	0	-2	0	-1	-4	$+4$
		$+3$	$+1$	$+1$	$+1$	-1	-2	$+1$	-4	-4
					$+2$	$+2$	$+1$	$+3$	-1	-7
								$+5$	$+5$	$+5$

n = number of levels of tested factor.

We illustrate the method by applying it to this situation: Samples of steel bars were prepared with different concentrations of manganese, each heated

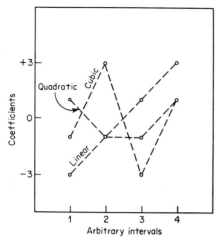

Fig. 11·3 Plot of coefficients for n = 4 (from Table 11·7).

for different times. The concentrations were 0.2, 0.4, 0.6, and 0.8 percent; the times of heating were ½, 1, ³⁄₂, 2, and ⁵⁄₂ hr. Each of the factors was quantitative, and tested at equally spaced levels.

The coded results are given in Table 11·8. If we were to run an ordinary ANOVA, the summary table would indicate that only the concentration of manganese was a significant factor. In running that analysis we make no use of our knowledge of the physical relationships of the factor levels (i.e., the steadily rising concentrations and times). Since we do have this information, we can now ask the question: What relationship exists between concentration and elongation? Since there are four levels of the concentration, we can by partitioning the SS term for concentration test for the three alternative

Table 11·8   Data from Experiment Testing Elongation of Steel Bars

Concentration	Heating time					$T_i.$	Mean
	H_1	H_2	H_3	H_4	H_5		
C_1	28.2	29.3	33.7	41.2	50.9	183.3	36.66
C_2	23.5	24.8	24.1	34.7	32.8	139.9	27.98
C_3	17.4	15.2	17.8	14.7	16.6	81.7	16.34
C_4	10.1	11.5	15.6	9.9	4.7	51.8	10.36
$T._j$	79.2	80.8	91.2	100.5	105.0	$T..$ 456.7	
Mean	19.80	20.20	22.80	25.13	26.25		

Table 11·9   Summary ANOVA for Data in Table 11·8

Source	SS	ν	MS	Calculated F ratio	Table F ratio
Concentration	2,077.07	3	692.36	20.4	$F_{3,12,0.025} = 4.4742$
Heating time	132.25	4	33.06	0.96	
Error	407.65	12	33.97		
Total	2,616.97	19			

functions. We apply the coefficients from Table 11·7 for $n = 4$ to the T_i. value for each level, as shown in Table 11·10. The divisor is again the sum of the squares of the coefficients in each row times the number of items in each total (here five because each total is made up of the result from five times of heating).

Table 11·10   Partitioned SS for the Factor Concentration (Data from Tables 11·7 and 11·8)

Concentration level	Total, T_i.	Coefficient			Product		
		L	Q	C	L	Q	C
C_1	183.3	-3	$+1$	-1	-549.9	$+183.3$	-183.3
C_2	139.9	-1	-1	$+3$	-139.9	-139.9	$+419.7$
C_3	81.7	$+1$	-1	-3	$+81.7$	-81.7	-245.1
C_4	51.8	$+3$	$+1$	$+1$	$+155.4$	$+51.8$	$+51.8$
	Divisor	20×5	4×5	20×5	-452.7	$+13.5$	$+43.10$ total

$$\text{Linear SS} = \frac{(-452.7)^2}{20 \times 5} = 2,049.37 \qquad \text{Cubic SS} = \frac{(+43.10)^2}{20 \times 5} = 18.58$$

$$\text{Quadratic SS} = \frac{(+13.5)^2}{4 \times 5} = 9.11$$

From the results, we prepare Table 11·11, a new summary ANOVA table. We test the partitioned sum of squares terms against the error mean square term, since the MS terms will equal the corresponding SS terms because of the single degree of freedom assigned to each. The calculated F ratios show that only the linear relationship is significantly different from error. A plot of elongation (using the means of the data for each level of manganese) against concentration, as shown in Fig. 11·4, indicates that the elongation is reduced with concentration. The results of the data analysis provide a basis for making decisions about the effect of manganese concentration, and about the desirability of changes in the composition of the steel.

Table 11·11 Revised ANOVA (Data from Tables 11·9 and 11·10)

Source	SS	ν	MS	Calculated F ratio	Table F ratio
Concentration: Linear	2,049.37	1	2,049.37	60.3	$F_{1,12,0.025} = 6.5538$
Quadratic	9.11	1	9.11	0.27	
Cubic	18.58	1	18.58	0.54	
Heating time	132.25	4	33.06	0.96	
Error	407.65	12	33.97		
Total	2,616.96	19			

The same techniques can now be applied to the factor "heating time." Since there are five levels of this factor, there are four functional relationships which can be tested. This time we would use the coefficients for $n = 5$ from Table 11·7. We need not compute all possible tests, for we can, if we care to, partition the SS for heating time into SS linear, and the rest into residual. Such a course of action would be proper if the SS linear were to utilize most of the existing SS for heating time, leaving very little to be partitioned into the other possible relationships.

In our example this is what happens. The SS linear term will consume almost all the 132.25 assigned to heating time. We therefore have insight into the functional relationship which exists between elongation and time of heating.

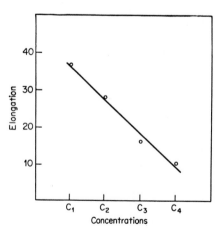

Fig. 11·4 Plot of mean response for levels of concentration (data from Table 11·8).

11·20   An investigation was carried out to discover whether a change in temperature or in running speed of an automatic processing machine would affect the yield of the product. Factor T (temperature) was run at 4 levels, and factor S (speed) was run at 5 levels, with results as given in Table 11·12.

Table 11·12   Results of Experiment Involving
Temperature and Running
Speed on Product Yield

Temperature	Speed, ft/min					Total
	18	20	22	24	26	
66°	7	9	11	12	15	54
68°	9	8	10	11	14	52
70°	7	7	11	12	14	51
72°	8	9	12	10	15	54
Total	31	33	44	45	58	211

(a) Complete the ANOVA for the experiment.
(b) Break down the sum of squares into components, and test for the kind of functional relationship that may exist.
(c) Graph the response variable against the levels of the significant factor, and sketch the line indicated by the answer to (b) above.

11·21   Using the data from Tables 11·8 and 11·11, partition the sum of squares for heating time into components and test, at a level of significance of 0.05, the functional relationships which might exist.

11·22   Perform the following experiment on a hard floor. Draw lines parallel to a wall at distances 6, 8, and 10 ft. Toss a coin toward the base of the wall from the different lines, and record, as your response variable, the distance from the wall that the coin stops. Use 4 sizes of coins and record 5 tosses from each distance.

11·23   Use the data gathered in answer to Prob. 11·22.
(a) List the questions you might answer from the experiment.
(b) Run an analysis of variance of the data.
(c) Using the factor "distance from the wall" test for possible functional relationships which might exist.

11·24   A quantitative factor with three levels A, B, and C was tested for possible functional relationships. Assume that the SS error term is equal to

1.00 with 21 degrees of freedom. One factor is being tested with 8 replications per level.

(a) Test for possible functional relationships if the averages for the factor levels, based upon 8 replicates per level, are $\bar{X}_a = 20.0$, $\bar{X}_b = 30.0$, and $\bar{X}_c = 25.0$.

(b) Test for possible relationships if $\bar{X}_a = 10$, $\bar{X}_b = 15$, and $\bar{X}_c = 50$.

11·7 TEST FOR FLYERS

It is customary in many analysis procedures to process three samples, and if one result appears unlike the other two, to discard the unusual result as a "flyer." An offhand judgment that the unusual value should be rejected is risky, especially when the sample size is small. For example, if the three results are 18, 19, and 25 percent, it is tempting, but dangerous, to consider the 25 percent value as incorrect.

An appropriate decision is based on the interval between the suspected number and the rest of the data. If the remaining numbers are very much alike, a relatively small difference between the suspected flyer and the other values is sufficient cause for rejection. If, on the other hand, the remaining values differ greatly, a considerable difference between the possible flyer and the remaining data must be observed. The necessary difference, furthermore, is related to the number of values in the set.

Table A·10 supplies ratios, based on the analysis above, which can be used to test suspected flyers and to reject unusual numbers with a specified α risk. To apply this test, we arrange the data in order of magnitude, with the suspect designated X_1. We calculate the ratio of two ranges, as required by the table formulas. If the calculated ratio is larger than the table value, we discard the suspected value, assuming an α risk as shown in the table. If the calculated ratio is less than the table value, we retain the number.

Consider these data, supposedly from the same population: 8, 7, 4, 9, 8, 7, 8. We suspect the value 4. We rearrange the data in order of size: 4, 7, 7, 8, 8, 8, 9. We calculate the ratio

$$r = \frac{X_2 - X_1}{X_n - X_1} = \frac{7 - 4}{9 - 4} = 0.60$$

The table value for $\alpha = 0.05$ is 0.507 ($n = 7$). Since the calculated ratio exceeds the table ratio, we may reject the suspect with a risk of error of less than 0.05.

The test for flyers should not be applied indiscriminately. If it is known that a mistake or a blunder was made in the experiment, or that a wrong analysis procedure was followed, the data resulting from that part of the experiment should be rejected, regardless of the nature of the data. A test for flyers is useful where the data themselves suggest its use, by the presence of a value that is seemingly out of place. However, one should look for reasons behind this "flyer." Is there a peculiarity of the experimental system?

Is the test method unstable? Much useful information can be lost by tossing out and ignoring such a point.

The method of testing for flyers is such that it should never be applied sequentially to the same set of data. If the test permits rejection of one value, the remaining data should be retained, without further test.

The test for flyers can be applied to sample means as well as to individual observations. In such a case, the terms used for the calculation of the test ratio are $\bar{X}$, rather than X, values. When the test for flyers is so used, it may be thought of as a quick test of hypothesis to discover whether one of a set of means is significantly different from the rest of the set.

PROBLEMS

11·25   Apply the test for flyers to each of the following:
(a) 6, 9, 9, 10, 11
(b) 6, 9, 9, 10, 10,
(c) 6, 9, 9, 10, 11, 11
(d) 6, 7, 9, 10, 10, 15
What conclusions do you draw from the results?

11·26   Two chemical analyses of the same coal sample gave ash contents of 21.3 and 24.2 percent. Find the extreme values which a third analysis could give without rejection by the test for flyers. Use $\alpha = 0.05$.

11·27   Using the data from Table 11·1, test to see if there is evidence that the value 20 is a flyer.

11·28   Using the data from Table 11·3, test the suspect value 20 as being a flyer.

11·29   The averages of students in a given subject area were recorded as follows: 93.5, 90.0, 89.7, 93.0, 67.0, 88.5, 89.2, 85.6, 90.3, 85.5. Test the suspect average to see if there is evidence of its being a flyer.

QUESTIONS

11·1   What is meant by the statement that ANOVA has failed to find a significant factor?

11·2   If we believe that a significant factor is present, but ANOVA does not support this belief, what explanations for this result are possible?

11·3   When ANOVA indicates that a factor is significant, what questions follow?

11·4   After data are collected from an appropriate experiment, what steps should be taken before ANOVA?

11·5   What are the implications involved in (a) accepting the null hypothesis; (b) rejecting the null hypothesis?

11·6   What is the relationship between least significant difference and the confidence limits on the mean?

11·7 What are the two rules for the orthogonality of a set of coefficients?

11·8 If a factor, tested and found to be significant, involves several degrees of freedom, why is it appropriate to subdivide the sum of squares associated with this factor?

11·9 If a significant factor has five levels, into how many components can the sum of squares be divided?

11·10 Is the numerical value of the coefficients used for making orthogonal comparisons of consequence? Why?

11·11 When we test a quantitative factor for a functional relationship, why must the intervals between the levels of the factor be equally spaced?

11·12 When we test a factor for a functional relationship, what conditions must be met (a) in the experiment; (b) in the set of coefficients used in the data analysis?

11·13 Under what conditions should we test for a flyer?

11·14 What is the probable basis of the test for flyers?

REFERENCES
25, 31, 35, 47, 59, 73, 77, 103, 104

12 Regression Analysis

We are often interested in a possible relationship between two or more variable quantities. We may suspect that when one of the quantities is changed, the other changes in some predictable manner. It is useful to express such a relationship in the form of a mathematical equation connecting the variables. If we can find the equation, we can then predict the value of one variable from a knowledge of the value of the other variable or variables.

12·1 REGRESSION LINES

Suppose that we have available two methods of testing a material for hardness. Method A is fast and inexpensive; method B is standard but is time-consuming and costly. We submit samples of the material to the two test methods and record the data, as in Table 12·1, for the pairs of measurements. We graph the data in Fig. 12·1, placing the data from method B on the X axis and those from method A on the Y axis.

We observe in Fig. 12·1 that the points tend to lie on a straight line. We may, for such data, draw by eye the line which appears to represent the data. From such a line we can obtain a reasonably good estimate of the expected result of the test method from a result found by method A and the line we have drawn. Thus, if we want a hardness of 325 by method B, we find that a value of about 78 should be obtained by method A.

Similarly, we may seek evidence of a relationship between fiber length and

Table 12·1	Hardness of Test Specimens as Measured by Two Different Test Methods

Test specimen	Method A	Method B
1	20	150
2	45	225
3	85	350
4	70	300
5	90	375
6	30	175
7	60	275
8	35	200
9	55	250

fabric strength, between the dosage of an analgesic drug and the alleviation of pain, between the temperature at which we run a process and the yield of the process. If we are successful in finding relationships between variables, such that a change in one variable is accompanied by a change in the other with a high degree of probability, we often obtain insight into the nature of the process. Furthermore, the detection of a relationship often eases control methods. We may find a relationship between the thickness of a paper and its resistance to breaking. If so, we may be able to substitute a simple, nondestructive measurement of thickness for a more difficult test of burst strength in the control of a process for manufacturing wrapping paper.

When we plot data as in Fig. 12·1, we call the line we use to represent the data a *regression line*. If, as in this case, we use the graph to estimate the value from method B (which we have plotted on the *Y* axis) from our knowledge of the value of method A (which we have plotted on the *X* axis) we call the line a line of regression of *Y on X*. Here we set the values of *X* and then measure the corresponding *Y*. Because these data seem to be fitted by a single line so well, we could also use this same graph to estimate values by method A if we had knowledge of the values by method B. In this latter

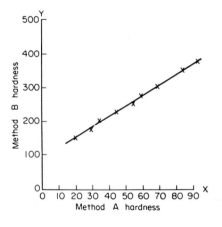

Fig. 12·1   Two tests of hardness, method A versus method B.

case, we would refer to the line as being the regression line of *X on Y*. Here we would set the value of *Y* and measure the corresponding value of *X*.

In Fig. 12·2 we have plotted data of a similar kind, but representing a situation in which we have obtained replicate values for the *X, Y* pairs. In Fig. 12·2a we have drawn a line approximating the "best fit" to the $\bar{Y}$ values. In Fig. 12·2b we have drawn a line approximating the "best fit" to the $\bar{X}$ values. We see that the lines differ in slope and intercept. This illustrates the concept of two different lines of regression. Even though the same data are

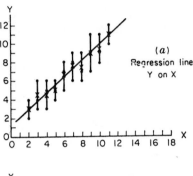

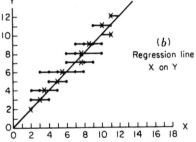

Fig. 12·2 Regression lines from given set of data, interchanging predictor and response variables.

plotted in each graph, the variability of the *X*'s and of the *Y*'s causes the lines of "best fit" to differ according to whether the averages of *Y* are plotted against fixed values of *X* (*Y* on *X*) or averages of *X* are plotted against fixed values of *Y* (*X* on *Y*).

Criteria for drawing a line by judgment are often stated as follows: (1) As many points as possible should fall on the line to be drawn. (2) As many points should fall above as below the line. (3) The deviations of the missed points should sum to zero, considering deviations above the line as positive and those below the line as negative. Such criteria are difficult to apply. Furthermore, they are insufficient. Many straight lines can be drawn which will meet the criteria of equal distribution of points and zero net deviation. For example, the net deviation of the mean values will be zero for any line whatever which passes through the point $(\bar{X}, \bar{Y})$, i.e., the point which lies at the mean of the two variables.

A more serious criticism of the "judgment" method of fitting a line to observed data is that it is difficult to test the extent to which the drawn line really represents the data. We have, as a better technique, an objective method of fitting lines to plotted data, known as the method of *least squares*. By this method, we define the line of "best fit" as that line which makes the *squares* of the deviations of the points from the line sum to a minimum.

PROBLEMS

12·1   The gloss of 10 different samples of paint was estimated by 2 different methods of rating. Use the data in Table 12·2.

Table 12·2   Rating of Gloss in Paints
by Two Different Rating Scales

Compound	Method A X	Method B Y
1	20	4.0
2	10	3.0
3	40	6.0
4	30	4.9
5	25	4.5
6	15	3.5
7	60	7.5
8	55	7.5
9	50	7.0
10	40	5.8

(a) Make a plot and draw by judgment the regression line of Y on X.
(b) Replot the data and draw by judgment the regression line of X on Y.
(c) Note the extent to which the two lines superimpose.
(d) On the X axis, plot a dot frequency distribution of the points, and find the mean and standard deviation.
(e) On the Y axis plot a dot frequency distribution of the points and find the mean and standard deviation. Compare the results with those in (d).

12·2   By interviewing 25 male persons, obtain their weight to the nearest pound and their height to the nearest inch.
(a) Plot the data, with weight as X and height as Y, and draw your estimate of the regression line.
(b) Replot the data with height as X and weight as Y and draw your estimate of the regression line.
(c) For each of the plots, assign + to deviations of the points above the

line and — to deviations of the points below the line. Find the algebraic sum of the deviations in both plots.

(d) What would it mean if the result in (c) above were zero? What would it mean if the result in (c) above were positive?

12·2 KINDS OF VARIABLES

When we plot graphs of experimental quantitative data, we are interested in (at least) two quantities which are subject to change, i.e., two or more variables. We usually can specify that the values of one or more of the variables are assignable at will; such variables we call *predictor* (or input) variables. Once we find the relationship connecting the predictor variable with the *response* (or output) variable, we can thereafter estimate the value of the response variable by knowing the value of the predictor variable.

If we test for the relationship between the strength of a metal bar and the temperature of the bar, we would in the experiment fix the levels of temperature and find the corresponding values of strength. In this case temperature is the predictor variable; strength is the response variable. If we find that the yield of a chemical product bears a relationship to the changing levels of four ingredients, we would have four predictor variables in the four ingredients; yield would be the response variable. If a relationship exists between the brittleness of metal and its elongation under stress, the distinction between predictor and response variables is determined by the situation, that is, which variable we wish to predict from the other.

It is conventional in plotting quantitative data to place the values of the predictor variable (if there is only one) on the horizontal axis of a two-dimensional plot, and the values of the response variable on the vertical axis. An important distinction between predictor and response variables is that in regression analysis we assume that the measurement of the predictor variable is without error. All error is attributable to the response variable.

12·3 REGRESSION MODELS

A line graph is more than merely a pictorial representation of data. To draw such a line represents our judgment that (1) a functional relationship really exists between the variables; (2) the line expresses this relationship in quantitative terms. Only if we draw a horizontal line do we say that no relationship exists between the variables and that Y is independent of the level of X.

If we perform regression analysis, the end result is a mathematical equation which describes the kind of relationship that exists between the variables we have tested. In ordinary regression, we deal with two variables, one predictor and one response. The resulting functional relation may plot as a straight line or a curved line. If it plots as a straight line, we have a *linear regression model*; if it plots as other than straight, we have a *curvilinear regression model*. If we are involved with more than one predictor variable, we may

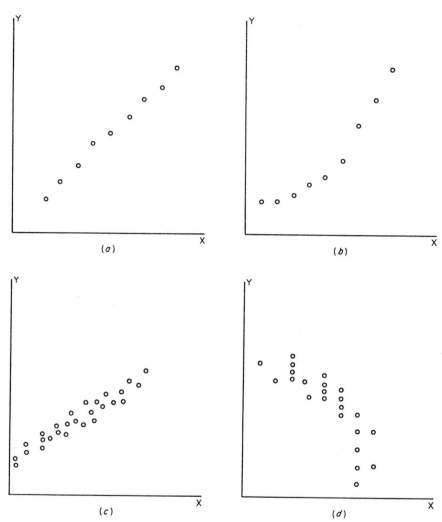

Fig. 12·3 Possible patterns in plots of the two-variable case.

have a multiple linear regression model or a multiple curvilinear regression model. In what follows we shall emphasize the two-variable case.

In Fig. 12·3 we show a few of the many possible patterns that may appear in plots of two-variable data. In (*a*) and (*b*) we have only one value of Y for each level of X. In (*c*) and (*d*) we have indicated that replicates were obtained for at least some levels of X. In a well-designed experiment, replicate values will be obtained to permit us to get an estimate of error.

Figure 12·4 indicates how a series of 21 experimental runs might be carried out in two different ways: in (*a*) the tests are made at numerous levels of X; in (*b*) only five levels of X were used. We will explain in the chapters

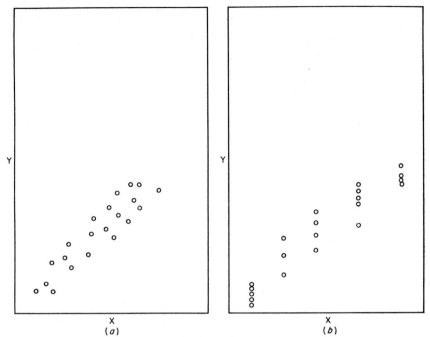

Fig. 12·4 Two designs for the two-variable case.

to follow (on experiment design) that the method which results in data such as in (*b*) is better. It is, however, sometimes necessary to use the former approach when we are working with industrial-plant data.

12·4 THE LINEAR REGRESSION MODEL

For the two-variable regression case, the correct approach is to make no hypothesis about the form of the relationship and to base a test for curvilinearity on the data. If we find no significant departure from a straight-line relationship, we can simplify the model to a linear one. Alternatively, we may have theoretical reasons for believing that the linear model is the appropriate one and restrict our analysis to this model. To avoid the complications in curvilinear regression analysis, we will begin with the linear model.

In the linear regression model, each set of paired data (X_i, Y_i) represents a fixed value (level) of X and a value of the response variable Y. The value of Y_i is dependent, if there is a line of regression, upon the level of X and the amount of random error present. In Fig. 12·5 we show a line of regression and a typical plotted point. We assume that the point is truly at the specified level of X. If the regression line is accurate, then that the point does not plot on the line indicates that error is present in the value of Y. The amount of error is the difference between the value $\hat{Y}$ (predicted from the line of regression) and the observed value of Y.

We assume that, for every level of X, the observed values of Y about the value of $\hat{Y}$ follow a normal distribution. Also, we assume that at every level of X the population of Y values about each $\hat{Y}$ has the same standard deviation $\sigma_{X.Y}$, the standard deviation of Y on X.

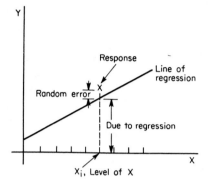

Fig. 12·5   Factors influencing location of the line of regression.

For each level of X we find the mean of the response values $\overline{Y}$ and find for these means a line of "best fit." We define this line of best fit as that line for which the sum of the squares of the deviations of the Y values is a minimum. This definition of the line of best fit is intuitively meaningful. Furthermore, it agrees with the concepts of ANOVA. We can use the sum of squares as a test of the extent to which the calculated line really represents the data. Actually, ANOVA is a short-cut least-squares analysis which we can use because of the balanced method of data collection.

If the $\overline{Y}$ values are linearly related to the X values, we hypothesize a "true" line of regression according to

$$Y = \beta_0 + \beta_1 X$$

where β_0 is the value of Y at the intercept, i.e., at the point where $X = 0$, and β_1 defines the contribution to the value of Y by the effect of the predictor variable X. For any single point X_i, Y_i there will be some contribution due to error, so that

$$Y_i = \beta_0 + \beta_1 X_i + \epsilon$$

The preceding equation is the mathematical model for linear regression analysis. It is based on the equation (from analytic geometry) for a straight line.

Since we have only sample data, we will estimate the coefficients in this equation by

$$\hat{Y} = b_0 + b_1 X$$

The task of linear regression analysis is that of determining the values of the coefficients b_0 and b_1 from the sample data, and evaluating the results by ANOVA for goodness of fit and error.

12·3   Plot the graph of the data in Table 12·3.

Table 12·3

Predictor X	Response Y
10	40
10	45
20	50
30	60
50	60
50	70
60	75
70	75
80	80
80	85

(a) By sight, draw a line of regression.
(b) Write the mathematical model that you believe to fit the data.
(c) List the amount of random error estimated at each plotted point.
(d) Square the error value found at each point in the graph in answer to (c) and sum the squared values.
(e) What is the proper name for the sum found in answer to part (d) above?

12·4   Plot the data in Table 12·4 and then answer parts (a) to (e) as given in Prob. 12·3.

Table 12·4

Predictor X	10	20	30	30	40	40	50	50	50	60	60	60
Response Y	20	25	25	30	30	35	40	45	50	70	80	90

12·5   Make sketches of the following:
(a) Y as a linear function of X
(b) R as a linear function of S
(c) R as a linear function of S and T
(d) Y as a quadratic function of X
(e) Z as a quadratic function of both X and Y

12·5   DATA COLLECTION FOR LINEAR REGRESSION ANALYSIS

We wish to determine the relationship between the quantity of a component (metol) actually present in a photographic developer and the quantity estimated by a prescribed method of chemical analysis. We believe that the probable relationship is a straight line.

We collect data by preparing samples with known amounts of metol and subjecting each of these samples to analysis. To estimate the variability of the analysis method, we test several samples at each known level, and make a total of 14 analyses. The results of the tests are shown in Table 12·5, where X_i symbolizes one of the chosen levels of the predictor variable (known amount of metol) and Y_i symbolizes the reported amount of the compound.

Table 12·5 Amount of Component Present and Found by a Chemical Analysis

X_i, Present	Y_i, Found
1	1.0
1	1.4
2	1.7
2	1.8
2	2.2
2	2.1
3	3.3
3	3.2
3	3.3
4	4.1
4	4.4
5	5.6
5	5.4
5	5.5

This problem is appropriate for regression analysis because we want to find the functional relationship between X and Y values, and because we assume no error in the X values. All the variability in the experiment associated with error is considered to be contained within the values of Y.

By making replicate runs at different levels of X, we will be able not only to estimate the error present in the tests but also to estimate the lack of fit. That is, we can estimate the extent to which the data indicate that our model (here, a linear one) was inappropriate. If no replicate runs are made at any level of X, we will not be able to test for lack of fit.

12·6 CALCULATION OF THE LINE OF LEAST SQUARES

We believe, on theoretical grounds, that the data in Table 12·5 are linearly related. The plot of these data, in Fig. 12·6, gives us no reason to disbelieve in the linearity of the data. We proceed to find the coefficients in the general equation for a straight line:

$$\hat{Y} = b_0 + b_1 X \tag{12·1}$$

where b_0 and b_1 are constants. We intend to find that single line (among many possible lines) which makes a minimum the sum of the squares of the deviations of the points from the line. What we do is first to use Eq. (12·1) to estimate the value of Y for each of the X values. Thus, for the first observation, X is 1; then, by Eq. (12·1), $\hat{Y} = b_0 + b_1(1)$. For the last point, $\hat{Y} = b_0 + b_1(5)$. We write an expression for the difference between the

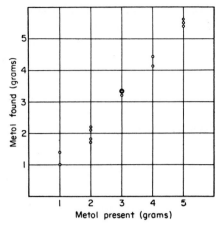

Fig. 12·6 Graph of amount of metol present versus metol found by analysis (data from Table 12·5).

actual value of Y from our sample data and the estimated value ($\hat{Y}$) of Y for this same X value found from Eq. (12·1). Thus, for the first point, the observed value of Y is 1.0, and the difference between this value and the estimated value is $1.0 - b_0 - b_1(1)$. We square all these differences, and now have the data for Table 12·6.

Table 12·6

Point	X_i	Y_i	$\hat{Y}_i$	$(Y_i - \hat{Y}_i)^2$
1	1	1.0	$b_0 + b_1$	$(1 - b_0 - b_1)^2$
2	1	1.4	$b_0 + b_1$	$(1.4 - b_0 - b_1)^2$
3	2	1.7	$b_0 + 2b_1$	$(1.7 - b_0 - 2b_1)^2$
4	2	1.8	$b_0 + 2b_1$	$(1.8 - b_0 - 2b_1)^2$
5	2	2.2	$b_0 + 2b_1$	$(2.2 - b_0 - 2b_1)^2$
6	2	2.1	$b_0 + 2b_1$	$(2.1 - b_0 - 2b_1)^2$
7	3	3.3	$b_0 + 3b_1$	$(3.3 - b_0 - 3b_1)^2$
8	3	3.2	$b_0 + 3b_1$	$(3.2 - b_0 - 3b_1)^2$
9	3	3.3	$b_0 + 3b_1$	$(3.3 - b_0 - 3b_1)^2$
10	4	4.1	$b_0 + 4b_1$	$(4.1 - b_0 - 4b_1)^2$
11	4	4.4	$b_0 + 4b_1$	$(4.4 - b_0 - 4b_1)^2$
12	5	5.6	$b_0 + 5b_1$	$(5.6 - b_0 - 5b_1)^2$
13	5	5.4	$b_0 + 5b_1$	$(5.4 - b_0 - 5b_1)^2$
14	5	5.5	$b_0 + 5b_1$	$(5.5 - b_0 - 5b_1)^2$

We want values for b_0 and b_1 that will make the sum of the squares of the deviations just found as small as possible. We find these values by taking the partial derivatives of the sum of squares with respect to b_0 and b_1:

$$\frac{\partial S}{\partial b_0} = -2(1 - b_0 + b_1) - 2(1.4 - b_0 - b_1) \cdots - 2(5.5) - b_0 - 5b_1)$$

$$\frac{\partial S}{\partial b_1} = -2(1 - b_0 - b_1) - 2(1.4 - b_0 - b_1) \cdots -2(5)(5.5 - b_0 - 5b_1)$$

We now set the partial derivatives equal to zero and solve simultaneously for b_0 and b_1. We find, from our example,

$$\frac{\partial S}{\partial b_0} = -90.0 + 28b_0 + 84b_1$$

$$\frac{\partial S}{\partial b_1} = -327.8 + 84b_0 + 304b_1$$

When we set the partial derivatives equal to zero and rearrange the terms we have

$$28b_0 + 84b_1 = 90$$

$$84b_0 + 304b_1 = 327.8$$

We multiply the first equation by 84/28 and subtract from the second to obtain

$$b_1 = 1.11$$

We substitute this value for b_1 in the first of the equations and find $b_0 = -0.12$. Hence the equation of the least-squares line for estimating Y from X is

$$\hat{Y} = -0.12 + 1.11X$$

We have now found the equation which is the line of best fit, in a least-squares sense, for the data in Table 12·6. We have drawn this line in Fig. 12·7. It is the regression line of Y on X.

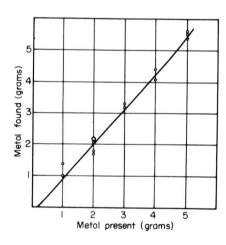

Fig. 12·7 Line of regression (data from Table 12·5).

12·7 LINEAR REGRESSION ANALYSIS—THE NORMAL EQUATIONS

We have used the calculus to find the coefficients for the line of regression. We now develop a computing method which is equivalent to the one we have just used. This method is useful for longhand computations and convenient when we use desk calculators.

In the general case,

$$\frac{\partial S}{\partial b_0} = -2 \sum (Y_i - b_0 - b_1 X) = -2(\sum Y_i - n b_0 - b_1 \sum X_i)$$

$$\frac{\partial S}{\partial b_1} = -2 \sum (X_i)(Y_i - b_0 - b_1 X_i) = -2(\sum X_i Y_i - b_0 \sum X_i - b_1 \sum X_i^2)$$

When we set each partial derivative equal to zero and divide by -2, we have

$$\sum Y_i - n b_0 - b_1 \sum X_i = 0$$

$$\sum X_i Y_i - b_0 \sum X_i - b_1 \sum X_i^2 = 0$$

By rearranging the terms, we have the *normal equations*

$$n b_0 + b_1 \sum X_i = \sum Y_i \tag{12·2}$$

$$b_0 \sum X_i + b_1 \sum X_i^2 = \sum X_i Y_i \tag{12·3}$$

Solving for b_0 and b_1 will produce

$$b_1 = \frac{n \sum X_i Y_i - (\sum X_i)(\sum Y_i)}{n \sum X_i^2 - (\sum X_i)^2} \tag{12·4}$$

$$b_0 = \frac{\sum X_i^2 \sum Y_i - \sum X_i \sum X_i Y_i}{n \sum X_i^2 - (\sum X_i)^2} \tag{12·5}$$

In Eq. (12·4) the numerator of the fraction is a measure of the relationship between Y and X. The denominator of the fraction is the sum of squares for X. The entire expression is a measure of the rate of change of Y with X and is thus an estimation of the slope of the line of best fit.

Equation (12·5) is equivalent to the statement that

$$b_0 = \bar{Y} - b_1 \bar{X} \tag{12·6}$$

Once we have found the value of b_1 by Eq. (12·4), we will use Eq. (12·6) to find the value of b_0.

From our experimental data we calculate by these equations the values of b_1 and b_0 in Table 12·7. We require the values of $\sum X_i$, $\sum Y_i$, $\sum X_i^2$, and $\sum X_i Y_i$. For the ANOVA to follow and for finding confidence limits on the calculated coefficients, we also need the value of $\sum Y_i^2$.

Substituting in Eq. (12·4), we have

$$b_1 = \frac{14(163.9) - 42 \times 45}{14(152) - 42^2} = 1.11$$

And by Eq. (12·6),

$$b_0 = \frac{45 - (1.11 \times 42)}{14} = -0.12$$

Table 12·7

Predictor, X_i	Response, Y_i	$X_i Y_i$	X_i^2	Y_i^2
1	1.0	1.0	1	1.00
1	1.4	1.4	1	1.96
2	1.7	3.4	4	2.89
2	1.8	3.6	4	3.24
2	2.2	4.4	4	4.84
2	2.1	4.2	4	4.41
3	3.3	9.9	9	10.89
3	3.2	9.6	9	10.24
3	3.3	9.9	9	10.89
4	4.1	16.4	16	16.81
4	4.4	17.6	16	19.36
5	5.6	28.0	25	31.36
5	5.4	27.0	25	29.16
5	5.5	27.5	25	30.25
42	45.0	163.9	152	177.30
ΣX_i	ΣY_i	$\Sigma X_i Y_i$	ΣX_i^2	ΣY_i^2

Thus the equation for the calculated line of regression is, as before, $\hat{Y} = -0.12 + 1.11X$.

If we code the values of X_i by subtracting $\bar{X}$ from each value of the predictor variable, then the sum of the X_i becomes zero, and the normal equations reduce to

$$b_1 = \frac{\sum X_i Y_i}{\sum X_i^2}$$

$$b_0 = \frac{\sum Y_i}{n} = \bar{Y}$$

Such a coding process is helpful, since it reduces the computing effort, especially in cases where large sets of data must be handled by pencil-and-paper methods.

PROBLEMS

(In each of these assume a linear model.)

12·6　An experiment was run to test for a relationship between the light level at an inspector's table and the number of defects the inspector found. The results were as in Table 12·8.

(a) Plot the data.

(b) Find and mark the average percent of defects at each light level; draw by sight the line of "best fit."

Table 12·8 Percentage of Defects Found by an Inspector at Varying Light Levels

Light level, foot-candles	Percentage of defects found					
50	82	79	80	83	80	81
75	87	83	85	84	86	86
100	89	90	87	87	89	90
125	89	92	91	91	91	92

(c) Calculate and draw in the line of best fit using the method of linear regression.

12·7   Use the data in Table 12·9.

Table 12·9

X	1	3	4	6	8	9	11	14
Y	1	2	4	4	5	7	8	9

(a) Plot the data.
(b) Sketch in the line of best fit freehand.
(c) Write the mathematical model for the line of best fit.
(d) Find the equation of the line of best fit.

12·8   Use the data in Table 12·10.

Table 12·10

Height, in.	70	63	72	60	66	70	74	65	62	67
Weight, lb	155	150	180	135	156	168	178	160	132	145

(a) Plot the data.
(b) Calculate the regression line of Y on X.
(c) Calculate the regression line of X on Y.
(d) Why do the two regression lines intersect at one point?

12·9   List five pairs of variables which you believe might have a functional relationship between them.
(a) For each pair identify which of the two variables is the predictor variable.
(b) Identify the type of functional relationship you believe might exist between each pair.

12·8 ANOVA FOR REGRESSION

Once we have calculated the coefficients for the line of regression, we must test to see whether or not the calculated equation describes, to a significant degree, a real relationship between the variables. We therefore run an analysis of variance on the data. We will be involved with only a single factor, the predictor variable. Our question is: Do the values of the response variable indicate that a genuine relationship exists between the levels of the predictor variable and the values of the response variable?

The null hypothesis is that the two variables Y and X are independent. We will accept the null hypothesis unless we have evidence that the value of the coefficient b_1 in Eq. (12·4) is other than zero. To reject the null hypothesis, and therefore to suppose that a real relationship exists, we must have evidence that b_1 is significantly different from zero.

There may be two different reasons for a failure to find a value other than zero for b_1: (1) Our mathematical model may be incorrect, and a relationship other than linear may exist between the variables. (2) The effect of random error may be large, making it impossible for us to detect the linear effect. To distinguish between these two possible reasons, we partition the total variability in Y (expressed as a sum of squares) into these divisions: (1) a sum of squares attributable to the value of the coefficient b_1, and associated with the relationship between the variables; (2) a sum of squares which estimates the effect of random error; (3) a sum of squares which estimates the extent to which the linear model is inappropriate.

By this analysis, we can answer these questions:

1 Does the model fit the data? To answer this question, we compare the mean square attributable to lack of fit with the error mean square.
2 If the answer to question 1 above is yes, then we ask, "Is there a real relationship between the variables?" We answer this question by comparing the mean square associated with the coefficient b_1 with the error mean square. (Note that if the answer to question 1 is no, we must assume a different model, no doubt a curvilinear one.)

In Table 12·11, we show the steps required for the ANOVA of the data from Table 12·7 and the resulting sums of squares and mean squares. In step 1, we have found the sum of the squares of the response variable about the origin. In step 2 we have found the value of the correction factor which is the contribution of $\overline{Y}$ about the origin. The value from step 3 is the total variability of the values of the response variable about the mean $\overline{Y}$. It is found by subtracting step 2 from step 1. It includes a possible variation in Y associated with the values of X (the regression relationship) and also any variability due to lack of fit and to error.

In step 4 we have calculated the variation in the Y values which is connected with changes in the levels of X. The difference, found in step 5, between the total sum of squares and the value of step 4 we call the residual

Table 12·11   Summary ANOVA for Regression

Steps	Source	Method	SS	v	MS
1	Crude SS (variability of Y about the origin)	$\Sigma\ Y_i^2$	177.300	14	—
2	SS due to b_0 (sum of squares of $\bar{Y}$ about the origin)	$\dfrac{(\Sigma\ Y_i)^2}{n}$	144.643	1	144.643
3	Total SS (total sum of squares of Y about $\bar{Y}$)	Step 1–step 2	32.657	13	
4	SS due to b_1 [sum of squares due to relationship of X and Y (regression)]	$\dfrac{b_1(n\ \Sigma\ X_i Y_i - \Sigma\ X_i \Sigma\ Y_i)}{n}$	32.079	1	32.079
5	SS residual (sum of squares still unexplained)	Step 3–step 4	0.578	12	0.048
6	SS error (sum of squares due to replication)	$\Sigma\ (Y_i - \bar{Y}_j)^2$	0.322	9	0.036
7	SS lack of fit (sum of squares still unexplained)	Step 5–step 6	0.256	3	0.085

sum of squares. This value contains the sum of squares due to error and to lack of fit. Note that the sums required for the value in step 4 were required for computing the value of b_1 by Eq. (12·4).

In step 6 we show the sum of squares for error, calculated as in Table 12·12. By subtracting this value from the residual sum of squares, we now have the value for the sum of squares due to lack of fit of the model since all other factors have been accounted for.

We had a total of $n = 14$ degrees of freedom in this experiment. We show in the ANOVA Table 12·11 that one degree of freedom is associated with the calculation of each of the two coefficients for the line of regression. Therefore, a total of 12 degrees of freedom remains for lack of fit and for error. The value of v for error we find in Table 12·12; for each set of observations obtained at a given level of X we subtract 1 from the value of n for that set. We find that a total of 9 degrees of freedom is associated with error. Therefore, a total of $v = 12 - 9 = 3$ remains for a test of lack of fit to the

Table 12·12 Calculation of Sum of Squares for Error in Regression

X_i	Y_i	$\overline{Y}_i$	$Y_i - \overline{Y}$	$(Y_i - \overline{Y}_i)^2$	D.F.
1	1.0	1.20	−0.20	0.0400	1
1	1.4		0.20	0.0400	
2	1.7		−0.25	0.0625	
2	1.8	1.95	−0.15	0.0225	3
2	2.2		0.25	0.0625	
2	2.1		0.15	0.0225	
3	3.3		0.0333	0.0011	
3	3.2	3.2667	−0.0667	0.0044	2
3	3.3		0.0333	0.0011	
4	4.1	4.25	−0.15	0.0225	1
4	4.4		0.15	0.0225	
5	5.6		0.10	0.0100	
5	5.4	5.5	−0.10	0.0100	2
5	5.5		0.00	0.0000	
Total				0.3216	9

SSE $= \Sigma (Y_i - \overline{Y}_i)^2 = 0.3216$ with 9 degrees of freedom

linear model. Table 12·7 shows that X was measured at five points. Since only two points are required to draw a straight line, it is reasonable that the other three would permit testing of a more complex model. If the latter is required it will be revealed by a significant "lack of fit" term.

We test first for lack of fit. To do so, we compare the MS for lack of fit to the MS for error. The null hypothesis we are now testing is $H_0: \sigma_{LF}^2 \leq \sigma_e^2$; the alternative hypothesis is $H_1: \sigma_{LF}^2 > \sigma_e^2$. To test these hypotheses we find an F ratio of the corresponding MS terms at an α risk of 0.05:

$$F = \frac{0.085}{0.036} = 2.36$$

The table value for $F_{3,9,0.05}$ is 3.86. We accept the null hypothesis and conclude that our data give us no reason to reject the linear model for this situation.

Now we test the MS for regression against the MS for error and obtain a ratio of 891. The table value of $F_{1,9,0.05}$ is 5.12. Thus we have good reason to accept the hypothesis that b_1 is real; i.e., we accept the hypothesis that the value of the response variable Y is genuinely related to the level of the predictor variable X.

We can approximate the fraction of the total variation in the values of Y that is accounted for by the line of regression by the value

$$r^2 = \frac{SS(b_1)}{SST} = \frac{32.079}{32.657} = 98\%$$

Thus about 98 percent of the total variation in the values of the response variable is attributable to the calculated regression relationship.

PROBLEMS

12·10   Using the data in Table 12·3, construct an ANOVA table as it would be used to test for linear regression. Is there evidence of a functional relationship in the data, using a level of significance of 0.05?

12·11   Test, at a level of significance of 0.10, the line of regression which was obtained from the data in Table 12·4.

12·12   Using the regression line of Y on X as computed from the data in Table 12·8, perform the analysis of variance to see if there is evidence (at the 0.05 level of significance) of a functional relationship as indicated by the line of best fit.

12·9 CONFIDENCE LIMITS FOR REGRESSION ANALYSIS

Once we have calculated the regression line from a set of data, we can use the equation to predict the value of the response variable $\hat{Y}$ for any assigned value of the predictor variable X_0. We would thus have a point estimate of the value of $\hat{Y}$.

As for other point estimates, the worth of the prediction is inversely related to the uncertainty in the data. Therefore, we should place confidence limits on the prediction. In addition, we should place confidence limits on the values of the regression coefficients themselves. The relationship between Y and X is best expressed as a band within which the values probably lie.

As we have said, in regression analysis we attribute all the uncertainty to the values of the response variable. Thus the standard error of the estimates we make from the line of regression is $s_{Y.X}$, the estimate of the standard deviation of the values of Y about X. This value is a measure of the variation in a set of calculated values of Y that is unexplained by the regression equation. Thus

$$s_{Y.X} = \sqrt{\frac{\sum (Y_i - \hat{Y}_i)^2}{n-2}} \tag{12·7}$$

based on the differences between the actual values Y_i and the calculated values $\hat{Y}_i$ for all the X_i we used. There are $n-2$ degrees of freedom because two parameters have been estimated in the regression equation. Since every $\hat{Y}$ comes from the regression equation,

$$\hat{Y}_i = b_0 + b_1 X_i$$

Eq. (12·7) may be rewritten

$$s_{Y.X} = \sqrt{\frac{\sum Y_i^2 - b_0 \sum Y_i - b_1 \sum X_i Y_i}{n - 2}} \qquad (12·8)$$

In Eq. (12·8) the expression under the radical is exactly the residual mean square which we found in the ANOVA. Thus

$$s_{Y.X} = \sqrt{MS(\text{residual})}$$

For the data in Table 12·11 we find that $s_{Y.X} = \sqrt{0.048} = 0.22$.

Confidence limits are based on the value of $s_{Y.X}$ and on the t_ν distribution for the number of degrees of freedom for the residual $(n - 2)$ and the desired significance level. Thus, for the value of $\hat{Y}$ at any desired X_0, the confidence limits are

$$\hat{Y} \pm t_{n-2,\alpha/2} s_{Y.X} \sqrt{\frac{1}{n} + \frac{(X_0 - \bar{X})^2}{\sum (X_i - \bar{X})^2}} \qquad (12·9)$$

For the data in Table 12·5, to place 95 percent confidence limits on the estimate of the response variable at any X_0, we have the table value for $t_{12,0.025} = 2.179$; $s_{Y.X} = 0.22$; $\bar{X} = \sum X_i/n$ or 3.00; $\sum (X_i - \bar{X})^2 = \sum X_i^2 - (\sum X_i)^2/n = 26$. If we select $X_0 = 2$ and make the substitutions in Eq. (12·9), we have

$$\hat{Y} \pm (2.179)(0.22) \sqrt{\frac{1}{14} + \frac{(2 - 3.00)^2}{26}}$$

Therefore, we obtain, as confidence limits at $X = 2$,

$$\hat{Y} \pm 0.16$$

Since $\hat{Y}$ at $X = 2$ is $-0.12 + 1.11(2)$, or 2.10, $\hat{Y}$ at $X = 2$ lies between the values 2.26 and 1.94 with 95 percent confidence. If we choose another value of X_0, say 5, and find the confidence limits at this point, we find that they are

$$\hat{Y} \pm 0.23$$

Thus, as the chosen value of X_0 is farther from the mean, the confidence band becomes wider.

We can also place confidence limits on the coefficients of the regression equation, as follows:
For β_1

$$b_1 \pm \frac{t_{n-2,\alpha/2} s_{Y.X}}{\sqrt{\sum (X_i - \bar{X})^2}}$$

For the intercept β_0

$$b_0 \pm t_{n-2,\alpha/2} \frac{s_{Y.X}}{\sqrt{n}}$$

The intercept is not dependent on the distance of X_i from X, while the slope is.

Note that in the confidence-limit formulas, three terms are common to all: they are the t_v value, the value of $s_{Y.X}$, and the sum of squares of X. Thus the confidence limits for the regression equation coefficients are

$$\beta_1 = 1.11 \pm \frac{2.179(0.22)}{\sqrt{26}} = 1.11 \pm 0.078$$

$$\beta_0 = -0.12 \pm \frac{(2.179)(0.22)}{\sqrt{14}} = -0.12 \pm 0.126$$

In Fig. 12·8 we show the data previously plotted in Fig. 12·7, with the confidence limits we have just found.

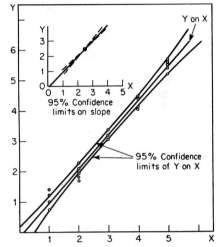

Fig. 12·8   Line of regression, with confidence limits.

PROBLEMS

12·13   In a single-factor experiment a quantitative factor was tested at four levels with the results of all runs as found in Table 12·13.

Table 12·13

Factor level	Replications				
2	159	148	152	152	139
4	241	247	248	256	264
6	335	347	351	339	350
8	443	453	465	460	436

(a) Make a graph of the results, and indicate the location of the average response per level.

(b) Calculate the line of best fit.

(c) Calculate, and graph, the 95 percent confidence limits on the line of regression.

12·14 In Table 12·14 are given the results of an experiment in which the weight of a chemical precipitate in grams (Y) is associated with the weight of a reagent (X) used in the preparation of the precipitate.

Table 12·14 Weight of a Precipitate in Grams (Y)
and the Weight of a Reagent (X)

Y	X	Y	X	Y	X	Y	X
9.4	8.2	9.3	7.1	14.2	8.6	6.5	5.8
10.6	7.8	13.8	9.0	7.3	6.4	11.4	8.1
11.3	8.1	7.9	5.9	12.2	7.5	8.1	5.7
9.0	6.5	13.3	7.9	6.1	5.3	10.8	7.7

(a) Plot the data, and draw a freehand line of best fit for the regression line Y on X.

(b) Calculate the equation of the line of best fit by the method of least squares. Plot the graph of this equation and compare with the line drawn in answer to part (a).

(c) If the weight of the reagent used is 8.0, what would be the value of the point estimate of the response value Y?

(d) Place 95 percent confidence intervals on the point estimate in answer to part (c) above.

(e) Place 95 percent confidence intervals on the coefficients of the regression equation found in answer to part (b).

12·15 Using the data from Prob. 12·14, place 90 percent confidence intervals on:

(a) the point estimate of Y at $X_0 = 8.0$

(b) the coefficients of the regression equation

12·10 MULTIPLE LINEAR REGRESSION

We can extend the concepts of this chapter to situations in which a linear relationship holds, but in which there is more than one predictor variable. For example, in a chemical analysis, the amount of one component found may be affected by (1) the amount of the component that is present and (2) the amount of a second component. Thus we have here two input variables.

The mathematical model for such a case is

$$Y = \beta_0 + \beta_1 X_1 + \beta_2 X_2 + \epsilon$$

where Y = value of the response variable
β_0 = value associated with the general level of Y
X_1 and X_2 = predictor variables
β_1 and β_2 = two slopes
The plot of such an equation is a plane, as suggested by Fig. 12·9.

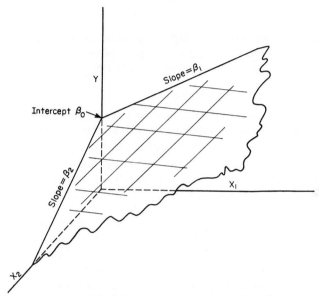

Fig. 12·9   Multiple linear regression, plane of best fit.

To estimate the values of the three coefficients in the model, we obtain from experiments in which we set the levels of X_1 and X_2 the corresponding values of the response variable.

The equation for which we calculate the coefficients is

$$\hat{Y}_i = b_0 + b_1(X_1)_i + b_2(X_2)_i$$

We can find the values of the coefficients by the simultaneous solution of three equations:

$$\sum Y_i = b_0 n + b_1 \sum (X_1)_i + b_2 \sum (X_2)_i$$

$$\sum (X_1)_i Y_i = b_0 \sum (X_1)_i + b_1 \sum (X_1)_i{}^2 + b_2 \sum (X_1)_i (X_2)_i$$

$$\sum (X_2)_i Y_i = b_0 \sum (X_2)_i + b_1 \sum (X_1)_i (X_2)_i + b_2 \sum (X_2)_i{}^2$$

The first of these three equations sums all the observations; the second equation is the first multiplied through by the values of X_1; the third equation is the first multiplied through by the values of X_2.

If by coding we make $\Sigma\ X_1 = 0$ and also $\Sigma\ X_2 = 0$, the equations simplify to

$$b_0 = \overline{Y}$$

$$b_1 = \frac{\Sigma\ (X_1)_i Y_i}{\Sigma\ (X_1)_i^2}$$

$$b_2 = \frac{\Sigma\ (X_2)_i Y_i}{\Sigma\ (X_2)_i^2}$$

Multiple regression analysis is especially useful in industrial experimentation in which it is risky to test for the effect of only one factor at a time. Details of the calculating methods are given in Chap. 17.

12·11 CURVILINEAR REGRESSION

As we said earlier in this chapter, unless we have good reasons for assuming linearity, we should, for the single-predictor-variable case, proceed with the regression calculations so as to test for curvilinearity. (In Chap. 11 we showed how to use ANOVA to test for linear, quadratic, and cubic patterns.)

The mathematical model for a second-degree equation is

$$Y = \beta_0 + \beta_1 X + \beta_2 X^2$$

For this model, although we have a single input variable, we can treat the data as if there were two predictor variables—one predictor variable being X and the other X^2. Thus X^2 takes the place of the second input variable X_2 in a multiple regression analysis similar to that used for the two-predictor-variable case. The equation for which we find coefficients is

$$\hat{Y}_i = b_0 + b_1 X_i + b_2 X_i^2$$

and the three equations to be solved simultaneously are

$$\sum Y_i = b_0 n + b_1 \sum X_i + b_2 \sum X_i^2$$
$$\sum X_i Y_i = b_0 \sum X_i + b_1 \sum X_i^2 + b_2 \sum X_i^3$$
$$\sum X_i^2 Y_i = b_0 \sum X_i^2 + b_1 \sum X_i^3 + b_2 \sum X_i^4$$

Details of the calculations are found in Chap. 17.

If the results of the calculations of the coefficients indicate that β_2 is not significantly different from zero, and that β_1 is significantly different from zero, we have no evidence of curvilinearity and can work with the simpler linear model. To test for curvilinearity is better than to assume a straight-line relationship.

Note that the mathematical model above assumes that a parabola is the curvilinear pattern. Such an assumption is risky, inasmuch as the real pattern may be an exponential or other curvilinear one.

To show what happens in a regression analysis if we assume linearity in a situation where a curvilinear relationship exists, we add to the data in Table 12·5 additional values (Table 12·15) found from analysis of greater

Table 12·15   Data Added to Those in Table 12·6

X_i, Present	Y_i, Found
6	10.0
6	10.1
6	10.5
7	14.0
7	14.2
7	14.4

amounts of the component being tested for. Now, for the original plus the added data, we have

$$n = 20$$
$$\sum X_i = 81$$
$$\sum Y_i = 118.2$$
$$\sum X_i^2 = 407$$
$$\sum Y_i^2 = 1,094.56$$
$$\sum X_i Y_i = 645.7$$

Substituting in Eqs. (12·4) and (12·5), we find that

$$b_1 = 2.115$$
$$b_0 = -2.656$$

and thus the least-squares straight-line equation that best fits these data is

$$\hat{Y} = -2.656 + 2.115X$$

In Table 12·16 we show the worked-out ANOVA for this set of data. When

Table 12·16   Analysis of Variance for Data in Tables 12·6 and 12·15

Source	SS	ν	MS
Crude sum	1,094.560	20	
SS b_0	698.562	·1	
Total SS	395.998	19	
SS b_1	353.184	1	
SS residual	42.814	18	
SS error	0.5416	13	0.042
SS lack of fit	42.272	5	8.454

we test the null hypothesis for lack of fit, i.e., $H_0\colon \sigma_{LF}^2 \le \sigma_e^2$, the F ratio for the MS terms is

$$F = \frac{\text{MS(lack of fit)}}{\text{MS(error)}} = \frac{8.454}{0.042} = 201$$

The occurrence of such a large F ratio compels us to reject the null hypothesis and to conclude that we have used a model which the data do not fit. Note the plotted data in Fig. 12·10.

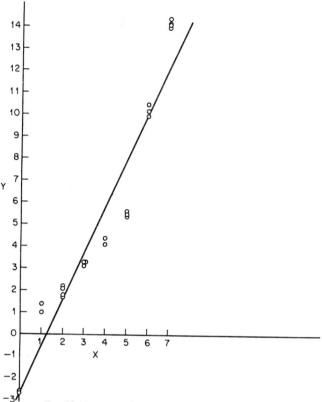

Fig. 12·10   Line of regression, showing lack of fit.

12·12   CORRELATION

We have so far in this chapter been considering the problem of regression, which involves the estimation of the value of one variable from knowledge of the value of another variable. We now discuss a different kind of problem in which we wish merely to measure the extent to which two variables are related. This measure is the correlation coefficient r.

In regression analysis the result is an equation. In correlation, the result is a number which indicates the degree of association of two variables (more than two in multiple correlation). Experiments testing the relationship between temperature T and yield of a process Y might give from a regression analysis an equation such as $Y = 0.08T^2 + 32.0$. To the extent that the data fit this equation we can predict the value of the yield from a measurement of temperature. If, on the other hand, we want only to measure the extent to which temperature and yield are related, we would find the correlation coefficient r to be, perhaps, 0.70. The value of r is a measure of the degree to which changes in the level of one variable are associated with changes in the level of the other variable.

In correlation calculations, we need not specify which of the variables is the predictor variable, since we are not interested in prediction in such problems. Furthermore, we need make no assumptions about the nature of the relationship between the variables (if one exists).

12·13 THE CORRELATION COEFFICIENT

The value of the correlation coefficient r lies between $+1$ and -1; If we find r to be nearly $+1$, it means that as one variable increases the other also increases, as in Fig. 12·11a. If we find the value of r to be nearly -1, it means that as one variable increases the other decreases, as in Fig. 12·11b. The

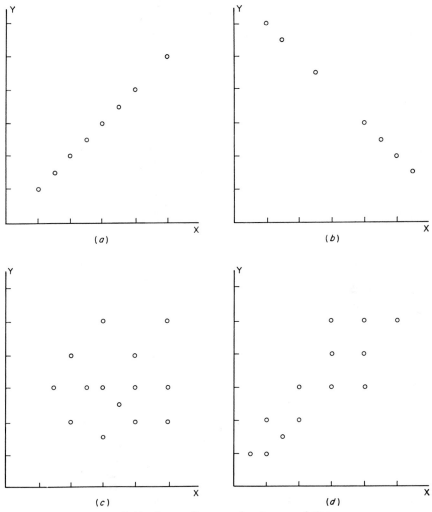

Fig. 12·11   Scatter diagrams of various correlations.

value of r will be near zero for data such as in Fig. 12·11c. Data which plot as in Fig. 12·11d would give a fraction for r.

The value of r^2 indicates approximately the percentage of the total variability in the data that is accounted for by a relationship between the variables. Thus, if $r = 0.70$, $r^2 = 0.49$. If we find such a value for r^2, we can say that only about half of all the variation in the data is attributable to a relationship between the variables. The rest of the variation is associated with error or untested factors.

We calculate the value of r by finding the ratio of two terms

$$r = \frac{\sum (X_i - \bar{X})(Y_i - \bar{Y})}{\sqrt{\sum (X_i - \bar{X})^2 \sum (Y_i - \bar{Y})^2}} \tag{12·10}$$

The numerator of the fraction in Eq. (12·10) is a measure of the variation associated with the relationship between the values of X and Y. The denominator of the fraction is the square root of the product of the X and the Y sum of squares. Thus r is the ratio (to the total variation) of the variation in the data connected with the changes of X with Y. If we divide the numerator and denominator by $n - 1$ we find

$$r = \frac{\sum (X_i - \bar{X})(Y_i - \bar{Y})/(n - 1)}{\sqrt{[\sum (X_i - \bar{X})^2/(n - 1)][\sum (Y_i - \bar{Y})^2/(n - 1)]}}$$

$$= \frac{\text{covariance } (XY)}{\sqrt{(\text{variance } X)(\text{variance } Y)}}$$

Thus

$$r = \frac{\sigma_{XY}}{\sigma_X \sigma_Y}$$

For calculating purposes, the terms in Eq. (12·10) are better transformed. We recall that

$$\sum (X_i - \bar{X})^2 = \sum X_i^2 - \frac{(\sum X_i)^2}{n} \tag{12·11}$$

Similarly,

$$\sum (Y_i - \bar{Y})^2 = \sum Y_i^2 - \frac{(\sum Y_i^2)}{n} \tag{12·12}$$

By analogy,

$$\sum (X_i - \bar{X})(Y_i - \bar{Y}) = \sum X_i Y_i - \frac{\sum X_i \sum Y_i}{n} \tag{12·13}$$

In each of these three expressions, we find a total sum of squares and subtract a correction factor, exactly as in the computations for ANOVA. To find the value of r from a set of paired observations, we first tabulate X_i, Y_i, X_i^2, Y_i^2, and $X_i Y_i$ and the sums of these sets. From these values and from $(\sum X_i)^2$ and $(\sum Y_i)^2$ we find the coefficient of correlation by Eq. (12·10).

12·14 CALCULATION OF r

In Table 12·17 the observations came from two different methods of estimating the potency of a hormone by biological techniques. We have made the computations preliminary to the calculation of r.

Table 12·17 Calculation of Correlation Coefficient

X_i	Y_i	X_i^2	Y_i^2	X_iY_i
8.2	9.4	67.24	88.36	77.08
5.8	6.5	33.64	42.25	37.70
6.4	7.3	40.96	53.29	46.72
5.9	7.9	34.81	62.41	46.61
6.5	9.0	42.25	81.00	58.50
7.1	9.3	50.41	86.49	66.03
7.8	10.6	60.84	112.36	82.68
8.1	11.4	65.61	129.96	92.34
7.5	12.2	56.25	148.84	91.50
7.9	13.3	62.41	176.89	105.07
8.6	14.2	73.96	201.64	122.12
9.0	13.8	81.00	190.44	124.20
8.1	11.3	65.61	127.69	91.53
5.7	8.1	32.49	65.61	46.17
5.3	6.1	28.09	37.21	32.33
$\Sigma X_i = 107.9$	$\Sigma Y_i = 150.4$	$\Sigma X_i^2 = 795.57$	$\Sigma Y_i^2 = 1{,}604.44$	$\Sigma X_iY_i = 1{,}120.58$

By Eq. (12·13),

$$\Sigma(X_i - \bar{X})(Y_i - \bar{Y}) = 1{,}120.58 - 107.9 \times \frac{150.4}{15} = 38.70$$

By Eq. (12·11),

$$\Sigma(X_i - \bar{X})^2 = 795.57 - \frac{107.9^2}{15} = 19.41$$

By Eq. (12.12),

$$\Sigma(Y_i - \bar{Y})^2 = 1{,}604.44 - \frac{150.4^2}{15}$$
$$= 96.43$$

And now, using the results of these three calculations,

$$r = \frac{38.70}{\sqrt{19.41 \times 96.43}}$$
$$= 0.895$$

The value of the correlation coefficient we have found indicates that there is a very pronounced positive corelationship between the two methods of estimating the hormone potency. The square of this value is 0.80, which implies that the association between the two variables accounts for 80 percent of the total variability of the data, leaving only 20 percent for error and for other sources of variation. We have plotted the observations in Fig. 12·12.

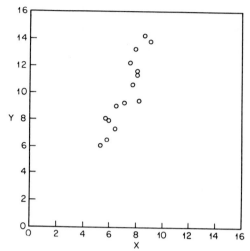

Fig. 12·12   Scatter diagram, data from Table 12·17.

12·15   TEST OF HYPOTHESIS FOR r

For small sample sizes, and where the calculated value of r is not close to ± 1, we may be uncertain whether or not we have demonstrated a real relationship between the tested variables. Table A·5 provides the basis for a test of hypothesis for the significance of r.

For various α risks (shown in the top row) and for various values of $\nu = (n - 2)$ (shown in the left-hand column) the body of the table gives that value of r which must be exceeded by the calculated value. The example used in this chapter involves a sample size of 15; therefore, $\nu = 13$. The calculated value of r exceeds all those given in the table for this sample size. Hence we can say that we run a risk of error (α risk) of less than 5 percent if we assume on the basis of this calculation that a real corelationship exists.

12·16   CONFIDENCE LIMITS FOR r

We can place confidence limits on the value of r by the use of Table A·9. The chart gives two curves for each of many different values of sample size. Interpolation may be used for sample sizes not included. The chart is applicable only for an assumed α risk of 0.05.

We enter the chart on the horizontal axis at the calculated value of r. On the vertical axis we read two other values of r, one from each of the two curves for the sample size. These two numbers are the confidence limits between which the true correlation coefficient lies to a probability of 0.95.

For the value of r calculated in the example above (0.895, sample size 15) the confidence limits are 0.71 and 0.91 approximately.

12·17 CORRELATION AND SAMPLE SIZE

Values of r lying between the extremes of 0 (no demonstrated correlation) and ± 1 (perfect correlation) are difficult to interpret. Specifically, a value of $r = 0.40$ would have very little meaning if obtained from a small sample size but might be a strong indication of corelationship if obtained from a large sample.

When we say that there exists a high degree of correlation between two variables, we usually mean that the value of r turned out to be a larger value than we expected, based on our experience. In part, this decision is based on our reasonable expectation of what the value of r is likely to be in various circumstances. This, in turn, is related to the sampling distribution of r, which involves this question: If we *know* the value of the correlation coefficient what values for r are we likely to find?

The sampling distribution for r is found to vary with the known value in this manner: If there is no correlation, and the known value of the coefficient is therefore 0, a set of samples from the population (all of the same large sample size) will give a set of r values that is nearly normally distributed. As the known value of the coefficient approaches either -1 or $+1$, the distribution of sample values of r will be more and more skewed. The skewness is negative (to the left) as the true coefficient approaches $+1$, and positive (to the right) as the value of the true coefficient approaches -1.

When there is no correlation between the variables, the near-normal sampling distribution of r has a standard deviation of

$$s_r = \frac{1}{\sqrt{n-1}}$$

Thus the observed values of r will cluster about the value 0; there will be some negative and some positive values of r, with frequency varying with the sample size. If the sample size is 10, the standard deviation of the distribution of the observed r values is

$$s_r = \frac{1}{\sqrt{10-1}} = \frac{1}{3} = 0.33$$

This value implies that, even with no correlation between the variables, we would expect to find a value of r with absolute value of 0 to 0.33 about 68 percent of the time; 32 percent of the time we would expect to find a value of r even larger than 0.33. It is for this reason that even moderately large values of r with small sample sizes cannot be considered as real. When we choose an α risk of 0.05, we must find a value of r greater in absolute value than 1.96 standard deviations above or below the true value (here assumed to be 0) before we can conclude that we have demonstrated a genuine corelationship.

If the sample size is 100,

$$s_r = \frac{1}{\sqrt{100 - 1}} = 0.100$$

Now, with a larger sample size, we can attach real significance to a much smaller observed value of r (still supposing that the assumed value is 0).

When the expected value of the coefficient is other than 0, the distribution is distinctly different from normal, and the same meaning no longer attaches to the value of the standard deviation of the sampling distribution. It is for this reason that Table A·5 is provided. It should always be consulted before one concludes that a corelationship exists.

In correlation problems, the hypothesis under test is again a null hypothesis. It is initially to be assumed that no correlation is present. Only after we have calculated a value of r and found it to be larger than the Table A·5 value at the indicated level of significance are we prepared to state that we have an indication of a significant correlation. We then should say that the correlation coefficient is other than zero, and specify the level of significance we have used.

12·18 SUMMARY OF CORRELATION TECHNIQUES

These are the steps in the kind of situation where we are concerned with two variables:

1 Find the sample size. To do this, we must specify the value of r which we need to find and the level of significance we wish to use. The required value of n is usually astonishingly large. In general, the larger the value of r we are seeking, the smaller will be the sample size. If we use a small sample we may easily miss a small but real correlation.
2 Take samples at random from the population, and record the data. It is often helpful to plot the observations.
3 Compute the value of r.
4 Test the calculated value of r against the value in Table A·5. If the null hypothesis is rejected, place confidence limits on the value of r, based on Table A·9.

PROBLEMS

12·16 From Table A·9, what sample size is needed to detect:
(a) $r = 0.70 + 0.10$
(b) $r = 0.40 \pm 0.10$
(c) $r = 0.70 \pm 0.20$
What inferences do you make from these answers?

12·17   Ten works of art were ranked by two judges with respect to quality; each assigned 1 to the best work and 10 to the poorest. The results were as given in Table 12·18.

Table 12·18   Ranks Assigned by 2 Judges to 10 Works of Art

Judge	Work									
	A	B	C	D	E	F	G	H	I	J
I	7	3	9	2	1	5	6	10	8	4
II	5	3	7	1	2	4	8	10	9	6

(a) Prepare a scattergram of the data. Is there an indication of correlation? Is it positive or negative?
(b) Calculate the correlation coefficient.
(c) At an α risk of 0.05, is there an indication of a statistically significant correlation?
(d) Place 95 percent confidence limits on the value of r.

12·18   Estimate and record your judgment of the length of a variety of objects. Measure the length of each one, and record. Plot a scattergram of the data. If there is an indication of correlation, calculate r, and place confidence limits on the value.

12·19   Stand 10 ft from a line on a floor or other flat surface. Toss one penny at a time at the line. Record the deviation of each attempt from the line. Is there a correlation between the sequence of the toss and the deviation from the line; i.e., do you learn by experience in this situation?

12·20   From among your acquaintances, select a group of about the same age. Record their weights (to the nearest pound) and their heights (to the nearest inch). Determine whether or not a statistically significant correlation exists for the two variables. State the hypothesis that is under test, and give the alternative.

12·21   From a sample size of 10, the following sums were calculated in a correlation problem:

$$\sum X = 71, \sum Y = 70, \sum X^2 = 555, \sum Y^2 = 526, \sum XY = 527$$

(a) Find the value of r.
(b) Is there evidence of a statistically significant correlation?
(c) Place 95 percent confidence limits on the value of r.
(d) Is it possible to sketch the scattergram from the information given about this situation? Why?

12·22   From a randomly selected sample of size 20, the following sums were calculated in a correlation problem:

$$\sum X = 210, \sum Y = 1611, \sum X^2 = 2,266, \sum Y^2 = 133,616, \sum XY = 17,366$$

What is your estimate of the value of the correlation coefficient for the population from which the sample was taken?

QUESTIONS

12·1   Under what circumstances is regression analysis an appropriate method of data analysis?

12·2   What is the mathematical model for:
(a) a two-variable linear regression analysis
(b) a three-variable linear regression analysis
(c) a two-factor curvilinear regression analysis

12·3   What are the differences between a problem suitable for correlation analysis and one suitable for regression analysis?

12·4   What assumptions are made in a two-variable linear regression analysis?

12·5   After a correlation analysis has been found to be significant, what information do we now have?

12·6   After a linear regression analysis has found a significant effect, what information do we now have?

12·7   Once a linear regression equation has been found, how do we test for significance?

12·8   Explain why, for any line drawn through the point $(\bar{X}, \bar{Y})$, the sum of the deviations of the points from the line will be zero.

12·9   Define the phrase "line of best fit."

12·10   What is the significance of the two coefficients found in a two-variable linear regression analysis?

12·11   Why must ANOVA follow the determination of the regression equation?

12·12   What is meant by "the regression sum of squares"?

12·13   Once ANOVA indicates that significance may be attached to a regression equation, why must we place confidence limits on the equation?

12·14   What factors determine the separation of the confidence limits placed on a line of regression?

12·15   How does multiple linear regression differ from the two-variable case?

12·16   Indicate for the following situations the kind of correlation you would expect to find: positive, negative, or none:
(a) the age of a car and its trade-in value
(b) the annual snowfall in inches and the death rate from heart disease
(c) years of education and annual income
(d) altitude above the earth's surface and air pressure
(e) success in love and luck in card playing
(f) years of experience and typing errors

12·17   List five pairs of variables that your experience suggests are correlated; for each pair indicate the kind of correlation that probably exists.

12·18   If a sample size of 2 is used in a correlation problem, what value of r will result? Why?

12·19   Is there a relationship between the inclination of the major axis of the ellipse enclosing the points in a scattergram and the value of r? Discuss.

12·20   If grouped data are used in preparing a scattergram, how will the plot look, as compared with one prepared from individual observations?

12·21   Why is it possible to obtain a finite value for r when in fact no correlation exists?

12·22   Why should we always test the null hypothesis in correlation problems?

12·23   How does the sample size influence the interpretation of a given value of r?

12·24   Once we have calculated the value of r, what precisely do we know?

REFERENCES
5, 10, 16, 23, 27, 30, 31, 35, 37, 44, 46, 47, 68, 71, 73, 78, 81, 82, 95, 97, 106, 123, 124

13 Planning Experiments for Statistical Analysis

Experiments need to be carefully planned so that, at the conclusion of the work, answers can be given to the problems which the experiment was intended to solve. We want to avoid the embarrassment of having accumulated a mass of data without having foreseen how the data were to be analyzed. We want to use available manpower and facilities in an economical manner.

Planned experiments are those for which we can formulate objectives with reasonable clarity, and for which we can foresee, at least in outline, the various alternatives that may result. There are, it is true, some kinds of experimentation which it is hardly possible to plan in this sense. An experiment in which we mix chemicals together without prior hypotheses in order to "see what will happen" cannot be planned. Some discoveries, typically those of Edison, have been made in this way. Most fruitful experimentation, however, is the result of detailed planning.

Thus we are here concerned not with free-ranging exploration but with specific problems to be attacked. This concept of experimentation, although it may seem restricted, is by no means to be belittled. Only a properly planned experiment can permit us to find answers to difficult questions with a minimum of effort and cost. Improperly planned experiments can be misleading as well as costly.

The plan of an experiment involves these major areas: (1) the formulation and clear statement of the objectives of the work, (2) the determination and clear statement of the methods to be used in the collection of data, (3) a decision on and a clear statement of the techniques of data analysis to be used. Such a plan will necessarily include the *design* of the experiment, which in statistics means the arrangement of a series of experimental runs in order to minimize the effects of time and trends and other factors, to reduce the amount of experimentation to achieve a desired effect, and to obtain good estimates of errors. We will describe some basic experimental designs in the chapters to follow; here we are concerned with examining the entire process of planning experimentation.

13·1 OBJECTIVES

If the objective of an experiment can be specifically stated, a major part of the planning has been done. To state an objective requires insight into the process being investigated and an understanding of how the data will be collected and treated. Although in a written report of an experiment the objective is normally one of the first sections of the report, that objective may have become clear to the experimenter only after he had acquired much familiarity with the process.

Although it is difficult to formulate meaningful objectives, this part of the planning must be most carefully done. Imprecise statements of objectives can lead to disaster. Such an objective as "To find out how well a milling machine works" cannot look forward to a definite answer. A similarly vague and too ambitious objective is "To design, fabricate, and test a system for detecting enemy missiles." Other poorly stated objectives are

1   To find the best method of making steel. How could it ever be known that any method is *best*? Optimization of a process is often a desirable goal, but it can be attained only through a long sequence of experiments.
2   To evaluate the effectiveness of an anticorrosion coating. The conditions of use and the method of evaluation must be specified to make this statement meaningful.
3   To investigate the factors that affect the operation of an automatic titrimeter. There may well be so many factors that they cannot all be evaluated. This objective as stated is too ambitious.
4   To improve the yield of a process for making nitrobenzene. There is no limit to the experimentation implied by this statement.

Examples of properly stated objectives are

1   Is there a difference in the average results obtained by two analysts A and B in the estimation by polarimetry of the sugar content of crude syrup? We can see that the result of this experiment will be the answer "yes" or "no" at a specified level of confidence. The method will use a *t* test of significance based on a series of analyses by each of the two persons.
2   Is there a difference in the variability of the analysts in the situation

described in 1 above? The answer will be of the same kind as before, but we will use an F test of significance.

3 Which of two drugs C and D is more effective in treating the pain of headache in male college students at a dosage rate of 20 grains per day? The answer will be "drug C" or "drug D." The phrase "more effective" must be spelled out in terms of a specific testing procedure and method of measurement. We would use a χ^2 test for correctly chosen samples of a suitable size.

4 To find whether or not there is a relationship between the dosage level of drug C and the degree of relief from headache of male college students. This objective may be met by the method of data analysis called "regression analysis," which requires data from testing at different levels of the drug and suitable sampling procedures.

We have left out the entire area of optimization, which is a long-range objective and not one likely to be met by any single set of experiments. It is true that we are not usually interested in significance tests only in themselves but rather want to obtain estimates of the relationship between variables and to use these estimates to tell us where to run further experiments, or if we should stop experimentation in a specific area. Thus the optimization of a process involves a sequence of experiments, each of which should have a clearly stated objective, and each of which is expected to contribute to the long-range objective.

If we intend to use tests of significance in data analysis, the best way of stating the objectives is in terms of the null and alternative hypotheses. When objectives are so stated, no question is left about the aims of the experiment.

PROBLEMS

13·1 Criticize the following experiment objectives:
(a) to find the most effective drug for headache
(b) to find out whether or not taking aspirin reduces the pain of headache
(c) to test the effects of 5 levels of a given fertilizer on the yield and strength of cotton fibers
(d) to compare the effects of feeding rations A and B on the amount and quality of milk from cows

13·2 A modern version of the divining rod is said to be useful in finding buried pipes of any kind. Two steel wires, about 18 in. long, are bent into right angles with equal legs. Each wire is held loosely in one hand, with the horizontal portions of the wires above, parallel and pointing forward. When the user carrying the wires walks over a buried pipe, the wires are said to spread apart and to take up positions parallel to the direction in which the pipe runs.
Write objectives for three different tests of the assertions implied in the

description of the divining rods. For each objective, indicate the method of data collection and the kind of analysis that will satisfy the objective.

13·2 CHOICE OF FACTORS TO BE TESTED

In almost every situation, the list of factors that might affect the outcome is nearly inexhaustible. In a chemical manufacturing process, there may be a score of steps; there may be several ingredients, each at any of many concentrations; factors such as pressure, temperature, and handling methods may all play a part. It is always most difficult to select those factors which should be investigated in an experiment.

Some statisticians believe that among the many factors which may affect a process, only a few are likely to have a major effect. The number of important factors is often said to be two or three or four, rarely more. To the extent that this belief is valid, a moderate familiarity with the process is thought to reveal those few factors for which an investigation is likely to be fruitful.

If it is, on the other hand, thought that a selection of factors is dangerous, screening designs such as "fractional factorials" can be used to obtain evidence about the influence of many factors. On the basis of such screening experiments, additional experimentation can be used to estimate the effects of the most important factors. We will introduce such experiments in Chap. 14. In that chapter we will, however, begin with smaller experiments involving only a few factors, on the assumption that prior knowledge is available about the relative importance of the factors of interest.

When only a few factors are chosen for testing, the problem of dealing with the possible effects of untested factors is the primary purpose of the "design" of experiments. We want to reduce the error associated with the effects of nontested factors so that we can maximize the efficiency with which the effects of tested factors can be estimated. If the influence of untested factors is great, the error mean square will also be large, and we will perhaps be unable to detect a small, but real, effect of the factors we wish to test.

When we choose factors for testing, we must distinguish between those which are fixed and those which are random (see Chap. 9). If "machine" is a factor, and we test three machines all of which (and only which) are used in production, this factor is fixed. If 50 machines are available, and we choose at random 3 of these for the experiment, the factor is random. We must take care to distinguish between these two different situations, inasmuch as the method of data analysis is dependent in part upon this distinction.

Similarly, the distinction between quantitative and qualitative factors affects the method of data collection and analysis. If we have only qualitative factors, such as with or without agitation, or at a high or low level of temperature, we make it impossible in the data analysis to test for linear versus quadratic effects. If, on the other hand, we can use quantitative factors and can therefore specify the levels of the response variable in numerical terms, we increase the variety of methods of data analysis we can use. It is well, therefore, to use quantitative levels of the input variable when we can.

PROBLEMS

13·3 Make a list of the factors that might be tested in each of the following situations:

(a) the gas mileage for a given automobile
(b) a comparison of the sharpness of two brands of razor blades
(c) the optimization of the method of preparing lemonade
(d) a comparison of the wearing properties of nylon versus wool rugs
(e) the effect on sales of a soft drink as affected by packaging in bottles or cans

13·4 For each of the situations in Prob. 13·3, assume that in a single experiment only three of the factors can be tested.

(a) Choose three of the factors you have listed and indicate the reasons for your choice.
(b) Explain, assuming that only three factors can be chosen, how you would proceed to handle the untested factors.

13·3 DEFINITION OF THE POPULATION BEING SAMPLED

No inference about a population can be made on the basis of an experiment unless that population has been properly sampled. An experimenter must never yield to the temptation to make inferences about populations other than those sampled. He also must take care that the sampling procedure does not omit any significant part of the population of interest.

If only college students are used as subjects for a psychological test, the results permit inferences only about college students. If all the tested students are men, the results can be applied only to male college students. If a drug is found to be without side effects when administered to rats, there is no reason from this result alone to suppose that the drug is safe for human beings. If the drug has produced no ill effects in human beings when it has been given over a period of 3 months, we cannot from this observation infer that the drug has no bad long-time effects. Recall again in this connection the distinction made between fixed and random factors in Chap. 9.

As part of the planning process for an experiment, the nature of the population to be sampled should be clearly defined. Furthermore, the techniques of choosing samples from the population should be such as to ensure that the sample is representative of the population. Random choice of the sample is often used for this purpose.

13·5 Refer to Prob. 13·4. For each situation, specify the populations being sampled.

13·4 CHOICE OF A RESPONSE VARIABLE

If we want to improve the yield or the purity of a product, we can measure the response variable with little difficulty. In other situations it may be hard to find a satisfactory measurement of the quality of a product. What, for example, should we choose as the response variable in an experiment intended to test a set of formulations used in cleaning windows? We would like the response variable to be well-defined and meaningful.

Here are some possibilities: (1) amount of formulation needed to clean a group of sample windows, (2) number of strokes with a cloth required to clean a group of sample windows, (3) cleanliness of a group of sample windows after we apply a specified amount of formulation in a specified manner. Each of these proposals suffers from the same defect: the quality "clean" is undefined. If now we define "clean" in terms of the appearance of the windows, we must specify both the technique of observation and the observers we will use. Each of these aspects of the situation presents new problems which we must solve.

In other situations, such qualities as hardness, abrasion resistance, gloss, color, and strength are difficult to define. Despite the difficulty of specifying these qualities, we must do so before we can engage in an experiment involving them. In any but routine situations, the ingenuity of the experimenter will be challenged to find appropriate response variables. Considerable familiarity with the product or process is required before intelligent decisions can be made.

It is usually easier to find suitable response variables for the description of a process than for the description of a product. If the manufacturer is the "customer," i.e., if what is wanted is that the process should give optimum yield or quantity, the selection of a response variable is usually without difficulty. If, on the other hand, the customer is the buyer of a manufactured item, the response variable should be related to the use to which the item is to be put. It is often necessary on this account to use panels of consumers to make judgments about the product.

Conflicts between seller and buyer of a product can be minimized by precise specification of response variables and by agreement on test methods. Disputes between production and inspection personnel within a plant arise when these matters are only vaguely understood.

PROBLEM

13·6 Refer to Prob. 13·4. For each situation, specify a suitable response variable.

13·5 OBJECTIVE VERSUS SUBJECTIVE MEASUREMENTS

Objective measurements such as length and weight are usually considered as preferable to subjective estimations of product quality such as taste, odor, or other perceptions. The distinction between subjective and objective estimations of product quality is, however, not absolute. Every measurement necessarily involves some judgment, if only a judgment of the position of a needle on a dial. Even in automatic machine readout, human judgment enters into the design of the response characteristics of the machine.

The important distinction is in the degree of variability with repeated measurements of the same quantity. "Objective" measurements are usually those for which the variation with repetition is small; "subjective" measurements are ordinarily characterized by large variability. In the presence of great measurement uncertainty, we are forced to use relatively large sample sizes. If the measurements are very repeatable, perhaps a single measurement will suffice.

It is most desirable that the values of the response variable be associated with a scale, that is, with a linearly arranged set of units. If measurements are scaled in this sense, we can be sure that if A is greater than B and B greater than C, A is surely greater than C. In tests of consumer preference, the responses are often not so scaled. Special problems arise when such response variables must be used; these problems can in general be solved only by the use of large sample sizes and "blind" test procedures.

13·6 EXPERIMENTAL ERROR

Nearly all statistical tests are in principle based on a comparison between effects associated with tested factors and an estimate of experimental error. The best estimate of error comes from replications within the experiment itself. When we replicate we attempt to make runs at constant levels of experimental conditions. The failure to obtain the same values of the response variable is our measure of error.

Completely identical treatment of different runs is impossible. There will be at the very least a time difference between replications. When we anticipate a large effect associated with time changes, such as day-to-day variations in conditions, we may randomize the sequence of runs within time to eliminate trends which may systematically affect the observed values.

The greater the number of replications, the more sensitive the statistical test; this increased sensitivity comes from the increased number of degrees of freedom associated with the error term. We discussed in Chap. 6 the methods for determining the sample size required for tests of hypotheses. Similar principles apply here. As part of the planning of an experiment, we require an estimate of the size of the effect we wish to detect, the risks of error we can tolerate, and the size of the experimental error. Only if we can establish these values in advance can we specify the minimum number of experimental runs we must make.

The α and β risks we decide to take are related to our estimate of the cost, in money and effort, connected with errors of the first and second kind. If we are testing rocket engines, it may be costly indeed to conclude that a new engine develops more thrust than a former model (when in reality there is no difference) since an error of this kind may lead to disastrous consequences. If, on the other hand, we test the acceptability to a customer of a food package, a false conclusion that red is preferred to blue may involve little loss. Thus we should set the levels of the α and β risks in accordance with our estimate of the costs of each kind of error.

In experiments in which we intend to use ANOVA for data analysis, we may use an interaction effect as a substitute estimate of error, thus making it unnecessary to replicate. The assumption we make when we do this is that the interaction effect is nonexistent. We should use an interaction effect as a measure of error only when on the basis of past performance or theory we have good reason to believe that no interaction will occur.

It is often useful, in order to avoid replication in an experiment, to perform a preliminary experiment with the sole objective of estimating error. This experiment involves replications of the procedure to be used in the full-dress experiment. The estimate of error we will thus obtain is valid only if the value of error is unaffected by the levels of treatment that will be used later. We may, for example, wish to test the effects of temperature and pressure on the yield of a complex process. We may estimate the value for error on the basis of the past history of the operation at a single level of temperature and pressure. If we use this estimate of error in the experiment, we must suppose that the error is not changed by alterations in the levels of temperature and pressure.

13·7 ASSUMPTIONS IN STATISTICAL EXPERIMENTS

An essential part of experiment planning is a statement of the assumptions that underlie the experimental process. Even in well-defined situations we find it necessary to make assumptions; we must be clear about them.

If a test of hypothesis rests on the assumption of normal distribution of the populations from which the samples are taken, this assumption should be recognized and tested following the methods of Chaps. 2 and 7. Often historical records of a process will provide evidence for or against normality. Tests which involve a comparison of means are often independent of the assumption of normality according to the Central Limit theorem. Chapter 18 deals with some test methods which require no assumptions about the population distributions.

When we write the mathematical model for ANOVA, we indicate that the effects of the tested factors and error are summed. Thus we are assuming that the effects are additive in a simple manner, not multiplicative. For example, if we are testing the effects of temperature and pressure on a process, we may assume that the effect of temperature is to increase yield by the same amount on the average for any pressure. This assumption is often correct only to a first approximation. Experimental designs such as "split-plot" and incomplete blocks do not necessarily involve this assumption. One type of nonadditivity is interaction, for which we can test in some of the designs to be described in Chap. 14.

A third assumption often is that error is independent of the levels of the tested factors. In some measuring instruments, the instrumental error varies with the level of the observation. In some experiments, there is a trend in time or in space. Often we encounter situations in which there is compensation for error. If the value of the error term varies with the treatment levels, we should obtain an estimate of error at the different levels. If there is a trend, we should randomize the sequence of runs. Any necessary restrictions placed on the process of randomization (within time, or within space or other factors) lead us to the selection of one of a number of experiment patterns which are called "experiment designs." The fundamental designs are the subject of Chap. 14.

In ANOVA, we assume that the error estimates have a common variance, since we pool the estimates of error from different subgroups. If we suspect that the error variances are different, we may find a separate error term for different comparisons.

Much experience indicates that the assumptions stated in this section are met well enough in most situations to justify the elementary methods we have described. It is well, however, to appreciate that there may be cases in which the assumptions are not justified. Familiarity with, and insight into, the situation are the bases for intelligent choices of statistical techniques.

The preferred method of expressing the assumptions involved in an experiment (and by implication the methods of data analysis that will follow the data collection) is to write the mathematical model for the experiment. If we can do this successfully, we will necessarily have thought through the experiment in detail.

If we write the model as

$$X_{ijk} = A_i + B_{j(i)} + \epsilon_{k(ij)}$$

we imply that we are testing only two factors A and B, that B is nested in A, and that all other sources of variation are included within the error term ϵ. The implication is that we will analyze the data by means of a two-factor nested ANOVA. If, on the other hand, we write the model as

$$Y_i = \beta_0 + \beta_1 X_i + \beta_2 X_i^2$$

we imply that the values of the response variable Y are determined only by the levels of a single predictor variable X, that a curvilinear functional relationship is plausible, and that we are therefore planning to run a curvilinear regression analysis.

PROBLEMS

13·7 Four thermometers are available for class use in a science laboratory. You intend to test these thermometers and also the class performance in the use of the thermometers.

(a) Write the mathematical model of an experiment you might conduct if you were to select randomly 5 students to use the thermometers.
(b) State the hypotheses which would be tested.
(c) Tell why it would be wise to replicate the tests.
(d) Outline the method of running the tests, showing the order in which the experimental runs would be made.
(e) What would the response variable be?

13·8 You are considering the purchase of a new automobile, a low-priced American economy car. You will test each brand of automobile, but testing must be kept to a minimum. For this situation, answer the following questions:

(a) List the possible factors for the test.
(b) From the list, choose the most important three. Justify your choice. How would you deal with the other factors?
(c) For each of these three selected factors, indicate the levels you would test.
(d) What response variable would you use?
(e) How would you obtain an estimate of the experimental error?
(f) How would you determine the appropriate sample size?

13·9 There is a traffic light at a busy intersection, and it is believed that the length of time for the red light causes traffic to back up too much in one direction. Prior to taking action on making changes, you are to investigate the whole problem and make suggestions.

(a) State the objectives of the experiment.
(b) List the possible factors to be investigated.
(c) Identify each factor as:
  (1) fixed or random
  (2) quantitative or qualitative
  (3) crossed or nested
(d) Write the mathematical model for the experiment you suggest be carried out.
(e) State the response variable.
(f) List the questions to which you hope to find answers.

13·10 It is suggested that the location of certain packaged foods in large supermarkets will contribute to the total sales of the foods. Apply to this situation the outline in Prob. 13·9.

13·8 FLEXIBILITY IN EXPERIMENT EXECUTION

Once an experiment has been well planned, a few final temptations must be avoided. One of these is to hew to the line of planned attack, regardless of evidence that the work has gone astray. It is often possible to examine data that are being accumulated in the course of an experiment and to see that a completely unexpected set of results is appearing. The data may be unexpected in that they show no effects whatever or that they show fantastically large effects. Sometimes obvious patterns begin to appear after the completion of only a small part of a planned elaborate experimental design. In all such cases, an intelligent appraisal of the data may indicate that further data accumulation is likely to be unprofitable and that a different approach is needed.

Often experimenters will subject data to exhaustive analysis when, from the practical point of view, any existing effects may be real but unimportant. It is necessary to appreciate the difference between statistical and practical significance. We may be able to demonstrate, with a sufficiently sensitive experiment, statistically significant differences where no practical difference exists. Conversely, it is pointless to carry out an elaborate analysis of variance when merely eyeing the data shows that all tested factors are clearly significant.

13·9 FLOW DIAGRAMS

The persons involved in the execution of an experiment must understand the different operations through which the product must go to reach the stage where the response variable can be obtained. A flow diagram is a graphical representation of the process which serves to display the required sequence in an orderly, understandable way. The person who does the actual experiment planning will often not carry out the test runs. There must be some method of communication between the planner and the performers. A flow diagram is useful for this purpose.

To produce a satisfactory flow diagram of an operation requires that all elements of the operation be placed in the correct sequence. This requirement often leads to the elimination of misunderstandings, since agreement must be reached among the personnel involved.

We are interested in an experiment in connection with a process which is used in the manufacture of chocolate fudge. Before we list the factors we might like to test, we make a flow diagram as in Fig. 13·1. From the flow diagram, we can now prepare a list of questions to be answered, some of which are:

1 Is the temperature important in the first step? If, so shall we include this as a factor in the experiment, or try to control it? If we control it, what should be the temperature?
2 Should we test, as factors in the experiment, the quantity of sugar and salt added at the second step?
3 Is the mixing process important?
4 What quantities of syrup and milk should be used?

5 Is the time needed for boiling off the water a factor?
6 What amounts of vanilla and butter should be used?
7 Is the cooling rate a factor?
8 How should the stirring be done? For what time?
9 What should be the response variables?

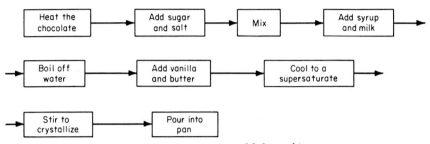

Fig. 13·1   Flow diagram of fudge making.

An expert fudge maker may well think that he has the answers to many of the questions. Other questions will be unanswerable on the basis of experience; these will serve as the possible factors to be tested. Furthermore, an experiment planner often mistrusts those who are expert solely because of experience, since if this experience gave rise to correct answers to all questions, there would be no reason for experimentation with the process. At any rate, once the flow diagram has been agreed upon, the factors to be tested can be determined.

Flow diagrams are particularly useful in the development of test methods which are to be used in different laboratories. Often, analysis and measurement methods require a series of manipulations that must be carried out according to a fixed procedure. The flow diagram is especially necessary in situations in which several laboratories cooperate in an extensive study.

In complex situations, the flow diagram may be branched if it is to show the required relationships. In Fig. 13·2 are shown the stages in the preparation of a sample. To describe the technique by means of a verbal statement would be difficult.

PROBLEMS

After any necessary research into the following processes, prepare flow diagrams for:

13·11   Cleaning house windows with a commercial preparation.
13·12   Painting a room.
13·13   Making bread.
13·14   Vaccinating a person for smallpox.
13·15   The lead-chamber process for making sulfuric acid.

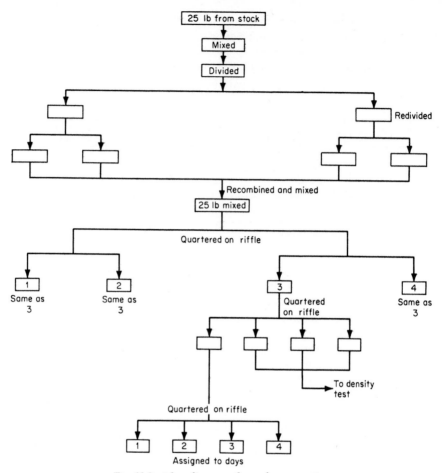

Fig. 13·2   *Flow diagram of sample preparation.*

13·16   Taking a favorite girl (or wife) to a formal dance.

13·17   Being sure to be on time for a very important crosstown appointment, a week from today.

13·10   CHECK LIST FOR PLANNING TEST PROGRAMS

The following outline[1] is a summary of the steps required for the successful completion of an experiment.

A. Obtain a clear statement of the problem.
    1. Identify the new and important problem areas.
    2. Outline the specific problems within current limitations.
    3. Define the exact scope of the test problem.

[1] C. A. Bicking, *Industrial Quality Control*, vol. 10, no. 4, pp. 20–23, January, 1954.

4. Determine the relationship of the particular problem to the whole research or development program.
B. Collect available background information.
  1. Investigate all available sources of information.
  2. Tabulate data pertinent to planning new program.
C. Design the test program.
  1. Hold a conference of all parties concerned.
    a. State the propositions to be proved.
    b. Agree on magnitude of differences considered worthwhile.
    c. Outline the possible alternative outcomes.
    d. Choose the factors to be studied.
    e. Determine the practical range of these factors and the specific levels at which tests will be made.
    f. Choose the end measurements which are to be made.
    g. Consider the effect of sampling variability and precision of test methods.
    h. Consider possible interactions of the factors.
    i. Determine limitations of time, cost, materials, manpower, instrumentation, and other facilities.
    j. Consider human relations angles of the program.
  2. Design the program in preliminary form.
    a. Prepare a systematic and inclusive schedule.
    b. Provide for step-wise performance or adaptation of schedule if necessary.
    c. Eliminate effect of variables not under study by controlling, balancing, or randomizing them.
    d. Minimize number of experimental runs.
    e. Choose the method of statistical analysis.
    f. Arrange for orderly accumulation of data.
  3. Review the design with all concerned.
    a. Adjust the program in line with comments.
    b. Spell out the steps to be followed in unmistakable terms.
D. Plan and carry out the experimental work.
  1. Develop methods, materials, and equipment.
  2. Apply the methods or techniques.
  3. Attend to and check details; modify methods if necessary.
  4. Record any modifications of program design.
  5. Take precautions in collection of data.
  6. Record progress of the program.
E. Analyze the data.
  1. Reduce the recorded data, if necessary, to numerical form.
  2. Apply proper mathematical statistical techniques.
F. Interpret the results.
  1. Consider all the observed data.

2. Confine conclusions to strict deductions from the evidence at hand.
3. Test questions suggested by the data in independent experiments.
4. Arrive at conclusions as to the technical meaning of results as well as their statistical significance.
5. Point out implications of the findings for application and further work.
6. Account for any limitations imposed by the methods used.
7. State results in terms of verifiable probabilities.

G. Prepare a report.
1. Describe work clearly, giving background, pertinence of the problem, and meaning of results.
2. Use tabular and graphic methods of presentation of data in good form for future use.
3. Supply sufficient information to permit reader to verify results and draw his own conclusions.
4. Limit conclusions to objective summary of evidence so that the work recommends itself for prompt consideration and decisive action.

QUESTIONS

13·1 Why should the method of data analysis be considered prior to the collection of the data?

13·2 In what sense, if any, is it possible to plan an experiment if the outcome is in no way foreseeable?

13·3 Why should the factor levels be chosen prior to the experimental runs? How should the factor levels be selected?

13·4 What methods can be used to minimize the effects of factors which are not under study?

13·5 What is the role of randomization in designed experiments?

13·6 What considerations affect the choice of response variables?

13·7 What is meant by "experimental error"? How can it be eliminated from an experiment? How can it be measured?

13·8 Consider the following statement: "A statistically designed experiment allows the experimenter to tailor the size of his experiment considered in terms of risks, size of difference which should be detected, and cost." How does each of the three items mentioned in the statement affect the size of the experiment?

REFERENCES

10, 11, 12, 13, 14, 20, 25, 30, 57, 59, 60, 70, 116, 127

14 Basic Experiment Designs

We discussed in Chap. 13 the principles that should be followed in planning any kind of experimentation. In this chapter we introduce a few of the experimental schemes called *experiment designs*. Although these designs are related to only a part of the necessary planning, they are important because they deal with the problem of ordering (in time or in space) the experimental runs. If we fail to arrange these runs appropriately, we may find the experiment to be inefficient or the results misleading.

Experiment designs differ primarily in the extent to which we randomize the sequence of runs, or in the restrictions which are placed on the randomization of this sequence. When we choose a specific design, we intend to accomplish these aims: (1) to reduce the variation associated with experimental error to an economic minimum, in order to distinguish potential real effects of the factors under study from random noise; (2) to detect the effects of tested factors with a minimum number of experimental runs.

Consider this objective: to discover whether or not there is a difference in the yield of four different varieties of wheat. The basic problem in experiment design arises from this: We believe that the time of planting, weather conditions, planting techniques, and especially the type of soil will be important factors in determining the level of the response variable. We are not interested in testing these other factors because we have from experience alone excellent reason to believe that they are highly significant. We fear the effects of the nontested factors to be so great that they may well mask a real effect associated with the factor of interest—variety.

An alternative way of stating the problem is this: Unless we design the experiment properly, we may find in the ANOVA to follow that the error term is so large that the factor "variety" will not prove to be significant even though the effect is present. Since, in ANOVA, we include in error all effects due to untested factors, the error term may be highly inflated.

An exactly similar problem appears in the following:

1 We wish to see whether or not three different kinds of steel rods differ in machining quality. We have good reason to believe that different lathes, different tool shapes, different speeds, different operators, etc., have a large effect.

2 We wish to test several different photographic emulsions for "speed." We expect different coating machines, different temperatures, different rates of application, etc., to have a large effect.

3 In a chemical analysis procedure we wish to find small differences in the concentration of an ingredient. The ingredient is unstable with time, as are some of the components of the analytical solutions. We fear, therefore, that the effect associated with time will cover the small differences in composition we are seeking.

In each of these situations, and in similar ones, the effort put into experimentation will be wasted unless we plan the work in such a way as to reduce the probability of an inconclusive outcome. In this chapter we discuss some of the basic methods of planning experimentation which lead to the economical use of effort in accumulating and analyzing data.

14·1 COMPLETE RANDOMIZATION

We return to the wheat experiment. Assume that we have available for this test a large plot of ground. To design the experiment means to decide on the manner in which the four varieties will be planted within this plot. For the moment we are primarily concerned with the effect of position on the yield. We consider these alternatives:

1 We will divide the plot into four equal parts, and assign each variety to one-fourth of the plot. We discard this possible method, since we will have only a single value of the response variable for each variety. Therefore, we will have no estimate of experimental error. More important, this technique gives us no way of distinguishing an effect due solely to position within the plot from the effect of variety. A similar defect would be present in an experiment on the machining quality of three samples of steel if we were to test each sample on a different lathe.

2 We will divide the entire plot of ground into 16 sections, and assign *at random* one variety to a section, using a total of four different sections for each of the four varieties. This method is far superior to the preceding one. We will now have four estimated values of the yield for each variety; from these we can arrive at an estimate of error. Furthermore,

and this is the point, this method will to some extent distribute the effect due to position among each of the four varieties. An experiment of this type is called "completely randomized."

We number the 16 sections of the plot as in Fig. 14·1a. We resort to a table of random numbers to assign the varieties to the plots. One such set of numbers is 13, 16, 5, 11, 10, 2, 12, 14, 15, 3, 6, 7, 9, 4, 1, 8. We let the first

I	2	3	4
5	6	7	8
9	10	I I	12
13	14	15	16

(a)

D 7	B 6	C 6	D 12
A 8	C I I	C 12	D 20
D 18	B 14	A 8	B 17
A 16	B 20	C 15	A 25

(b)

Fig. 14·1 Random assignment to four treatments (varieties) to an experiment. (a) Plot section numbers. (b) Assignments of position.

four numbers specify the positions for planting the four samples of variety A, the next four those positions for variety B, etc. Thus we arrive at the arrangement of Fig. 14·1b, to which we have added the coded values representing the yields for each plot.

One of the virtues of completely randomized experiments is flexibility. We have a wide choice of the number of replications. Here, we have used four plantings of each variety. We might have, had we liked, used other numbers merely by making a different division of the plot. Similarly, if it will take 5 days to complete a set of experiments of the same kind, we may completely randomize the experiment within the 5 days; we may, as practical matters require, run two, or three, or any other number of experiments within a day.

Note, however, that this accidental choice of random numbers has by no means distributed the samples evenly among the plot sections. For example, three of the samples of variety C are juxtaposed; for both varieties A and B three of the samples are in the lower half of the plot; all the samples of variety C are in the center two columns.

It is a disadvantage of a completely randomized experiment that we cannot be confident that we will have equally distributed the effect associated with the nontested factor (here position within the plot). We have, when we use this design, no way to estimate the effect of this untested factor if it exists. Any contribution to variability due to position will appear as experimental error.

Other experimental designs, to be discussed later in this chapter, overcome in part the defect just mentioned. These other designs, however, are less flexible in nature, but where we suspect the presence of patterns within the nontested factors, they are most valuable.

The completely randomized experiment for which we have the data of Fig. 14·1b can be analyzed by a one-factor ANOVA. Thus we arrange the observations in Table 14·1 where we have grouped in columns the yields for each of the four varieties. The rows of this table have no significance.

Table 14·1   Data from Fig. 14·1b

Variety	A	B	C	D
Yield	8	6	6	7
	8	14	11	12
	16	17	12	20
	25	20	15	18

The mathematical model for this experiment is

$$X_{jr} = \mu + F_j + \epsilon_{r(j)}$$

where X_{jr} = any observed value of response variable
μ = general level of yield
F_j = a possible effect of factor indicated by any j column
$\epsilon_{r(j)}$ = error effect of replications nested within columns

The null hypothesis is $H_0: F_j = 0$; the alternative is $H_1: F_j \neq 0$. We will test these hypotheses by means of an analysis of variance in which we will determine

$$SST = \sum X_{rj}^2 - \frac{T..^2}{n}$$

$$SSC = \frac{1}{r} \sum T._j^2 - \frac{T..^2}{n}$$

$$SSE = SST - SSC$$

Here we use the customary notation:

SST = total sum of squares
$T..$ = grand total of response variables
SSC = sum of squares due to columns (tested factor)
$T._j$ = total for a specified column over all rows
r = number of entries in each column (number of rows)
SSE = sum of squares due to error

The values of v are total $n - 1$; for the tested factor, $c - 1$ (the number of columns less one); and for error the remainder.

PROBLEMS

14·1   Complete the ANOVA for the data from the wheat experiment.

14·2   It is desired to test the hypothesis that the means of four normal populations are equal. Assuming the populations have equal variances, and we decide to take three independent samples from each population, answer the following:

(a) How many factors are being considered for investigation?
(b) How many levels of each factor are there?
(c) Lay out a table having 3 rows and 4 columns.
(d) Using the columns of the table as the levels of the factor, and the numbers 1, 2, 3 as the samples from each level, show how the samples should be collected if the experiment is a completely randomized one.
(e) If the samples were to be obtained one at a time, as the process produces them, show the order of running if a completely randomized order is followed.

14·3   In a sporting event such as broad jumping, each contestant is given the same number of attempts. The usual method of setting the order of jumps is to select at random the first set of jumps, and then to repeat them in the same order for a definite number of times.

(a) If there are 5 contestants, and each one is given 4 jumps, show how the order of jumps would be made. (Use A, B, C, and D as the jumps, and numbers for the contestants.)
(b) If all jumps were assigned by a complete randomization, how might the jumps be carried out?
(c) Repeat the order of jumps for the contestants, by once more using complete randomization, and compare the answer with that obtained in answer to part (b).
(d) In the case of the sporting event, should the answer to (a) or (b) be used? Explain.

14·4   An investigation into the effect of conductivity on TV tubes was made by testing four types of coating on the tubes. If it is agreed that four observations per coating should be adequate, design a completely randomized experiment to carry out the investigation.

14·5   Set up the ANOVA table for the experiment in answer to Prob. 14·4, and write the mathematical model.

14·2   MISSING DATA IN COMPLETE RANDOMIZATION DESIGN

In a complicated experiment, it often happens that an observation is lost through accident. When this occurs, we have the alternatives of performing the whole experiment over again or of salvaging what we can from the remaining data. For ease in analysis, it is convenient to be able to operate with a complete set of data.

We can, by methods that vary somewhat with the type of experiment design, calculate a dummy value to be used instead of the missing one; thus we can complete an ANOVA in the customary manner. The dummy value is calculated so as to leave unchanged the mean of the data and is assumed to contain no error. (Thus it is based on a "least-squares" method.) Whenever we calculate a dummy value, therefore, we lose one degree of freedom in the experiment. This loss has the effect of reducing by one the value of v for the error term. If several pieces of data are lost, the experiment becomes insensitive, and we often do better to rerun the experiment.

For a completely randomized experiment, to find the dummy value we merely use the mean of the remaining data in the column to which the missing value belongs. The ease with which the dummy can be calculated is an additional virtue of the completely randomized experiment design.

14·3 RANDOMIZED BLOCK DESIGN

We have seen that in a completely randomized experiment there is a danger of obtaining a large error term by the accidental grouping of data when there are extraneous assignable effects in the experimental space. We can reduce this danger, remove the effect of the extraneous factors from the error term, and also estimate their influence, by randomizing in sections rather than over the whole experiment.

In applying this technique to the tests of the wheat varieties, we would divide the entire plot of ground into four *blocks*, subdivide each block into four sections, and randomize the position of the varieties within the blocks. We would thus make sure that each variety of wheat would be tested equally within each block.

A randomized block design is also appropriate in this situation: A complete experiment requires 3 runs of each of 4 treatments; each run takes 2 hours. The whole experiment will require 3 days, and we fear a large day-to-day difference in results. We would therefore consider each day as a block, and randomize the testing sequence within each day, arranging the runs so as to have one treatment each day. We say that we are "blocking on days."

In similar fashion, an experiment testing 3 raw materials may require the use of 4 processing machines. We would block on machines, testing each material on each machine in random sequence.

In general, we block on effects that we feel sure will be real. Since they are factors we usually are not interested in per se, we do not especially wish to test for them. We want, however, to take them into account in order to reduce the size of the error term and thus to increase the sensitivity of the experiment with respect to the factors we do want to test.

In the wheat experiment, assume that we decide to block in the up-down direction in the plot. Thus we have an arrangement as in Fig. 14·2 in which we have assigned positions within each block by using a table of random numbers. Each variety appears once within each block. We have placed the same yield values in the array as in Fig. 14·1*b*, but we have ordered them in such a way as to indicate an among-block effect which we anticipated when we planned the randomized block design.

The mathematical model for this experiment is

$$X_{ijr} = \mu + B_i + F_j + \epsilon_{r(ij)}$$

The value of each observation is attributable to four causes: a mean; an effect associated with blocks (we believe this to be significant); an effect due to the factor of interest; error.

Fig. 14·2 Randomized block arrangement of an experiment.

Block				
I	D 7	B 6	A 8	C 6
II	C 11	A 8	D 12	B 14
III	B 17	A 16	C 12	D 18
IV	C 15	B 25	A 25	D 20

For the purposes of ANOVA, we order the data from Fig. 14·2 as in Table 14·2. For the values of *v*, we have 3 each for the 2 factors (one fewer than the number of rows or columns); for residual we have 9. The randomized block design (as compared with the completely randomized design) has withdrawn from error three degrees of freedom in order to include the block effect.

Table 14·2   Data from Fig. 14·2

Block	Variety			
	A	B	C	D
I	8	6	6	7
II	8	14	11	12
III	16	17	12	18
IV	25	20	15	20

Therefore, we now add to the sums of squares calculated in the previous ANOVA a sum of squares due to rows:

$$\text{SSR} = \frac{1}{c}\sum T_{i\cdot}{}^2 - \frac{T_{\cdot\cdot}{}^2}{n}$$

We will now be able to complete the two-factor analysis. Note that we fully expect the effect of rows (associated with blocks) to be significant. Otherwise we would not have taken the trouble to block on position and thus to have sacrificed three degrees of freedom from the error term.

When we compare the randomized block design with a completely randomized design, we see that we lose degrees of freedom for error and thus suffer a loss in sensitivity for the F test. In effect, the error term is the block-factor interaction. An essential assumption in all blocking designs is that there are no block-factor interactions and thus that this term measures only error. We usually make the blocks homogeneous enough to make this assumption reasonable. If not, then we must run a factorial experiment (see Chap. 15) in order to be able to test for interactions.

Whether on the whole we gain or lose by introducing the restriction needed for randomized block designs depends on the influence the block factor exerts on the response variable. If that influence is great, we will gain by separating it from the error term, in spite of the loss of degrees of freedom. If that influence is small, we will lose. Thus our decision about which design to use is based on our belief concerning the effect of that factor on which we may block. A similar comment applies to the designs described in the rest of this chapter.

PROBLEMS

14·6   Complete the analysis of variance for the data in Table 14·2.

14·7   Consider the sport of broad jumping, with 5 contestants, each one being allowed 4 jumps. Each jumper is to complete his turn before any contestant takes an additional jump.
   (a) If the random order of jumping has been established as Mr. B, Mr. C, Mr. A, Mr. E, and Mr. D, lay out a table with 5 columns and 4 rows, and show the order of the jumps as they are carried out at the event.
   (b) If after each jump, the order of jumping is once again randomized, we then have a randomized block design.
       (1) What have we blocked on?
       (2) Show how the order of jumping might be assigned by the randomized block technique.

14·8   We have 4 tests we wish to make. We will repeat each test 4 times, calling for 16 tests in all. To speed up the running of the tests, we will use four similar testing machines A, B, C, and D. The tests to be made are the W, X, Y, and Z tests, and we will use each test 4 times.
   (a) Lay out a 4 × 4 table and assign the machines to the 4 columns.
   (b) We wish to block on machines, as we feel they might be enough different to influence the results. Randomly place the 4 tests in each of the 4 machines.
   (c) If each test is to be carried out one at a time how might we decide on the order of running the tests?

(d) If the tests were to be carried out 4 at a time, by using all 4 machines at once, what problems might be encountered?

14·9 In an investigation concerning a chemical compound, it is required to test whether different blends of the compound give different yields. Five different blends are made and three batches of product are prepared from each blend. If the blends are tested one at a time, any variation from blend to blend might be due to a time trend in the process and not to any real difference between blends. To eliminate this effect from the experiment, design a randomized block experiment for the investigation.

14·10 To test the effects of 5 levels of application of potash on the yield and properties of cotton, an experiment was arranged in 3 randomized blocks of 5 plots each. The results of the experiment are given in Table 14·3.

Table 14·3 Strength Index of Cotton in a Randomized Block Experiment

Lb K$_2$O per acre	Replication			Total
	1	2	3	
36	7.62	8.00	7.93	23.55
54	8.14	8.15	7.87	24.16
72	7.76	7.73	7.74	23.23
108	7.17	7.57	7.80	22.54
144	7.46	7.68	7.21	22.35
Total	38.15	39.13	38.55	115.83

(a) Show how the field might have been laid out to carry out the investigation.
(b) Write the mathematical model for the experiment.
(c) Complete the analysis of variance for the experiment.
(d) List the hypotheses which are being tested.
(e) Complete the analysis of the data from the experiment.

14·4 MISSING OBSERVATION IN RANDOMIZED BLOCK EXPERIMENT

If, in an experiment based on randomized blocks, one observation is missing or must be discarded because of a blunder in the experimentation, we can estimate a dummy value by

$$X = \frac{tT + bB - S}{(t - 1)(b - 1)} \tag{14.1}$$

where X = dummy

t = number of treatments

T = sum of observations for same treatment as missing term

b = number of blocks

B = sum of items in same block as missing term

S = sum of all available data

In Table 14·4 we have selected data such that inspection indicates what the missing value must be if the data are to be consistent.

Table 14·4   One Missing Observation

Block	Treatment		
	A	B	C
I	3	7	11
II	4	?	12
III	5	9	13
IV	6	10	14

Here $t = 3$, $T = 26$, $b = 4$, $B = 16$, $S = 94$. Thus

$$X = \frac{(3 \times 26) + (4 \times 16) - 94}{(3 - 1)(4 - 1)} = \frac{48}{6} = 8$$

In completing the ANOVA based on the use of a calculated dummy value, we reduce by one the value of ν for the error term, since in calculating the dummy value we assume that it contains no error.

14·5 LATIN SQUARE EXPERIMENTS

We reconsider the randomized block arrangement of Fig. 14·2 on this assumption: There may well be a variation associated with right-left position within the plot. The random choice used to place the varieties within the blocks has resulted in these accidental patterns: All the samples of A are in the central two columns; all the samples of B are in the outside positions; three of the samples of D are in the right-hand half of the plot. Any effect that we have associated with variety may instead have been in part associated with variation due to these accidental patterns of position.

As compared with a completely randomized experiment, the use of randomized blocks imposed one restriction on the planting sequence. We may now wish to adopt a second restriction in order to let us remove a second effect from the error term (in addition to that of the up-down position). The new restriction is this: We will arrange the planting sequence so that each variety appears first in one up-down block, second in another, etc., but with no arrangement repeated in the four blocks. Such an array is called a Latin square; it is sometimes referred to as "double blocking." In this case, the second block is the right-left position in which each treatment now appears once and only once.

One of the 576 possible arrangements of a 4 × 4 Latin square is shown in Fig. 14·3, in which the small letters above the array indicate the order of planting from right to left. Since each variety occupies once each of the possible positions in each of the four blocks, any effect due to a general

Position	a	b	c	d
Block I	A 8	B 6	C 6	D 7
II	C 11	A 8	D 12	B 14
III	B 17	D 18	A 16	C 12
IV	D 20	C 15	B 25	A 25

Fig. 14·3 Latin square arrangement of an experiment.

	Temp. I	2	3
Machine I	A	C	B
II	C	B	A
III	B	A	C

Fig. 14·4 Latin square experiment design.

change in level with position is in all likelihood balanced out. By this design, we will be able to take out from the error term two block effects—up-down and right-left—and also to test for the effect of variety, which is the factor of interest.

A Latin square design is useful in situations like this: We wish to test whether or not temperature affects the yield of a process. The process involves three similar machines and three similar lots of raw material, but we believe that there are real differences among machines and real differences among the lots of raw material, We want to arrange the experiment so that we can find changes in yield due to temperature alone. We will use a Latin square, necessarily testing at three temperatures 1, 2, and 3. We will block on Machines I, II, and III and assign the order to the use of raw materials A, B, and C so as to complete the Latin square array. We obtain the arrangement shown in Fig. 14·4. We have made sure that each raw material is tested on each machine, that each temperature is used with each machine, and that each lot of raw material is tested at each temperature, without duplicating any runs. Note the economy of this design: Only 9 runs are needed to acquire the data, in comparison with the 27 runs we would need to test every machine with every lot of raw material at every temperature. As usual, we have paid a price: We must assume no block-factor or block-block interactions. In a Latin square design the failure to satisfy this assumption will confuse the treatment estimates and inflate the error term.

The mathematical model for this design is

$$X_{ijk} = \mu + B_i + P_j + T_k + \epsilon_{ij(k)}$$

The value of each observation is affected by perhaps five contributing causes: the general level μ, the up-down block B, the right-left block P, the treatment T (variety of wheat), and error ϵ.

In the ANOVA we will find, as usual, a total sum of squares SST, a sum of squares for columns SSC, a sum of squares for rows SSR, and now one more: a sum of squares for treatments SSTr. The last is found by lifting out of the table of data first the total for the cells marked A in Fig. 14·3, then the total for the cells marked B, etc. Now,

$$SSTr = \frac{1}{t} \sum T_{\cdot k}^2 - \frac{T_{\cdot \cdot}^2}{n}$$

The row, column, and treatment sample sizes are necessarily the same; in our example $n = 4$. The value of v for each block and treatment factor is $n - 1 = 3$. From the total number of degrees of freedom $(n - 1)$, or 15, there remain 6 for the error term.

Latin square designs are frequently used. They are economical, inasmuch as two possible assignable causes may be removed from the error effect. Latin square designs are, however, inflexible. The number of blocks (and of blocks within blocks) must equal the number of treatments to be tested. If the number of treatments is necessarily large, the size of the experiment is great. By withdrawing from the SS term for error the effects of two blocks, we reduce the value of v for error, and by this reduce the sensitivity of the F test. On the other hand, we increase the sensitivity of the experiment by generating a smaller sum of squares for error. The most serious criticism of a Latin square is that when we use it we cannot test for interaction. We use this design, therefore, only when we have good reason to believe that interactions are not likely to be influential in the process being investigated.

PROBLEMS

14·11 Four kinds of gasoline (A,B,C,D) were tested to see whether or not they differed in performance. The response variable was miles per gallon. It was supposed that different cars and different drivers would have a significant effect; therefore, a Latin square design was used, as indicated in Table 14·5, where the cell numbers are the response-variable data. All cars were driven over the same course on the same day to minimize environmental factors.

(a) Write the mathematical model for this experiment.
(b) State the hypothesis that is under test.
(c) Complete the analysis of variance.

(d) Do the gasolines differ with respect to the variable used?
(e) What assumptions were made in this experiment?

Table 14·5 Latin Square Experiment—Test of Mileage for Four Different Gasolines

Car	Driver				Total
	I	II	III	IV	
1	A 21.2	B 20.9	C 25.0	D 24.4	91.5
2	C 28.3	D 26.2	A 20.7	B 21.1	96.3
3	D 26.5	C 24.8	B 23.2	A 22.2	96.7
4	B 21.8	A 22.3	D 25.3	C 26.7	96.1
Total	97.8	94.2	94.2	94.4	380.6

14·12 In a soapbox derby, the final race of the day was to be run in 3 parts, as there were 3 lanes in the racecourse. Each of the three finalists was to race once in each of the three lanes. The cars were Yellow, Red, and Blue. The lanes were No. 1, No. 2, and No. 3. Lay out a 3 × 3 table and show how the race might be carried out. ·

14·13 List some of the advantages in conducting the final race in the way described in Prob. 14·12.

14·14 In the testing of a rubber-covered fabric, a machine with 4 testing surfaces is used to measure the loss in weight of the fabrics due to abrasion. There are slight differences between the 4 positions on the machine, and experience has shown that replicates from the same position differ slightly. Design a Latin square experiment to test 4 fabrics for loss of weight due to abrasion. It can be assumed that the 3 factors (position, runs, and materials) act independently, so that if one position gives a higher rate of wear than another it will do so in every run and on any material.

14·6 CALCULATION OF MISSING OBSERVATION FOR LATIN SQUARE

If in running a Latin square design one observation is lost, a dummy value can be calculated by

$$X = \frac{m(R + C + T) - 2S}{(m - 1)(m - 2)} \qquad (14·2)$$

where m = number of rows (same as number of columns and treatments)

R = sum of observations in same row as dummy

C = sum of observations in same column as dummy

T = sum of observations for same treatments as for dummy

S = sum of all available observations

As before, in the ANOVA, the value of ν for the error term must be reduced by 1.

14·7 BALANCED INCOMPLETE BLOCK DESIGN

Latin square designs and modifications of them are restrictive in the sense that the number of treatments must equal the number of blocks. There are experimental situations in which this restriction is undesirable, or for practical reasons impossible to accept.

Assume that we have to test five varieties of wheat but that we have a sufficient sample of each for only four plantings. We cannot use a square design. Instead, we use a rectangular array called a balanced incomplete block design, as we show in Fig. 14·5.

Here we have placed four of the five varieties in each block, with the positions so chosen that each variety appears the same number of times in

Position	a	b	c	d
Block I	A	B	C	D
II	B	C	E	A
III	D	E	A	B
IV	E	A	D	C
V	C	D	B	E

Fig. 14·5 Balanced incomplete block arrangement of an experiment.

each position and each variety is associated equally with each other so that every pair of varieties occurs the same number of times. The arrangement shown here is only one of many possible such arrangements; we might have randomized the positions rather than use this systematic placement.

In Table 14·6 we give the values of the response variable such as we might have obtained from the randomized block design. The table indicates that we have the equivalent of a 5 × 5 design, but with a value missing from each block and from each set of treatments. As the data stand, we cannot directly compare, for instance, the total for treatment A with that for treatment B, because the sums are associated with different blocks. Therefore, the differences among the totals for treatments are affected, to some extent, by the differences in the block combinations that appear in each set of treatments.

The method of correcting the treatment means so that they are independent of the block differences is analogous to the calculations of missing data, but somewhat more involved since we have more missing values. What we require is a correction term to be applied to each treatment total.

Table 14·6   Values of the Response Variable for an Incomplete Random-ized Block Design

Treatment	Block					Sum	Q	P	Corrected treatment mean
	I	II	III	IV	V				
A	17	20	17	19	...	73			
B	21	23	22	...	23	89			
C	15	15	...	16	15	61			
D	25	...	26	24	23	98			
E	...	20	22	22	23	87			
Sum	78	78	87	81	84	408	0	0	

In Table 14·6, the first row sum is composed of the results of four runs, each of which includes the effect of treatment A. The effect of block I is included in the first cell of this row, of block II in the second, etc. Thus the row sum comprises the A effect four times, and each of the blocks I, II, III, and IV once, or $4A + I + II + III + IV$. We would like to isolate the effect of treatment A.

We now look at the column totals: the first column includes $4I + A + B + C + D$; the second includes $4II + A + B + C + E$. We show in Table 14·7 the composition of all the row and column sums. If now we add the effects for the first four columns, we have $4I + 4II + 4III + 4IV + A + (3A + 3B + 3C + 3D + 3E)$. The term in parentheses is equal to zero, inasmuch as the effects are deviations from the grand average, and therefore $A + B + C + D + E$ must be zero.

We multiply the effects included in the first row sum by 4, and obtain $16A + 4I + 4II + 4III + 4IV$. When we subtract from this term the sum of the effects for the first four columns we eliminate the block effects and have as a remainder $15A$. By dividing by the coefficient 15 we have found, as we wished to, the effect of A alone.

In general, we find the individual treatment totals using the following notation: n the number of observations (20); t the number of treatments (5); k the number of treatments per block (4). The correction term for each

	I	II	III	IV	V	Sum
A	X	X	X	X	· · ·	$4A + I + II + III + IV$
B	X	X	X	· · ·	X	$4B + I + II + III + V$
C	X	X	· · ·	X	X	$4C + I + II + IV + V$
D	X	· · ·	X	X	X	$4D + I + III + IV + V$
E	· · ·	X	X	X	X	$4E + II + III + IV + V$
Sum	4I $+A$ $+B$ $+C$ $+D$	4II $+A$ $+B$ $+C$ $+E$	4III $+A$ $+B$ $+D$ $+E$	4IV $+A$ $+C$ $+D$ $+E$	4V $+B$ $+C$ $+D$ $+E$	

treatment total involves a sum and a divisor. The sum $Q_i = kT_{i.} -$ (sum of
block totals containing the ith treatment). The divisor is $n(k - 1)/(t - 1)$.
The result is applied to the grand mean.

Thus, for the first treatment A, we multiply the sum for that treatment (73)
by 4, and subtract the block totals for those blocks that contain treatment A;
these are blocks I, II, III, and IV, for which the treatment totals are 78, 78,
87, and 81. Thus $Q_A = (4 \times 73) - (78 + 78 + 87 + 81) = -32$. The
divisor is $20(4 - 1)/(5 - 1)$, or 15. The correction for treatment A is now
$-32/15 = -2.13$. The grand mean is 408/20, or 20.40. The corrected treatment mean for treatment A is $20.40 - 2.13$ or 18.27. The other values of
corrected treatment means are computed in a similar manner, giving Table
14·8.

Table 14·8   Corrected Treatment Means for Data in Table 14·6

Treatment	Block					Sum	Q	P	Corrected treatment mean
	1	2	3	4	5				
A	17	20	17	19	· · ·	73	-32	-2.13	18.27
B	21	23	22	· · ·	23	89	$+29$	$+1.93$	22.33
C	15	15	· · ·	16	15	61	-77	-5.13	15.27
D	25	· · ·	26	24	23	98	$+62$	$+4.13$	24.53
E	· · ·	20	22	22	23	87	$+18$	$+1.20$	21.60
Sum	78	78	87	81	84	408	0	0	

For the ANOVA, we find the total sum of squares in the usual way. We also find the sum of squares for the actual block totals as usual: SS blocks = $1/k(\Sigma T_{.j}^2) - T_{..}^2/N$. Note, however, that this sum of squares for blocks is not a true estimate of the influence due to blocks, since each block contains a different set of treatment totals. The mean square for blocks is likely to have little significance (especially if it happens that we reject the null hypothesis for treatments), and therefore it need not be calculated.

For the sum of squares for treatments we use the corrected sums:

$$\text{SS treatments} = \frac{\sum Q_i^2}{nk(k-1)/(t-1)}$$

For the ANOVA, therefore,

$$\text{SST} = (17^2 + 20^2 + \cdots + 23^2) - (408^2/20) = 232.80$$
$$\text{SS blocks} = \tfrac{1}{4}(78^2 + 78^2 + 87^2 + 81^2 + 84^2) - (408^2/20) = 15.30$$
$$\text{SS treatments} = [(-32)^2 + 29^2 + (-77)^2 + 62^2 + 18^2]/60 = 199.37$$

The sum of squares for error is obtained by subtraction of the SS blocks and the SS treatments from SST; it is 18.13. We have as values for degrees of freedom 4 for treatments $(c-1)$, 4 for blocks (number of blocks -1), and

Table 14·9   Completed ANOVA for Data in Table 14·8

Source of variation	SS	D.F.	MS
Between corrected treatment totals	199.37	4	49.84
Between actual block totals	15.30	4	
Residual (error)	18.13	11	1.65
Total	232.80	19	

19 for total. Thus we have 11 left for error. We complete the ANOVA in Table 14·9. The ratio of the treatment mean square to the error mean square is 49.84/1.65 = 30.21. Based on 4 and 11 degrees of freedom, this ratio is highly significant, since $F_{4,11,0.01} = 5.67$.

Now that we have found a significant effect associated with the treatment factor we can reconsider the data in Table 14·8 and pose this question: Which of the treatments contribute to the effect which we have found? If we arrange the corrected treatment means in order of magnitude, we have

C	A	E	B	D
15.27	18.27	21.60	22.33	24.53

Which of these are to be considered as different from their neighbors is determined by (1) the difference among them and (2) the estimate of the standard deviation of the difference between the means. Our estimate of the standard deviation of the difference is related to the error term mean square, the square root of which is $\sqrt{1.65}$ or 1.28. We can interpret this value as an estimate of the standard deviation between replicates. The standard deviation of the difference in means will be smaller than this by a factor dependent on sample size for means, or in this case by the factor

$$\sqrt{\frac{2k(t-1)}{n(k-1)}}$$

The factor here is 0.730, which when multiplied by 1.28 gives 0.93. The difference between means required for significance at the 0.05 level is connected with the value of $t_{11,0.05}$, which is 2.20. Thus $2.20 \times 0.93 = 2.46$, the required difference in means for significance at the 0.05 level.

This value of the required difference indicates that treatments E and B have not been shown to be different but that a significant difference is found among the remaining treatments.

PROBLEMS

14·15   In the testing of automobile tires, large experimental errors can be reduced by building up a tire of 3 part treads. Thus each tire will have a complete tread, one third of which is from material A, a second part from material B, and the third part from material C. Design a balanced incomplete block experiment to test 4 materials using 4 three-part tread tires. Tires: 1, 2, 3, 4. Materials: A, B, C, D.

14·16   Five cars are to race to determine the fastest car on a given course. The course is only a 4-lane track with each lane of the same length. If each car must remain in the lane assigned to it, design a balanced incomplete block experiment to show how the 5 cars might be raced to determine the winner. Each race must be run with 4 cars at the same time, but more than one run for each car may be required.

14·17   Complete the analysis of the data from the balanced incomplete block design contained in Table 14·10.

QUESTIONS

14·1   Under what circumstances may we omit the determination of experimental error in an experiment? What is the justification for your answer?

14·2   Under what circumstances is the experimental error estimate likely to be inflated? What is the consequence?

Table 14·10   Balanced Incomplete Block Design

Treatment	Block				Sum
	1	2	3	4	
A	5	7	3	· · ·	15
B	5	8	· · ·	9	22
C	6	· · ·	9	10	25
D	· · ·	8	7	11	26
Sum	16	23	19	30	88

14·3   In some experimental designs, the estimate of error is found by the residual remaining when the sums of squares of the tested factors are subtracted from the total sum of squares. What is the composition of this residual sum of squares?

14·4   Explain the difference between the residual sum of squares and the sum of squares for error.

14·5   Summarize the advantages and disadvantages of the completely randomized design.

14·6   If we must make a large number of runs in order to replicate treatment combinations, can we also test for factors other than treatments in a completely randomized design? Explain.

14·7   We wish to run four treatment combinations (a, b, ab, and 1) and we want to use four similar machines to speed the experimentation. Prepare a completely randomized design for this purpose.

14·8   Set up a completely randomized experiment including 5 replicates of each of 5 treatments. Place in the cells values selected at random from Table A·1. Now remove one of the values and calculate the number that would be placed in the cell by the missing-data calculation. What do you conclude?

14·9   What is meant by blocking? Under what circumstances do we block on time? on machine? on raw material?

14·10   When we analyze the data from an experiment, we find that the blocking effect is highly significant. How should we interpret this result? What should we conclude if the blocking effect is not significant?

14·11   If we do not remove the block effect, where will it appear in the ANOVA table for a completely randomized experiment?

14·12   In what sense is a randomized block design restricted, as compared with a completely randomized experiment?

14·13   When we calculate a value by the missing-data technique, is the value intended to estimate the missing number? Explain.

14·14   What is sacrificed if ANOVA is carried out with one observation missing?

14·15   Why is a Latin square design sometimes referred to as "double blocking"?

14·16   What restrictions on randomness are involved in a Latin square design?

14·17   If the value for the uppermost left-hand cell had been missing from Table 14·4, what value would you have used in its stead?

14·18   Why do we not test for interactions in Latin square designs?

14·19   In what kinds of circumstances are we forced to use an incomplete block design? What disadvantages are associated with this design?

14·20   What information is required in order to permit us to decide which type of experimental design should be used?

REFERENCES
4, 13, 20, 25, 28, 30, 42, 56, 59, 68, 70, 73, 82, 126, 127

15 Factorial Experiments

We are interested in the effects of some of the conditions which may affect the rate at which a specified type of bacteria multiplies on a culture medium. Of the many conditions, we believe that temperature, humidity, and the concentration of agar in the medium are exceptionally important, and we intend to test these three factors.

The classical "controlled" experiment would be carried out by preparing samples with the "same" concentration of agar, holding them at the "same" temperature, and testing these controlled samples at two or more different humidities. We would thus control at some level the factors other than the single one to be tested, and we would make a decision about the effect of humidity on the basis of such a set of tests. Another similar "controlled" experiment would be performed to find out the effect of temperature, and still another to find out the effect of agar concentration.

Such a group of "controlled" experiments suffers from several serious defects:

1 The effect of humidity (and the other factors as well) is known at only *one arbitrarily chosen* level of each of the other factors. To find out whether this effect holds at other temperatures and other concentrations would require an additional group of experiments.
2 Only a few observations are made for any single set of conditions. Experimental error may under these circumstances outweigh the effect being sought. In order to estimate the effect of error we must perform several experiments under the same conditions.

3 We cannot really control those factors which are not for the moment being tested. We delude ourselves if we believe that there is no variation whatever in temperature or humidity for those tests in which we intend to change only the agar concentration.

Factorial experiment designs are superior to controlled experiments in many respects:

1 We can study the effects of several factors in the same set of experiments.
2 We can test for the effect of each factor at all levels of the other factors and can discover whether or not this effect changes as the other factors change.
3 We can test not only for the effects of the factors separately—the main effects—but also for interactions—joint effects of two or more factors combined.
4 Every judgment we make about the effects of the factors is based on all the observations accumulated in the entire set of experiments, not merely on a few of the observations. Thus factorial experiments are more sensitive in the detection of small effects.

In general, as compared with "controlled" experiments, factorial designs are more efficient in the utilization of experimental time and effort. They permit us to make judgments of greater validity and they are more sensitive.

15·1 FACTORIAL DESIGNS

A factorial experiment is one by which we obtain the same number of observations (one as a minimum, more if we desire more) for each level of the tested factor. In the illustration we are using, suppose that we are interested in 2 temperatures (70 and 80°F), 2 humidities (20 and 60 percent), and 2 concentrations of agar (2 and 5 percent). We have here a *two-level three-factor* experiment. We specify such an experiment as $2 \times 2 \times 2$ or a 2^3 experimental design. If we want one observation at each combination of levels we must make $2^3 = 8$ runs.

Factorial experiments may involve 3 levels of 1 factor and 2 levels of each of 2 factors: $3 \times 2 \times 2 = 12$ runs. When all factors have the same number of levels, we have a special, simple case; e.g., 3^4, involving 3 levels of each of 4 factors and a total of 81 runs for an experiment in which every possible combination of factor levels is used once—a complete factorial experiment. A complete factorial which requires testing many factors at many levels may require the running of an impractical number of tests. We will see in Chap. 16 that it is possible, with some minor sacrifices, to reduce the total number of runs by the use of fractional factorials that are called half replicates or quarter replicates, etc. In such experiment designs, only some of the runs of a complete factorial experiment are needed.

A 2^5 factorial design requires 32 runs. Sometimes we will wish to make the experiment more sensitive and will make two or more runs for each combination of factor levels. The effect of replicating is to increase the number of degrees of freedom for experimental error and to increase the sensitivity of the F test in the ANOVA. More often than not, we will want to improve our test sensitivity by means other than this type of replication, which is often too inefficient.

Factorial experiments can be designed for any number of levels of any number of factors. The number of levels need not be the same for all factors. We will emphasize the 2^n factorial series in which we use only two levels of each factor. Such designs are useful for preliminary, or screening, experiments. With only two levels of each factor we can find only linear effects. Further experimentation would be needed to find curvilinear effects, by methods some of which we described in Chap. 11. Experiments based on composite designs discussed in Chap. 17 permit us to supplement 2^n factorials for this purpose.

15·2 TREATMENTS

A treatment is a specific combination of specific levels of the factors used in a single run of a factorial experiment. The systematic analysis of a factorial experiment requires that we have some method of designating different treatments. Three methods are commonly used:

1 The factors themselves are designated by appropriate capital letters. In the present illustration we use T for temperature, H for humidity, and C for agar concentration. Each of these is tested at each of two levels—high or low. In designating the treatment combination, we use the corresponding small letter to indicate that the treatment involves the high level of the factor; the absence of the letter in the designation signifies the low level of the factor. Thus the treatment combination thc indicates a run at the high level of all factors, i.e., at 80°F, 60 percent humidity, and 5 percent agar concentration. The treatment symbol t specifies a run at the high level of temperature but at the low level of both humidity and concentration (since the letters h and c are missing from the treatment symbol). The special symbol (1) designates the low level of all three factors. This method is suitable only for a 2^n design.

2 Another common system uses -1 to designate the low level of a factor and $+1$ to designate the high level. Thus the symbols represent a shift of one experimental level above or below the average level. This notation is particularly useful when regression techniques are used for data analysis. For example, the symbol $-1, +1, +1$ designates the low level of the first factor, and the high level of the second and third factors.

3 A third system, probably not used so often as the other two, uses the symbols 0, 1, 2, etc., to designate factor levels. This notation can be used

Table 15·1 Treatment Combinations for 2^3 Factorial Design

	Temperature, low		Temperature, high	
	Humidity, low	Humidity, high	Humidity, low	Humidity, high
Concentration, low	(1) −1, −1, −1 000	h −1, +1, −1 010	t +1, −1, −1 100	th +1, +1, −1 110
Concentration, high	c −1, −1, +1 001	hc −1, +1, +1 011	tc +1, −1, +1 101	thc +1, +1, +1 111

for designs including any desired number of levels. In this notation, the symbol 211 designates this treatment combination: third level of the first factor, second level of the second factor, second level of the third factor.

In Table 15·1 we show how we might record the data for a complete 2^3 factorial experiment. In the cells of the table we have placed the treatment combinations as symbolized by each of the methods described above. In Table 15·2 we give the results of the experimental runs; the table data are counts of bacterial colonies.

Although we record the data in this systematic fashion, the runs themselves, depending on the situation, would have been made in completely randomized sequence, or in another of the designs described in Chap. 14. Assume that we can in 1 day prepare 8 samples but that we fear an effect associated with the order of preparation. We could, in a singly replicated experiment, randomize the order of preparation of the samples within the day. Alternatively, we might decide to prepare 2 replicates of each run, to

Table 15·2 Data from 2^3 Factorial Experiment

	Temperature, low		Temperature, high	
	Humidity, low	Humidity, high	Humidity, low	Humidity, high
Concentration, low	36	43	17	77
Concentration, high	22	68	28	76

use 2 days for the experiment, and block on days. In any case, we would attempt, by the proper design, to reduce the error term. The point is that the treatment table, such as Table 15·1, is merely a convenient way of displaying the set of experimental conditions we intend to use and is independent of the specific experimental design that we believe is appropriate.

15·3 CONTRASTS

In our singly replicated 2^3 experiment, we have a total of 8 runs; 4 of these were at the high level of temperature and 4 at the low level of temperature. Each of these sets of 4 was made at both high and low levels of the other 2 factors; thus we can estimate the effect of temperature by the average difference between these sets of 4 data.

The same 8 runs can be differently regarded: 4 were at a high level of humidity and 4 at a low level. Each of these sets of 4 runs included both levels of the other factors, and we can estimate the effect of humidity by the average difference between these sets of 4 data.

With the same 8 runs we can make a total of 7 independent different effect estimates, depending on how we group the data. These 7 estimates correspond to the total of 7 degrees of freedom that we have in the experiment design. Each such estimate of an effect is called a *contrast*.

The contrast for the temperature effect is, in symbols,

$$\tfrac{1}{4}[t + th + tc + thc - (1) - h - c - hc]$$

The multiplier $\tfrac{1}{4}$ comes from there being 4 sets of data for each level of temperature. The first four terms in brackets specify treatment combinations all at the high level of temperature; the last four specify treatment combinations all at the low level of temperature. We insert the observed data from Table 15·2: $\tfrac{1}{4}(17 + 77 + 28 + 76 - 36 - 43 - 22 - 68) = \tfrac{1}{4}(29) = 7.25$. Thus the *total* effect associated with a change in temperature from 70°F to 80°F is 29; the *average* effect or contrast is a little more than 7 colonies of bacteria.

Similarly we can estimate the effect of humidity by the contrast

$$\tfrac{1}{4}[h + th + hc + thc - (1) - t - c - tc]$$

Here we are comparing 4 observations all at a high level of humidity with 4 observations all at a low level of humidity. We have

$$\tfrac{1}{4}(43 + 77 + 68 + 76 - 36 - 17 - 22 - 28) = \tfrac{1}{4}(161) = 40.25$$

Thus the increase in humidity gives a total increase of 161 in the number of colonies, an average of over 40.

In the same manner we estimate the effect of the change in agar concentration as a total of 21, an average of a little more than 5.

Two-factor interaction effects can be found by similar computations. For the interaction between temperature and humidity, we find the contrast between the contrast of temperature at high humidity and at low humidity: $\frac{1}{4}[(th + thc - h - hc) - (t + tc - (1) - c)]$. The first of the two terms in this expression estimates the effect of a temperature difference at the high level of humidity; the second of these terms estimates the effect of temperature difference at the low level of humidity. The entire expression estimates the average effect of the difference in these two differences. We may write this expression as $\frac{1}{4}[th + thc + (1) + c - t - h \doteq tc - hc]$. In the treatment combinations th and thc both factors are at a high level; for the treatment combinations l and c both factors are at a low level. The remaining treatment combinations involve a high level of one of the two factors and low level of the other factor. Using the same data as before, we have

$$\frac{1}{4}(77 + 76 + 36 + 22 - 17 - 43 - 28 - 68) = \frac{1}{4}(55) = 13.75$$

The effects of the interactions between temperature and agar concentration, and between humidity and agar concentration, are found in the same way.

Finally, the three-factor interaction effect is the difference between the two-factor interactions at the high and low levels of the third factor. This interaction is estimated by the contrast

$$\frac{1}{4}\{[(thc - hc) - (tc - c)] - [(th - h) - (t - (1))]\}$$
$$= \frac{1}{4}[t + h + c + thc - (1) - th - tc - hc]$$

Insertion of the data gives $\frac{1}{4}(-51) = -12.75$.

Thus we can find from the 8 observations secured by the 2^3 factorial design a total of 7 contrasts. Three of these are the main effects, associated

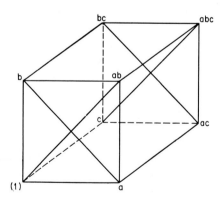

Fig. 15·1 Diagram of a 2^3 factorial experiment.

with single factors; three more are associated with two-factor interactions; one is associated with the three-factor interaction.

A 2^3 design can be represented by a cube, as in Fig. 15·1, where the treatment combinations are represented as the corners of the cube. In the analysis of the data, we compare one-half of the data against the other half; each set of four pieces of data is represented by a plane.

The A effect is found by comparing the four values for the treatment combinations on the right side of the cube with the data from the treatment combinations on the left side of the cube. Observe that the four on the right all involve the high level of factor A; the four on the left all involve the low level of factor A.

The estimation of the B effect involves a comparison of the upper surface of the cube against the lower surface. The estimation of the C effect involves a comparison of the front surface of the cube against the back surface of the cube.

The two-factor effects also are related to planes in the cube, but the planes intersect. One such pair is shown in the figure; the treatment combinations (1), c, abc, and ab lie at the corners of one plane; the treatment combinations b, bc, ac, and a lie at the corners of the other. A comparison of the data from these sets of treatment combinations gives the estimation of the AB effect.

PROBLEMS

15·1　How many experimental runs are required in the following factorial experiments:
(a) $3 \times 4 \times 2$
(b) 2^5
(c) 2×2^3
(d) A twice replicated 2^4

15·2　Using the factors A, B, C, D, and E, each at two levels:
(a) How many treatment combinations are called for?
(b) Show how the experimental runs might be made if we choose to run a completely randomized experiment.
(c) How might the experiment be carried out if we wished to run 8 treatments at the same time?

15·3　List the treatment combinations in a 2^4 factorial experiment:
(a) using the lowercase letters a, b, c, and d to signify the high level
(b) using -1 for low levels of the factors and $+1$ for the high levels of the factors
(c) using 0 for low levels and 1 for high levels

15·4　A 2^4 factorial experiment is planned to investigate the factors A (agitation), B (burst strength), C (cooling rate), and D (duration of stress). Show how the treatment combinations should be compared to estimate the following effects:
(a) B (burst strength)
(b) AB interaction
(c) all factors at the high level

15·5 In a 2^3 full factorial, randomly assign the digits 1 through 8 to the treatment combinations as response values. Then find the total effects and average effects for all factors and interactions.

15·6 Prepare a table similar to Table 15·1 for a 3^3 full factorial experiment. Indicate the treatment combinations.

15·4 TWO-WAY TABLES

A convenient way of displaying the main effects and the first-order inter-actions for two of the factors at a time is by the use of two-way (summing-over) tables derived from one such as Table 15·2. To show the effects attributable to temperature and concentration (independent of humidity) we prepare a 2×2 table using sums of pairs of observations, as follows: We sum the data for the treatment combinations (1) and h: $36 + 43 = 79$. This is the total effect at a low level of temperature and concentration summed over both values of humidity. We enter this value in the upper left cell of Table 15·3.

Table 15·3   One of the Three Two-way Tables Derivable from Table 15·2

	Temperature, low	Temperature, high
Concentration, low	$(1) + h$ 79	$t + th$ 94
Concentration, high	$c + hc$ 90	$tc + thc$ 104

Similarly, we sum the data for the treatment combinations t and th to get 94. This is the total effect for a high level of temperature and a low level of concentration summed over both levels of humidity. We enter this sum in the upper right cell of Table 15·3. We continue in this manner to complete the table.

In Table 15·3, the two row totals compare the total effects of the two levels of concentration. The two column totals compare the total effects of the two levels of temperature. The two diagonal sums compare the inter-action effects of these two factors.

Similar tables can be prepared to show the effects of humidity and con-centrations, and of humidity and temperature.

The reason for the preparation of two-way ANOVA tables is that when there are more than two factors in the experiment, the row-column treatment of Chap. 14 is inadequate. If there are more than two factors, we have not only rows and columns but also rows within rows, and perhaps (in a large experiment) columns within columns. By preparing a series of two-way

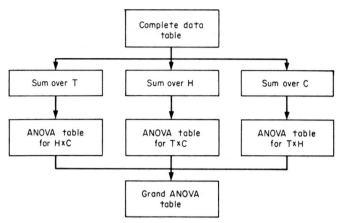

Fig. 15·2 Flow diagram for the preparation of ANOVA table for a 2^3 factorial experiment.

tables from the entire set of data, we can continue to use the technique of rows and columns.

Figure 15·2 shows the required steps in block form. For the 2^3 factorial we are using, it will be necessary to make three summing-over tables, one for each pair of factors. For each two-way table, we run an analysis of variance, with sources of variance as factor A, factor B, interaction, and total.

After the three two-way tables are made, the results are transferred to a grand ANOVA table, and the total sum of squares is computed for the entire set of data. The analysis of variance is completed by finding mean square values and F ratios in the usual manner. If we have only one observation for each treatment combination, the residual found by subtraction will be used for the estimate of error. If we believe that the three-factor interaction may be important, replication of runs will be necessary unless an outside estimate of error is available.

15·5 ANOVA FROM SUMMING-OVER TABLES

We illustrate the use of two-way summing-over tables, beginning with the data in Table 15·4. The numbers in the cells of the table represent coded data from a factorial experiment.

Table 15·4 Data for ANOVA Using Summing-over Tables

A_1				A_2			
B_1		B_2		B_1		B_2	
C_1	C_2	C_1	C_2	C_1	C_2	C_1	C_2
3	5	4	7	4	2	3	7

We first sum over the factor C: We add the first two numbers in the array, thus finding the sum of the observations for both levels of C at the first level of B and the first level of A. The sum (8) appears in the upper left-hand cell of the summing-over Table 15·5a. (The 2 in the corner of this cell is to remind us that the value in this cell is composed of two observations; we will need this number in the ANOVA. We do not place a 2 in every cell merely to avoid unnecessary repetition.) Next, we find the sum of the next two values in the array in Table 15·4; this sum (11) is the total of the observations for both levels of C at the second level of B and the first level of A. This sum appears in the upper right-hand cell of Table 15·5a.

Table 15·5 (Data from Table 15·4) Summed over Factor C

(a) Two-way Table

	B_1	B_2	Total
A_1	8 $\boxed{2}$	11	19
A_2	6	10	16
Total	14	21	35

(b) ANOVA for Factors A and B

Source	SS	ν	MS
A	1.125	1	1.125
B	6.125	1	6.125
AB	0.125	1	0.125
Total	7.375	3	

Continuing in this way, we find 6 as the sum of the observations for both levels of C at B_1 and A_2, and 10 as the sum of the observations for both levels of C at B_2 and A_2. We have thus completed the cell entries in the summing-over table. These entries indicate the observations for changing levels of factors A and B, independent of the factor C.

We complete Table 15·5a by finding row and column totals of the cell entries. We now have the numbers needed for an analysis of variance of the two factors A and B:

$$SST_{AB} = \tfrac{1}{2}(8^2 + 6^2 + 10^2 + 11^2) - 35^2/8 = 160.500 - 153.125 = 7.375$$
$$SSA = \tfrac{1}{4}(19^2 + 16^2) - 35^2/8 = 154.25 - 153.125 = 1.125$$
$$SSB = \tfrac{1}{4}(14^2 + 21^2) - 35^2/8 = 159.25 - 153.125 = 6.125$$
$$SS(AB) = SST_{AB} - SSA - SSB = 7.375 - 1.125 - 6.125 = 0.125$$

We have just found a total sum of squares for the factors A and B (SST_{AB}), a sum of squares for A alone (SSA), a sum of squares for B alone (SSB), and a sum of squares for the interaction of A and B [$SS(AB)$]. Each of these is independent of the effect of factor C.

In Table 15·5b we show the data to this point in the form of an ANOVA table. We do not continue the analysis to find F ratios, since these data will soon be transferred to the grand ANOVA table.

We prepare the two-way table summed over B by a similar method. We first add the observations in the first and third cells of Table 15·4; this sum (7) is for both levels of B at C_1 and A_1. We enter this value as the first in Table 15·6a. The next entry in this table is 12; it is the sum of the values 5 and 7 for both levels of B and C_2 and A_1. Similarly we complete the two-way summing-over table.

Table 15·6   (Data from Table 15·4) Summed over Factor B

(a) *Two-way Table*

	C_1	C_2	Total
A_1	7	12	19
	2		
A_2	7	9	16
Total	14	21	35

(b) *ANOVA for Factors A and C*

Source	SS	ν	MS
A	1.125	1	1.125
C	6.125	1	6.125
AC	1.125	1	1.125
Total	8.375	3	

The sums of squares are found as before. They are:

$$SST_{AC} = 8.375$$
$$SSA = 1.125$$
$$SSC = 6.125$$
$$SS(AC) = 1.125$$

In Table 15·6b we show these data as in a summary ANOVA table. Observe that we find here the same SSA value that we found in the summing-over table for A and B.

A similar procedure gives the summing-over and the two-way ANOVA tables for the factors B and C. The results are given in Table 15·7a and b. By these calculations we obtain only two sums of squares (for the BC interaction and for the total) that we did not already know.

Table 15·7   (Data from Table 15·4) Summed over Factor A

(a) *Two-way Table*

	C_1	C_2	Total
B_1	7	7	14
	2		
B_2	7	14	21
Total	14	21	35

(b) *ANOVA for Factors B and C*

Source	SS	ν	MS
B	6.125	1	6.125
C	6.125	1	6.125
BC	6.125	1	6.125
Total	18.375	3	

To complete the analysis of variance, we return to the original table of data, find a total sum of squares for these values, and subtract the sums of squares we found in the summing-over tables. The result is the ABC interaction; the value is 3.125. We transfer the data to the grand ANOVA summary Table 15·8.

Table 15·8 Grand ANOVA Summary (Data
from Tables 15·4 to 15·7 and from
External Error Estimate)

Source	SS	ν	MS	F ratio calculated
A	1.125	1	1.125	1.12
B	6.125	1	6.125	6.12
C	6.125	1	6.125	6.12
AB	0.125	1	0.125	0.12
AC	1.125	1	1.125	1.12
BC	6.125	1	6.125	6.12
ABC	3.125	1	3.125	3.12
Total	23.875	7		
Error, external	24.100	24	1.004	

$F_{1,24,0.05} = 4.26$

Since we did not replicate the runs in this experiment, we have two alternatives for the error term: We may use the highest-order interaction MS value, in this case for the ABC interaction; or we may use an external estimate of error. Let us suppose that we have from an earlier experiment of the same type, but with four replications, a mean square for error of 1.0, with $\nu = 24$. We include this information in the ANOVA Table 15·8. Based on this external error estimate, the F ratios indicate that factors B and C and the BC interaction are statistically significant.

PROBLEMS

15·7 Using the data in Table 15·2, construct all the two-way summing-over tables and compute the three ANOVA tables for the summing-over tables.

15·8 Using the answer to Prob. 15·7, complete the grand ANOVA table for the complete experiment, if the runs were made in a completely randomized order. (Assume the three-factor interaction effect to be nonexistent.)

15·9 We wish to investigate three factors R, S, and T. Each factor is to be tested at two levels, and we are to make two runs at each treatment combination.

(a) Construct a treatment table showing the entire set of treatment combinations to be run.

(b) Write the mathematical model for this experiment if all three factors are crossed and fixed.

(c) Using the digits from 1 to 16 as the response values, randomly assign the response values to the treatment combinations.

(d) Since we have randomly assigned the digits to the treatments, what results should we expect in the grand ANOVA table of this experiment?

(e) Complete the analysis of variance for the experiment.

15·6 STANDARD ORDER OF ANALYSIS

For facilitating the process of finding effects and for the ANOVA to come, a standard order of tabular arrangement of data is useful, particularly so for experiments more complicated than the 2^3 design we are considering.

The standard order is found as follows: We begin with treatment combination (1); we multiply this term by the first treatment factor to get t; we then multiply both of these by the second treatment factor h to get h and th; finally we multiply these four by the third factor c to get c, tc, hc, and thc. Thus we develop the sequence of treatment combinations shown in the left-hand column of Table 15·9. Across the top of the table we place the desired effects in the same order as in the column. Main effects are indicated by single capital letters; interactions are indicated by products of capital letters.

Table 15·9   Signs for Calculating Effects for 2^3 Factorial Experiment

Treatment	Total	T	H	TH	C	TC	HC	THC
(1)	+	−	−	+	−	+	+	−
t	+	+	−	−	−	−	+	+
h	+	−	+	−	−	+	−	+
th	+	+	+	+	−	−	−	−
c	+	−	−	+	+	−	−	+
tc	+	+	−	−	+	+	−	−
hc	+	−	+	−	+	−	+	−
thc	+	+	+	+	+	+	+	+

The signs in the body of the table are those which we will apply to the observations (from each treatment combination) to calculate the effects. The signs are placed in the table column by column. For the main-effect columns, we mark − if the factor is at a low level in the treatment combination; we mark + if the factor is at a high level. For the interaction effects, the sign is the algebraic product of the signs already indicated in the appropriate main effects. Note the tie-in here with the −1, +1 type of notation for factor levels. The 1's do not appear in Table 15·9 but are understood.

For example, opposite (1) we place − in each of the main-effect columns because the treatment combination symbol (1) signifies the low level of every factor. In this same row under TH we place +, the product of the two negative signs already associated with the separate factors T and H. Under THC in this row we place −; the negative sign is the triple product of the signs attached to the three main factors.

The remainder of the table is completed in the same manner with the exception of the first column marked total. For this column we assign + to every row entry. The sum of the data will in this column give us the grand total of the observations which we will need for the ANOVA.

The application of this table is this: The signs indicate how we are to combine the observations in order to calculate each of the seven total effects we can estimate from this 2^3 factorial design. Once we find the total effects, we find the average effect by dividing the total effect by the number of comparisons involved in the effect calculation. In this case, the number of comparisons is four.

15·7 BINOMIAL METHOD

An alternative way of determining how the data from the experiment are to be combined to find effects is this: For each factor in the experiment form a binomial of the pattern $(a \pm 1)$ and find the product of these binomials. The sign in the binomial is − if the factor letter is *present* in the desired effect; the sign in the binomial is + if the factor letter is missing. The presence of − implies that we are interested in the *contrast* associated with the factor; the presence of a + implies that we are *averaging* the factor effect.

Thus, to find the required method of combining data to get the A effect in a factorial involving the factors A, B, C, we write the binomials

$$(a - 1)(b + 1)(c + 1)$$

Since we are for the moment interested in only the A effect, we have inserted a negative sign in the binomial containing a, and a positive sign in the other two binomials. Multiplying the three binomials together gives $+a +ac +ab +abc -(1) -b -c -bc$. Combining the observations from the experiment according to these signs will give the A effect total.

Similarly, to get the AB effect, we form the binomials $(a - 1)(b - 1)(c + 1)$. The product of the binomials is $+abc +c +ab +(1) -bc -ac -b -a$; the signs direct us how to combine the experiment observations in order to estimate the total AB interaction effect.

The binomial method outlined above can be used for a factorial experiment involving any number of factors; all factors must be tested at two levels only.

PROBLEM

15·10 If the 3 factors to be investigated in a 2^3 factorial are R, S, and T, what is the standard order for the treatment combinations?

15·11 If a fourth factor Z is added to the experiment in Prob. 15·10:
(a) what is the standard order of the treatment combinations if we run all factors at 2 levels
(b) what is the mathematical model for the experiment if all factors are run at 2 levels

15·12 In a 2^4 full factorial, the 4 factors are C, B, N, and S.
(a) List the treatment combinations in standard order.
(b) Build a plus and minus table, like Table 15·9, to show how the total effects are to be estimated.
(c) From the table in answer to (b) above, indicate how the total effect CBN would be estimated.
(d) Using the binomial method, show how the treatment combinations should be combined to calculate the CB total effect.
(e) Using the binomial method, how would the average effect of S be calculated?

15·8 YATES METHOD

A systematic method for calculating effects in a factorial experiment is useful because it reduces the chance for errors, especially in large experiments. The Yates method requires listing the treatment combinations in standard order, and beside them the corresponding data—the observations (or the sum of the replicates). Additional columns complete the table, the number of added columns being equal to the exponent of the factorial. A 2^3 experiment requires 3 added columns; a 2^5 experiment requires 5.

Each column of numbers is generated from the preceding column in the same manner. The first four numbers in the new column are the *sums* of the successive pairs of numbers in the preceding column; the next four numbers in the new column are the successive *differences* in these same pairs. To find the difference, we always subtract the *first* member of each pair from the *second*. The final column of numbers is the desired set of effects totals.

We have applied the Yates method to the data from Table 15·2; the results are given in Table 15·10.

15·9 CALCULATIONS FOR ANOVA IN FACTORIAL DESIGNS

We have described four somewhat different methods of finding the total effects attributable to the experiment factors and interactions. Regardless of the method used, if we square each of the factor total effects and divide by the number of items making up each total (in this case 8) we obtain as a result the SS term for that factor or interaction. The SS term so found for

Table 15·10 Application of Yates Method to a 2³ Factorial

| Treatment | Observation | Factor columns | | | Effect |
		I	II	III	
(1)	36	53	173	367	Total
t	17	120	194	29	*T*
h	43	50	15	161	*H*
th	77	144	14	55	*TH*
c	22	−19	67	21	*C*
tc	28	34	94	−1	*TC*
hc	68	6	53	27	*HC*
thc	76	8	2	−51	*THC*

the "Total" in the Yates calculation (Table 15·10) is the correction factor used in the ANOVA calculations. In general,

$$SS = \frac{(\text{effects})^2}{pk^n}$$

where p = number of replications
n = number of factors
k = number of levels of each factor

The total sum of squares must yet be found in the usual way: We square the individual observations, sum these squares, and subtract the correction factor.

We apply the preceding rules to the data in Table 15·10. We have in the last column the number 29 as the total effect for factor T. Then

$$SS_T = \frac{29^2}{8} = 105.125$$

where SS_T is the sum of squares for the factor "temperature." Similarly:

$SS_H = 161^2/8 = 3240.125$
$SS_{TH} = 55^2/8 = 378.125$
$SS_C = 21^2/8 = 55.125$
$SS_{TC} = -1^2/8 = 0.125$
$SS_{HC} = 27^2/8 = 91.125$
$SS_{THC} = -51^2/8 = 325.125$

In each of the SS terms, the subscript indicates the main factor or interaction estimated by the calculated numbers.

15·10 ANOVA FOR SINGLY REPLICATED FACTORIALS

We find that the sums of squares we have found for the three main factors, the three two-factor interactions, and the three-factor interaction together exhaust the total sum of squares for the whole experiment. We have therefore no residual and no SS term for error. We cannot as a consequence perform the F tests against the usual estimate of error.

This situation arises from our having only one observation for each of the treatment combinations. The total of 8 observations gives us 7 degrees of freedom. One degree of freedom is associated with each factor and with each interaction, leaving none for error.

In any singly replicated experiment, we must find a substitute for the error term in order to run F tests on the data. We have two alternatives: to obtain an external estimate of error from another experiment; to use high-order interaction SS terms. There are two justifications for the second of these procedures. First, as we have designed the experiment we have no suitable alternative. Second, experience shows that interactions (especially those involving more than two factors) are only rarely of practical significance. Apparent high-order interactions are usually attributable to error.

In Table 15·11 we show the values of the response variable from a 2^4 factorial. We use the Yates method to find the effects as shown in Table 15·12; from this we have prepared the ANOVA Table 15·13. We now pool the three-factor effects and call the resulting MS term "residual" in Table 15·14; we use this residual MS as the substitute for the error term in the calculation of the F ratios. Thus we attach statistical significance to the C and the D effects and to the AB interaction. With an α risk of 0.10 we find the CD interaction also significant.

Table 15·11   Data for a 2^4 Factorial
Experiment

		A_1		A_2	
		B_1	B_2	B_1	B_2
C_1	D_1	4	2	2	1
	D_2	9	5	8	8
C_2	D_1	3	1	2	4
	D_2	11	9	9	11

Table 15·12   Yates Method Applied to Data
from Table 15·11

Treatment	Response	I	II	III	IV
(1)	4	6	9	19	89
a	2	3	10	70	1
b	2	5	30	−1	−7
ab	1	5	40	2	13
c	3	17	−3	−3	11
ac	2	13	2	−4	3
bc	1	20	2	5	7
abc	4	20	0	8	3
d	9	−2	−3	1	51
ad	8	−1	0	10	3
bd	5	−1	−4	5	−1
abd	8	3	0	−2	3
cd	11	−1	1	3	9
acd	9	3	4	4	−7
bcd	9	−2	4	3	1
abcd	11	2	4	0	−3

Table 15·13   ANOVA Summary for Data
from Table 15·12

Source	SS		MS
A	0.06	1	0.06
B	3.06	1	3.06
C	7.56	1	7.56
D	162.56	1	162.56
AB	10.56	1	10.56
AC	0.56	1	0.56
AD	0.56	1	0.56
BC	3.06	1	3.06
BD	0.06	1	0.06
CD	5.06	1	5.06
ABC	0.56	1	0.56
ABD	0.56	1	0.56
ACD	3.06	1	3.06
BCD	0.06	1	0.06
ABCD	0.56	1	0.56
Total	197.90	15	

Table 15·14 ANOVA for Data from Table 15·11, Three-factor Effects Pooled

Source	SS		MS
A	0.06	1	0.06
B	3.06	1	3.06
C	7.56	1	7.56*
D	162.56	1	162.56**
AB	10.56	1	10.56*
AC	0.56	1	0.56
AD	0.56	1	0.56
BC	3.06	1	3.06
BD	0.06	1	0.06
CD	5.06	1	5.06
Residual	4.80	5	0.96
Total	197.90	15	

$F_{1,5,0·10} = 4.06$

$F_{1,5,0·05} = 6.61$

PROBLEMS

15·13 Using the data in Table 15·2, find the sum of squares of all sources by Yates technique.

15·14 Compare the sum of squares as found in answer to Prob. 15·13 with the values as found by the summing-over tables and the grand ANOVA table.

15·15 The data in Table 15·15 are the result of a 2^3 factorial experiment, where all factors are fixed and all tests were run in a completely randomized order. Using the Yates technique, find the effects and then the sum of squares for this experiment.

Table 15·15 Data from a 2^3 Factorial Experiment

Material	Running speed			
	35 ft/min		50 ft/min	
	Percent humidity		Percent humidity	
	40	50	40	50
Type A	(1) 37.4	h 39.2	r 36.6	hr 40.0
Type B	t 35.5	ht 37.7	rt 32.0	hrt 38.8

15·16   The data in Table 15·16 are the result of a twice-replicated 2^3 factorial experiment. Calculate the mean squares for all sources using the Yates technique.

Table 15·16   Data from a Twice-replicated
2^3 Factorial Experiment

Type	35 ft/min		50 ft/min	
	40%	50%	40%	50%
A	37.4 36.9	39.2 40.1	36.6 36.4	40.0 39.9
B	33.5 33.2	37.7 38.5	32.0 31.8	38.8 39.3

(NOTE: Use the totals of each treatment combination as the response variable, and then increase the divisor of the sum of squares by a factor equal to the number of replicates per treatment combination.)

15·11   ANALYSIS OF 3^3 FACTORIAL BY THE YATES METHOD

A three-level experiment in many situations is superior to a two-level experiment because in the data analysis we can not only test for main effects and for two-factor interactions; we can also test for linear and quadratic effects of each of the main factors by the use of the table of orthogonal coefficients as described in Chap. 11. The data analysis is only slightly more laborious, and the results are far more comprehensive.

As in the previous description of the Yates method, we list the treatments in standard order and follow a strict procedure, based on the sum and difference of the response variables, to determine the effects associated with the factors and their interactions. By applying a divisor to the effects, we find values for sums of squares which make possible the ANOVA.

For the listing of the treatment combinations, we will use the alternative symbolization in which the symbol 000 represents the low level of all three factors, the symbol 100 represents the middle level of factor A and the low level of the other two factors, the symbol 211 represents the highest level of factor A and the middle level of the other two factors, etc. Furthermore, the subscript 1 represents an effect which is linear with respect to the designated factor, and the subscript 2 represents an effect which is quadratic with respect to the designated factor.

Thus, in the tabulation of the computations by the Yates method, we will find in the effects column not only the effects and the interactions but also the division of the effects into linear and quadratic terms. The analysis should be followed by reference to Table 15·17, in which we show the results of the following procedure:

Table 15·17 *Application of the Yates Method to a 3^3 Factorial Experiment*

Treatment	Yield	I	II	III	Effect	Divisor	SS
(1) 000	5	16	34	137	Total		
a_1 100	6	4	52	−7	A_1	18	2.72 ⎫
a_2 200	5	14	51	17	A_2	54	5.35 ⎬ A
b_1 010	2	16	−7	13	B_1	18	9.39 ⎫
a_1b_1 110	1	15	−3	5	A_1B_1	12	
a_2b_1 210	1	21	3	1	A_2B_1	36	⎬ B
b_2 020	8	11	1	23	B_2	54	9.79 ⎭
a_1b_2 120	4	19	7	−13	A_1B_2	36	
a_2b_2 220	2	21	9	11	A_2B_2	108	
c_1 001	9	0	−2	17	C_1	18	16.05 ⎫
a_1c_1 101	5	−1	5	10	A_1C_1	12	
a_2c_1 201	2	−6	10	8	A_2C_1	36	
b_1c_1 011	5	−7	−6	12	B_1C_1	12	
$a_1b_1c_1$ 111	4	1	10	7	$A_1B_1C_1$	8	
$a_2b_1c_1$ 211	6	3	1	−9	$A_2B_1C_1$	24	⎬ C
b_2c_1 021	6	0	4	−28	B_2C_1	36	
$a_1b_2c_1$ 121	6	2	2	1	$A_1B_2C_1$	24	
$a_2b_2c_1$ 221	9	1	−5	17	$A_2B_2C_1$	72	
c_2 002	5	−2	22	−19	C_2	54	6.68 ⎭
a_1c_2 102	1	1	7	2	A_1C_2	36	
a_2c_2 202	5	2	−6	−4	A_2C_2	108	
b_1c_2 012	5	1	−4	−2	B_1C_2	36	
$a_1b_1c_2$ 112	7	3	−6	−25	$A_1B_1C_2$	24	
$a_2b_1c_2$ 212	7	3	−3	−5	$A_2B_1C_2$	72	
b_2c_2 022	7	8	−2	2	B_2C_2	108	
$a_1b_2c_2$ 122	6	−2	−2	5	$A_1B_2C_2$	72	
$a_2b_2c_2$ 222	8	3	15	17	$A_2B_2C_2$	216	

1 In the first column are listed the treatment combinations in standard order. Observe carefully the sequence of the symbols and their significance.

2 In the second column are listed the values of the response variable.

3 The tabulation is divided into sections of three rows each, associated with the three levels of each of the three factors.

4 In the next column we place first the successive sums of the sets of three in each row section. There will be nine such sets, filling in the first nine rows of the table.

5 Following the sums, we calculate and enter the differences between the first and third item in each of the sets of the original data (we subtract the first from the third). These differences fill in the next nine rows.

6 We complete the column by calculating and entering the sum of the first and third values in each of the original row sections minus twice the second. These values complete the third column of the table. (Note: Steps 4, 5, and 6 involve the application of the orthogonal coefficients in Table 15·18 to the data we collected.)

Table 15·18   Orthogonal Coefficients

Observations	First	Second	Third
Sum	+1	+1	+1
Linear	−1	0	+1
Quadratic	+1	−2	+1

7 We repeat steps 4 through 6 twice more, applying the process to the calculations in the third column to generate the values in the fourth column, and then to those values to generate the values in the fifth column. Three successive sets of calculations are needed because this is a three-factor experiment. The numbers in the last of these columns are the effects indicated by the letter designations in the next column.

8 We enter the value of the divisor according to

$$\text{Divisor} = 2^m 3^{n-p}$$

where m = number of terms in effect symbol
$\quad\ n$ = number of factors in experiment
$\quad\ p$ = number of linear terms designated 1 in treatment combinations

For example, to find the divisor for the effect $A_1 B_1$, substitute 2 for m (since two factors are listed in the effect term), 3 for n (since this is a three-factor experiment), and 2 for p (since both factors are designated 1 in the treatment combination 110): $2^2 \times 3^{3-1} = 4 \times 3 = 12$; this divisor appears in the appropriate column opposite $A_1 B_1$. The other divisors are obtained by the same procedure.

9 Now we find the SS terms as before, applying the divisors just found to the effects squared. Thus we have the values in the final column.

10 We identify the SS terms for the desired effects as follows: Those designated A_1, B_1, and C_1 are the *linear* effects for each of the main factors; those designated A_2, etc., are the quadratic effects for the same factors. For the interactions, we have various combinations: $A_1 B_1$ specifies the linear two-factor interaction; $A_2 B_1$ specifies an interaction

quadratic in A and linear in B; etc. Unless we have reason to do otherwise, we would probably sum the separate two-factor interactions for A and B, for B and C, and for A and C. This we have done in preparing Table 15·19, giving the summary ANOVA for the data from Table 15·17.

Table 15·19 Summary ANOVA for Data in Table 15·17

Source	SS		v		MS	
Main effects:						
A Linear	2.72	8.07	1	2	2.72	4.04
Quadratic	5.35		1		5.35	
B Linear	9.39	19.18	1	2	9.39	9.59
Quadratic	9.79		1		9.79	
C Linear	16.05	22.73	1	2	16.05	11.36
Quadratic	6.68		1		6.68	
Two-factor interactions:						
$L_A L_B$	2.08		1		2.08	
AB $Q_A L_B$	0.25	8.13	1	4	0.25	2.04
$L_A Q_B$	4.69		1		4.69	
$Q_A Q_B$	1.11		1		1.11	
$L_A L_C$	8.33		1		8.33	
AC $Q_A L_C$	1.78	10.37	1	4	1.78	2.59
$L_A Q_C$	0.11		1		0.11	
$Q_A Q_C$	0.15		1		0.15	
$L_B L_C$	12.00		1		12.00	
BC $Q_B L_C$	21.80	33.95	1	4	21.80	8.49
$L_B Q_C$	0.11		1		0.11	
$Q_B Q_C$	0.04		1		0.04	
Three-factor interactions (residual)	41.41		8		5.18	
Total	143.84		26			

The analysis we have demonstrated applies only to quantitative, equally spaced levels of each factor. If any one, or all, of the factors were qualitative, we would carry out the procedure as we have described, but we would not attach any significance to the linear-quadratic division of the sums of squares and would add them for each factor and each interaction effect, to generate single terms for the ANOVA.

15·17   An experiment was conducted to study the effect of three different drying temperatures (equally spaced increments), three different types of washing methods, and three transport systems, on the dimensional stability of a new high-resolution photographic film. The amount of film change with respect to length, as measured between two reference points on a standard, is shown in Table 15·20.

Table 15·20   Data from an Experiment on Dimensional Stability of Film

	Temperature 1			Temperature 2			Temperature 3		
	W_1	W_2	W_3	W_1	W_2	W_3	W_1	W_2	W_3
Transport I	3	5	10	2	4	11	3	6	9
Transport II	4	6	10	4	4	12	5	7	13
Transport III	6	7	12	7	8	12	6	7	14

(a) Using the Yates technique, calculate the sum of squares for the sources of variation.
(b) Which of the three factors can be considered as quantitative?
(c) Find the linear and quadratic sum of squares for the quantitative factor(s).

15·18   To study the effects of various temperatures and various humidities on the resistance of a coil, three temperatures T_1, T_2, and T_3 were used in tests with three levels of humidity H_1, H_2, and H_3. The results of the investigation are given in Table 15·21.

Table 15·21   Data from an Experiment
on Resistance of Resistors

	T_1	T_2	T_3
H_1	26	36	28
H_2	40	49	38
H_3	35	40	36

(a) Using the Yates technique, complete the analysis of the data, treating each factor as a qualitative factor.
(b) Complete the analysis of the data, treating each factor as a quantitative factor with equally spaced levels.
Assume all interactions to be nonexistent.

QUESTIONS

15·1 Define a "factorial" experiment.

15·2 What are the advantages of a factorial design, as compared with other designs? What are the disadvantages?

15·3 How many runs would be required for a factorial experiment for each of the following: (a) 3^4; (b) 4^3; (c) 3^{4-1}; (d) 4^{3-1}; (e) 2^{5-2}?

15·4 In a fully replicated 2^3 factorial experiment, what is the source of the error estimate?

15·5 What are the advantages of a 3^3 design as compared with a 2^3 design? What are the disadvantages?

15·6 List the treatment combinations for a factorial experiment using 4 factors each at 2 levels. Call the factors D, G, T, and W.

15·7 For Question 15·6, which treatments would be used to obtain the DG effect? How would they be compared?

15·8 How would the average DG effect be computed in Question 15·6?

15·9 List the treatment combinations for a 3^4 design, using as factors A, B, C, and D.

15·10 Write the mathematical model for a fully replicated 2^4 factorial design.

15·11 Prepare a table of signs for combining the treatments for a 2^4 factorial design.

15·12 Use the products of binomials to determine the methods of finding the effects for Question 15·11.

15·13 Discuss the analysis of a completely crossed experiment in which 3 factors are used, each with a different number of levels.

15·14 If an experiment with 3 or more factors is broken down into two-way tables, why is an ANOVA table prepared from each of the two-way tables?

15·15 (a) Using 3 factors A, B, and C, each at 2 levels, arrange the treatment combinations in standard order as required for the Yates technique of data analysis.

(b) Repeat (a) above, but for 4 factors A, B, C, and D.

(c) Compare your answers for (a) and (b).

15·16 When the Yates technique is used, how are the total effects converted to sums of squares? How is the total sum of squares found?

15·17 Under what circumstances may MS terms be pooled?

15·18 What are the advantages of replication in factorial experiments? the disadvantages?

REFERENCES

3, 13, 47, 56, 58, 59, 68, 72, 123, 126, 127

16 Confounding and Fractional Factorials

When we are involved with experimentation with many factors, we may have more treatment runs than we can manage in a reasonable period of time. In a 2^3 factorial we will need 8 runs; if each treatment requires 2 hours, and we work an 8-hour day, the experiment will require at best a total of 2 days. The question to be answered at this point is this: Which four treatments should we assign to each day's work? Any day-to-day difference, if it exists, will influence the results of the second day's work. If we use a completely randomized experiment the error term will be inflated by day-to-day variability.

If we believe that the day-to-day variability is important, we will block on days to remove this source of variation from the error term. By blocking in this way, we will use one degree of freedom. There will remain six degrees of freedom from the seven with which we began. These remaining six degrees of freedom will permit an estimation of the three main effects and the three two-factor interaction effects, but there will be no possible way of obtaining an independent measure of the three-factor effect.

16·1 CONFOUNDING

When we find it impossible to run all the treatment combinations required for a full factorial in a single block, we select certain effects (usually high-order interactions) that we are willing to sacrifice in order to remove a block effect. Such a situation is called *confounding*. Confounding means "to combine in an indistinguishable manner." Confounding is not at all the same as "confusion"; we can control the effect to be confounded with the block effect. The main effects in the experiment and the other interactions will remain free and clear of any block effect; the experimental error will by this procedure be a smaller, purer error term because it too will be free of the block effect.

We find the technique we are about to describe especially useful in screening experiments, in which we wish to find the important factors in a process when there are many factors that might be important. We are usually interested in main effects and sometimes in low-order interactions, but we often have no reason to believe that high-order interactions are of practical significance. We are therefore willing to sacrifice an estimate of high-order interactions in order to remove an effect on which we can block.

In a 2^3 factorial experiment, assume that the three-factor interaction is the one we can best afford to lose. We proceed by first forming a series of binomials, using a negative sign in each binomial which contains the letter included in the interaction we are willing to sacrifice. In this case, we have $(a - 1)(b - 1)(c - 1)$. When we expand the product of these binomials, we obtain $+abc$, $-bc$, $-ac$, $+c$, $-ab$, $+b$, $+a$, $-(1)$. We have four positive terms and four negative terms. We now assign to day 1 all the positive terms and to day 2 all the negative terms, or the contrary.

We may now block on days as follows:

Day 1: abc, a, b, c
Day 2: (1), ab, ac, bc

When we analyze the experimental runs, we will be able to calculate the day-to-day variability by working with the day totals in the usual way. The A effect will be computed from the total of the four treatments with A at the high level compared with the total from the four treatments with A at the low level. Each day contains two treatments with A at the high level and two with A at the low level. Even if there is a day-to-day difference, the effect due to A will still be free of its influence. The same will also hold for the other two main effects, and for the three two-factor interactions.

Suppose that each result on the second day was increased by 10 units of response because of the day-to-day difference. In the calculation of the effect due to A, we would have the following:

Day 1: abc, a, b, c
Day 2: $(1) + 10$, $ab + 10$, $ac + 10$, $bc + 10$

$$\begin{aligned}
\text{Effect sum } A &= abc + a + (ab + 10) + (ac + 10) - b - c \\
&\qquad\qquad - (bc + 10) - [(1) + 10] \\
&= abc + a + ab + ac + 20 - b - c - bc - (1) - (20) \\
&= abc + a + ab + ac - b - c - bc - (1) \\
&= a(bc + 1 + b + c) - (bc + 1 + b + c) \\
&= 4A, \text{ or total effect of } A, \text{ obtained from four comparisons.}
\end{aligned}$$

When we compute the three-factor interaction, using the same 10-unit increment on the second day, we see that the effect sum ABC does not reduce as did the A effect sum, since the 10 units of increase does not drop out of the calculations:

$$\text{Effect sum } ABC = (1) + 10 + ab + 10 + ac + 10 + bc + 10$$
$$- abc - a - b - c$$
$$= (1) + ab + ac + bc + 40 - a - b - c - abc$$
$$= 4ABC + 40$$

This last calculation is also the calculation for the block effect. If we had not known that the second-day increment was 10, the results obtained would demonstrate a change in level. We would be unable to determine if the change was due to the blocking or to the three-factor interaction. *Thus we have confounded the ABC interaction with the blocking effect.*

The three main factors and the three two-factor interactions are free of the block effect; only the ABC interaction is confounded with the blocking. The ANOVA for this experiment will be as in Table 16·1.

Table 16·1   ANOVA for 2^3 Experiment, ABC Effect Confounded with Blocks

Source	SS	D.F.	MS
A		1	
B		1	
C		1	
AB		1	
AC		1	
BC		1	
Blocks (ABC)		1	
Total		7	

There is another way to find out which treatments should be assigned to the different blocks, based on a table of $+$ and $-$ values such as Table 16·2. The effects (capital letters) are the column headings; contained in each column is a listing of the eight treatment combinations. If the ABC effect is to be confounded with blocks, then by looking under the ABC column, the eight treatments are assigned the plus and minus values as listed, the same as we determined by the algebraic technique of binomial multiplication.

If there were some reason for our wanting to confound AB interaction with blocks, then from Table 16·2 we would group the eight treatments as follows:

Day 1: (1), ab, c, abc (the positive signs in column AB)
Day 2: a, b, ac, bc (the negative signs in column AB)

Table 16·2   Plus-Minus Values for 2^3 Factorial

Treatment combinations	T	A	B	AB	C	AC	BC	ABC
(1)	+	−	−	+	−	+	+	−
a	+	+	−	−	−	−	+	+
b	+	−	+	−	−	+	−	+
ab	+	+	+	+	−	−	−	−
c	+	−	−	+	+	−	−	+
ac	+	+	−	−	+	+	−	−
bc	+	−	+	−	+	−	+	−
abc	+	+	+	+	+	+	+	+

Table 16·3   Twice-replicated 2^3 Factorial, Blocked on Days (Four Runs per Day)

(a)

Replicate	Block		Day
1	1	(1) ab ac bc	1
	2	a b c abc	2
2	1	(1) ab ac bc	3
	2	a b c abc	4

(b) ANOVA Table

Source	SS	ν	MS
Main effects:			
A		1	
B		1	
C		1	
Interactions:			
AB		1	
AC		1	
BC		1	
Blocks:			
ABC		1	
Replicates		1	
Replicates × ABC		1	
Residual, error		6	
Total		15	

Consider now the case where we wish to run a 2^3 factorial with two replicates per treatment combination. This is a 2×2^3 which requires 16 runs in place of the 8 as discussed so far. One of the main reasons for considering a twice-replicated experiment is to secure an internal estimate of the experimental error. This is always best done by replicates of the same treatment combination. (To complete the analysis of the data in the 2^3 factorial, either we must have an external estimate of the error or we must have the right to assume that the two-factor interactions are not existent, and use them as the estimate of error.)

Now we have 16 runs to make. Four days will be needed, requiring four blocks and causing us to use three degrees of freedom for blocks. The ANOVA now takes the form of Table 16·3. The analysis of a confounded experiment is done in the same manner as an experiment that has no confounding; the difference is that the confounded terms do not appear in the ANOVA table since their sums of squares will be included in the term for blocks.

The use of the Yates technique will produce the sums of the various effects, and the sums of squares can then be obtained from these in the usual way. We can also use the method of Chap. 8 to arrive at the sums of squares and to complete the analysis of the data.

16·2 PARTIAL CONFOUNDING

It is possible to design an experiment so as to include several replicates of the complete factorial. When this is the case, it is desirable to secure some information about all the treatment combinations and their effects. If the same interaction is used for the confounding in every replicate set, then we are not able to obtain any insight into the influence of that interaction as it will be confounded with the blocks. Partial confounding involves the confounding of a different term in each replication, or at least two different ones, so that it is possible to secure some information about every term—main factors and all interactions.

Consider a 2^3 factorial experiment where we wish to make four replicates, and the restriction of only four runs per day has been imposed. This experiment will then require 32 runs, or a total of 8 days. Since we believe that there is a significant day-to-day variation, we intend to block on days. Furthermore, we would like some estimate of every main-factor effect and interaction effect. We will use a different term for confounding in each of the four replicates, as shown in Table 16·4.

The ABC effect is confounded in replication I but not in II, III, or IV. Since three of the four replicate sets do not confound ABC with the block effect, we may use these three to obtain an estimate of the ABC sum of squares. The ABC term is calculated from the second, third, and fourth

Table 16·4   Confounding on Different Effects

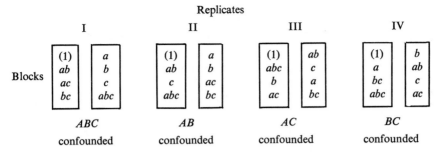

Replicates

	I		II		III		IV	
Blocks	(1) ab ac bc	a b c abc	(1) ab c abc	a b ac bc	(1) abc b ac	ab c a bc	(1) a bc abc	b ab c ac
	ABC confounded		*AB* confounded		*AC* confounded		*BC* confounded	

replicates by the following:

$$ABC = \frac{1}{4}(a-1)(b-1)(c-1)$$
$$= \frac{1}{4}[(abc + a + b + c) - (ab + ac + bc + 1)]$$
$$= \frac{1}{4}(S_1 - S_2)$$

where S_1 is the sum of abc, a, b, and c; and $S_2 = $ sum of ab, ac, bc, and (1).

(a) From replicate II, calculate $(S_1 - S_2)$.
(b) From replicate III, calculate $(S_1 - S_2)$.
(c) From replicate IV, calculate $(S_1 - S_2)$.

Sum the answers to (a), (b), and (c), and denote by $S(ABC)$. Then $ABC = S(ABC)/(3 \times 4)$ and sum of squares of $ABC = S(ABC)^2/24$. The estimate of the interaction is the average of the estimates from each of the three replicates. The divisor of the sum of squares is the number of observations involved in the expression $S(ABC)$.

The other three interactions which have been confounded can be calculated in a similar way by using the three replicates where the confounding does not take place.

All main effects are free from block effects. All interactions are said to be partially confounded, since the interaction effects may be estimated by using three-fourths of the total set of runs. While all effects are estimated by such an experimental design, the main effects are estimated more precisely than are the interactions. Thus confidence limits about these estimates would be tighter for the main effects (based on all the data) than for the interactions (using only part of the data).

PROBLEMS

16·1   Factors R, S, and T are to be investigated using a 2^3 factorial experiment. If the RS interaction is known not to occur because of physical reasons, and we wish to run our experiment in 2 blocks, which treatment combinations should be run in the 2 blocks?

16·2   Set up the ANOVA table for the experiment given in Prob. 16·1, and fill in the degrees of freedom.

16·3   In an investigation involving the factors R, S, T, and Z, a 2^4 factorial experiment is run. Of the 16 runs to be made, only 8 can be made each day, and we decide to block on days.

(a) Assign the treatments to 2 blocks if we choose to confound on the highest-order interaction.

(b) Assign the treatments to 2 blocks by confounding on the RST interaction.

16·4   A 2^4 factorial experiment is to be run in 4 blocks of 4 runs each. Assign the treatments to the 4 blocks if each block is to be confounded with one of the three-factor interactions.

16·5   A twice-replicated 2^4 factorial experiment is to be run in 4 blocks. If the factors are A, B, C, and D,

(a) Assign the treatments to 4 blocks by confounding on the 4 three-factor interactions.

(b) Set up the ANOVA table for the experiment and fill in the degrees of freedom.

16·3   CONFOUNDING 2^n IN MORE THAN TWO BLOCKS

When it becomes necessary to use more than two blocks in a factorial experiment, more than one term will be confounded with the block effect. If a 2^4 factorial is to be run in four blocks, with four treatments per block, there will be three degrees of freedom for blocks. Thus we will have to sacrifice three of the terms in the blocking action. The first two terms to be sacrificed may be selected as we desire, but the third term is determined by the selection of the first two.

If we choose ABC and BCD to be confounded with blocks, then we obtain the third term to be confounded by multiplying the first two terms together, and dropping any letter that occurs twice. In this case, ABC times BCD will produce AB^2C^2D; when we drop the B^2C^2, the remaining term is AD, the third term to be confounded.

In general, the number of blocks for a 2^n experiment will be equal to 2^p, where p is the number of independent terms to be selected for confounding. The p interactions selected from the 2^p blocks are called independent *defining contrasts*. In the example here, ABC and BCD are the independent defining contrasts with AD the additional defining contrast required for the block confounding.

After we have made sure that the proper terms are involved in the confounding by obtaining the defining contrasts and the dependent terms involved, it remains for us to place the treatments into the four blocks. We can use the defining contrasts to help us in this task. If we select any treatment combination that has an even number of letters (or no letters) in common with each defining contrast, it will belong in the first block, which is called the *principal block*; this block will always contain the treatment combination (1).

The principal block is a "group closed to multiplication": any two members when multiplied together will produce a member which is contained in the group, after each letter which has occurred twice has been dropped. Our defining contrasts are *ABC* and *BCD*, since the term (1) has no letters in common with them; so we know it belongs in the principal block. The term *bc* has two letters in common with both defining contrasts; so it also belongs in the principal block. As four terms are required for the principal block, we must select another term that belongs, and then by using the multiple of closed groups we can generate the rest. Selecting the *abd* term, we find that it also contains two letters in common with the defining contrasts, and therefore add it to the principal block. Then by multiplying *abd* by *bc*, the fourth and last term of the principal block is found to be *acd*. Thus the principal block contains (1), *bc*, *abd*, and *acd*.

The three remaining blocks can now be generated by selecting a term *not* contained in the principal block and multiplying the entire principal block by that term with the resulting terms comprising the next block. Selecting for the second block the term *a*, since it is not in the principal block, by multiplication, we get for the second block *a*, *abc*, *bd*, and *cd*. The same can be done for the third and fourth block, as long as we can select a term not already used in the preceding blocks. The final result for the treatment assignment in the four blocks is shown in Table 16·5. The order of running within each block remains to be randomized.

Table 16·5 Confounding with Defining Contrasts ABC, BCD, and AD

Principal block	Block 2	Block 3	Block 4
(1)	a	b	ab
bc	abc	c	ac
abd	bd	ad	d
acd	cd	abcd	bcd

PROBLEMS

16·6 A 2^4 factorial is to be run in 2 blocks of 8 treatment combinations.
(a) If the term *abc* is confounded with blocks, assign the treatment combinations to the 2 blocks.
(b) Confound the *abcd* term with blocks, and assign the treatment combinations to the 2 blocks.

16·7 A 2^4 factorial is to be twice replicated and run in 4 blocks of 8 treatment combinations.
(a) Confound the *abcd* term with blocks, and assign the treatment combinations to the 4 blocks.
(b) Show the form of the ANOVA table.

16·8   A 2^5 factorial is to be run in 4 blocks of 8 treatment combinations.
(a) Show the treatment combinations which would be assigned to each of the 4 blocks.
(b) Show the ANOVA table for this experiment, showing the source and degrees of freedom.

16·9   A 2^3 factorial is to be replicated 3 times and run in blocks of 4.
(a) Show how it is possible to assign the treatment combinations to be blocks if it is desirable to get information about all the main effects and interactions.
(b) Show the form that the ANOVA table would take.

16·10   A 2^5 factorial is to be run in blocks of 4.
(a) How many blocks will be required?
(b) How many terms will constitute the defining contrast?
(c) Assign the terms to the principal block.
(d) List the terms to be assigned to the entire set of blocks.

16·4   FRACTIONAL FACTORIALS

To run a 2^3 factorial requires 8 runs, a 2^4 requires 16 runs, and a 2^5 requires 32 runs, even without any replications. As the number of factors increases, the total number of runs required for a full factorial experiment increases at a very rapid rate. Often in the early stages of an investigation, there are so many factors to be considered that to run a full factorial, even with all factors at two levels, would require too many runs. In addition to the excessive number of experimental runs called for in multifactored factorials, a large number of higher-order interactions show up as possibilities. In most cases, even if we found a four-factor interaction to be significant, we would find it very hard to explain what is meant. Usually, we are interested only in main effects and in two-factor interactions and, if we were to run the complete factorial of five or more factors, would find that we were wasting time calculating the effects of high-order interactions.

To reduce the number of runs required in the case where a large number of factors are involved, only a fraction of a full factorial is run. When this is done according to prescribed rules, we have what is known as a *fractional factorial*. If seven factors are to be investigated at two levels each, a full factorial would call for 128 runs. If the economics of the situation is such that only about 60 runs can be made, then by running a one-half replicate of the full factorial, only 64 runs would be required. In symbols, this would be written as a 2^{7-1} factorial. If only one-fourth of the full factorial were run, we would have a quarter replicate, or 2^{7-2}.

For running a fractional factorial of any size, the methods discussed in this chapter (under confounding) are used. If a one-half replicate is to be run, then the treatments must be placed into two blocks, and only one of the blocks of treatments is run. Which of the two blocks is to be run can be determined by choice or by chance, since the results from either will give the same insight into the problem. Just as in blocking for a full factorial, the term used to divide the treatments will be sacrificed. When only half the tests are run, some other changes must be made in the analysis of the data.

16·5 ALIASES

Consider a one-half replicate of a 2^3 or a 2^{3-1}. Here there are three factors A, B, and C, each at two levels, calling for eight treatment combinations, of which only four are to run. If we decide to sacrifice the ABC term, we split the eight treatments into the following two blocks:

Block 1: (1) ab bc ac
Block 2: a b c abc

We can now run either block 1 or block 2; if we have no preference, we can decide by chance. Suppose we run only block 2, giving us a one-half

Table 16·6   2³ Factorial Experiment in Two
Blocks (Half Replicates)

(a) Block 1 (Principal Block)

Treatment	Effect						
	A	B	AB	C	AC	BC	ABC
(1)	−	−	+	−	+	+	−
ab	+	+	+	−	−	−	−
bc	−	+	−	+	−	+	−
ac	+	−	−	+	+	−	−

(b) Block 2

Treatment	Effect						
	A	B	AB	C	AC	BC	ABC
a	+	−	−	−	−	+	+
b	−	+	−	−	+	−	+
c	−	−	+	+	−	−	+
abc	+	+	+	+	+	+	+

replicate of the 2^3 design. Table 16·6 shows which treatments are to be used to calculate the various effects. Table 16·6 lists only the treatments which are to be run but contains all the effects of the full 2^3 factorial. The A effect is found by $(a + abc) - (b + c)$; the BC effect is found by $(a + abc) - (b + c)$. Hence both the A effect and the BC effect are found by the same set of differences. Had we run block 1 instead, the two effects would have been found by the same set of differences, but the sign of the differences would have been opposite (see Table 16·6a). In terms of the absolute difference of the two sets of terms, we cannot distinguish between the A and BC effects because they are both calculated in the same way when we use a half-replicate design. When two or more effects have the same numerical value in a partial-replicate design, they are called *aliases*.

For every effect we want to find in a one-half replicate, there will be an alias. In this case, B and AC are aliases; so also are C and AB. By choosing a specific defining contrast, we can determine which paired terms will be aliases. We indicate the aliases in the ANOVA table and assign one degree of freedom to each pair. By confounding properly, we can arrange the design so that all main factors will have high-order interactions as their aliases. Usually we can reasonably suppose that high-order interactions do not exist (have no physical meaning) and therefore that the full effect shown must be due to the main factor. Under these circumstances, we will not have lost much information by running only half the full experiment, and we will have saved considerable effort.

If we choose to use the AB effect as the defining contrast, the aliases of our 2^{3-1} experiment become A with B, C with ABC, and AC with BC. Now, two main factors are confounded and we will be unable to distinguish their effects. This choice of defining contrast loses much information and is therefore a poor one. When we run fractional replicates, we should choose the defining contrasts and list the aliases to ensure that important terms have not been confounded with each other. Although we have used a 2^3 factorial to explain the method of obtaining a fractional replicate, such as the 2^{3-1}, the real advantage of fractional factorials is best seen when used with large factorials. For the 2^{3-1}, two-way interactions were aliased with main effects and we are not usually willing to assume these to be nonexistent. With a larger number of factors, high-order interactions can be ruled out in most cases.

16·6 A HALF-REPLICATED 2^5 EXPERIMENT

We believe that five factors A, B, C, D, and E may affect the yield of an industrial process. In a preliminary investigation we wish to test the effects of these factors at each of two levels. The required treatment combinations for the fully replicated 2^5 experiment are given in Table 16·7; in such an experiment 32 different runs are needed. We assume now that time does not permit us to run the entire set of 32; hence we consider a half-replicate 2^{5-1} design for which only 16 runs will be needed.

We choose the highest-order interaction $ABCDE$ as the defining contrast. Multiplying out the binomials

$$(a - 1)(b - 1)(c - 1)(d - 1)(e - 1)$$

Table 16·7 Treatment Combinations for a Fully Replicated 2^5 Experiment

			A, low		A, high	
			B, low	B, high	B, low	B, high
C, low	D, low	E, low	(1)	b	a	ab
		E, high	e	be	ae	abe
	D, high	E, low	d	bd	ad	abd
		E, high	de	bde	ade	abde
C, high	D, low	E, low	c	bc	ac	abc
		E, high	ce	bce	ace	abce
	D, high	E, low	cd	bcd	acd	abcd
		E, high	cde	bcde	acde	abcde

and grouping the terms according to sign, we have the following treatment combinations, all with positive sign:

a, b, c, d, e, abc, abd, abe, acd, ace, ade, bcd, bce, bde, cde, abcde

and the following, all with negative sign:

(1), ab, ac, ad, ae, bc, bd, be, cd, ce, de, abcd, abce, abde, acde, bcde

We now have two sets of treatment combinations; each is an appropriate set for a half-replicate design.

Let us choose the principal block [containing (1)] because we happen to be interested in the effect for all factors at the low level. In Table 16·8 we show the 16 treatment combinations for the principal block, and the response data that we obtained in the actual experiment.

We find the aliases for the treatment combinations in the principal block by multiplying each factor in turn by the defining contrast $ABCDE$ and dropping any squared terms. Thus the alias for the AB effect is

$$AB \times ABCDE = A^2B^2CDE$$

Dropping the two squared terms, we have CDE, the alias for the two-factor interaction AB. The aliases are given in Table 16·9. For an odd number of

Table 16·8   *Treatment Combinations and Data for a Half-replicated 2^5 Design*

			A, low		A, high	
			B, low	B, high	B, low	B, high
C, low	D, low	E, low	(1) 10	...	...	ab 15
		E, high	...	be 12	ae 14	...
	D, high	E, low	...	bd 11	ad 13	...
		E, high	de 9	...	...	abde 16
C, high	D, low	E, low	...	bc 12	ac 14	...
		E, high	ce 11	...	...	abce 16
	D, high	E, low	cd 10	...	...	abcd 16
		E, high	...	bcde 11	acde 13	...

letters in the defining contrast, the signs in the plus and minus table would have to be reversed for the principal block; so the aliases are the reverse too. For an even number of letters in the defining contrast, the principal block has positive aliases.

We might have chosen a different defining contrast; then the aliases would be differently paired. By using $ABCDE$ as the defining contrast, we have confounded the main effects with the four-factor interactions which are least likely to exist.

We use the Yates method for calculating effects as shown in Table 16·10, as if we were involved with a 2^4 experiment. We arrange the treatment combinations in the first column in standard order, ignoring all e's for this

Table 16·9   *Aliases for a Half-replicated 2^5 Factorial Design*

Effect	Alias
A	$\pm BCDE$
B	$\pm ACDE$
C	$\pm ABDE$
D	$\pm ABCE$
E	$\pm ABCD$
AB	$\pm CDE$
AC	$\pm BDE$
AD	$\pm BCE$
AE	$\pm BCD$
BC	$\pm ADE$
BD	$\pm ACE$
BE	$\pm ACD$
CD	$\pm ABE$
CE	$\pm ABD$
DE	$\pm ABC$

— For principal block
+ For other block

purpose but indicating their presence by the use of parentheses. In the second column we have placed the data from Table 16·8, and we complete the table in the customary fashion.

We find the average effect by dividing the effect totals by 8, the number of comparisons entering into the effects. We find the factor sums of squares by squaring each effect total and dividing by 16, the number of runs in the experiment. We find the total sum of squares from the observations themselves. From these calculations, we prepare the summary ANOVA Table 16·11.

We have no a priori reason to suspect that interactions are important; the data in Table 16·11 do not suggest that any of the interactions have a large effect. Especially since we are here performing a screening experiment, we pool all the interaction effects to obtain a residual effect with a sum of squares of 3.1250, 10 degrees of freedom, and a mean square of 0.3125 as shown in Table 16·12. When we construct the F ratios using the residual MS term as the denominator, we find that factors A, B, and D are statistically significant, with factor D having a lower probability. Any further experimentation should concentrate on these three factors in order to obtain added insight into the process as economically as possible.

Table 16·10   Estimates of Effects by Yates Method from Data in Table 16·8

Treatment Combination	Response	(1)	(2)	(3)	(4)	Effect	Alias
(1)	10	24	51	104	203	Total	$-ABCDE$
a(e)	14	27	53	99	31	A	$-BCDE$
b(e)	12	25	49	14	15	B	$-ACDE$
ab	15	28	50	17	3	AB	$-CDE$
c(e)	11	22	7	6	3	C	$-ABDE$
ac	14	27	7	9	−1	AC	$-BDE$
bc	12	23	9	0	−1	BC	$-ADE$
abc(e)	16	27	8	3	3	ABC	$-DE$
d(e)	9	4	3	2	−5	D	$-ABCE$
ad	13	3	3	1	3	AD	$-BCE$
bd	11	3	5	0	3	BD	$-ACE$
abd(e)	16	4	4	−1	3	ABD	$-CE$
cd	10	4	−1	0	−1	CD	$-ABE$
acd(e)	13	5	1	−1	−1	ACD	$-BE$
bcd(e)	11	3	1	2	−1	BCD	$-AE$
abcd	16	5	2	1	−1	ABCD	$-E$

Greater fractionation is not only usable but is also desirable for screening experiments. For a quarter replicate that is perhaps 2^{5-2}, only 8 of the full 32 runs are required. A quarter replicate implies four blocks of runs, only one of which is actually performed. Thus we have three defining contrasts and each effect has three aliases. Such confounding is not costly in comparison with the economy of the experiment.

Table 16·11 Summary of ANOVA, Data from Table 16·10

Source of variation	Total effect	Average effect	SS	ν	MS
A	31	3.875	60.0625	1	60.0625
B	15	1.875	14.0625	1	14.0625
C	3	0.375	0.5625	1	0.5625
D	−5	−0.625	1.5625	1	1.5625
E	+1	+0.125	0.0625	1	0.0625
AB	3	0.375	0.5625	1	0.5625
AC	−1	−0.125	0.0625	1	0.0625
AD	3	0.375	0.5625	1	0.5625
AE	+1	+0.125	0.0625	1	0.0625
BC	−1	−0.125	0.0625	1	0.0625
BD	3	0.375	0.5625	1	0.5625
BE	+1	+0.125	0.0625	1	0.0625
CD	−1	−0.125	0.0625	1	0.0625
CE	−3	−0.375	0.5625	1	0.5625
DE	−3	−0.375	0.5625	1	0.5625

Table 16·12 ANOVA, Data from Table 16·11, Interaction Pooled

Source	Total effect	Average effect	SS	ν	MS	F ratio
A	31	3.875	60.0625	1	60.0625	192**
B	15	1.875	14.0625	1	14.0625	45**
C	3	0.375	0.5625	1	0.5625	1.8
D	−5	−0.625	1.5625	1	1.5625	5.0*
E	1	0.125	0.0625	1	0.0625	
Residual			3.1250	10	0.3125	
Total			79.4375	15		

$F_{1,10,0.05} = 4.96$
$F_{1,10,0.01} = 10.0$

PROBLEMS

16·11 Assume an experiment carried out under normal industrial conditions to test the effect of 5 factors on the yield of a process. Each factor was run at two levels—low and high. The factors were A (agitation), T (temperature), S (speed), O (operator), and M (material).

(a) Lay out a 2^5 factorial design for a crossed experiment.

(b) Enter the treatment combination symbols in the cells of the design.

(c) Indicate which runs should be made for a half-replicate experiment, confounding on the $ASOM$ effect.

(d) List the tested effects and their aliases.

(e) Assume that the experiment in answer to (c) has been run and that the resulting yields were the numbers 20 through 35 in sequence in the 16 cells required for the half-replicated experiment. Using these numbers, complete the ANOVA.

(f) If the same data had been assigned in a random way to the same cells, what result would you have expected from the data analysis? Why?

16·12 For the same factors given in Prob. 16·11, use the highest-order interaction $ATSOM$ as the defining contrast and run a half-replicate experiment using the principal block. For the response values, randomly assign the numbers from 20 to 35 to the 16 selected treatments.

QUESTIONS

16·1 Why do we block on factors which we suspect are large influences in the response variable?

16·2 If an experiment requires more than one day to complete the experimental runs and there is a day-to-day variability, where will this variability show up in the ANOVA table if we do not block on days?

16·3 What effect does blocking have on the degrees of freedom of the entire experiment?

16·4 What is meant by the term "confounding"?

16·5 Why are factorial experiments used as screening experiments?

16·6 How is it possible to determine the number of runs required for a factorial experiment?

16·7 If a factorial experiment is to be run in two blocks, how do we decide which block will be run first?

16·8 If a factorial experiment is run in two blocks, why must we sacrifice one of the treatment effects?

16·9 How can a treatment effect be changed to a sum of squares?

16·10 In generating a plus and minus table for factorials, how is it possible to generate all the signs of the table from just the treatment combinations?

16·11 What is the desirability of partial confounding?

16·12 What are the independent defining contrasts?

16·13 How are the other defining contrasts generated from the independent defining contrasts?

16·14 Why should all the defining contrasts be generated prior to starting the experimental runs?

16·15 What do we mean by a "group closed to multiplication"?

16·16 Which block is called the principal block?

16·17 Should the order of running the treatments within each block be randomized? Why?

16·18 What is a 2^{5-1} factorial experiment?

16·19 Define an "alias."

16·20 Under what conditions should a fractional factorial be considered as the proper experiment to be run?

16·21 Why is it desirable to know the aliases of all the confounded effects?

16·22 It is sometimes incorrect to choose the highest-order interaction as the defining contrast. Why?

16·23 If we choose the principal block to be run in a half replicate of a factorial, why are the aliases assigned minus values in some cases?

16·24 Why is it important to maintain the correct algebraic sign in front of the total effect of the treatments?

16·25 In 2^5 factorial experiments, where does the estimate of experimental error come from if there is no external estimate of error?

REFERENCES
25, 30, 35, 47, 58, 59, 64, 68, 73, 111

17 Determination of Optimum Conditions

In almost any research or industrial situation, it may be important to find the conditions under which we can obtain a result which is best in terms of quality or cost or some combination of these two measures. A systematic method which permits us economically to find these conditions is often of great importance in the success or failure of an enterprise. Inefficient methods not only may be costly but also may actually mislead us into accepting an inferior set of conditions.

The hardness of iron is related to the percentage of carbon which it contains. Wrought iron contains almost no carbon; it is very soft. Cast iron, which contains a few percent of carbon, is much harder. No doubt a batch of iron containing 50 percent carbon would be fragile. Thus, if we want maximum hardness, we expect that there exists an optimum percentage of carbon.

Ignoring all the other variables which affect the response variable, it is reasonable to suppose that a plot of hardness versus percent of carbon would look something like Fig. 17·1, with a single maximum. We wish to find this maximum.

We have a choice of experimental methods. One could be this: we make a large number of batches of iron, of slightly different carbon content. We would obtain many closely spaced points on a graph. If the optimum value lay within the set of carbon levels we chose for the experiment, we could thus find at least the approximate position of the assumed maximum. This

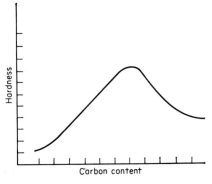

Fig. 17·1 Plot of iron hardness versus carbon content.

method would perhaps be costly, because we might make many runs at a level of carbon far from the maximum.

An alternative method is to make only a few experimental runs and to infer from these few how to run a following experiment. This second method is called *response surface methodology*. It is especially useful in the research and development stage of a process. An economical strategy combines regression analysis and factorial designs (or modifications of these designs). We will in this chapter discuss some of the simpler techniques in the study of response surfaces.

In the manufacturing state of a process, following the research and development stage, at the same time that the process turns out acceptable product, it can furnish information for its own improvement. A technique known as *evolutionary operations* is based on making slight modifications of the process —so slight that the product will still be satisfactory. By the use of an appropriate experimental design, we can shift the process to better and better levels. In this chapter we will describe an application of this technique.

The basis of response surface methodology is intuitively as follows: we are trying to find the optimum concentration of carbon in iron; we make only two runs at different carbon levels. Assume that the experimental error is known to be small. If the two runs give far different hardnesses, we will shift the next experiments in the direction of that level of carbon which gave the better result. If the two batches give nearly the same hardness, we do not know whether we are far from the maximum or have bracketed the maximum. If the first inference is correct, we should change the carbon level considerably, and we do not know in which direction to go. If the second inference is correct, we should make runs next at levels of carbon between those we tested first. Although factorial designs do not completely relieve us of making such judgments, in general their superior efficiency makes it easier to decide on the next stage of a sequential experiment.

17·1 MATHEMATICAL CONCEPTS

We are interested in the yield of a process, i.e., in the response variable Y. We believe that the yield is related to a set of predictor variables, perhaps pressure, temperature, rate of agitation, concentration of ingredients. We symbolize the predictor variables by $X_1, X_2, X_3, \ldots, X_n$. The methods we will describe are based on the concept that Y is a function of the predictor variables, meaning that for a set of different levels of the predictor variables we will obtain a set of different Y values. A plot of the relationship between the Y values and the levels of the various predictor variables is called a *response surface*.

If there are only two predictor variables, the relationship is generalized as $Y = f(X_1, X_2)$, which is to say the value of Y is determined by some relationship between the response variable and the levels of the two predictor variables. The response surface can be visualized as a three-dimensional graph, as in Fig. 17·2. The response surface may be a plane, as in Fig. 17·2a, but is more often a curved surface, perhaps as in Fig. 17·2b.

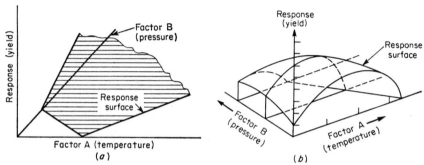

Fig. 17·2   Response surfaces. (a) Plane. (b) Curved.

The process of optimization includes two aspects: (1) the identification of the nature of the surface; (2) the location of an optimum if it exists. We will identify the surface by assuming a plane, for example, and testing the fit to that plane. If we find a lack of fit we will hypothesize a curved surface and test for that surface. We will find the presence of an optimum (or lack of an optimum) on the basis of the calculated slopes for the surface. The simplest model we can use is based on the belief that the values of Y are related to the levels of a single factor X_1. The mathematical model would be of the form $Y = \beta_0 X_0 + \beta_1 X_1 + \epsilon$ if the functional relationship were linear. The mathematical model would have the form $Y = \beta_0 X_0 + \beta_1 X_1 + \beta_2 X_1^2 + \epsilon$ if the relationship were quadratic.

For the linear model, we would expect the data to plot as in Fig. 17·3, and from an analysis of the data we would expect to conclude that a higher level of X gives a higher level of Y. For the curvilinear model, we would expect a plot of data as in Fig. 17·4, and a conclusion that maximum Y would be found in the vicinity of $X_1 = 0$. An efficient experiment design requires that we select the levels of X_1 in advance and that we use sound experimental designs.

Note that in the one-predictor-variable case, the response "surface" is a line. We want by experimentation to learn where we are on what line. To optimize means to find a maximum (or minimum) point on the line.

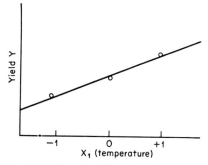

Fig. 17·3 Plot of yield versus temperature, linear model.

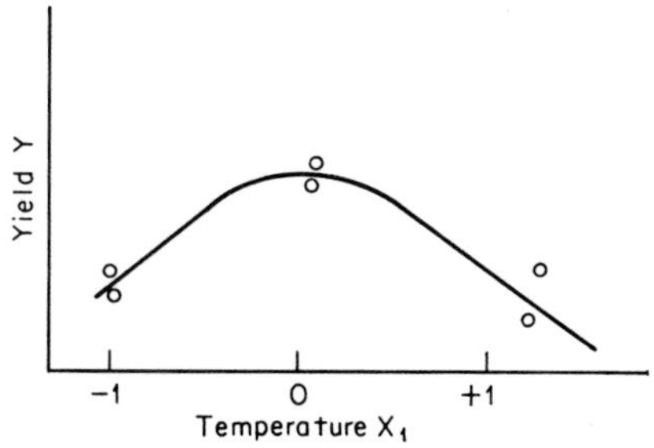

Fig. 17·4 Plot of yield versus temperature, curvilinear model.

17·2 ONE-VARIABLE FIRST-ORDER MODEL

In the simplest case, we have reason to believe that only one predictor variable is important and that a linear relationship exists between the input and the response variables. Suppose that we believe that only temperature is important in affecting the yield of a process. The mathematical model for the linear case is

$$Y = \beta_0 X_0 + \beta_1 X_1 + \epsilon$$

In this model, β_0 specifies the Y intercept and β_1 specifies the slope of the assumed straight line. Experimental error is symbolized by ϵ. We will estimate the coefficients in the mathematical model from our sample data, and approximate the model by

$$\hat{Y} = b_0 X_0 + b_1 X_1$$

where $\hat{Y}$ is the value of the response variable (yield) estimated from the calculated equation, and b_0 and b_1 are the calculated equation coefficients estimating the actual coefficients β_0 and β_1.

In both the mathematical model and the equation found from sample data, the term X_0 appears. This term represents a "dummy" variable, having no physical significance and always assigned the value 1. This dummy X_0 is put into the equations as a device to facilitate the calculations now to be explained.

We use here, as before, a least-squares method to find the equation coefficients. We developed the normal equations in Chap. 12, and pointed out that with an appropriate coding method the estimates of the equation coefficients can be found from

$$b_0 = \frac{\sum X_0 Y}{\sum X_0^2} \tag{17·1}$$

$$b_1 = \frac{\sum X_1 Y}{\sum X_1^2} \tag{17·2}$$

By "appropriate" coding we mean that the rules for orthogonality are obeyed. We must code so that $\sum X_1 = 0$ and $\sum X_0 X_1 = 0$. For the temperatures 180 and 200, we can code by

$$X_1 = \frac{T - 190}{10}$$

by which the coded value of 180° is -1 and the coded value of 200° is $+1$. We can always code two values in the same way. For more than two values of the input variable, we should select them so that they can be coded so as to conform to the rule just above; that is, they should be equally spaced. Since the dummy factor X_0 is made equal to 1, the second orthogonality rule just above is also fulfilled.

If we now run a one-factor two-level factorial experiment with two replicates per treatment, we would require $2 \times 2^1 = 4$ runs. The data resulting from the experiment are shown in Table 17·1.

Table 17·1 Data from 2×2^1 Factorial Experiment

X_0	X_1	Y
1	$+1$	54
1	$+1$	56
1	-1	59
1	-1	61

The computations for finding the estimates of the equation coefficients and the terms needed for ANOVA are systematically carried out as in Table 17·2. In Table 17·2a we show the generalized identification of the sums of squares and cross products we need; in Table 17·2b are shown the results for the data in Table 17·1. The value 230, for example, which appears in the upper right cell of Table 17·2b is the value for $\sum X_0 Y$, i.e., the sum of the products of the dummy X_0 and each of the set of response values. $\sum X_0$ is the equivalent of the sample size.

We now take from Table 17·2 the values needed to compute the equation coefficients:

Table 17·2 Tables of Sums and Cross Products for Data in Table 17·1

(a)

	X_0	X_1	Y
X_0	$\Sigma X_0{}^2$	$\Sigma X_0 X_1$	$\Sigma X_0 Y$
X_1	$\cdots$	$\Sigma X_1{}^2$	$\Sigma X_1 Y$
Y	$\cdots$	$\cdots$	ΣY^2

(b)

	X_0	X_1	Y
X_0	4	0	230
X_1	$\cdots$	4	-10
Y	$\cdots$	$\cdots$	13,254

by Eq. (17·1),

$$b_0 = \frac{230}{4} = 57.5$$

by Eq. (17·2),

$$b_1 = \frac{-10}{4} = -2.5$$

Therefore,
$$\hat{Y} = 57.5 - 2.5 X_1$$

When we uncode,

$$\hat{Y} = 57.5 - 2.5 \ \frac{T - 190}{10}$$

and
$$\hat{Y} = 105 - 0.25T$$

From the calculated values in Table 17·2b we take the numbers needed for the summary ANOVA Table 17·3. Note that we have available all the numbers we require except the residual term. This we can find by subtraction or as indicated in the second column of Table 17·3.

Table 17·3 Summary ANOVA Based on Data in Tables 17·1 and 17·2

Source	SS	ν	MS
Crude SS $\cdots$	$\Sigma Y^2 = 13,254$	4	
Due to b_0	$b_0 \Sigma X_0 Y = 57.5 \times 230 = 13,225$	1	
Due to b_1	$b_1 \Sigma X_1 Y = -2.5 \times -10 = 25$	1	25
Residual by difference	4	2	2

By the methods of Chap. 12, we can now partition the residual mean square into two terms which estimate error and lack of fit, and follow this procedure by F tests of significance.

We can conclude that (1) a linear relationship is probable over the range of temperatures tested; (2) an increase in temperature in this range produces a decrease in yield. Therefore, in the next attempt to optimize the yield (for the factor temperature) we should run a second experiment at temperatures below those of this test. Note especially the economy of this method of experimentation, as shown by the inferences we can make from only four runs.

PROBLEMS

17·1   In an investigation into the resulting strength of joints in a plastic, the rate of drying of the cement was the major factor affecting the strength of the joint. Three levels of the factor "drying rate" were investigated, and replicate runs were made at each level. The results of the investigation are given in Table 17·4.

Table 17·4   A Twice-replicated One-factor Experiment Using Three Levels, to Investigate Strength of Joint as a Function of Drying Rate

Drying Rate, min	Strength of Joint, psi
2.0	132
2.0	134
2.5	140
2.5	139
3.0	155
3.0	154

(a) Set up the design matrix for the experiment.
(b) If we believe the relationship between drying rate and strength of the joint to be a linear one, what is the mathematical model under test?
(c) Code the treatment combinations.
(d) Set up a table of sum of squares and cross products for this experiment.
(e) Calculate the coefficients for the line of best fit.
(f) Using ANOVA, test the assumption of linear relationship.
(g) If the line of best fit is linear and the data from the experiment do not suggest lack of fit of the linear model, where should we run the next experiment to determine the maximum strength of a joint as a function of drying rate?

17·2   If we believe we are near a maximum strength of a joint in plastic, with respect to the drying rates under test (see Prob. 17·1), what experimental design should we consider?

17·3 ONE-VARIABLE SECOND-ORDER MODEL

If we suspect that a linear model may be inadequate for a situation involving a single predictor variable, we may consider a second-order model. We would in fact expect to find a curvilinear relationship near a maximum or minimum point. If temperature and yield are again the factors of interest, we could run a twice-replicated 3^1 factorial design for a total of 6 runs. We must select levels of the predictor variable that are equally spaced, such as 180, 200, and 220 degrees. If we code by the rule $X_1 = (T - 200)/20$, we have the coded values -1, 0, and $+1$.

The mathematical model now is

$$Y = \beta_0 X_0 + \beta_1 X_1 + \beta_{11} X_1^2 + \epsilon$$

and we will estimate the coefficients by

$$\hat{Y} = b_0 X_0 + b_1 X_1 + b_{11} X_1^2$$

where again X_0 is a dummy variable set equal to 1.

We run the experiment, and record the levels of the predictor variables and the response variable in Table 17·5. X_1 is orthogonal to both X_0 and X_1^2; by such an arrangement we ease the calculations to follow.

Table 17·5 Design Matrix for Regression Experiment with One Predictor Variable, Second-order Model

X_0	X_1	X_1^2	Y
1	$+1$	1	54
1	$+1$	1	56
1	0	0	59
1	0	0	61
1	-1	1	59
1	-1	1	61

In Table 17·6 we show the generalized tables for finding sums of squares of cross products for this case, and also the resulting terms. The following equations are those for calculating the estimates of the coefficients:

$$b_0 = \frac{(\sum X_1^4)(\sum X_0 Y) - (\sum X_0 X_1^2)(\sum X_1^2 Y)}{(\sum X_0^2)(\sum X_1^4) - (\sum X_0 X_1^2)^2}$$

$$= \frac{(4 \times 350) - (4 \times 230)}{(6 \times 4) - 4^2} = 60$$

$$b_1 = \frac{\sum X_i Y}{\sum X_i^2} = -\frac{10}{4} = -2.5$$

$$b_{11} = \frac{(\sum X_0^2)(\sum X_1^2 Y) - (\sum X_0 X_1^2)(\sum X_0 Y)}{(\sum X_0^2)(\sum X_1^4) - (\sum X_0 X_1^2)^2}$$

$$= \frac{6 \times 230 - 4 \times 350}{6 \times 4 - 4^2} = -2.5$$

Thus the calculated coefficients give the equation

$$\hat{Y} = 60 - 2.5 X_1 - 2.5 X_1^2$$

which may be decoded to give the equation estimated to relate yield and temperature.

Table 17·6 Tables of Sums of Squares and Cross Products for Data in Table 17·5

(a)

	X_0	X_1	X_1^2	Y
X_0	ΣX_0^2	$\Sigma X_0 X_1$	$\Sigma X_0 X_1^2$	$\Sigma X_0 Y$
X_1	$\cdots$	ΣX_1^2	ΣX_1^3	$\Sigma X_1 Y$
X_1^2	$\cdots$	$\cdots$	ΣX_1^4	$\Sigma X_1^2 Y$
Y	$\cdots$	$\cdots$	$\cdots$	ΣY^2

(b)

	X_0	X_1	X_1^2	Y
X_0	6	0	4	350
X_1	$\cdots$	4	0	-10
X_1^2	$\cdots$	$\cdots$	4	230
Y	$\cdots$	$\cdots$	$\cdots$	20,456

The data we found in Table 17·6 also serve for the preparation of the summary ANOVA in Table 17·7. An F test shows significance for the coefficient b_1, which is the first-order slope of the line, but at the 0.10 confidence level we cannot attach significance to the coefficient b_{11}, which is the second-order coefficient and which therefore is associated with curvilinearity. We have not in this experiment demonstrated a curvilinear relationship in the tested range of temperatures. We should now plan a new experiment at reduced temperatures since we have found as before that an increase in temperature is associated with reduced yield, as shown by the negative sign for the coefficient b_1.

Table 17·7 Summary ANOVA for Data in Table 17·6

Source	SS	ν	MS
Crude SS $\cdots$	$\Sigma Y^2 = 20,456.0$	6	
Due to b_0	$b_0 \Sigma X_0 Y = 20,416.7$	1	
Due to b_1	$b_1 \Sigma X_1 Y = 25.0$	1	25.0
Due to b_{11}	$b_{11}\left(\Sigma X_1^2 Y - \dfrac{\Sigma X_0 X_1^2 \Sigma X_0 Y}{\Sigma X_0^2}\right) = 8.3$	1	8.3
Residual	6.0	3	2.0

$F_{1,3,0.10} = 5.5$

PROBLEM

17·3 Using the data from Prob. 17·1, assume a quadratic relationship exists between the two variables.
(a) Write the mathematical model which is now under test.
(b) Set up the design matrix for a quadratic relationship.
(c) Using regression techniques, solve for the line of best fit by the sum of least squares.

(d) Using ANOVA, test to see if there is an indication of a lack of fit of the curvilinear model.

17·4 TWO-VARIABLE FIRST-ORDER MODEL

The methods of response surface experimentation are especially fruitful in cases in which we wish to test for the effects of more than one factor. If we believe that temperature and pressure affect process yield, and we wish to maximize the yield, we may run a twice-replicated 2^2 factorial design. From only eight runs we can find the coefficients of the regression equation, and from the same data we can calculate the terms needed for ANOVA.

If we assume a linear relationship, our mathematical model is

$$Y = \beta_0 X_0 + \beta_1 X_1 + \beta_2 X_2 + \epsilon$$

which is represented by a plane surface as shown in Fig. 17·2a. We code the levels of the predictor variables so that they follow the orthogonality rules, as shown in Table 17·8. In the same table at the left are the treatment combinations, and also the values of the response variable Y. In Table 17·9 are the sums of squares of cross products.

Table 17·8   Design Matrix for Two-variable
First-order Model

Treatment combinations	X_0	X_1	X_2	Y
ab	1	+1	+1	54
	1	+1	+1	56
b	1	−1	+1	59
	1	−1	+1	61
a	1	+1	−1	65
	1	+1	−1	67
(1)	1	−1	−1	66
	1	−1	−1	68

Table 17·9   Sums of Squares of Cross Products for Data in Table 17·7

(a)

	X_0	X_1	X_2	Y
X_0	ΣX_0^2	$\Sigma X_0 X_1$	$\Sigma X_0 X_2$	$\Sigma X_0 Y$
X_1	$\cdots$	ΣX_1^2	$\Sigma X_1 X_2$	$\Sigma X_1 Y$
X_2	$\cdots$	$\cdots$	ΣX_2^2	$\Sigma X_2 Y$
Y	$\cdots$	$\cdots$	$\cdots$	ΣY^2

(b)

	X_0	X_1	X_2	Y
X_0	8	0	0	496
X_1	$\cdots$	8	0	−12
X_2	$\cdots$	$\cdots$	8	−36
Y	$\cdots$	$\cdots$	$\cdots$	30,948

We calculate the coefficients in the equation

$$\hat{Y} = b_0 X_0 + b_1 X_1 + b_2 X_2$$

by means of the normal equations

$$b_0 = \frac{\sum X_0 Y}{\sum X_0^2} = \frac{496}{8} = 62.0$$

$$b_1 = \frac{\sum X_1 Y}{\sum X_1^2} = -\frac{12}{8} = -1.5$$

$$b_2 = \frac{\sum X_2 Y}{\sum X_2^2} = -\frac{36}{8} = -4.5$$

Thus in coded form

$$\hat{Y} = 62.0 - 1.5X_1 - 4.5X_2 \tag{17·3}$$

The summary ANOVA in Table 17·10 indicates that both coefficients b_1 and b_2 are significant at the 0.05 confidence level. That these coefficients are both negative indicates that the optimum yield, if it exists, will be found at a temperature and pressure less than those we tested. Further experimentation should lie in this direction.

Table 17·10   Summary ANOVA for Data in Table 17·8

Source	SS	v	MS
Crude SS	$\sum Y^2 = 30{,}948$	8	
Due to b_0	$b_0 \sum X_0 Y = 30{,}752$	1	
Due to b_1	$b_1 \sum X_1 Y = 18$	1	18
Due to b_2	$b_2 \sum X_2 Y = 162$	1	162
Residual	$\sum (Y - \hat{Y})^2 = 16$	5	
Error	From pairs of observations $= 8$	4	2
Lack of fit	By subtraction $= 8$	1	8

$F_{1,4,0.05} = 7.7$

17·5 CONTOUR LINES

For the one-factor predictor-variable case, we can plot on rectangular coordinates the graph of the regression equation found by the methods of this chapter. When we have two input variables, the plot requires three dimensions. An alternative to such a plot is to show, by contour lines of equal values of the response variable, the projection of the response surface on a graph whose axes are the predictor variables.

Such contour lines are shown for Eq. (17·3) in Fig. 17·5. They are calculated from the regression equation by substituting for $\hat{Y}$ any desired numerical value and thus developing an equation in the two predictor variables X_1 and X_2. The resulting equation is then plotted.

For example, to find the equation of the contour line for $\hat{Y} = 62$, we write

$$62 = 62 - 1.5X_1 - 4.5X_2$$

Simplifying, and solving for X_2, we have

$$X_2 = (-\tfrac{1}{3})X_1$$

This equation is that of the straight line passing through the origin and with a slope of $-\tfrac{1}{3}$. The graph of this line is marked 62 in Fig. 17·5. It represents

Fig. 17·5 Contour lines for Eq. (17·3).

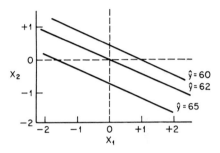

a plot of the values of the predictor variables (in coded form) that we would expect to give, within the limits of the experimental error, always a yield of 62.

Similarly, to find the equation for a yield of 65, we write

$$65 = 62 - 1.5X_1 - 4.5X_2$$

which can be simplified and rearranged to give

$$X_2 = (-\tfrac{1}{3})X_1 - \tfrac{2}{3}$$

This equation plots as a line with the same slope as the previous one because we are dealing with a plane surface. The intercept of this line is at $-\tfrac{2}{3}$.

The appearance of the contour lines gives us a clue to the direction in which we should make further experimentation in our search for the optimum levels of temperature and pressure. In most cases, we should move in a direction perpendicular to the contour lines. This is based on the principle of "steepest ascent." These contour lines indicate that we will do well in the next experiment to change the level of X_2 more than that of X_1, inasmuch as a line perpendicular to the contour lines lies in such a direction.

PROBLEMS

17·4 An investigation was carried out to determine the effect of two factors on the strength of joints of a specific type of plastic. The factors investigated were pressure per square inch during drying and temperature. The results of the investigation are presented in Table 17·11.

Table 17·11   Joint Strength as a Function of Two Factors

Pressure X_1	Temperature X_2	Strength Y
15	68	151
15	68	149
15	70	144
15	70	143
16	68	143
16	68	144
16	70	139
16	70	141

(a) What is the design of the experiment?

(b) Assume a linear relationship. Write the mathematical model for the function $Y = f(X_1, X_2)$.

(c) Using regression techniques, solve for the plane of best fit. (Assume a linear relationship.)

17·5   Use the data in Prob. 17·4 and the equation arrived at through regression to answer the following: If the glue joint is held with 15 lb pressure per square inch and the drying temperature is 70.0° F, what would you estimate the strength of the joint to be?

17·6   From your answer to Prob. 17·4, draw the contour lines for a joint strength of 150 and 140.

17·6 FORWARD DOOLITTLE SOLUTION
OF THE NORMAL EQUATIONS

We have described the calculating method for finding the estimated values of the coefficients in a regression model. A second method for accomplishing the same purpose is known as the "forward Doolittle solution." This technique has an additional advantage as compared with the one we have used so far. By the Doolittle technique we can, at the same time we find the equation coefficients, calculate the sums of squares for these coefficients. We need these sums of squares for testing the validity of the regression analysis of the data.

We begin with a first-order model in one variable

$$Y = \beta_0 X_0 + \beta_1 X_1 + \epsilon$$

The results of five experimental runs were as shown in Table 17·12. From this table we prepare a table of sums of cross products, shown in Table

17·13. This table is much like Table 17·2, except that we now do not need the third row.

Table 17·12   Data from Regression Experiment

X_0	X_1	Y
1	2	8
1	3	7
1	4	5
1	5	3
1	6	1

Table 17·13   Table of Sums and Cross Products for Doolittle Solution of Data in Table 17·10

	X_0	X_1	Y
X_0	$\Sigma X_0{}^2$	$\Sigma X_0 X_1$	$\Sigma X_0 Y$
X_1	$\cdots$	$\Sigma X_1{}^2$	$\Sigma X_1 Y$

As the first step in the Doolittle solution we replace the symbols in the cross-products table with the numerical values. As the second step we bring down and repeat the values found in the first row. As the third step we divide each member of this row by its leading elements. We now have Table 17·14.

Table 17·14   Early Stages in the Doolittle Solution of Data in Tables 17·12 and 17·13

		X_0	X_1	Y
Step		5	20	24
			90	78
2		5	20	24
3		1	4	4.8

As step 4, the second element—the *pivot multiplier*—in the row just formed is multiplied by the number immediately above it, and this product is subtracted from the number in the second row. Thus 4×20 is subtracted from 90. The same multiplier is used with the last element in the third row, and the product is subtracted from the last number in the second row: 4×24 is subtracted from 78. This process is equivalent to multiplying one row of a matrix by a scalar value (in this case $\frac{4}{5}$) and subtracting the results from a second parallel row of the matrix.

Finally, as step 5, we divide the numbers in the row just formed by the leading element of the row. We have now completed the forward Doolittle solution in Table 17·15. The results of these calculations are as follows:

Table 17·15   Complete Doolittle Solution for Data in Tables 17·12 and 17·13

Step	X_0	X_1	Y
	5	20	24
		90	78
2	5	20	24
3	1	4	4.8
4	$\cdots$	10	-18
5	$\cdots$	1	-1.8

1 The last line is read as (1) $b_1 = -1.8$.
2 The third line from the bottom is read as (1) $b_0 + 4b_1 = 4.8$. Hence $b_0 = 12$, and the estimate of the regression equation is $\hat{Y} = 12.0 - 1.8X$.
3 The product of the two Y terms in the bottom pair of rows is the sum of squares due to $b_1 = 32.4$.
4 The product of the two Y terms in the next two rows above is the sum of squares due to $b_0 = 115.2$.

We have shown how the Doolittle method can be applied to a simple model. The same techniques are applicable, regardless of the number of coefficients we need to estimate. We consider now the data in Table 17·16

Table 17·16   Data for Second-order Model of Single Predictor Variable, in Regression Analysis

X_0	X_1	X_1^2	Y
1	1	1	6.4
1	1	1	5.6
1	1	1	6.0
1	2	4	7.5
1	2	4	6.5
1	3	9	8.3
1	3	9	7.7
1	3	9	11.7
1	4	16	10.3
1	4	16	17.6
1	5	25	18.0
1	5	25	18.4

for the one-variable second-order model

$$Y = \beta_0 X_0 + \beta_1 X_1 + \beta_{11} X_1^2 + \epsilon$$

We will use the Doolittle method to estimate the values of the three coefficients. The solution is shown in Table 17·17. The steps in this solution are as follows:

Table 17·17   Doolittle Solution for the Data in Table 17·16

	X_0	X_1	X_1^2	Y
X_0	12	36	136	124
X_1	...	136	576	452
X_1^2	...	...	2,584	1,920
Step 1	12	36	136	124
Step 2	1	3	11.33	10.33
Step 3	...	28	168	80
Step 4	...	1	6	2.857
Step 5	...	...	34.67	34.67
Step 6	...	...	1	1

1   Bring down the first row.
2   Divide this row by its leading element.
3   Use 3 as a pivot multiplier, and perform the following operations:
    3×36 is subtracted from 136
    3×136 is subtracted from 576
    3×124 is subtracted from 452
    The results of these operations form the next row.
4   Divide the newly formed row by its leading element.
5   We are now down *two* pairs of rows. We have *two* pivot multipliers—
    6 and 11.33. We find the terms for the next row by these operations:
    $(6 \times 168) + (11.33 \times 136)$ subtracted from 2,584
    $(6 \times 80) + (11.33 \times 124)$ subtracted from 1,920
6   Divide the newly formed row by its leading element.

From the last line of the Doolittle solution, $b_{11} = 1.0$. In the third row from the bottom, (1) $b_1 + 6.0 b_{11} = 2.857$. By substituting the now known value of b_{11}, we find that $b_1 = -3.14$. In the fifth row from the bottom, (1) $b_0 + 3 b_1 + 11.33 b_{11} = 10.33$. Thus $b_0 = 8.43$. Our estimate of the regression equation is

$$\hat{Y} = 8.43 - 3.14X + X^2$$

The products of the terms in the Doolittle solution, taking the rows in pairs, also furnish us with the sums of squares:

SS due to $b_0 = 124 \times 10.33$
SS due to $b_1 = 80 \times 2.857$
SS due to $b_{11} = 34.67 \times 1$

After we have obtained the estimates of the coefficients of the mathematical model and have incorporated them in the model, we must still test to see if the mathematical model we used is the proper one. This we do by running an analysis of variance of the results. The completed ANOVA table is given in Table 17·18.

Table 17·18 ANOVA for Second-order Model Regression Problem (Data from Table 17·16)

Source	SS	ν	MS
Crude SS	ΣY^2	12	
Due to b_0	$(10.33)(124)$	1	
Due to b_1	$(2.857)(80)$	1	
Due to b_{11}	$(1)(34.67)$	1	
Residual	$\Sigma Y^2 - (SSb_0 + SSb_1 + SSb_{11})$	9	
Lack of fit	By subtraction	2	
Error	$\Sigma (Y_i - \bar{Y}_i)^2$	7	

The residual term in the ANOVA consists of two members: the experimental error, and the influence due to lack of fit of the data to the mathematical model we have used. To separate the two parts of the residual, we require an estimate of the error term. The error can be estimated from repeated experimental runs. From the data list in Table 17·16 we see that there were three runs made at X_1 equal to 1, two at X_1 equal to 2, three at X_1 equal to 3, two at X_1 equal to 4, and two at X_1 equal to 5. These replicated runs will permit us to obtain an estimate of error that is independent of the proposed mathematical model.

Error is obtained by averaging the replicated runs, and summing the squares of the deviations of each run from the average value at each level. For $X_1 = 1$, the average response value was $Y = 6.00$. The deviations from this average were 0.4, -0.4, and 0.0. The squares of these deviations are 0.16, 0.16, and 0.0. The SS for error from the replicated runs at $X_1 = 1$ was therefore 0.32. The sum of the error from all sets of replicated runs is the sum of squares for the error term. When the error sum of squares is subtracted from the residual term, the result is the sum of squares for lack of fit. We first test to see if the lack of fit term is significant. To do this we run an F test against the error term. If we find significance, we have evidence that the mathematical model used is not the correct one for the data. If we do not find significance, we accept the model as an appropriate explanation of the data.

PROBLEMS

17·7   Using the data in Table 17·4, fit a first-order model to the data with the forward Doolittle technique.

17·8   With the information from the forward Doolittle solution in answer to Prob. 17·7, set up the ANOVA table for the data.

17·9   From the data in Table 17·11, and assuming a linear relationship, fit a first-order model for the two-variable case using the Doolittle technique.

17·10   From the information found in the forward Doolittle solution in answer to Prob. 17·9, set up the ANOVA table for the experiment.

17·7   TWO-VARIABLE SECOND-ORDER MODEL

If we suspect a nonlinear relationship involving two predictor variables, we may by only a 3^2 factorial design test for the six coefficients in the following model:

$$Y = \beta_0 X_0 + \beta_1 X_1 + \beta_2 X_2 + \beta_{11} X_1^2 + \beta_{22} X_2^2 + \beta_{12} X_1 X_2 + \epsilon$$

This experiment will require nine runs at the treatment combinations shown in Table 17·19. It will enable us to find the coefficients of the equation

$$\hat{Y} = b_0 X_0 + b_1 X_1 + b_2 X_2 + b_{11} X_1^2 + b_{22} X_2^2 + b_{12} X_1 X_2$$

Table 17·19   Two-variable Second-order Model (Using a 3^2 Factorial Design)

Treatment combination	X_0	X_1	X_2	X_1^2	X_2^2	$X_1 X_2$	Y
2, 2	1	+1	+1	1	1	+1	54
0, 2	1	−1	+1	1	1	−1	59
2, 0	1	+1	−1	1	1	−1	65
0, 0	1	−1	−1	1	1	+1	66
1, 1	1	0	0	0	0	0	63
1, 2	1	0	+1	0	1	0	57
1, 0	1	0	−1	0	1	0	66
2, 1	1	+1	0	1	0	0	62
0, 1	1	−1	0	1	0	0	64

he values of the predictor variables must be coded as before, to give the ·thogonal coefficients in Table 17·18. From the values of the response .riable in the same table, we can by methods similar to those previously ·scribed arrive at the equation

$$\hat{Y} = 63.23 - 1.33 X_1 - 4.50 X_2 - 0.33 X_1^2 - 1.83 X_2^2 - 1.00 X_1 X_2$$

From this equation we can find contour lines as shown in Fig. 17·6. For $\hat{Y} = 60$, we have

$$60 = 63.23 - 1.33X_1 - 4.50X_2 - 0.33X_1{}^2 - 1.83X_2{}^2 - 1.00X_1X_2$$

which on rearrangement gives, for $X_1 = 0$,

$$1.83X_2{}^2 + 4.50X_2 - 3.23 = 0$$
$$X_2 = 0.58$$

Thus we have the point on the contour line at $X_1 = 0$, $X_2 = 0.58$. Similarly, when we substitute $X_1 = -1$ we find that $X_2 = 0.82$, and when we substitute

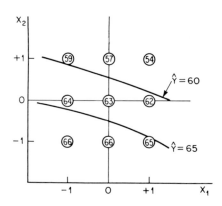

Fig. 17·6 Contour lines for two-variable second-order model.

$X_1 = +1$ we find that $X_2 = 0.24$. In the same way we find the points for the contour line at $Y = 65$.

The shapes of the contour lines, like those on a physical map, show that the maximum lies below our chart and slightly to the left, thus indicating where the next experiments should be run.

PROBLEMS

17·11   Thickness and Brinell hardness affect the number of folding operations which a cardboard carton can withstand. To find the combination of the two factors which would produce the maximum number of folds a carton could withstand, an experiment was run, with results as in Table 17·20.

(a) What was the experimental design used in the treatment combinations?
(b) If we assume that a second-order model should be used, what is the mathematical model which shows the functional relationship between the response variable and the two factors?
(c) Using the Doolittle technique, solve for the coefficients of the mathematical equation which estimates the functional relationship. [Use the mathematical model in answer to part (b).]

Table 17·20 Number of Folds as a Function of Thickness and Hardness

Thickness, X_1	Hardness, X_2	Folds, Y
0.260	50	154
0.255	50	161
0.250	50	164
0.260	45	150
0.255	45	153
0.250	45	159
0.260	40	148
0.255	40	156
0.250	40	161

(d) Using the information from the forward Doolittle solution, set up the ANOVA table for the experiment.

(e) Draw the contour lines for 155 folds and 160 folds.

(f) In which direction does this experiment suggest the maximum is located?

17·12 Refer to the data in Table 17·11. Answer the same questions as listed in parts (a) to (d) in Prob. 17·11.

17·8 COMPOSITE DESIGNS

Suppose that we are interested in testing a situation in which we believe that two predictor variables are important in determining yield. We have no a priori reason to believe that the effects of the predictor variables are linear. If we run a 2^2 factorial experiment, we will need to make only four tests. We will, however, have no internal error estimate, and we will be unable to test for the curvilinearity which we suspect. If we fully replicate, we will require eight runs, and we still will only be able to test for linear effects. Furthermore, half our experimental effort will have been put into obtaining an estimate of experimental error.

If we run a 3^2 experiment, for a total of 9 runs, we will have no internal error estimate, though we will be able to test for curvilinear effects. A fully replicated 2×3^2 experimental design will take 18 runs, which may be costly.

An experimental design which is more economical is one of a class called "composite designs." Such designs are based on factorial experiments with added runs, as indicated in Fig. 17·7. The four positions marked with triangles are those of a 2^2 factorial. To those we add (1) four replicates at the center, which will provide us with an estimate of error and avoid the necessity of making an F test with only $v = 1, 1$; (2) the four axial points marked

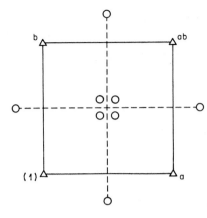

Fig. 17·7 Composite design, 2^2 factorial with four center points and axial points.

with circles. By adding these four axial points, giving us a total of 12 runs, we have in effect five levels along each of the axes X_1 and X_2. This situation permits us to make good tests for curvilinearity, since we have increased our chances of finding nonlinear effects with more levels of the factors. The saving of experimental effort is not great for only two factors, but it is very great for more complicated situations.

What we are doing here is to explore the response variable over a circular field, as indicated by the dotted line which forms the envelope of the positions in Fig. 17·8. In order to make the positions of the axial points equivalent to the others (in terms of their distance from the central point) we assign

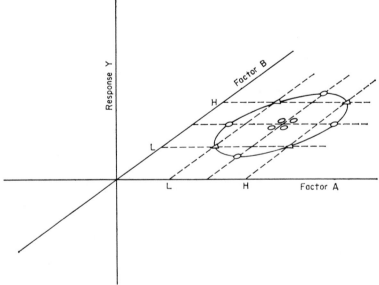

Fig. 17·8 Region of investigation with composite design.

Table 17·21   Design Matrix for 2² Factorial Composite with Four Center Points and Added Axial Points

Treatment combination	X_0	X_1	X_2	X_1^2	X_2^2	$X_1 X_2$	Y
(1)	+1	−1	−1	+1	+1	+1	66
a	+1	+1	−1	+1	+1	−1	65
b	+1	−1	+1	+1	+1	−1	58
ab	+1	+1	+1	+1	+1	+1	56
Center	+1	0	0	0	0	0	64
Center	+1	0	0	0	0	0	64
Center	+1	0	0	0	0	0	63
Center	+1	0	0	0	0	0	63
Axial	+1	0	+1.4	0	2	0	55
Axial	+1	0	−1.4	0	2	0	66
Axial	+1	+1.4	0	2	0	0	59
Axial	+1	−1.4	0	2	0	0	64

Table 17·22   Forward Doolittle Solution

	X_0	X_1	X_2	X_1^2	X_2^2	$X_1 X_2$	Y
X_0	12	0	0	8	8	0	743
X_1	⋯	8	0	0	0	0	−10
X_2	⋯	⋯	8	0	0	0	−32.4
X_1^2	⋯	⋯	⋯	12	4	0	491
X_2^2	⋯	⋯	⋯	⋯	12	0	487
$X_1 X_2$	⋯	⋯	⋯	⋯	⋯	4	−1
	12	0	0	8	8	0	743
	1	0	0	$\frac{2}{3}$	$\frac{2}{3}$	0	61.917
		8	0	0	0	0	−10
		1	0	0	0	0	−1.25
			8	0	0	0	−32.4
			1	0	0	0	−4.05
				6.67	−1.33	0	−4.333
				1	−0.2	0	−0.65
					5.6	0	−9.2
					1	0	−1.643
						4	−1.0
						1	−0.25

them the coded value ± 1.4, since they lie on the diagonals of squares of sides 1. Therefore, the design matrix is as shown in Table 17·21. The mathematical model for this experiment design is

$$Y = \beta_0 X_0 + \beta_1 X_1 + \beta_2 X_2 + \beta_{11} X_1^2 + \beta_{22} X_2^2 + \beta_{12} X_1 X_2 + \epsilon$$

The forward Doolittle solution is shown in Table 17·22. Working upward in the table, $b_{12} = -0.25$ and $b_{22} = -1.643$. Moving now to the fifth line from the bottom, $b_{11} = -0.65 - 0.2(1.643)$, from which $b_{11} = -0.9786$, or approximately -1. Similarly, working upward, $b_2 = -4.05$, $b_1 = -1.25$, and $b_0 = 63.678$. Thus the calculated equation is

$$\hat{Y} = 63.678 - 1.25 X_1 - 4.05 X_2 - X_1^2 - 1.643 X_2^2 - 0.25 X_1 X_2$$

The analysis of variance is shown in Table 17·23; we conclude from this that we have obtained evidence for a significant influence due to the first- and second-order effects of both variables.

Table 17·23 Analysis of Variance Table

Source	SS	ν	MS	F_0
Crude SS	$\Sigma\, Y^2 = 46{,}169.00$	12		
Due to b_0	$(743)(61.917) = 46{,}004.33$	1		
Due to b_1	$(-10)(-1.25) = \quad 12.25$	1	12.25	36.75*
Due to b_2	$(-32.4)(-4.05) = \quad 131.22$	1	131.22	393.6*
Due to b_{11}	$(-4.33)(-0.65) = \quad 2.82$	1	2.82	8.46(?)
Due to b_{22}	$(-9.2)(-1.643) = \quad 15.12$	1	15.12	45.36*
Due to b_{12}	$(-1)(-0.25) = \quad 0.25$	1	0.25	0.75
Residual	By subtraction $= \quad 3.01$	6		
Lack of fit	2.01	3	0.67	
Error	1.00	3	0.33	

* Significant
$F_{1,3,0.05} = 10.1$
$F_{1,3,0.10} = 5.54$

We find the contour lines in Fig. 17·9 by assigning any desired value for $\hat{Y}$, and in the resulting equation substituting successive values of X_1 to obtain the corresponding values for X_2. For example, if $\hat{Y} = 65$,

$$65 = 63.678 - 1.25 X_1 - 4.05 X_2 - X_1^2 - 1.643 X_2^2 - 0.25 X_1 X_2$$

Simplifying,

$$1.25 X_1 + 4.05 X_1 + X_1^2 + 1.643 X_2^2 + 0.25 X_1 X_2 - 1.322 = 0$$

In this simplified equation, we substitute successively values for X_1 of 0, $+1$, and -1 and find the corresponding values of X_2, which are -0.39, imaginary, and -0.33. If we substitute -0.5 for X_2 we find that $X_1 = 0.63$.

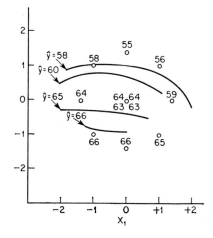

Fig. 17·9 Contour lines for composite design.

We ignore the point which gave the imaginary value, and sketch the contour line marked 65 in Fig. 17·9. Similarly, we compute points permitting us to sketch the other contour lines of the figure.

The shape of these lines suggests that the optimum lies below and to the left of the area explored by the experimental runs.

PROBLEMS

17·13 A central composite design is used to investigate the response surface of yield as a function of factor A and factor B. The design consists of a 2^2 factorial with replicated center points and axial points. The design matrix and the yields are listed in Table 17·24.

Table 17·24 A Two-factor Composite Design

Trial	X_0	X_1	X_2	Percent yield
1	1	-1	-1	71.7
2	1	$+1$	-1	80.1
3	1	-1	$+1$	76.3
4	1	$+1$	$+1$	75.8
5	1	0	0	81.5
6	1	0	0	81.2
7	1	0	0	81.7
8	1	-1.4	0	75.2
9	1	$+1.4$	0	79.1
10	1	0	-1.4	79.2
11	1	0	$+1.4$	80.2

(a) Sketch a diagram of the experimental treatment combinations.
(b) Using the data in Table 17·24, solve for the mathematical equation which represents the response surface.

(c) Draw the contour lines for 75 percent yield and for 80 percent yield.

(d) Test to see if the equation arrived at in answer to part (b) above shows evidence of lack of fit.

17·14 Consider the information in Table 17·25 from an experiment which was run to determine optimum conditions.

Table 17·25   Determination of Optimum Conditions

Trial	X_0	X_1	X_2	Response
1	1	−1	−1	77.2
2	1	+1	−1	77.3
3	1	−1	+1	80.1
4	1	+1	+1	76.6
5	1	0	0	78.2
6	1	0	0	78.1
7	1	0	0	78.3
8	1	0	0	78.2
9	1	−1.4	0	78.0
10	1	+1.4	0	76.2
11	1	0	−1.4	76.8
12	1	0	+1.4	78.0

(a) What design was used?

(b) Fit a mathematical model of second degree in two variables to the data, and solve for the equation.

(c) Using ANOVA, test the significance of the coefficients of the equation in answer to part (b), and test for lack of fit of the equation.

(d) Draw the contour lines on the experimental results.

17·15 Consider the design matrix in Table 17·26 (see top of next page).

(a) From the design matrix, what type of experiment is used?

(b) Make a sketch of the design, showing the location of the treatment combinations.

(c) Write the mathematical model for a second-degree equation to be fitted to the response data.

(d) Comment on how an estimate of error might be obtained.

17·9 EVOLUTIONARY OPERATION

Consider the problem of improving a manufacturing process which is already producing a satisfactory product. It may be costly to experiment on the process because we fear that any consequential change in the process conditions may lead to the production of unusable items. If, nevertheless, we have no reason to believe that the process is operating at an optimum

Table 17·26   *Design for Determination of Optimum Conditions*

Trial	X_1	X_2	X_3
1	-1	-1	-1
2	1	-1	-1
3	-1	1	-1
4	1	1	-1
5	-1	-1	1
6	1	-1	1
7	-1	1	1
8	1	1	1
9	$-a_1$	0	0
10	a_1	0	0
11	0	$-a_2$	0
12	0	a_2	0
13	0	0	$-a_3$
14	0	0	a_3
15	0	0	0

level, we may want to try to improve the process, in terms of either quality, yield, or cost.

Evolutionary operation (EVOP for short), developed by G. E. P. Box, is a technique of statistical experimentation which is particularly appropriate in such a situation. EVOP has these characteristics:

1 There is no interruption of the process, and no increase in defective production (over the normal level) occurs.
2 To ensure continuation of useful production, changes in the process conditions are deliberately made very small, unlike other experimentation in which we often use large changes in factor levels in order to detect significant effects. In EVOP, if temperature is known to have an effect on the process with a change of 5°F, we may make a change of only 1°F in order to avoid drastic changes in product.
3 EVOP is based on a factorial design (fully or partly replicated) and with randomization of the sequence of treatment combinations. The design may involve any number of factors (usually two or three) at two levels, to which we add one additional treatment combination—the present operating conditions. Thus, for a fully replicated 2^2 EVOP, there will be a total of five treatment combinations, as indicated in Fig. 17·10. The changes in process conditions are centered about the present operation which is designated (1) in the figure.
4 The experimentation is carried out in *cycles*, each cycle comprising a complete set of treatment combinations. Since the changes in the factor levels are small, several cycles will be needed to detect a significant change.

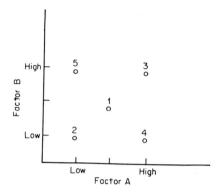

Fig. 17·10   EVOP treatment combinations.

EVOP therefore requires considerable time, which is not harmful since the process continues to produce satisfactory material.

5   Once evidence is acquired that one or more of the tested factors has a significant effect, we shift the process conditions in the indicated direction and begin a new *phase*. We start a new set of cycles centered about a new (better) set of operating conditions. Experimentation continues thus until we are satisfied that no improvement of this process is possible by further changes of the tested factors.

6   The nature of EVOP requires that we make no change in process conditions until we are quite sure that a desirable effect has been found. Our test of hypothesis will be based on an error estimate, as usual. In EVOP, we estimate the standard deviation s from differences between values of the response variable for the same treatment combination in successive cycles. To obtain any estimate of error by this method requires

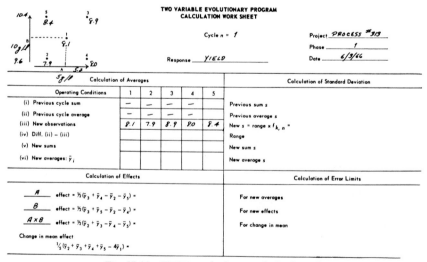

Fig. 17·11   EVOP work sheet, cycle 1, phase 1.

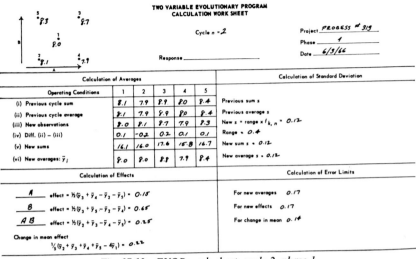

Fig. 17·12   EVOP work sheet, cycle 2, phase 1.

two cycles; to find an average value of s requires at least three cycles. Therefore, three is the minimum number of cycles we can safely run before a decision is made to change the process conditions for the second phase. Since to change the process implies a major decision, we often continue the first phase for several cycles more, or until we have a strong indication of the changes to be made in the process levels, and the direction to take.

We carry through the calculations for a typical application. Reference should be made to the calculation work sheets (Figs. 17·11 to 17·13).

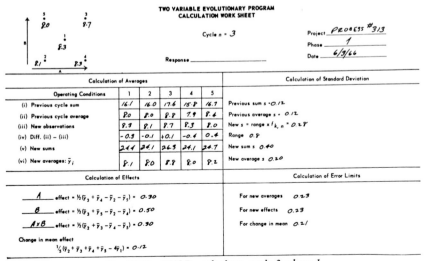

Fig. 17·13   EVOP work sheet, cycle 3, phase 1.

In a chemical manufacturing operation, we are interested in improving the yield of a process which is already operating at an economical level. We know from a history of the process that the concentrations of two ingredients A and B are important factors; we intend to optimize the process with respect to these two factors only. We will therefore run a $2^2 + 1$ EVOP. The process is now operating at 5 grams per liter of ingredient A and 10 grams per liter of ingredient B. We believe that a 10 percent change in each concentration has a real effect on the yield, and we will therefore change the concentrations by only 2 percent. The levels of ingredient A are therefore 4.8 and 5.2 grams per liter, and of ingredient B 9.6 and 10.4 grams per liter, giving us four treatment combinations, to which we add the fifth treatment combination at the present level of both factors.

As a result of a complete experimental cycle, in which we test in random order all five treatment combinations, we have as the coded yields 8.1, 7.9, 8.9, 8.0, and 8.4 in order of treatment combination. Since the completion of only one cycle gives us little information about the effects of the changes we introduced, we carry out another cycle. In Fig. 17·12, we have entered the data so far collected as follows: The values of the response variable from cycle 1 are listed in rows (i) and (ii) under the heading "Calculation of averages." The data from cycle 2 are entered in row (iii). In row (iv) we place the differences (with due regard for sign) between the data for cycle 1 and cycle 2. These will provide us with an estimate of experimental error.

In row (v) are found the sums of the observations for each treatment combination, and in row (vi) the average, found by dividing the data in row (v) by the number of cycles n.

The calculations of effects at the bottom of the work sheet are straightforward. The subscripts for the values of $\bar{y}$ (the average values of the response variable) refer to the column numbers under calculation of averages. Treatment combinations 3 and 4 represent (see Fig. 17·10) runs at the high level of factor A; treatment combinations 2 and 5 represent runs at the low level of factor A. Thus the difference in the pairs of averages estimates the total effect associated with the factor; we apply 2 as a divisor to find the average effect, since we have 2 pairs of data. A similar analysis applies to the calculation of the B effect and the AB interaction. The term designated "Change in the mean effect" compares the average of all the new data with that of the former data by a simplification of $\frac{1}{5}(\bar{y}_1 + \bar{y}_2 + \bar{y}_3 + \bar{y}_4 + \bar{y}_5) - \bar{y}_1$. The value of the change in the mean will be small (near zero) if the factors have linear effects over the range studied. Near a maximum the value will be negative, and near a minimum the value will be positive.

The right-hand side of the calculation work sheet is used for the computation of the error estimate. At cycle 2, we leave blank the first two lines, inasmuch as the first (preceding) cycle gives us too little information to estimate s. We enter on the fourth line the range of the differences found in row (iv) of the calculations of averages; the range is here 0.4. An estimate of the value of s is found from the range multiplied by a factor $f_{k,n}$ taken from a table of such factors as given in Table 17·27. The subscript k identifies the number of treatment combinations (here 5) and the subscript n identifies the cycle (here 2). Thus the factor from the table is the first value in column 5 or 0.30. The application of this factor to the range gives 0.12, which is identified as the "New s" on the work sheet. Since this is our first estimate of the value of s in this phase of the experiment, we enter this same value as the "New sum s" and the "New average s." Hereafter, the value of the new average s will be the new sum s divided by $n - 1$.

Table 17·27   Values of the Constant $f_{k,n}$

Number of cycles, n	k = number of sets of conditions in block								
	2	3	4	5	6	7	8	9	10
2	0.63	0.42	0.34	0.30	0.28	0.26	0.25	0.24	0.23
3	0.72	0.48	0.40	0.35	0.32	0.30	0.29	0.27	0.26
4	0.77	0.51	0.42	0.37	0.34	0.32	0.30	0.29	0.28
5	0.79	0.53	0.43	0.38	0.35	0.33	0.31	0.30	0.29
6	0.81	0.54	0.44	0.39	0.36	0.34	0.32	0.31	0.30
7	0.82	0.55	0.45	0.40	0.37	0.34	0.33	0.31	0.30
8	0.83	0.55	0.45	0.40	0.37	0.35	0.33	0.31	0.30
9	0.84	0.56	0.46	0.40	0.37	0.35	0.33	0.32	0.31
10	0.84	0.56	0.46	0.41	0.37	0.35	0.33	0.32	0.31

SOURCE: G.E.P. Box, *Applied Statistics*, vol. 6, no. 2, pp. 3–23, 1957.

The remaining section of the calculation work sheet permits us to place confidence limits on the effects we previously found. The limits are usually set at the 95 percent level of confidence. Thus, for the error limits for new averages and for new effects, we use $2\bar{s}/\sqrt{n}$, and for the change in mean we use $1.78\bar{s}/\sqrt{n}$, the constant 1.78 being associated with the 5-point design we used in this example. The error limits are here 0.17 for the averages and the effects, and 0.14 for the change in the mean.

The application of these error limits to the effects we have found suggests that we have not demonstrated an A effect, but we have found a B effect, an AB interaction effect, and a change in the mean. Therefore, we might well make a process adjustment at this point. Let us suppose that we continue the experiment, giving the results shown in Fig. 17·13.

In completing the work sheet for cycle $n = 3$, we transfer the appropriate data from the previous sheet, add the new observations, and perform the computations as for the previous cycle. To find the standard deviation, we use a new $f_{k,n} = f_{5,3} = 0.35$. This gives us a new value for $s = 0.28$. When we average this value with the one found for cycle 2, we have a new average s of 0.20. Our error limits are now changed to ± 0.23 for the averages and for the effects, and 0.21 for the change in mean.

If we were to make a decision on the basis of the data to this point, we would say (from the error limits and the effects we have found) that both factors have an effect, and so also does the interaction. Thus we would change the process to the conditions under which the best result was obtained, that is, treatment combination 3; and we would begin phase 2 of the EVOP with treatment combinations centered about this new set of operating conditions.

The relative ease with which EVOP can be applied to a working process, and the concept that continued operation under the same conditions is not justified without evidence that optimal results are being obtained, suggest that experimentation of this type can be exceptionally useful. The methods are explainable to personnel untrained in statistics, and therefore appeal to production supervisors.

17·10 EVOP INFORMATION BOARD

When setting up an evolutionary operation program, we must provide for systematic small changes in the levels of the factors under examination. Such changes should be introduced in a routine way, and made on the actual process, not on a pilot-plant operation or small laboratory setup. It is

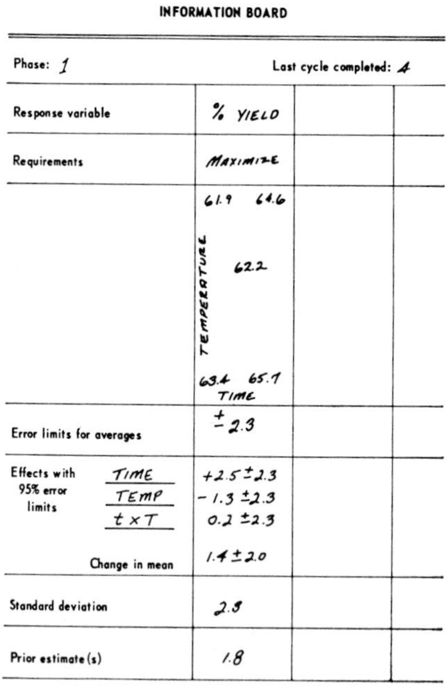

INFORMATION BOARD

Phase: 1		Last cycle completed: 4	
Response variable	% YIELD		
Requirements	MAXIMIZE		
	61.9 64.6 TEMPERATURE 62.2 63.4 65.1 TIME		
Error limits for averages	± 2.3		
Effects with 95% error limits — TIME TEMP $t \times T$	$+2.5 \pm 2.3$ -1.3 ± 2.3 0.2 ± 2.3		
Change in mean	1.4 ± 2.0		
Standard deviation	2.5		
Prior estimate(s)	1.8		

Fig. 17·14   EVOP information board.

important that the results of these small changes be fed back to the persons responsible for the process. The feedback of results can be aided by an information board as illustrated in Fig. 17·14.

A wall-mounted blackboard, or large bulletin board, with proper identifying features may be used. After each cycle has been completed and the results have been worked out on the work sheets, the results are posted on the information board for study by the evolutionary operation committee. Such a committee will continually review the boards and suggest new action to be taken or to be introduced at some later phase.

The information board may include several response variables on the same problem, all posted side by side. The decision by the committee is made with respect to all important variables and not just one alone. An increase in yield might be achieved, but if the cost of production has increased, it may be advisable to continue the process at the reduced yield but less expensive operation. The cost of operation may have been decreased and the yield increased, but the visual characteristics of the product may have become such that the product will not appeal to the buyer. It is important that the response variables be well chosen, and all major ones investigated, before any decision is made to change the process.

PROBLEMS

17·16   Continue the problem begun in Figs. 17·10 to 17·13. Complete the work sheet for cycle 4 if the results are as follows: (1) 7·8; (2) 7·9; (3) 8.6; (4) 8.0; (5) 8.3.

17·17   Complete the information board as it would appear:
(a) after cycle 3
(b) after cycle 4

17·18   A two-factor two-level evolutionary operation program was carried out with the following results in grams per liter of yield:

Cycle	Conditions				
	1	2	3	4	5
I	45	46	43	45	44
II	46	46	42	47	43
III	48	46	40	49	44
IV	49	45	41	49	43

(a) Complete the work sheets for the 4 cycles of the EVOP.
(b) Complete the information board for the last 3 cycles of the operation.
(c) What action would you suggest be taken at the end of the fourth cycle, if any?

17·1   If Y is a function of X, and the graph of the function is a quadratic relationship, how is the maximum or minimum value of Y found by methods of the calculus?

17·2   What is meant by "response surface methodology"?

17·3   In evolutionary operations, where does the experiment take place?

17·4   Define sequential experimentation.

17·5   When Y is a function of X_1 and X_2, why do we investigate the response as a surface?

17·6   In the $Y = f(X_1, X_2)$ case, what is the nature of the response surface if we find a linear relationship to be present?

17·7   When we are experimenting with two or more predictor variables, why do we resort to the use of contour lines to describe the response surface?

17·8   In the one-variable first-order model, are we concerned with a response surface? Explain.

17·9   In the solution of the normal equations, why is it useful to code the treatment combinations to orthogonal coefficients?

17·10   Is it possible to solve the normal equations without coding the treatment combinations? Explain.

17·11   When we define X_0 as being equal to 1, what value do we obtain in the table of sums of squares of cross products for the term ΣX_0^2?

17·12   After the equation for the response surface has been found by regression techniques, why do we use ANOVA?

17·13   When the levels of a factor are chosen, what considerations should be given to their selection if we are interested in the response surface?

17·14   When we are interested in the response surface of a two-variable case, what is the mathematical model to use:

(a) for a linear relationship

(b) for a curvilinear situation

17·15   After we have found the equation of the response surface how do we find the contour line for a given value of the response in the two-variable case?

17·16   How do the contour lines show the location or direction of the location of an optimum?

17·17   What is the forward Doolittle technique used for?

17·18   What is the backward Doolittle technique used for?

17·19   Can a technique other than the forward Doolittle method be used to solve the normal equations? Explain.

17·20   What are some of the advantages of using the Doolittle technique in the solution of the normal equations?

17·21   What are the major advantages of a composite design as compared with a full factorial or replicated factorial?

17·22   Why do we replicate one or more treatment combinations in the composite design when we have no outside estimate of experimental error?

17·23   In the composite design, if we are using a two-level factorial, what role do the axial points play?

17·24   What advantages does evolutionary operations have over the pilot-plant versus production-plant type of experimentation?

17·25   Why are relatively small changes made in the variables in evolutionary operations?

17·26   When a cycle has been run in evolutionary operations and none of the variables has been found to be significant, how does the running of more cycles increase the possibility of finding significance, should one of the variables under test affect the response?

REFERENCES
13, 15, 17, 28, 32, 58, 65, 97, 113, 124

18 Nonparametric Statistics

Nearly all the statistical methods we have so far discussed are based on the assumption that the populations of interest are normally distributed. These methods have for the most part involved the assumption of the existence of two population parameters—the mean and the standard deviation. Not all the preceding methods require that we know the parameters, but most of them do presuppose that the parameters could be found if needed.

We consider now situations in which we cannot safely assume anything about the population parameters, hence the term *nonparametric* statistics. We apply nonparametric methods in situations in which we have no reason to suppose that the populations are normal. We also apply them as short-cut techniques when we know the distribution shape. In fact, one of the real advantages of these techniques is the ease and speed of application. They are often used as rough and ready approximations when calculators are not readily available. Such methods are therefore useful in preliminary investigations to which later we may find it possible to apply conventional parametric techniques. If we are necessarily restricted to small sample sizes, we may prefer to use nonparametric methods, because the validity of most parametric methods is suspect unless the nature of the population is well known. Furthermore, nonparametric methods are applicable in experiments in which only limited qualitative data can be secured. As a rule, nonparametric methods have a smaller efficiency than the corresponding parametric methods.

18·1 NONPARAMETRIC VERSUS PARAMETRIC METHODS

In general, nonparametric methods are based on the median as the measure of central tendency, whereas parametric methods are based on the mean. We use the median here because it is less liable to be influenced by a few abnormal observations and is therefore more stable for nonnormal distributions.

Nonparametric methods have these advantages:

1 Data can usually be secured easily and at low cost.
2 Calculations are relatively simple.
3 The usefulness of conclusions from tests of hypothesis does not depend on assumptions about the nature of the parent populations. As usual, however, we do not get something for nothing. We must select the nonparametric test to run, and since some are more powerful under certain conditions than others, we have to assume the circumstances. We trade assuming the form of the distribution for selecting a nonparametric distribution with power in our area of interest.
4 The basic assumption we make for nonparametric tests is the independence of data.

Nonparametric methods have these deficiencies:

1 Nonparametric methods are generally less efficient (i.e., they require a larger sample size to obtain the same precision) than parametric tests if the distribution is known. This is particularly true for the normal distribution.
2 They are generally weaker than parametric methods in that we can make fewer inferences about the population of interest. Often all that we can conclude is that the populations are the same or different. If a difference can be demonstrated, we are often not able to specify the nature of the difference.
3 Until fairly recently, statisticians have paid less attention to these methods, hence they are not so well elaborated as are parametric techniques.

18·2 RUN TEST

It is intended that two machines should produce the same quality of goods. The null hypothesis we will test is that the two populations of goods are not different. We will make no assumptions about the population distributions and will therefore use a nonparametric test.

We collect data from the two processes by a suitable random method, ensuring independence of the two sets of data. A representative set of data is given in Table 18·1; here we have arbitrarily designated one set of observations as X_i and the other as Y_i. We now arrange the entire set of data in order of magnitude of the observations; we underscore the X values and overscore the Y values, thus:

2 $\overline{3}$ 4 $\overline{7}$ 8 $\underline{10\ 14\ 21\ 32}$ $\overline{33\ 48\ 67\ 79}$ 87 $\overline{98}$ $\underline{114}$ 118 120 $\overline{136\ 146}$

$X\ Y\ X\ Y\ X\ \ X\ X\ X\ X\ \ Y\ Y\ Y\ Y\ \ X\ \ Y\ \ Y\ \ X\ \ X\ \ Y\ \ Y$

Table 18·1   Performance Data
for Two Machines

X_i	Y_j
2	3
4	7
8	33
10	48
14	67
21	79
32	98
87	114
118	136
120	146

Each sequence of X_i values is a *run* of X; each sequence of Y_i values is a run of Y. The test of hypothesis we are about to make is based on the distribution of *runs*.

If the two populations from which the data were taken are *not* different, it is reasonable to expect that we will find many runs, since the data will intermix. Also, we expect that the runs will be well distributed throughout the combined set of observations. If there are few runs, and these are found mostly near the center of the combined set, giving a pattern as in

$\underline{X\ X\ X\ X\ X\ X}\ \overline{Y}\ \underline{X}\ \overline{Y\ Y}\ \underline{X}\ \overline{Y\ Y\ Y\ Y\ Y\ Y}$

it is probable that the medians of the two sets of data differ. If there are few runs, mostly near the center of the data, but the pattern is as

$\overline{Y\ Y\ Y\ Y}\ \underline{X\ X\ X}\ \overline{Y}\ \underline{X\ X\ X}\ \overline{Y}\ \underline{X\ X}\ \overline{Y\ Y\ Y\ Y}$

it is probable that the data differ in variability.

In the data from Table 18·1, there are 10 runs, well distributed throughout the combined set.

In addition to inspecting the data, we may calculate a test statistic, based on this principle: For data treated as we have done here, the number of runs for many such sets has approximately a normal distribution (regardless of the distribution of the data themselves) if the two populations are similar. Thus we can use a test of hypothesis based on the comparison of a calculated and a table t value.

If we let v = observed number of runs, μ_v = mean number of runs, σ_v = standard deviation of number of runs, and n_1 and n_2 = two sample sizes,

$$\mu_v = \frac{2n_1n_2}{n_1 + n_2} + 1 \qquad (18 \cdot 1)$$

$$\sigma_v = \sqrt{\frac{2n_1n_2(2n_1n_2 - n_1 - n_2)}{(n_1 + n_2)^2(n_1 + n_2 - 1)}} \qquad (18 \cdot 2)$$

As in Chap. 2,

$$t = \frac{v - \mu_v}{\sigma_v}$$

Note that the calculations in Eqs. (18·1) and (18·2) are concerned solely with the sample size. The values of the observations themselves are inconsequential.

In the example, the sample sizes are both 10. We apply Eqs. (18·1) and (18·2):

$$\mu_v = \frac{2 \times 10 \times 10}{10 + 10} + 1 = 11$$

$$\sigma_v = \sqrt{\frac{2 \times 10 \times 10(2 \times 10 \times 10 - 10 - 10)}{(10 + 10)^2(10 + 10 - 1)}} = 2.2$$

From these values,

$$t = \frac{10 - 11}{2.2} = -0.455$$

That the calculated value of t is so small indicates that the observed number of runs is close to the average number of runs for such sample sizes. Hence this test indicates that we should accept the null hypothesis. Our data give us insufficient reason to suppose that the two sets of observations differ, as indicated by the run test.

If ties are found when the data are arranged in order of magnitude, we may break the ties by a random method, such as coin tossing. If many ties appear, we have reason to suppose that the two sets of data are representative of the same kind of population. Coin tossing will, on the average, assign the observations to the higher group 50 percent of the time.

Alternatively, where ties occur, we may compute the t value twice, once on the assumption that all the ties should be resolved in favor of one sample, and the other time on the assumption that all ties should be resolved in favor of the other sample. Very often we find that the hypothesis test requires us to accept (or reject) the null hypothesis for both calculations.

PROBLEMS

18·1 Two brands of packaged popcorn were tested by similar treatment in a popcorn machine. The response variable is the number of unpopped kernels, recorded as follows:

$$Brand\ H.... \quad 12 \quad 15 \quad 20 \quad 14 \quad 15 \quad 13 \quad 13 \quad 15 \quad 16 \quad 20$$
$$Brand\ R.... \quad 9 \quad 11 \quad 14 \quad 15 \quad 14 \quad 12 \quad 22 \quad 11 \quad 10 \quad 9$$

Use a run test at a significance level of 0.05 to test for a difference in the two brands.

Table 18·2 Counts of Ragweed Pollen in New York City for 2 Years

Date	1963	1962
Aug. 27	5	7
Aug. 28	9	1
Aug. 29	12	3
Aug. 30	3	9
Aug. 31	10	5
Sept. 1	27	10
Sept. 2	8	12
Sept. 3	33	11
Sept. 4	28	2
Sept. 5	7	7
Sept. 6	4	9
Sept. 7	10	12
Sept. 8	19	1
Sept. 9	12	
Sept. 10	9	2
Sept. 11	7	53
Sept. 12	15	25
Sept. 13	20	5
Sept. 14	2	14
Sept. 15	3	4
Sept. 16	4	2
Sept. 17	0	7
Sept. 18	0	3
Sept. 19	1	4

18·2 In Table 18·2 are given the data for ragweed-pollen count for 2 years. On the basis of a run test, is there a difference in the ragweed-pollen count for the 2 years?

18·3 TEST OF THE MEDIAN

We can test the hypothesis that there is no difference between the median of a population and any desired number on this basis. The probability is 0.5 that any observation will fall on either side of the true median. The test therefore involves counting the number of observations above and below the assumed value of the median, and then applying a χ^2 test as in Chap. 5. The expected value here is half the sample size.

In Table 18·1 the median of the entire set of data is 40.5. We now will test the hypothesis that the two separate sets of observations are consistent with 40.5 as the median for each set.

For the samples designated as X_i, 3 items lie above the supposed median and 7 below; the expected value is 5. As in Chap. 5,

$$\chi^2 = \sum \frac{(O - E)^2}{E}$$
$$= \frac{(3 - 5)^2 + (7 - 5)^2}{5} = 1.6$$

We have only one degree of freedom for this test. The table values for $\chi^2_{1,\alpha}$ are 2.71 for an α risk of 0.10 and 3.84 for an α risk of 0.05. We would there-fore accept the null hypothesis for the X values.

PROBLEM

18·3   Apply a χ^2 test to the Y values in Table 18·1.

18·4 TESTS OF MORE THAN ONE MEDIAN

A modification of the preceding rule permits us to test two or more popula-tions. The test is based on a count, for each of the several sets of data, of the observations lying above and below the median of the combined set of observations.

$$\chi^2_{m-1} = \frac{\sum_1^m t_a^2/n - (T_a)^2/N}{T_a T_b/N^2} \tag{18·3}$$

where t_a = number of observations above the median for each m set
   n = number of samples in each m set
   N = total sample size
   T_a = total number of observations above the median
   T_b = total number of observations below the median

From a completely randomized experiment we have obtained the data in Table 18·3. The data represent counts of a specific wing defect in 5 different

Table 18·3   Number of Wing Defects in Three
Strains of Fruitflies

A	B	C
3	9	1
7	12	2
7	11	6
6	8	4
2	5	7

samples of fruitflies of 3 different strains. The null hypothesis is that of no difference in the median number of observed defects for the 3 strains. The median of all the data is 6. We prepare contingency Table 18·4 and from this acquire the data to apply Eq. (18·3):

$$\chi_2{}^2 = \frac{2^2/5 + 4^2/5 + 1^2/5 - 7^2/15}{7 \times 8/15^2} = 3.75$$

We have only two degrees of freedom for this test, inasmuch as we can enter data in only two of the cells of the contingency table in an unrestricted way.

Table 18·4   Contingency Table (Data from Table 18·3)

	Strain			Total
	A	B	C	
Above median (t_a)	2	4	1	$7(T_a)$
Below median (t_b)	3	1	4	$8(T_b)$

The rest of the entries are determined by the knowledge of the sample sizes and the totals. The table value of $\chi^2_{2,0.10}$ is 4.61 for an α risk of 0.10 and $\nu = 2$. Therefore, we accept the null hypothesis and conclude that we have not by this test demonstrated a difference in the medians of the three sets of observations.

In applying the preceding test, we take values equal to the median as below the median; we are interested for the purposes of this test in values above the median or not. By the use of this classification method, we make a binomial transformation of the data and thus can make use of the methods for dealing with binomial populations. Because we have made such a transformation of the data, it is well if large sample sizes are used, preferably at least 10.

PROBLEM

18·4   Test the hypothesis that the medians are equal for the population samples in Table 18·5.

18·5 THE SIGN TEST

An exceptionally simple nonparametric test may be performed if we pair samples from two populations, test the pairs under appropriate conditions, and then count the number of positive or negative differences in the results.

Table 18·5 Data for Test of Equality of Medians

A				B				C			
61	54	60	65	60	67	61	56	55	51	54	54
60	52	51	58	60	54	58	54	58	60	62	56
64	59	60	57	61	70	56	59	51	52	55	54
53	58	53	54	55	64	56	51	54	51	54	62
53	58	60	55	51	55	56	56	55	71	53	54

These differences are found by subtracting the second observation from the first and disregarding the magnitude of the deviations. Only the sign is used. This test is applicable to tests of preference. The null hypothesis is that of no difference. If this hypothesis is correct, the number of negative differences should be about the same as the number of positive differences.

If n is the number of pairs of observations and X is the number of occurrences of the more frequently observed sign, then the extent of the departure of X from $\frac{1}{2}n$ is given by

$$\frac{2X - 1}{\sqrt{n}} - \sqrt{n}$$

The departure from $\frac{1}{2}n$ is given in terms of standardized units and thus can be compared with the value 1.96 (if a level of significance of 0.05 is used).

If we test six pairs of samples, and all six members of population A give a larger response than the six from population B, the differences would give six signs that would be the same (either positive or negative depending on the order of comparison). If no difference exists between the two populations, the probability of occurrence of either sign is $\frac{1}{2}$, and the probability of occurrence of six identical signs is 1/32, or about 0.031.

Application of the formula to the above six-pair sample gives

$$\frac{2 \times 6 - 1}{\sqrt{6}} - \sqrt{6} = 2.04$$

Since the calculated value exceeds 1.96, we would reject the null hypothesis at the 0.05 level of significance.

Two types of paint are compared for weather resistance under different conditions of exposure. A sample of paint A is paired with a sample of paint B and exposed to condition 1. The test is repeated for each of 20 different test conditions, and two pairs are used under each condition. We thus have a total of $n = 40$ test pairs. The observations show that type B was preferred 30 times. We wish to test the hypothesis of no difference between the paints.

Applying the formula for deviation from $\frac{1}{2}n$ we have

$$\frac{2 \times 30 - 1}{\sqrt{40}} - \sqrt{40} = 3.00$$

The outcome of the test indicates that with an α risk of 0.05 we should reject the null hypothesis.

If the probability of observing a positive or a negative difference is $\frac{1}{2}$ (where the null hypothesis holds) the cumulative binomial distribution enables us to calculate the probability of obtaining less than any specified number of differences of one sign for different numbers of sample pairs. Dixon and Mood have published[1] tables giving the maximum allowable number of the less frequent sign for several probability levels. Table 18·6, a condensed form of these tables, includes sufficient data to cover most situations.

Table 18·6 Maximum Allowable Number of the Less Frequent Sign

n	Probability level		
	0.10	0.05	0.01
8	1	0	0
9	1	1	0
10	1	1	0
12	2	2	1
14	3	2	1
16	4	3	2
18	5	4	3
20	5	5	3
25	7	7	5
30	10	9	7
35	12	11	9
40	14	13	11
45	16	15	13
50	18	17	15
75	29	28	25
100	41	39	36

For the paint example, $n = 40$, and the number of observations of the less frequent sign was 10. At the 0.05 level, the maximum allowable number of occurrences for $n = 40$ is 13. Since we found only 10, the null hypothesis is rejected.

The sign test is most useful under the following conditions:

1 There is a series of pairs of observations of two possibly different objects or events.
2 For each pair, the observations are made under similar conditions.
3 For different pairs, different conditions may be involved.

In a sign test, we assume that the differences between paired observations are independent, so that the results from one pair are not influenced by the

[1] *Journal of the American Statistical Association*, vol. 41, p. 557, 1946.

results for another pair. We ignore the size of the difference between pairs; we must suppose that a small difference is as consequential as a large difference.

The sign test does not permit ties. If ties do occur, we omit these pairs from the test, and so reduce the sample size.

If we are interested in detecting small differences reflected by a narrow difference in the occurrence of the signs, we must be prepared to use a large sample size. A small sample size is sufficient if only large differences must be found.

18·6 χ^2 TEST OF SIGNS

Where data can logically be arranged in pairs, we can test the hypothesis of no difference between two populations by testing the hypothesis that the signs of the pair differences are symmetrically distributed. We use a χ^2 test to see whether or not the number of plus and minus signs is significantly different from a chance result. At least 10 pairs of numbers are needed if this is to be reasonably sensitive.

Table 18·7 Number of Defective Items
for Two Shifts

Day	Shift A	Shift B	Sign of difference, $A - B$
1	8	7	+
2	6	7	−
3	5	7	−
4	0	2	−
5	5	3	+
6	2	3	−
7	4	4	
8	3	5	−
9	9	8	+
10	6	8	−
11	11	12	−
12	8	7	+
13	5	6	−
14	8	9	−
15	10	7	+
16	3	5	−

We apply the method to the data in Table 18·7 which represent the number of defective items produced on each of 16 working days by two shifts A and B. The null hypothesis is that of no difference between the number of defective items. The last column of the table gives the sign of the difference $A - B$.

We have 15 pairs of data; we ignore day 7 because the two values are the same. The expected number of plus signs (and of minus signs) is half the sample number, or $7\frac{1}{2}$. We have recorded 5 plus signs and 10 minus signs. We apply the usual formula for the χ^2 test:

$$\chi_1{}^2 = \frac{\sum (O - E)^2}{E} = \frac{(5 - 7.5)^2 + (10 - 7.5)^2}{7.5} = 1.67$$

When we compare the calculated value with the table value for $\chi^2_{1,0.05} = 3.84$, we conclude that the data do not justify the inference that there is a difference in the performance of the two shifts on the basis of the data.

PROBLEMS

18·5 The usefulness of an aid to weight loss was tested by administering the preparation to one of each of 12 sets of twins. Each other member of the pair received a placebo. The weight loss for each person was as recorded in Table 18·8.

Table 18·8   Weight Lost by Twins

	Twin number											
	1	2	3	4	5	6	7	8	9	10	11	12
Loss with medication	6	3	0	5	4	8	3	12	5	2	11	4
Loss with placebo	4	5	3	5	3	2	6	5	6	3	4	2

(a) State the null hypothesis.
(b) Test the hypothesis on the basis of deviations of number of signs from $n/2$.
(c) Test the hypothesis by the use of χ^2.

18·6 Two types of glue were tested on 15 kinds of materials. The response variable was breaking strength of the joint. In Table 18·9, a $+$ sign indicates a greater strength. Are the glues different?

Table 18·9   Data for Test of Glue Strength

Glue	Test number														
	1	2	3	4	5	6	7	8	9	10	11	12	13	14	15
A	+	−	+	+	−	−	+	+	+	−	+	+	−	+	−
B	−	+	−	−	+	+	−	−	−	+	−	−	+	−	+

18·7   Use a sign test for the data in Table 18·10. The entries are the coded values for miles per gallon for 2 gasolines in 15 makes of automobiles; one pair of each kind of automobile was driven over similar routes.

Table 18·10   Data for Test of Mileage for Two Gasolines

Gasoline	Test number														
	1	2	3	4	5	6	7	8	9	10	11	12	13	14	15
A	6	6	3	6	8	4	5	6	1	4	6	7	4	3	8
B	4	7	2	3	9	2	4	7	4	3	3	8	5	2	6

18·7   RANK CORRELATION

In situations in which two judges are asked to make subjective evaluations (as of taste, odor, appearance) we wish to test the hypothesis of independence of the judges. Rejection of the hypothesis means that there is evidence of agreement between the judges. We are here interested in a one-sided test.

We can, without making any assumptions about the distributions of the ranks assigned by each judge, find a coefficient of correlation r_s. This coefficient is similar to that described in Chap. 12. A value of $+1$ for r_s indicates complete agreement between the judges; a value of -1 indicates complete disagreement; a value of 0 indicates no relationship. The coefficient of rank correlation is found by

$$r_s = 1 - \frac{6 \sum d_i^2}{n(n^2 - 1)}$$

The values d_i are the differences in the ranks assigned by each of the judges to each tested item; n is the number of items tested.

In Table 18·11 are shown the results obtained when 10 photographic prints were ranked for quality by each of two judges A and B. The value of the coefficient of rank correlation is

$$r_s = 1 - \frac{6 \times 54}{10(100 - 1)} = 0.673$$

If there had been perfect agreement between the judges, the value of d_i^2 would have been 0, giving a value of 1 for r_s. That we have found an appreciable value for $\sum d_i^2$ has reduced the value of r_s below unity.

Some agreement between the judges is indicated. We now need a method of deciding what value of r_s will be sufficient to permit us to reject the hypothesis of independence.

One method employs Table A·9. The value of r_s needed for rejection of the null hypothesis at $\alpha = 0.025$ is found from the curve for the appropriate sample size at the vertical line representing $\rho = 0$. For the sample size of 10 in our illustration, we find that the table value is $+0.60$. Since our calculated

*Table 18·11   Results of Ranking 10 Photographic
Prints by 2 Judges*

Judge A	Judge B	d_i	$d_i{}^2$
1	2	1	1
2	1	1	1
3	3	0	0
4	7	3	9
5	4	1	1
6	9	3	9
7	10	3	9
8	6	2	4
9	5	4	16
10	8	2	4
			Total 54

value exceeds the table value, we reject the hypothesis of independence of the judges and conclude that there is evidence of agreement between them.

An alternative method of finding a minimum required value for r_s is based on the use of the t_v table. The minimum value is

$$\frac{t_\alpha}{\sqrt{t_\alpha{}^2 + n - 2}}$$

where t_α is the table value for $(n - 2)$ degrees of freedom. Using the value from the t_v table for a sample size of 10 and an α risk of 0.05, we have

$$\frac{t_{8,0.05}}{\sqrt{t_{8,0.05}^2 + 10 - 2}} = \frac{1.86}{\sqrt{1.86^2 + 8}} = \frac{1.86}{3.385} = 0.549$$

Since the calculated value of 0.673 exceeds this minimum value, we again have reason to reject the hypothesis of independence.

PROBLEMS

18·8   Two high school English teachers ranked 10 compositions, with results as given in Table 18·12. Is there a lack of independence in the rankings?

*Table 18·12   Ranks Assigned to 10 Compositions by 2
Teachers*

Teacher A	9	4	3	7	2	1	5	8	10	6
Teacher B	7	6	4	9	2	3	8	5	10	1

18·9   The top 10 song hits of the day were ranked by 2 students, based on their own ideas of what constitutes good music. If the results were as given in Table 18·3, is there any indication of agreement on the part of the 2 students?

Table 18·13   Ranks of Ten Songs by Two Students

Student	Song									
	1	2	3	4	5	6	7	8	9	10
A	5	3	8	9	1	2	7	6	10	4
B	7	4	6	2	8	3	9	10	5	1

18·8 THE RANK-SUM TEST

We may test the hypothesis of no difference between two populations by finding the rank-sum statistic T'. To find T', we intermix the samples from the two populations in question, and arrange them in order of size. If the rank score of 1 is assigned to the smallest value, the score of 2 to the next larger, and so on, then T' is the sum of the ranks of the observations in the sample of smaller n. (If the sample sizes are the same, either may be used for finding T'.) For ties, the observations that have the same value are assigned the mean of the ranks for which they tie. Thus, if there is a tie for ranks 3, 4, and 5, all three observations are assigned a value of 4.

When n_1 and n_2 each exceed 10, the sampling distribution of T' is approximately normal, with mean

$$\mu_{T'} = \frac{n_1(n_1 + n_2 + 1)}{2} \tag{18·4}$$

and variance

$$\sigma_{T'}^2 = \frac{n_1 n_2(n_1 + n_2 + 1)}{12} \tag{18·5}$$

We use these formulas in the customary way. We select an α risk and calculate a t value

$$t = \frac{T_0' - \mu_{T'} + \frac{1}{2}}{\sigma_{T'}} \tag{18·6}$$

T_0' is the observed value of T'. The fraction $\frac{1}{2}$ is the Yates correction term.

In Table 18·14 we give the results of counting the number of defects found when a process was run at two different speeds. The numbers in the columns headed "Rank" indicate the order of magnitude of each count when the data are combined. Here $n_1 = 10$, $n_2 = 12$, and $T' = 69$. Applying Eqs. (18·4) and (18·5) we have

$$\mu_{T'} = \frac{10(10 + 12 + 1)}{2} = 115$$

$$\sigma_{T'}^2 = \frac{10 \times 12(10 + 12 + 1)}{12} = 230$$

$$\sigma_{T'} = 15.17$$

Table 18·14 Ranks of Defects for Two Different Speeds of a Process

Defects at speed 1	Rank	Defects at speed 2	Rank
212	3	233	20
220	10	224	14
215	5	216	6
200	1	222	12
217	7	230	18
221	11	228	16
218	8	225	15
205	2	214	4
219	9	235	21
223	13	231	19
		229	17
	$T' = 69$	237	22

$n_1 = 10$

$n_2 = 12$

Substituting the observed value T_0' in Eq. (18·6) gives

$$t = \frac{69 - 115 + \frac{1}{2}}{15.17} = -45.5$$

The table value for $t_{0.025} = -1.96$; thus the calculated value (-45.5) is significant at this α risk. We conclude that we have evidence to reject the hypothesis that the speed of the process has no effect on the number of defects.

18·9 RANK-SUM TEST FOR SEVERAL SAMPLES

We can use the method of ranks to test the hypothesis that k samples of sizes $n_1, n_2, \ldots, n_k$ are randomly drawn from k identically distributed populations. Again, all observations are arranged in order of size and assigned ranks as for the two-sample case. The test statistic is now

$$H = \frac{12}{N(N+1)} \sum \frac{R_i^2}{n_i} - 3(N+1)$$

where N is the sum of all the sample sizes, and R_i is the sum of the ranks of the ith sample.

If the n_i values are not small, and if the hypothesis of no difference is true, the sampling distribution of the statistic H is approximately χ^2 with $(k-1)$ degrees of freedom. If ties occur in the data, the rank for each tied observation is replaced by the mean of the ranks for which it ties.

18·10   Test the hypothesis that the data in Table 18·15 are random samples from identically distributed populations. Use a rank-sum test.

Table 18·15   Data for Rank-sum Test Prob. 18·10

Sample A	42	53	48	49	52	53	49	50	59	57
Sample B	41	46	40	45	39	42	44	45	42	40

18·11   Test the hypothesis that the data in Table 18·16 are randomly selected from populations of identical distribution.

Table 18·16   Data for Rank-sum Test Prob. 18·11

Sample A	1	3	6	6	3	3	5	7	8	4	3	2	4
Sample B	8	7	8	7	9	10	9	7	9	5	8		
Sample C	4	5	6	4	7	8	6	7	5	8			
Sample D	11	5	5	6	9	12	7	8	3	4	1		

18·10   PAIRED-REPLICATE RANK TEST

A method of testing for a significant difference between pairs of data is called the Wilcoxon Paired-replicate Rank test. The absolute differences between the pairs are ranked without regard to sign, and then each rank is assigned the mathematical sign of the difference when set B is consistently subtracted from set A. The sum of the ranks for each sign is found, and the smaller total is used for test purposes. If any pair has a difference of zero, this pair is excluded from the test. In case of ties between two differences (regardless of sign) the differences are assigned the mean score of the ranks to be assigned.

If there is no consistent difference between pairs, the sum of the ranks with positive sign should be close to the sum of the ranks for negative sign, and each should be about half the sum of the total ranks. Table 18·17 gives a minimum sum of ranks for either sign that may be expected at the indicated level of significance. When the smaller sum of ranks of the same sign is less than the table value, we have evidence of a significant difference between the sets.

It is intended that the values of Table 18·17 are to be used for a two-sided test. They may be applied to a one-sided test if the sign of the smaller rank total is set in advance of the test. In this case, the probability levels will be one-half the values given in the table.

Number of replicates, n	Probability		
	0.05	0.02	0.01
6	0		
7	2	0	
8	4	2	0
9	6	3	2
10	8	5	3
11	11	7	5
12	14	10	7
13	17	13	10
14	21	16	13
15	25	20	16
16	30	24	20
17	35	28	23
18	40	33	28
19	46	38	32
20	52	43	38
21	59	49	43
22	66	56	49
23	73	62	55
24	81	69	61
25	89	77	68

Table 18·17 Minimum Value of Smaller Sum of Ranks

We apply this method to the data of Table 18·18. The observations represent the results of an investigation intended to test whether a trained weight guesser has a tendency to guess consistently high or low when he is asked to estimate the weights of packages of various sizes. The hypothesis of no difference between the actual weights and the estimated weights would, if accepted, justify the continued use of this approach to estimating weights in a noncritical situation. If the null hypothesis is rejected, then we will have reason to believe that the man has a tendency to estimate weights either too high or too low, based on the sign of the smaller sum of ranks.

The first column of Table 18·18 lists the estimated weights of nine different packages; the second column lists the actual weights. The differences are shown in the third column. In the fourth column are the ranks of the differences in order of absolute value; the ranks are assigned the signs of the corresponding differences. Note that there were three scores with absolute difference of 1; since they tie for the first three ranks (1, 2, and 3) we assign each of them the mean of this set of three ranks, or 2.

The smaller rank sum is $+17$. When we compare this value with those in Table 18·17 for $n = 9$, we find it to be far larger than any of those given in the table. We are therefore constrained to accept the null hypothesis. The test does not permit us to decide whether or not the weight guesser comes close to the actual value; it merely indicates that he does not consistently guess high or low.

Table 18·18 Estimated and Actual Weights of 10 Items

Guessed weight, A	Actual weight, B	Difference, $A - B$	Rank
33	35	−2	−5
17	20	−3	−7
104	100	+4	+8
20	25	−5	−9
4	5	−1	−2
7	6	+1	+2
26	25	+1	+2
38	40	−2	−5
100	100	0	
12	10	+2	+5

PROBLEM

18·12   Fifteen samples of different fabrics, wet from a dyeing operation, were divided into equal portions. Each portion was subjected to a different method of drying. After a standard time, the weight loss was recorded as in Table 18·19. Analyze the data by the use of a paired-replicate rank test.

Table 18·19 Weight Loss of 15 Specimens under Each of Two Conditions

Pair number	Condition A	Condition B
1	23	27
2	35	36
3	48	42
4	55	53
5	62	64
6	72	78
7	86	83
8	95	94
9	26	24
10	33	37
11	45	47
12	58	52
13	64	64
14	76	75
15	87	84

18·11 THE CORNER TEST

To test the hypothesis that two continuous variables are independent, a graphical test may be used. This corner test provides a very simple substitute for a more formal correlation study. The test is based on a plot of the data, to which are added lines at the X and Y medians, as in Fig. 18·1. The data are thus divided into four quadrants. The upper right and lower left quadrants are given positive signs, and the others negative signs. (These signs are related to the concept of positive and negative correlations.)

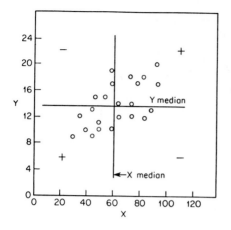

Fig. 18·1 Scattergram for corner test.

We now examine the points on the left-hand side of the X median, and draw a vertical line to the left of which will lie as many points as possible for one of the two quadrants lying there, without also separating from the main body of data any points in the other quadrant. We see that we can draw such a line at $X = 46$, which leaves to the left five points in the left positive quadrant, and no points in the left negative quadrant.

Similarly, examining the points on the right side of the X median, we can draw a vertical line at $X = 92$ which leaves to the right two points in the right positive quadrant, and none in the right negative quadrant. We now have Fig. 18·2.

In the same manner, by considering the points above and below the Y median, we draw two horizontal lines. Each of these lines separates out as many points as possible in one quadrant, without doing so for the other quadrant lying on the same side of the Y median. We now have Fig. 18·3.

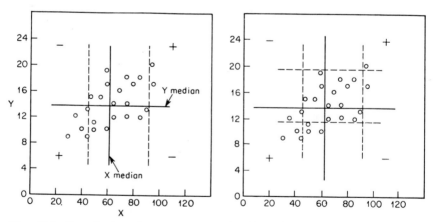

Fig. 18·2 Second stage in corner test.   Fig. 18·3 Completed plot for corner test.

The test of significance is based on a count of the number of points falling beyond the added parallel lines. To the number of points is assigned the sign of the quadrant in which they lie. Points falling beyond two of the guide lines are counted twice. If a guide line passes through two points, one on either side of the perpendicular median line, each point is counted as $\frac{1}{2}$.

The algebraic sum of all the outlying points is the test statistic S. If the value of S from the data is equal to or greater than the value in Table 18·20, the hypothesis of no association between the variables may be rejected at the indicated level of significance. If S is positive, a positive association is indicated; a negative sign is an indication of negative association.

Table 18·20   Table of S Values for Corner Test

Significance level	Quadrant sum
0.01	15
0.02	13
0.05	11
0.10	9

We apply the method to Fig. 18·3; two points lie in the positive right-hand quadrant to the right of the guide line; one point lies in this quadrant above the horizontal uppermost line; five points lie in the left-hand positive quadrant to the left of the leftmost parallel line; six points lie in this same quadrant below the lowest horizontal line. We thus have the sum $+2, +1, +5, +6 = 14$. In Table 18·20 the required value at the 0·05 level of significance is 11. Since the calculated value is larger, we have reason to reject the hypothesis of independence; that is, we conclude that conditions A and B are correlated.

PROBLEM

18·13   Test the hypothesis of independence of the two variables B and G using the corner test for the data in Table 18·21.

18·12   TCHEBYCHEFF'S INEQUALITY

A rule that can be applied to any distribution whatever is this: The relative size of the area for the section deviating from the mean by k or more units of standard deviation is less than $1/k^2$. Thus the area farther than two standard deviations from the mean is always less than $\frac{1}{2}^2$ or 0.25 of the total area. (For a normal distribution the actual value is far less than this: approximately 0.045.)

Since this rule—*Tchebycheff's inequality*—applies to any distribution, it can be used in situations where we are completely ignorant of the shape of the parent distribution. If we know that we are dealing with a symmetrical

Table 18·21　Data for Corner Test

Test number	Variable B	Variable G
1	5	11
2	7	9
3	8	10
4	6	12
5	9	14
6	7	11
7	9	11
8	10	13
9	7	14
10	11	11
11	10	12
12	13	12
13	15	16
14	13	14
15	12	15
16	10	15
17	14	17
18	12	10
19	5	9

distribution with a single mode, the Camp-Meidell inequality states that the total area k or more units of standard deviation from the mean will be less than $(\frac{2}{3}k)^2$ or $1/2.25k^2$. Note that the more we know (symmetry for Camp-Meidell, normality for the normal distribution) the more precisely we can state our probabilities.

PROBLEMS

18·14　The form of the distribution of a product is not known and is believed to be skewed, but there are good estimates of its mean and standard deviation. The population average is 90 and the standard deviation is 11. What is the probability that an observation will fall outside of 3-sigma limits?

18·15　If the distribution of observations is known to be unimodal, but it is not safe to assume a normal shape, find the probability that a single observation will deviate from its mean:
(a) by more than 3 times the standard deviation
(b) by more than 2 times the standard deviation
(c) by more than 1 time the standard deviation

18·16　If the distribution is unimodal and if the population mean is 100 with a standard deviation of 10:
(a) what is the probability that X will be above 130 or below 70
(b) what is the probability that X will be above 120 or below 80

18·1   List the differences between parametric and nonparametric statistical methods.

18·2   What advantages and disadvantages are involved in the use of nonparametric statistics?

18·3   List five situations in which nonparametric methods are appropriate. What characteristics of the situations suggest that such methods would be useful?

18·4   If we apply a run test, what would it mean if:
    (a) all the X values lie on one side of all the Y values
    (b) the X and Y values alternate
    (c) the X values were clustered in the center of the Y values

18·5   In a run test, should the data be questioned if
    (a) there is a very small number of runs? Explain.
    (b) there is a very large number of runs? Explain.

18·6   List three situations in which a run test would be appropriate.

18·7   Why is the χ^2 probability distribution used in a test for the median?

18·8   When the sign test is used, what hypothesis is being tested?

18·9   When we are using a sign test to compare populations A and B, how would the total number of sign changes be influenced by each of these situations:
    (a) A and B have the same mean; B has the larger variance
    (b) A and B have the same variance; the mean value of A is larger than the mean value of B
    (c) A and B have the same mean value and the same variance

18·10   What hypothesis is under test when a rank correlation method is being used?

18·11   What is the difference between a conventional correlation technique and the rank correlation technique?

18·12   What is the hypothesis under test when the rank-sum test is being used?

18·13   Can we use more than one kind of nonparametric test on the same set of data? Illustrate.

18·14   If the null hypothesis is rejected by the rank-sum test, what do we infer about the populations involved?

18·15   What hypothesis is tested by the paired-replicate rank test?

18·16   List three situations in which a paired-replicate test would be useful.

18·17   Show the relationship between the corner test and correlation analysis.

REFERENCES

8, 16, 100, 107, 108, 114, 119, 122, 123

19 Probability

In Chap. 1 we defined probability as the limit of the relative frequency of the occurrence of a specific event. In that chapter and in succeeding ones, we made use of the concept of probability in statistical inference. A somewhat more formal approach to probability is desirable for application to sampling procedures and control charts. This approach is discussed in this chapter.

19·1 PROBABILITY AS THE BASIS FOR DECISIONS

Every rational decision we make is based on the consideration of the possible alternative results of any projected course of action. Of the alternatives, we attempt to determine which will be favorable in their consequences for us, and what is the probability of this favorable outcome.

Such a procedure applies to games of chance. It was, in fact, in connection with these games that the theory of probability found its early applications. Winning play depends upon the player's knowledge of the possible alternatives in any given situation. The player determines his bet and his strategy by considering the possible reward and the probability of success. Note that in these games (assuming that they are fairly played) the probabilities are based not simply on experience but on an analysis of the *expected* results based on mathematical laws.

More complicated decisions are necessary in research, in business practice, in the determination of production methods, and in military operations. The assessment of the possible outcomes is far more difficult in such situations than in games of chance. Nevertheless, the intellectual process is similar; statistical methods often supply a basis for the evaluation of even complex industrial operations. A variety of methods has been developed to assist in the decision-making process; some of these will be discussed in the chapters to follow.

Care is needed in dealing with problems where probability is involved. As a trivial example of the dangers of a "common-sense" approach, consider this problem involving pairs of socks. Suppose that each of three boxes is known to contain just two socks—either a pair of black or a pair of white or one black and one white. If we take one sock from one of the boxes and find it to be black, what is the probability that the other sock in the box is also black? A "common-sense" solution might be this: The second sock in the box is surely either black or white, and the chances appear to be equally balanced.

The correct solution can be found by an examination of a table showing the possible cases (Table 19·1). Since a black sock was given as the first pick, we cannot have opened box 2. In fact, we must be involved with only one of the cases 1, 2, and 5. Of these cases, two (cases 1 and 2) will yield a black sock on the second pick, and only one of them (case 5) will yield a white sock. Therefore, the chances are 2 to 1 in favor of black as the remaining sock, not an even bet as "common sense" might suggest.

Table 19·1   Possible Arrangements of Socks

Box	Case	Sock chosen	Sock remaining
1	1	Black A	Black B
	2	Black B	Black A
2	3	White A	White B
	4	White B	White A
3	5	Black	White
	6	White	Black

Another example: Suppose there are 30 persons at a meeting; what is the probability that at least 2 of them have the same birthday (day and month only)? Since there are 365 days in a year (except for leap year) and only 30 persons, the chances seem remote. As a matter of fact, the chances are 7 in 10 that there will be at least one coincidence of birthdays. The odds are about even if only 23 persons are in the group.

19·2 METHODS OF DETERMINING PROBABILITY

There are two distinct methods of finding the probability associated with the occurrence of an event. The methods differ because of the inherent nature of the events with which we are involved.

1 The probability may be obtained in advance, before the data for the situation are collected. Such a probability is called *a priori*, meaning "from something before." A priori probabilities apply for the most part to games of chance, where all the possible events and outcomes can be enumerated or mathematically described.

2 The probability must be estimated by an actual count of events. Such a probability is called *empirical*, meaning "as a result of experience." Empirical probabilities are all that can be obtained in most industrial and experimental situations, because we rarely know in advance the underlying nature of the events. It is therefore necessary to collect data in order to estimate these probabilities. We were dealing with empirical probabilities in, for example, the setting of confidence limits in Chap. 4.

19·3 A PRIORI PROBABILITY

In games of chance, we can usually determine the probability of a specific event because we know the rules of the game and because we know by examination the nature of the objects used in the game. In the game of dice, we know that each die has six sides. In a bridge deck of cards, we know that there are 52 items marked in different ways.

We find the probability of occurrence of an event by forming a ratio of two terms: one term is the possible number of favorable events; the other term is the total number of all events. Thus the a priori probability is

$$P = \frac{E}{N} \tag{19·1}$$

where P = a priori probability
E = number of cases in which desired event occurs
N = total number of possible cases

What is the probability of drawing an ace of spades from a deck of 52 different cards? There are 52 possible results, only one of which is favorable. The probability is therefore 1:52, or about 0.019. What is the probability of throwing a 6 on a single roll of a die? There are 6 faces, and 6 possible results, only one of which is favorable. The probability is therefore 1:6, or about 0.17.

It is essential to understand the assumptions underlying these calculations. They are that the game is honest, i.e., that there are really 52 different cards in the deck, that the deck is well shuffled, that the choice of cards is not forced.

For the die, the assumptions are that there are truly six different faces, that the die is cubical, not weighted, and fairly rolled.

It is also important to understand the meaning of the probability found by this method. Suppose the game we play is to try to draw the ace of spades. That the probability of drawing the ace of spades is said to be 1:52 does not mean that the card will be drawn exactly once in a game of 52 attempts. Because the probability is the *limit* of the relative frequency, the ratio 1:52 will hold only for a very large number of games; that is, it will be true only in terms of the long run. In fact, we never expect to find the ratio of 1:52 to hold exactly for *any* specific number of games; we do expect to obtain a result closer to this ratio as the number of games increases toward infinity.

For situations more complicated than we have so far mentioned, a systematic examination of the possible outcomes is needed to calculate a priori probabilities. For rolling two dice, one red and one green, we make a tabulation as in Table 19·2. The red die may fall in one of six ways; so also may the

Table 19·2 Totals of Two Dice

Green Die	Red die					
	1	2	3	4	5	6
1	2	3	4	5	6	7
2	3	4	5	6	7	8
3	4	5	6	7	8	9
4	5	6	7	8	9	10
5	6	7	8	9	10	11
6	7	8	9	10	11	12

green die. In the body of the table appear all the possible sums of the two faces.

There are 36 possible outcomes; hence N is 36. The sum 2 appears only once in the table; for this event, E is 1. Therefore, the probability of obtaining the sum of 2 from a throw of two dice is 1:36 or about 0.028. Since there are 6 occurrences of the sum 7, the probability of throwing a 7 with two dice is 6:36, or about 0.17. The probabilities of obtaining other sums are found in the same way. If the probabilities of all possible outcomes are added, the total necessarily is unity.

For problems more involved than those associated with a pair of dice, the tabulation method we have used becomes tedious. Rules for calculation methods are, however, suggested by this example.

19·1   A deck of cards used for bridge contains 4 suits of 13 cards each. What is the probability that a single card drawn at random from the complete deck will be:

(a) of a specified suit

(b) a specified card

(c) an honor card, if each suit contains 5 honor cards

19·2   What is the probability of tossing a 3 on the first toss of an honest die? of obtaining 6 as the sum when a pair of honest dice are thrown?

19·3   Make a table like Table 19·2 for the various sums possible when 3 dice are thrown. What is the probability of obtaining the most frequently occurring sum?

19·4   If a process has produced 1,000 parts of which 10 are known to be defective, what is the probability of getting a defective part as the first part selected from the lot, on the assumption of random selection?

19·5   A box contains 3 red balls, 4 blue ones, and 7 yellow ones. What is the probability that you will be able to select a blue ball from the box without looking? What is the probability that the ball will not be yellow?

19·6   If it has rained 20 times on the fourth of July in the past 25 years, what is the probability of rain on July 4 this year? Defend your answer.

19·4 EMPIRICAL PROBABILITIES

We have defined empirical probabilities as those which we estimate from experience. As an example, consider this question: If a number of carpet tacks fall on a flat surface, what is the probability that one tack will fall point uppermost? It hardly seems possible to calculate this probability from a physical examination of a carpet tack. Hence it would be necessary to make actual trials to estimate the probability.

A series of 10 trials with 10 carpet tacks from the same box gave this group of favorable outcomes: 0, 2, 1, 2, 2, 4, 0, 3, 2, 1. The total number of tacks lying point upwards is 17. The probability based on this series of 100 events is therefore 0.17.

No doubt trials with thumbtacks would give a different empirical probability, since thumbtacks are constructed differently from carpet tacks. The point here is that an empirical probability applies only to a specific group (or set) of events and cannot be applied to other sets.

A somewhat different kind of empirical probability is found in a statement such as: The probability of snow in Rochester, N.Y., on Christmas is 0.80. This probability would have come from a series of observations over a period of years; perhaps it has snowed on 40 occasions in the last 50 years. Here there is no possibility of repetition of the trials, as could be done at pleasure with the tacks.

Industrial probabilities are very often empirical. They are estimated from data recorded for a process or a production line. A machine makes parts, some of which are defective. To calculate the number of defective parts that the machine will make in any period of time requires an estimate of the probability that the machine will produce a defect. This probability is estimated by examining a large number of parts and counting the number of defective parts in the lot. The probability is then expressed as the ratio of defective to total number of parts.

In finding such empirical probabilities, and in applying them to real situations, we must observe several precautions. Inasmuch as it is an empirical probability, the reliability of the calculation depends in part on the number of items that we examine. It will be recalled that a probability is a *limit* of a relative frequency.

Furthermore, the inspected items must all come from the same group, or set, of items. The parts must all have been made by the same machine if we want to apply the result to future production by that machine. We must not suppose without trials that another supposedly "identical" machine is really producing items of the same kind. The process from which we select the items for examination must be operating in a normal way, and nothing must be done to alter the performance during the time that the sample is collected.

Finally, it may well be that we cannot possibly examine *all* the parts produced by a machine, simply because so many of them are produced in a relatively short time. Under these circumstances we select only a portion of the items for inspection. It is essential that the subgroup of items be chosen in a random manner. This problem was discussed in Chap. 1.

19·5 PROBABILITIES COMBINED

There are many circumstances in which the probabilities of two or more events have been found, either a priori or empirically, and we are interested in the probability that both events will occur, or that either will occur. There are several different cases, depending on the nature of the situation. First, we distinguish between events which are *mutually exclusive* and those which are not. Two events are mutually exclusive if they cannot possibly occur together. For example, in a dice game, the occurrences of the two sums 6 and 8 are mutually exclusive events. In a production process, a machine part is either satisfactory or defective; these are mutually exclusive events. Second, we distinguish between events which are *independent* and those which are not. In tossing two dice, the fall of one die is independent, supposedly, of the fall of the second.

19·6 FIRST ADDITION THEOREM

Here we know the probabilities of two mutually exclusive events, and we wish to find the probability of the occurrence of *either* of them. The rule is

$$P(A \text{ or } B) = P(A) + P(B) \qquad (19\cdot2)$$

where P means "the probability of" and A and B are two mutually exclusive events.

For example, from Table 19·2, the probability of obtaining the sum of 7 with 2 dice is 6/36, or $P(7) = 6/36$. Similarly, $P(11) = 2/36$. Then the probability of obtaining either 7 or 11 when two dice are thrown is $P(7 \text{ or } 11) = P(7) + P(11) = 6/36 + 2/36 = 8/36$, about 0.22. The first addition theorem may be extended to include any number of events:

$$P(A \text{ or } B \text{ or } C \cdots \text{ or } N) = P(A) + P(B) + P(C) + \cdots + P(N)$$

The theorem implies for mutually exclusive events that the sum of the probabilities of all possible events is unity.

If the probability of a machine's producing a defective part that is too small is 0.05 and the probability of the machine's producing a defective part that is too large is 0.10, the probability that a part is either too small or too large is

$$P(S \text{ or } L) = P(S) + P(L) = 0.05 + 0.10 = 0.15$$

These two events are mutually exclusive, since a part cannot be simultaneously too small and too large.

19·7 MULTIPLICATION THEOREM

We know the probability of two *independent* events, and we wish to know the probability that *both* will occur. The rule is

$$P(A \text{ and } B) = P(A) \times P(B) \tag{19·3}$$

This rule may be extended to include any number of independent events.

The probability of obtaining a head on a single toss of a coin is 1/2. What is the probability of obtaining two heads when two coins are tossed? Since these events are independent, we use the multiplication theorem:

$$P(H \text{ and } H) = P(H) \times P(H) = 1/2 \times 1/2 = 1/4$$

If the probability of finding a defective piece in a specified lot of items is 0.2, what is the probability of finding two defective items from the same lot? Assuming that the selection of pieces is random, the finding of one defect does not affect the finding of another; hence the events are independent. The probability of finding two defective pieces is therefore $0.2 \times 0.2 = 0.04$. If $P(d)$ represents the probability of finding one defective part and $P(s)$ the probability of finding one satisfactory part,

$P(d) = 0.2$
$P(s) = 0.8$ by the first addition theorem
$P(2d) = P(d \text{ and } d) = P(d) \times P(d) = 0.2 \times 0.2 = 0.04$
$P(2s) = P(s \text{ and } s) = P(s) \times P(s) = 0.8 \times 0.8 = 0.64$
$P(d \text{ and } s) = 0.2 \times 0.8 = 0.16$
$P(s \text{ and } d) = 0.8 \times 0.2 = 0.16$

The last four calculations above exhaust the possible arrangements of a sample of two items from the lot. Since they are mutually exclusive possible outcomes they necessarily sum to unity.

PROBLEMS

19·7   The probability of getting a head in the single toss of a coin is known to be 0.5. Toss a single coin a great number of times, recording the number of heads obtained, and calculate the empirical probability after each set of 5 tosses. How do your probabilities compare with the theoretical one?

19·8   Using 10 thumbtacks of like design, find the empirical probability of having 5 or fewer thumbtacks falling with the points in an upward direction. (After every 5 tosses of the thumbtacks, recalculate the probability.)

19·9   If, in a large lot of inspected items, the probability of selecting a defective item is 0.1, tabulate the probability of obtaining each of the possible arrangements of four items selected from the lot at random.

19·10   An industrial process has been operated for a long period of time and the fraction of defective items has remained constant at 0.03.
   (a) What is the probability of obtaining, by random sampling, a defective item on the first sample taken from the process?
   (b) What is the probability of obtaining 3 defectives in a row, provided the machine maintains a constant rate of defectives?
   (c) What is the probability of getting a defective item in either the first or second selection of single units from the process?

19·11   If the probability of a person's being a blond is 0.4, and the probability of a person's wearing black shoes is 0.3, what is the probability of the following:
   (a) a person's being blond and wearing black shoes? (Assume the two events to be independent events.)
   (b) a person's either being blond or wearing black shoes?
   (c) three people all being blonds?
   (d) three people all wearing black shoes?
   (e) two people being either blond or wearing black shoes?

19·8   SECOND ADDITION THEOREM

If two probabilities do not apply to mutually exclusive events, the probability of the occurrence of either of the events is not merely the sum of the probabilities as in the first addition theorem. Consider, for example, the probability that you cannot complete a telephone call, if the probability that your line is busy is 0.4 and the probability that the called line is busy is 0.7. Here the probabilities sum to more than unity, which is an impossible answer. The probabilities sum to more than unity because they pertain to non-mutually-exclusive events. The lines can both be busy at the same time.

For non-mutually-exclusive events, to find the probability of occurrence of either of them, we find the sum of the two probabilities and subtract the probability that both events may occur. The number to be subtracted is found by multiplication theorem. Thus, for non-mutually-exclusive events,

$$P(A \text{ or } B) = P(A) + P(B) - P(A \text{ and } B) \tag{19.4}$$

In this example, the probability is

$$0.4 + 0.7 - (0.4 \times 0.7) = 0.82$$

An electrically operated piece of apparatus will not function if either of two relays is open. If the first relay is open at random 0.2 of the time and the second is open at random 0.3 of the time, what is the probability that the apparatus will not function? These two possible events are not mutually exclusive, since either or both relays may be open. Hence we use the second addition theorem

$$P(A \text{ or } B) = 0.2 + 0.3 - (0.2 \times 0.3) = 0.44$$

The probability that the apparatus is functioning is then $1 - 0.44 = 0.56$. This last answer is the same as that obtained by considering the probabilities that the relays are closed and applying the multiplication theorem to these probabilities.

19.9 CONDITIONAL PROBABILITY

When we discussed the multiplication theorem, we said that it applied only for independent events. There are, however, many situations in which the probability of a second event is dependent on whether or not the first event has occurred. In these situations, it is essential that the determination of the probabilities take this dependence into account.

The probability that a machined part will have the correct size and the correct surface finish involves the multiplication theorem. The empirical probability of the first event may be 0.8 and that of the second may be 0.7. It can easily be, however, that the probability of securing a good surface is dependent upon the size of the part. If the part has the right size, the probability of obtaining a good surface may be 0.9. We speak of such a dependent probability as a *conditional* probability, and use for it the symbol $P(B \mid A)$, which represents the probability of the occurrence of B if A has occurred.

19.10 SUMMARY OF THE RULES OF PROBABILITY

  I The probability of an event is equal to, or lies between, zero and unity.
  II Addition theorems.
    1 For mutually exclusive events,

$$P(A \text{ or } B) = P(A) + P(B)$$

This theorem implies that the sum of all possible mutually exclusive events of a specific class is unity.

2   For non-mutually-exclusive events,

$$P(A \text{ or } B) = P(A) + \mathbf{P}(B) - P(A \text{ and } B)$$

III   Multiplication theorem.

$$P(A \text{ and } B) = P(A) \times P(B)$$

$P(B)$ may be conditional on the occurrence of A, in which case we write it $P(B \mid A)$.

We illustrate the application of these rules in the following situation: Two companies A and $\bar{A}$ (the bar signifying "not") are suppliers of similar parts. We inspect a number of parts from each supplier and classify them as defective d or satisfactory $\bar{d}$. The results of the inspection are presented in the two-way frequency Table 19·3. On the assumption that the data are representative, we can calculate probabilities as follows:

Table 19·3   Frequency of Defective and Satisfactory Parts Supplied by Two Companies

	Parts		
	d	$\bar{d}$	Total
Company A	10	40	50
Company $\bar{A}$	20	130	150
Total	30	170	200

1   The probability that a part is defective, $P(d)$: Of the total of 200 parts 30 are defective; therefore, $P(d) = 30/200 = 0.15$.
2   The probability that a part was supplied by A, $P(A)$: Of the total of 200 parts, 50 were from company A; therefore, $P(A) = 50/200 = 0.25$.
3   The probability that a part is from company A if it is defective, $P(A \mid d)$: Of the total of 30 defective parts, 10 were from A; therefore, $P(A \mid d) = 10/30 = 0.33$. This is a conditional probability.
4   The probability that a part is from A and is defective, $P(A \text{ and } d)$: Of the total of 200 parts, 10 appear in the table in the A row and the d column; therefore, $P(A \text{ and } d) = 10/200 = 0.05$. This value may also be found by the multiplication theorem:

$$P(A \text{ and } d) = P(d) \times P(A \mid d) = 0.15 \times 0.33 = 0.05$$

5   The probability that a part was supplied by company A or is defective involves non-mutually-exclusive possibilities; hence

$$P(A \text{ or } d) = P(A) + P(d) - P(A \text{ and } d) = 0.25 + 0.15 - 0.05 = 0.35$$

19·12   From the data in Table 19·3, find the following:

(a) $P(\bar{d})$ (e) $P(A \text{ or } \bar{A})$

(b) $P(\bar{A})$ (f) $P(A \text{ and } \bar{A})$

(c) $P(A \mid \bar{d})$ (g) $P(d \text{ or } \bar{d})$

(d) $P(\bar{A} \mid \bar{d})$ (h) $P(\bar{A} \text{ or } \bar{d})$

19·11   APPLICATION

The preceding illustration used counts of numbers found by inspection. The same methods may be applied to probabilities, inasmuch as probabilities are the limits of relative frequencies. We consider now a situation in which a motor drives an electric generator. We assume that we have knowledge of the probabilities that each piece of equipment is in or out of operation, as in Table 19·4. We note that these events are independent but not mutually exclusive.

Table 19·4   Probabilities Associated with the Operation of a Motor and a Generator

	Out	In
Motor	$(\bar{M})$ 0.02	(M) 0.98
Generator	$(\bar{G})$ 0.03	(G) 0.97

The probability that the equipment will be working requires the use of the multiplication theorem:

$$P(M \text{ and } G) = P(M) \times P(G) = 0.98 \times 0.97 = 0.9506$$

Similarly, the probability that the motor is operating and the generator not operating is

$$P(M \text{ and } \bar{G}) = P(M) \times P(\bar{G}) = 0.98 \times 0.03 = 0.0294$$

By the same theorem,

$$P(\bar{M} \text{ and } G) = 0.02 \times 0.97 = 0.0194$$
$$P(\bar{M} \text{ and } \bar{G}) = 0.02 \times 0.03 = 0.0006$$

The probability that the combination will not operate at any specific time requires the use of the second addition theorem, since we are dealing here with non-mutually-exclusive events:

$$P(\bar{M} \text{ or } \bar{G}) = P(M \text{ and } \bar{G}) + P(\bar{M} \text{ and } G) - P(\bar{M} \text{ and } \bar{G})$$
$$= 0.0294 + 0.0194 - 0.0006 = 0.0494$$

Our first calculation was the probability that the equipment will operate; our last was that it will not operate. The sum of these probabilities 0.9506 + 0.0494 is unity, as the first rule of probability requires. The probabilities just found are tabulated in Table 19·5.

Motor	Generator		Total
	In (G)	Out ($\bar{G}$)	
In (M)	0.9506	0.0294	0.9800
Out ($\bar{M}$)	0.0194	0.0006	0.0200
Total	0.9700	0.0300	1.0000

Table 19·5   *Combined Probabilities of Operation of Motor and Generator*

A final illustration will be based on the data in Table 19·5. If the generator is in operation, what is the probability that the entire apparatus is functioning? We are involved only with the first column of the table. Of the total probability 0.97 we have a probability of success of 0.9506. Hence the probability desired is the ratio of these two probabilities 0.9506:0.97 or 0.98. But this answer follows from the original data: If the generator is working, the probability that the equipment is in operation depends on the motor alone, and we knew at the beginning that the probability that the motor operates is 0.98.

PROBLEMS

19·13   In Table 19·6 are given the results of inspection of parts supplied by 3 different companies.

Table 19·6   *Classification of Parts Supplied by Three Different Companies*

Parts	Company A	Company B	Company C	Total
Correct	3,900	2,400	1,400	7,700
Too long	400	250	200	850
Too short	700	350	400	1,450

(a) What is the probability that a part selected at random from the entire lot will be defective?
(b) What is the probability that the defective part will be too long?
(c) What is the probability that company A supplied a defective part?
(d) What is the probability that a part is supplied by either company A or company B?
(e) What is the probability that a part is both defective and from company C?

19·14 If you toss a coin onto a piece of paper which has red and green sections, what is the probability that the coin will land with a head upward, and also fall on the red section, if the red section of the paper covers 30 percent of the total area?

19·15 Company A supplied 100 units of which 10 were defective; company B supplies 300 units of which 20 were defective, and company C supplies 100 units of which 30 were defective. All the units are placed in storage in such a way as to cause them to be intermixed in a random fashion.

  (a) What is the probability that a single unit (selected at random from the entire lot of 500) came from company A?
  (b) What is the probability that it came from company B?
  (c) If the first selected unit is found to be defective, what is the probability that it came from company C?
  (d) From the information above, from which company would you prefer to buy additional units?
  (e) If 1,000 additional units were purchased from company B, how many defectives would you expect to find in the new purchase, if the rate of defectives for that company remains constant?

19·16 A pH meter contains 5 electronic tubes; the meter will not operate unless all tubes are operative. If the probability of failure of each tube is 0.1, what is the probability of failure of the meter?

19·17 To start a car, both the starter motor and the battery must function. If the probability that the starter motor will operate is 0.95, and the probability that the battery will operate is 0.97, what is the probability that the car will not start if all other reasons for not starting are not considered?

19·12 ARRANGEMENTS OF EVENTS

Sampling plans used in the inspection of large lots of items are based on the probabilities of observation of samples of specified characteristics. The next sections of this chapter discuss the fundamental concepts on which such plans are built.

Suppose that we have a large box of tags, of which one-third are marked A, one-third B, and one-third C. We pick these two at a time, recording the identifying letters in the order of their selection. We replace the tags, mix them thoroughly, and repeat the drawing. If we ignore combinations previously found and recorded, we can obtain nine different arrangements, as in Table 19·7.

We see that each position in any pair can be filled with three different letters. To compute the total number of possible pairs, we multiply the number of ways the first position can be filled by the number of ways the second position can be filled. Here, $3 \times 3 = 3^2 = 9$.

If the number of ways is the same for each position,

$$A_r{}^n = n^r \tag{19·5}$$

Table 19·7 Possible Arrangements of	First Draw	Second draw		
Two Tags Drawn from a		A	B	C
Set of Equal Numbers of	A	A, A	A, B	A, C
Three Different Tags	B	B, A	B, B	B, C
	C	C, A	C, B	C, C

where A_r^n = number of possible arrangements
n = number of ways each position can be filled
r = number of items we select

The rule applies to throwing two dice. There are 6 different ways for each die to fall and 2 dice; hence the number of different outcomes is $6^2 = 36$.

If we throw a die and toss a coin, the number of different outcomes requires multiplication of the two different sets of results: $6 \times 2 = 12$.

The a priori probability of observing a specific event of the kind we are discussing is found from the total possible outcomes and the application of Eq. (19·1). Thus the probability of drawing a specific pair of tags from the box containing equal numbers of A's, B's, and C's is 1/9.

PROBLEMS

19·18 What is the number of possible outcomes, order considered, when 4 dice are thrown? What is the probability of obtaining
    (a) 1, 1, 1, 1
    (b) 1, 1, 2, 1
    (c) all 4 dice showing the same value

19·19 There are 13 spades in 1 pile of cards and 13 hearts in another pile. How many different groups of 2 can be observed if 1 card is taken from each pile? If the piles are combined into 1, how many different groups of 2 can be observed if the cards are chosen 1 at a time and returned to the pile?

19·20 From a box of 16 bolts and 16 nuts which fit the bolts, how many arrangements can be made using a bolt and nut at random?

19·21 There are 7 locked doors and 7 keys, each of which fits 1 door. Is it possible that you might require 7 attempts before you could unlock a single door? If the key must be returned to the others after each trial, and mixed with them, is it possible that you might require more than 50 trials to unlock a single door? How many door-key arrangements are possible?

19·22 Considering a "word" to be any arrangement of 5 letters (for example, XYXXA), how many can be formed?

19·13 PERMUTATIONS

When we speak of *permutations* we mean arrangements of items exclusive of repetitions of the same item. If, in the tabulation of pairs of lettered tags in Table 19·7, we omit the repetitions A, A; B, B; and C, C, we are left with 6 sets of 2 tags. That is to say, the number of permutations of 3 items taken 2 at a time is 6.

We use the symbol P_r^n for "the number of permutations of n things taken r at a time." Thus P_2^3 means the number of permutations of 3 items taken 2 at a time. We find the number of permutations by

$$P_r^n = \frac{n!}{(n-r)!} \tag{19·6}$$

The sign (!) is read "factorial"; for any integer it stands for the product of that integer and all integers smaller than the number preceding the sign (!). Thus $4! = 4 \times 3 \times 2 \times 1$, and $8! = 8 \times 7 \times 6 \times 5 \times 4 \times 3 \times 2 \times 1$. $0!$ is defined as having the value 1.

To find the number of permutations of three items taken two at a time, we substitute in Eq. (19·6):

$$P_2^3 = \frac{3!}{(3-2)!} = \frac{3!}{1!} = \frac{3 \times 2 \times 1}{1} = 6$$

This result agrees with the count we made in Table 19·7. It may assist the recall of Eq. (19·6) to note that n is the upper number in the symbol P_r^n and that $n!$ is the numerator of the fraction; in the symbol r lies below n, suggesting the subtraction in the denominator of the fraction.

PROBLEMS

19·23   How many permutations are there of the faces of 3 dice?

19·24   If 5 different inks are available, in how many ways can a 2-color job be printed? a 3-color job?

19·25   An inspector visits 7 production lines. How many different orders of inspection can he use?

19·14 COMBINATIONS

By combinations of items we mean arrangements neglecting repetition (as we did in permutations) and also neglecting sets which differ only in order but contain the same items; i.e., combinations are permutations without regard to order of occurrence. If in Table 19·7 we omit all pairs which repeat letters (as we did in counting permutations) and we also omit every pair which contains the same letters as another, we are left with three pairs only: A, B; A, C; B, C. Hence there are three combinations of three items taken two at a time.

In general, if C_r^n symbolizes the number of combinations of n things taken r at a time

$$C_r^n = \frac{n!}{(n-r)!\,r!}$$ (19·7)

Equation (19·7) is similar to the equation for the number of permutations but with the inclusion of $r!$ in the denominator of the fraction. Thus

$$C_r^n = \frac{P_r^n}{r!}$$

The division by $r!$ has the effect of eliminating one of each of the pairs which are identical except for order.

We apply Eq. (19·7) to the lettered tags:

$$C_2^3 = \frac{3!}{(3-2)!\,2!} = \frac{3 \times 2 \times 1}{1 \times 2 \times 1} = 3$$

Consider this problem: We know that in a lot of 10 manufactured items 3 are defective; a sample of 5 items is chosen at random from the lot. What is the probability that we will find exactly 1 defective item in the sample? We intend to apply Eq. (19·1); thus we need to know the total number of different ways we can select our sample of 5 items out of a lot of 10. Then we need to know how many of these will contain one defective item. Here we cannot be concerned with repetition, since an item can be found only once in each sample of 5. Furthermore, order of occurrence is not consequential. Therefore, we are involved with combinations of 10 items 5 at a time.

We first find the total number of combinations:

$$C_5^{10} = \frac{10!}{(10-5)!\,5!} = \frac{10 \times 9 \times 8 \times 7 \times 6 \times 5 \times 4 \times 3 \times 2 \times 1}{(5 \times 4 \times 3 \times 2 \times 1)(5 \times 4 \times 3 \times 2 \times 1)} = 252$$

Thus there are 252 possible sets of 5 items from a lot of 10 items, any one set of which may be chosen at random.

We now find the number of combinations that will contain just 1 defective item. A sample of 5 must contain 4 satisfactory items if only 1 is to be defective. The 4 satisfactory items will be one of the possible combinations from the 7 good items in the lot of 10. The single defective item must be one of the original three defectives. Since we can combine any set of good items with any of the defective items, the total possible number of combinations fulfilling the conditions will be the product of the number of combinations of good and bad items separately.

For the satisfactory items,

$$C_4^7 = \frac{7!}{(7-4)!\,4!} = 35$$

For the defective items,

$$C_1^3 = \frac{3!}{(3-1)!\,1!} = 3$$

The total number of combinations of the 35 possible sets of good items and the 3 sets of defective items is now $35 \times 3 = 105$; this is the number of combinations of 5 items that may contain just 1 defective item.

The probability of finding exactly 1 defective item in the sample of 5 is now found from Eq. (19·1):

$$P = \frac{E}{N} = \frac{105}{252} = 0.417$$

Thus, if we select at random 5 items from a lot of 10, and if the lot contains 3 defective items, we have a probability of a little more than 41 percent of finding 1 defective item.

Calculations such as the preceding form the basis for inspection sampling procedures, which will be further discussed in later chapters. Sampling plans are constructed from calculations of the probabilities of discovering defective items in lots submitted for inspection. If we know the quality of the lot submitted, we can calculate the probability of finding a specified number of defective items in any specified sample size. Conversely, from an examination of a sample, we can make inferences about the probable fraction defective of the lot from which the sample was taken.

PROBLEMS

19·26 In a game of chance where the throwing of a pair with 2 dice is a loss, how many of the 36 possible arrangements constitute a losing arrangement? What is the probability of losing in this game? of winning?

19·27 How may different 5-card hands can be dealt in the game of poker?

19·28 In how many ways can an inspector choose a sample of 3 items from a lot of 10 items?

19·29 If a sample of 6 items is taken from a lot of 20, and there are 3 defective items in the lot,

  (a) In how many ways can the sample of 6 be taken?
  (b) In how many ways can a sample of 6 be taken, if the sample is to include 5 good pieces and 1 defective piece?
  (c) What is the probability of getting a sample from the lot with exactly 1 defective item and 5 satisfactory items?

19·30 In a box of 16 parts, there are 4 defective parts. What is the probability of taking a sample of 5 parts from the box and getting all good parts?

19·31   An assembled part consists of 3 items. There are 8 pieces of each item; 1 of each set of 8 is defective.
  (a) How many ways are there for assembly of the item?
  (b) How many of these will be completely satisfactory?
  (c) What is the probability that the assembled part will be defective in some respect?

19·32   Given the digits 1, 2, 3, 4, and 5, how many 4-digit numbers may be formed if:
  (a) repetitions are allowed
  (b) no repetitions are allowed
  (c) the number must be odd, without repetition
  (d) the number must be even, but repetition is allowed

19·33   How many different license-plate designations can be made if each plate has 1 letter, followed by a 4-digit number?

19·34   Johanssen gauge blocks are made in different thicknesses and may be stacked to different totals. If there are available 4 blocks of the following sizes: 0.5, 1.0, 1.5, and 2.0 in., how many different totals can the blocks be arranged to produce?

19·35   Of a lot of 10 items, 2 are scratched.
  (a) How many different samples of 4 can be obtained from the lot of 10 items?
  (b) How many of these samples of 4 will contain no scratched items?
  (c) How many of the samples of 4 will contain just 1 scratched item?
  (d) What is the probability that a sample of 4, selected at random from the lot of 10, will contain 1 or fewer scratched items?

QUESTIONS

19·1   Why do we define probability as the *limit* of the relative frequency of occurrence of an event?

19·2   Of the two methods of determining probabilities, which one applies to games of chance? What are the assumptions which are involved in estimating such probabilities?

19·3   Of the two methods of determining probabilities, which one applies to the prediction of defects in the items from a production line? What are the assumptions which are involved in estimating such probabilities?

19·4   If both methods of finding probabilities are applied to a situation, why will the two results often not agree? If the results differ, which is more to be trusted? Explain.

19·5   Distinguish between independent and mutually exclusive events. Must independent events be mutually exclusive? Explain. Must mutually exclusive events be independent? Explain.

19·6   Under what conditions may the addition theorem be used?

19·7   Under what conditions is the multiplication theorem used?

19·8   Under what circumstances are conditional probabilities needed?

19·9   For the same number of items taken 5 at a time, how will the number of combinations be related to the number of permutations?

19·10   What is the role of the formula for combinations in the calculation of simple, single-sampling plans?

19·11   What is the role of basic probability theory in taking samples from a population of finite size?

REFERENCES

16, 21, 22, 27, 33, 35, 37, 53, 62, 75, 78, 80, 82, 115, 121, 125

20 Acceptance Sampling

The elements of probability discussed in Chap. 19 provide the basis for understanding the principles upon which modern inspection methods rely. The problem arises in situations like this: A manufacturer buys subassemblies in lots of considerable size, perhaps from several different suppliers. He wishes to know whether or not any specific lot is of satisfactory quality. He would like all the parts to be perfect, but he knows that in spite of all precautions some defective parts will almost surely be found in any lot. For a specific level of quality, the cost of finding and removing the defective items will be prohibitively expensive, compared with their detection and removal during a later stage of manufacture. The manufacturer must decide what is the permissible percentage of defective items that he is willing to accept before this situation no longer holds true. This level is the acceptable quality level (AQL). Each lot must now be examined to discover whether or not it is at least as good as he desires.

One possible method of inspecting the lots submitted for purchase is to examine every piece in the lot and to classify it as defective or satisfactory. Such 100 percent inspection is usually a poor technique, on the following grounds: (1) 100 percent inspection is usually prohibitively expensive, especially for large lots. (2) If inspection methods involve destructive testing 100 percent inspection is impossible. (3) Even the most careful 100 percent inspection methods are not in themselves perfect, since mechanical inspection devices have some margin of error and human inspectors are subject to fatigue and other sources of mistakes.

Better inspection methods involve sampling the lot. We select a number of pieces at random, and on the basis of the nature of the sample decide whether or not to accept the whole lot. Sampling is less costly than 100 percent inspection. Furthermore, good sampling plans are as effective as 100 percent inspection and in many instances even more effective because they supply analytical information in addition to discriminating between good and poor material.

Sampling plans are designed to answer this question: Shall we accept or reject the population being sampled, based on our sample results? To accomplish this for a specified acceptable fraction defective, we must define what is the number of defective items in a sample of a given size to cause us to accept or reject the lot. The answer to this is based on probability considerations.

20·1 INSPECTION BY ATTRIBUTES

In the discussion to follow, we will necessarily suppose that we can make a decision whether or not a piece is defective. We are thus involved with a "go, no-go" situation, in which only two classifications of the inspected items exist. (Sometimes defects are classified as major or minor; different sampling plans may be used for different classifications of defects.) In such situations, we say that we are inspecting by *attributes*. Some examples of attribute inspection would involve the classification of items according to good or bad finish; right or wrong size; satisfactory or unsatisfactory color; the part functions or does not function.

Attribute sampling plans are based on a consideration of the different number of defectives which may be found in a sample of a given size from a population with some specific percent defective. Sampling variability will yield different values. When there are only two possible categories (i.e., defective and nondefective) into which any item may be classified, the basis for such sampling plans is the binomial expansion.

20·2 THE BINOMIAL EXPANSION

We toss four similar coins. We wish to compute the probability of occurrence of any combination of heads and tails. Let $p =$ the probability of obtaining a head; let $q =$ the probability of obtaining "not heads," i.e., tails. (Note that here $p = q = \frac{1}{2}$ and that in any case $p + q = 1$ since there are only two possibilities.) Then if we expand the binomial term $(p + q)^n$, where n is the number of coins, we will find all the ways in which a sample of n coins can combine. Here,

$$(p + q)^4 = p^4 + 4p^3q + 6p^2q^2 + 4pq^3 + q^4$$

The coefficients of the terms of the expansion are, in order, 1, 4, 6, 4, 1; the sum of these is 16. This means that there are 16 different ways in which the sample of 4 coins can combine; the coefficient of each term indicates the number of ways each different combination can occur. The kind of combination is shown by the exponents of the letters in the term. The first term of the expansion contains only the letter p, with exponent 4. Since the coefficient of this term is 1, there is only one way (out of the total of 16) that 4 heads can appear. The probability of obtaining 4 heads is then 1:16, the ratio of events of a specific kind to the total number of possible outcomes.

Similarly, the second term $4p^3q$ shows that there are 4 ways of getting a sample with 3 heads and 1 tail. The third term of the expansion shows that there are 6 ways of obtaining a combination of 2 heads and 2 tails. The probability of obtaining such a result is 6:16.

The binomial expansion may be generalized to provide for cases of any sample size n, where n is an integer, thus:

$$(p + q)^n = \frac{n!}{n!\,0!}p^n + \frac{n!}{(n-1)!\,1!}p^{n-1}q + \frac{n!}{(n-2)!\,2!}p^{n-2}q^2$$
$$+ \cdots + \frac{n!\,q^n}{0!\,n!} \quad (20\cdot1)$$

As in the example involving coins, the coefficients of the terms in the expansion are the basis for the calculation of probabilities of occurrence of any specified combination, provided that $p = q = \frac{1}{2}$. In the more general case, p and q will be unequal; in such a case, the values of p and q enter into the calculation, as indicated in Eq. (20·1).

A rule for finding the probability of occurrence of any possible combination and for any sample size is

$$P(E;n) = C_E{}^n p^E q^{n-E} \quad (20\cdot2)$$

In this equation, $P(E;n)$ represents the probability of finding E events of a specified kind in a set of n trials (such as 3 heads in 20 tosses), and $C_E{}^n$ represents the number of combinations of n items, E at a time, as in Chap. 19. Therefore, the term $C_E{}^n$ indicates the number of cases in which a specified arrangement can occur, and the term $p^E q^{n-E}$ the probability of occurrence of each of these cases. The product of the two terms gives the probability of occurrence of all the cases of a specified kind.

20·3 APPLICATION TO SAMPLING
We have illustrated the binomial expansion by means of a trivial coin-tossing example. The same concept applies to inspection and other situations in which these conditions hold:

1   There is a fixed number n of repeated trials.
2   Each of the results of the trials can be classified as falling into one of two classes, as in attribute inspection.
3   The probability p of an event of one specified kind is the same on each trial.
4   The trials are independent.

5   We wish to determine the probability of finding a specified number of events of one kind in the set of n trials.

Assume that a lot of 100 items contains 10 defective items. Assume further that the conditions above hold. What is the probability that a random sample of 5 items from the lot will contain exactly one defective item? We apply Eq. (20·2), with these substitutions: $n = 5$; $E = 1$; $p = 0.1$; and $q = 0.9$. Then

$$P(1;5) = C_1{}^5 \times 0.1^1 \times 0.9^4 = 0.33$$

Thus there is a chance of about 1 in 3 that a random sample of 5 from the lot of 100 will contain exactly 1 defective item.

PROBLEMS

20·1   Find the probability that a sample of 5 from a lot of 100 (containing 10 defectives) will contain the following numbers of defective items: (a) 0; (b) 2; (c) 3; (d) 4; (e) 5.

20·2   Plot a graph of probability of occurrence versus number of defectives from the results obtained in answer to Prob. 20·1.

20·4 BINOMIAL DISTRIBUTIONS

In Fig. 20·1 we have plotted *frequency* of occurrence of various numbers of heads when we toss various numbers of coins—2, 6, 8, and 10. On the vertical axis we have plotted the coefficients only, found by the binomial expansion illustrated in Eq. (20·1). All the plots are symmetrical; as n increases, the graphs approach the appearance of the normal curve.

In Fig. 20·2 we show plots in which we now demonstrate the distribution of *probabilities*. Here we have plotted the results of calculations based on Eq. (20·2).

If p is the probability that an event will happen in any single trial, then the probability that the event will happen exactly E times in n trials is given by the binomial expansion as written in the formula given in Eq. (20·2). The use of this formula will produce the probability of occurrence for a given E in n trials; to produce the entire distribution of probability values would require the formula to be used for each possible value of E.

When the binomial $(p + q)^n$ is used, the expansion will reveal the entire set of frequencies of occurrences of all possible outcomes. When the set of frequencies is graphed as in Fig. 20·1, a distribution is obtained. A frequency distribution obtained from the expansion of the binomial $(p + q)^n$ will have a mean value of np and a variance of $np(1 - p)$. When the value of p is equal to 0.5, the binomial distribution is symmetrical around its mean, while a value of $p < 0.5$ will produce a distribution which is skewed to the right, with the right-hand tail of the distribution being longer then the left-hand

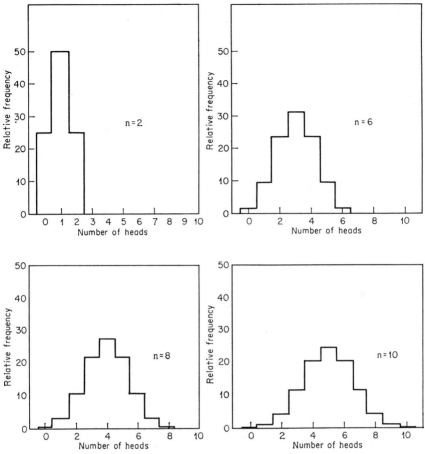

Fig. 20·1 Binomial distributions—coin-tossing frequency plots.

tail. A value of $p > 0.5$ will produce a distribution skewed to the left. The standard deviation of the binomial distribution is a maximum when $p = 0.5$.

When the value of n is large, the expansion of the binomial $(p + q)^n$ becomes a time-consuming task, and often only a few of the terms of the expansion are of interest. In such a case, the probability values of the terms of interest are computed by using Eq. (20·2) and omitting the rest of the terms in the distribution.

The probability of getting two or fewer heads in six tosses of a fair coin would be, using Eq. (20·2),

$$[C_0^6(\tfrac{1}{2})^0(\tfrac{1}{2})^{6-0}] + [C_1^6(\tfrac{1}{2})^1(\tfrac{1}{2})^{6-1}] + [C_2^6(\tfrac{1}{2})^2(\tfrac{1}{2})^{6-2}]$$
$$= \tfrac{1}{64} + \tfrac{6}{64} + \tfrac{15}{64} = \tfrac{22}{64}$$

That is, the probability of getting two or fewer heads in six tosses of a coin (for which the population probability $p' = 0.5$) is equal to 22/64 or 0.34.

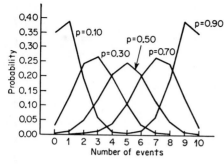

Fig. 20·2 Binomial distributions—coin-tossing probability plots.

For calculations of probabilities involving values of p' other than 0.5, the same procedure is followed. In the case of coin tossing, the value of p' is a known value. In industrial applications, p' is the value of the population fraction defective. If a lot of p' value is submitted, we wish to know the probability of its being accepted when we use a specific sampling plan. To answer this, we calculate the probability of acceptance for all conceivable values of p' and present them as points on the *operating characteristic* curve. Each specific sampling plan will have its own operating characteristic curve, which will show the probability of acceptance of the lot having p' fraction defective, as long as the specific sampling plan is used. If the sampling plan is changed, the operating characteristic curve is also changed. The construction of such curves for given sampling plans is covered later in this chapter.

If a given process, whose product was submitted for sampling, was actually running steady at a p' value of 0.02, the probability of finding three or fewer defectives in a sample of $n = 10$ could be computed by the same method we used to find the probability of two or fewer heads in the coin problem:

$$[C_0^{10}(0.02)^0(0.98)^{10}] + [C_1^{10}(0.02)^1(0.98)^9] + [C_2^{10}(0.02)^2(0.98)^8]$$

$$+ [C_3^{10}(0.02)^3(0.98)^7]$$

The required computations are often long and involved. Fortunately, statisticians have worked out tables of such cumulative probabilities.[1]

The binomial distribution is a discrete probability distribution; for $E = 0, 1, 2, \ldots, n$ it corresponds to successive terms in the binomial expansion. It is possible to compute the mean and the standard deviation (or variance) of the binomial distribution and to concern ourselves only with the probabilities of occurrences we are interested in knowing. Generally, we compute the probabilities of a range of possible p' values, with a given sample size n and an allowable number of defectives. If we find more defectives in a sample of size n than we choose as our rejection number, we

---

[1] "Tables of the Binomial Probability Distribution," Applied Mathematics Series 6, Government Printing Office, Washington, D.C.

should investigate or reject the lot from which the sample was taken. This is the basis for industrial sampling plans for attributes.

20·5 c VALUES IN INSPECTION

In attribute sampling we want to know the probability of accepting a lot based on taking a sample of size n and using a decision rule that says: Accept if the number of defectives found is some number c or fewer, and reject if the number of defectives is greater. This probability will be the sum of the probabilities of finding the individual number of defective items up to and including the specified number c. For example, we return to a consideration of samples taken from a lot of 100 which is 10 percent defective. We want to know the probability of obtaining a sample of 5 that will contain one or fewer defectives. This probability is the sum of two probabilities: the probability of obtaining a sample with no defectives and the probability of obtaining a sample with exactly one defective item.

When we say that we will accept a lot on the basis of finding one or fewer defective items in a sample of a specified size, we specify that $c = 1$. If we specify that $c = 3$, we mean that the sample may contain three or fewer defective items and still be acceptable.

A common procedure in acceptance sampling is to consider each submitted lot of product on its own merits and to base the decision on acceptance or rejection of the lot on the evidence of samples taken at random from the lot. When the decision is made on the evidence of only one sample of size n, the acceptance plan is called a single sampling plan. Any systematic plan for single sampling requires that three values be specified. The first is the value of N, the number of articles in the lot; the second is the value of n, the sample size to be used; and the third is the value of c, the acceptance number. Acceptance sampling is used to decide whether the lot in question is to be accepted or rejected. The specified sampling plan is the decision-making tool used in the acceptance or rejection of the lot.

PROBLEMS

20·3   From the probabilities found previously for samples of $n = 5$ taken from a lot of $N = 100$, with 10 defectives (see Prob. 20·1), find the probabilities for (a) $c = 0$, (b) $c = 1$, (c) $c = 3$, (d) $c = 5$. $c = $ #of defectives

20·4   An inoperative radio is by accident grouped with 5 working radios. What is the probability that a random sample will contain only operative radios if the sample size is (a) 1, (b) 2, (c) 3, (d) 4, (e) 5, (f) 6?

20·5   If the situation is as in Prob. 20·4 and a set of 3 radios is chosen, what is the probability that the inoperative radio will be one of the three?

20·6   A game of chance is played as follows: 2 players each contribute 3 coins of the same kind; the coins are mixed and then tossed.

(a) If player A wins when 3 heads appear and loses when any other number of heads appears, what should be the odds on the game if player A is to break even?

(b) How should the game be played so that both players have an equal chance of winning with no odds?

20·7 Ten similar coins are mixed and tossed.

(a) What is the probability of obtaining just 3 heads?

(b) What is the probability of obtaining 3 heads or fewer?

20·8 Over an extended period of time, a process yields 1 reject for every 10 items produced. What is the probability of finding exactly 2 rejects in a sample of 50 items from the process?

20·9 A process which has been performing in the same fashion for a long time produces on the average a process fraction defective of 0.05. If a random sample of 15 is taken from the production line, what is the probability that:

(a) There will be no defective units in the sample?

(b) There will be 2 or fewer defective units in the sample?

20·10 A supplier ships lots that are 4 percent defective. If random samples of 12 items are taken from large lots, what is the probability of finding:

(a) no defectives

(b) one or fewer defectives

(c) two or fewer defectives

20·6 CUMULATIVE PROBABILITIES

Calculations based on the binomial expansion, if carried out for many different values of n, c, and fraction defective, lead to binomial probability tables. The use of such tables makes it unnecessary to calculate the probabilities, inasmuch as the necessary data can be found from the tables instead.

A separate table is needed for each different sample size, since the coefficients of the terms in the binomial expansion vary with the value of n. In Table 20·1 are given data for a sample size of 10. In the body of the table

Table 20·1 Cumulative Binomial Probabilities for $n = 10$

c \ p'	0.01	0.05	0.10	0.20	0.30	0.40	0.50	0.60	0.70	0.80	0.90	0.95	0.99
0	0.904	0.599	0.349	0.107	0.028	0.006	0.001	0.000	0.000	0.000	0.000	0.000	0.000
1	0.996	0.914	0.636	0.376	0.149	0.046	0.011	0.002	0.000	0.000	0.000	0.000	0.000
2	1.00	0.988	0.930	0.678	0.383	0.167	0.055	0.012	0.002	0.000	0.000	0.000	0.000
3	1.00	0.999	0.987	0.879	0.650	0.382	0.172	0.055	0.011	0.001	0.000	0.000	0.000
4	1.00	1.00	0.998	0.967	0.850	0.643	0.377	0.166	0.047	0.006	0.000	0.000	0.000
5	1.00	1.00	1.00	0.994	0.953	0.834	0.623	0.367	0.150	0.033	0.002	0.000	0.000
6	1.00	1.00	1.00	0.999	0.989	0.945	0.828	0.608	0.350	0.121	0.013	0.001	0.000
7	1.00	1.00	1.00	1.00	0.998	0.988	0.945	0.833	0.617	0.322	0.070	0.012	0.000
8	1.00	1.00	1.00	1.00	1.00	0.998	0.989	0.954	0.851	0.623	0.364	0.086	0.004
9	1.00	1.00	1.00	1.00	1.00	1.00	0.999	0.994	0.972	0.893	0.651	0.401	0.096

are found values of P; P represents the cumulative probability of observing c or fewer events of a specified class (usually defective items in an inspection procedure). The various possible values of c are found in the left-hand vertical column. Some of the various possible values of p' (fraction defective of the population from which the sample is drawn) are found in the uppermost horizontal line.

Cumulative probability tables are used to answer questions such as this: What is the probability of finding 3 or fewer defective items in a sample of 10 taken at random from a lot which contains 30 percent defective items? We note that $c = 3$; $p' = 0.30$. We find the table value to be 0.650. Thus we would expect, in a large series of samples of 10 taken from this lot, that slightly fewer than two-thirds would contain 3 or fewer defectives. Similarly, the table value in the same column opposite $c = 4$ is 0.850, the probability of finding a sample from the same lot with 4 or fewer defective items. Note that the difference between the two probabilities we have just found $(0.850 - 0.650)$ is 0.200; this is the probability of finding a sample with exactly 4 defective items. In this manner, the cumulative probability tables may also be used to find the probability of occurrence of a specific kind of sample.

PROBLEMS

20·11　If 20 percent of the items produced by a machine are defective, determine the probability that out of a sample of 10 items chosen at random,
　(a) 5 or fewer items will be defective
　(b) 4 or fewer items will be defective
　(c) exactly 5 items will be found to be defective

20·12　If the probability that an entering college freshman will graduate from the college 4 years later is 0.30, determine the probability that out of 10 randomly selected students,
　(a) 4 or fewer will graduate
　(b) at least 1 will graduate
　(c) 4 will graduate

20·13　From Table 20·1, binomial probability of $n = 10$, find the following:
　(a) The probability of finding no defectives in a sample of 10 when the lot process fraction defective p' is 0.01; 0.05.
　(b) If p' is 0.05, what is the probability of finding 3 or fewer defectives in a sample of $n = 10$?
　(c) If $p' = 0.10$, what is the probability of finding just 1 defective item in a sample of $n = 10$?

20·7 THE POISSON DISTRIBUTION

For ease of calculations, the binomial distribution may be approximated by the Poisson distribution. The larger the value of n, and the smaller the value of p' (the lot fraction defective), the closer the approximation becomes.

The Poisson term for the probability of 0 occurrences of a defect is $e^{-np'}$, where e is the base of the system of natural logarithms and np' is the average value of the expected number of occurrences (defects). Often c' is substituted for np'.

The expression for the probability of exactly r occurrences is given by the binomial in the following form:

$$P = \frac{n!}{r!\,(n-r)!}\,(p')^r(q')^{n-r} \tag{20·3}$$

In this expression, the sample size is n, the fractional defective is p', the fractional nondefective is q'. The corresponding expression for the Poisson term is

$$P = \frac{(np')^r}{r!}\,e^{-np'} \tag{20·4}$$

Very often for devising sampling plans, it is necessary to know the probability of finding less than a given number of occurrences; this involves the summation of the terms of the binomial or Poisson distributions. Tables have been prepared which give these summations; one such table (Table A·8) gives these summations for the Poisson distribution. In this table, the number of occurrences is represented by the symbol c; the table value is the summation of all terms with c or fewer occurrences. Although all values of p' are not given in Table A·8, linear interpolation will usually give values very close to the correct value.

The normal curve can also be used as an approximation to the binomial and gives a more rapid answer than does the Poisson approximation when the value of np' is greater than any table value. The normal curve is best used with approximations of cumulative terms rather than with individual terms of the binomial. The approximation as computed by the normal curve will become better as the value of n increases and as p' approaches 0.5.

For values of np' less than 10 (large n, small p) the greatest timesaver in approximating the binomial distribution is the use of Table A·8, the summation of terms of Poisson's exponential. In most industrial sampling problems, these Poisson approximations are so close to the actual binomial values that they are used in place of the binomial values.

If $p' = 0.02$ and we use $n = 170$, the probability of 4 or fewer occurrences of defectives would be 0.745 if computed by the binomial, and would be 0.744 as listed in the cumulative Poisson table, Table A·8. The calculations for exactly 4 defectives as worked out using the binomial would be

$$\frac{170!}{4!\,166!}\,(0.98)^{166}(0.02)^4$$

while the same value as computed by the Poisson approximation would be

$$\frac{(np')^4}{4!} e^{-np'}$$

$$\frac{(3.4)^4}{4!} e^{-3.4}$$

Just as with the binomial distribution, the particular population values of np or $c(np'$ or $c')$ are substituted in the formula to obtain the appropriate probability values. The Poisson distribution has a standard deviation equal to $\sqrt{c'}$ and the average is the value c'. It is a distribution of importance in its own right as well as serving as an approximation to the binomial.

PROBLEMS

20·14   Find the probability of getting 3 or 4 heads in 10 tosses of an honest coin by using:
(a) the binomial distribution
(b) the normal approximation to the binomial distribution

20·15   If the probability that a football player will suffer a serious injury is given as 0.001, determine the probability that out of 2,000 players, two or fewer players will suffer such an injury.

20·16   If 3 percent of the items produced by a company are defective, find the probability that in a sample of 100 items,
(a) none will be defective
(b) one will be defective
(c) two or fewer will be defective

20·8   OPERATING CHARACTERISTIC CURVES

Data from binomial probability tables may be graphed by plotting cumulative probability P against fraction defective p'. There will be a different set of graphs for each sample size n and a different graph in the set for each acceptance number c. In Fig. 20·3 we show five such curves for data taken from Table 20·1. Figure 20·4 shows similar curves based on the Poisson distribution.

Such graphs have a special significance in their relationship to sampling plans. For a given sample size (here 10) the numbers found on the vertical axis express the probability that we will accept a lot of quality described by the value found on the horizontal axis, when we make the decision that we will judge the quality of the lot by the c value assigned to the curve.

Suppose that we decide to accept a lot if we find 3 or fewer defective items in a random sample of 10. The curve in Fig. 20·3 marked $c = 3$ applies. If the lot is really 50 percent defective ($p' = 0.50$) we find from the curve that $P = 0.17$. Therefore, about 17 percent of the time we would accept lots that

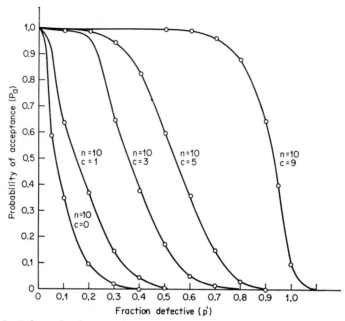

Fig. 20·3 Relationship between fraction defective and cumulative probability for n = 10 and different values of c based on binomial probability.

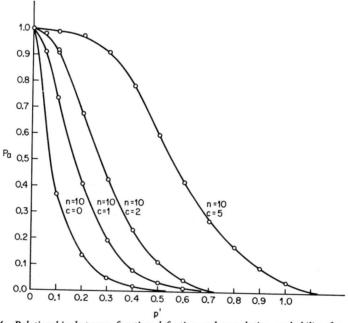

Fig. 20·4 Relationship between fraction defective and cumulative probability for n = 10 and different values of c based on the cumulative Poisson distribution.

were 50 percent defective on the basis of our accepting 3 or fewer defectives in the sample of 10. From the same curve, we see that lots as bad as 60 percent defective would give us samples with 3 or fewer defectives about 5 percent of the time.

If we now change our criterion for accepting a sample, we change the value of c. If we decide to accept the lot only if the sample of 10 contains 1 or fewer defectives, we use the curve marked $c = 1$. If the lot fraction defective is 50 percent, the probability of finding samples of 10 with 1 or fewer defectives is very small indeed. Thus, a sampling plan based on $n = 10$

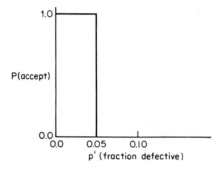

Fig. 20·5 Ideal operating characteristic curve for a sampling plan.

and $c = 1$ would almost always reject lots that were as bad as 50 percent defective or worse. From the same curve, if the lot were 30 percent defective, this sampling plan would accept the lot about 15 percent of the time.

Operating characteristic (OC) curves such as these are used to judge the worth of acceptance sampling plans. The ability of a sampling plan to choose from among "good" and "bad" lots of items is shown by the OC curve, particularly by its shape. A plan that would discriminate perfectly would be shaped as in Fig. 20·5, where the vertical axis now has the meaning: probability of acceptance. The ideal curve indicates that we would always reject a lot of undesirable quality as defined by the values of p' lying to the right of the foot of the curve and would always accept a lot of satisfactory quality as defined by the values of p' lying to the left of the foot of the curve. This curve indicates that we would invariably accept a lot of 5 percent defective or better and invariably reject a lot of poorer quality.

Only with 100 percent inspection of lots could we possibly realize the ideal OC curve, and since even 100 percent inspection is not infallible, we rarely could achieve this. We shall see that we can approach such a curve, at the cost of a laborious sampling procedure, but we cannot reach it.

20·9 CONSTRUCTION OF AN OC CURVE

Since the adequacy of any sampling plan is based on the nature of the OC curve, such a curve should be constructed for every proposed sampling plan. The curve then should be examined to determine whether or not the plan will succeed in distinguishing satisfactory from unsatisfactory lots of items. The following method is used to construct an OC curve based on the binomial distribution.

Let us assume that we are inspecting lots of items as follows: We examine a sample of three items from each lot; we accept any lot as satisfactory if in the sample we find no more than one defective item. Thus $n = 3$ and $c = 1$.

We expand the binomial formula:

$$(p' + q') = p'^3 + 3p'^2q' + 3p'q'^2 + q'^3$$

As before, p' represents the fraction defective of the *lot*. Since our plan will allow either one defective item or none, we are concerned with the last two terms of the expansion, where p' has, respectively, the exponent 1 and 0 (since p' does not appear in the last term).

We obtain the data required for plotting the OC curve by substituting various values of p' and q' in these terms and summing the results. [We recall that $(p' + q') = 1$.] Thus, if we let $p' = 0.1$ and $q' = 0.9$, then $3p'(q')^2 + (q')^3 = 3(0.1 \times 0.9^2) + 0.9^3 = 0.243 + 0.729 = 0.972$. The number 0.972 is P_a; that is, it represents the probability of accepting a lot containing 10 percent defective items (since p' was 0.1) on the basis of the assumed n and c values. Similar calculations for other p' values give the data in Table 20·2.

Table 20·2   Calculation of Data for OC Curve
from Binomial Distribution
$(n = 3, c = 1)$

p'	q'	$3p'(q')^2$	$(q')^3$	$P_a = 3p'(q')^2 + (q')^3$
0.0	1.0	0.000	1.000	1.000
0.1	0.9	0.243	0.729	0.972
0.2	0.8	0.384	0.512	0.896
0.3	0.7	0.441	0.343	0.784
0.4	0.6	0.432	0.216	0.648
0.5	0.5	0.375	0.125	0.500
0.6	0.4	0.288	0.064	0.352
0.7	0.3	0.189	0.027	0.216
0.8	0.2	0.096	0.008	0.104
0.9	0.1	0.027	0.001	0.028
1.0	0.0	0.000	0.000	0.000

We now plot the OC curve (Fig. 20·6) by using the values of p' (on the horizontal axis) and the values of P_a on the vertical axis. Examination of this curve indicates that the chosen sampling plan (sample size of 3 and c value of 1) is a poor one. We would by this plan, for example, accept 50 percent of the time lots which are 50 percent defective. This inference comes from the point $p' = 0.5$, $P_a = 0.5$. Over 20 percent of the time we would accept lots

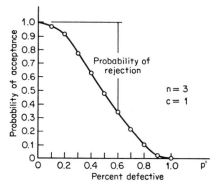

Fig. 20·6 Operating characteristic curve for n = 3 and c = 1 (data in Table 20·2).

which are 70 percent defective. Methods of improving the discrimination among poor and good lots of items can be inferred from an examination of the factors that determine the shape and position of the OC curve.

The two numbers n and c completely determine the OC curve:

1 Sample size n: For the same acceptance number c, increasing the sample size causes the OC curve to assume a more nearly vertical position as in Fig. 20·5. A sample size of 5 permits only very poor discrimination among lots of different quality; increasing the sample size greatly improves the discrimination. The magnitude of this change in power to discriminate is large for relatively small sample sizes. There is a great improvement when the sample size is increased from 5 to 10, but a much smaller improvement when the sample size is increased from 20 to 40. In addition, increasing sample size adds to the cost of inspection. It becomes, then, a management problem to decide whether to inspect larger samples and therefore to have a more efficient sampling plan, or to risk accepting lots of poorer quality than is really desired. Perhaps the cost of handling smaller lots is less than the added cost of efficient inspection.

2 The acceptance number: For the same sample size, increasing the c value shifts the steepest portion of the OC curve to the right, thus changing the region of fraction defectives where the sampling plan discriminates well. These changes may be seen in the curves in Fig. 20·3.

Since the effects of the two factors of sample size and acceptance number are to a considerable degree independent, it is possible by the appropriate selection of the values for n and c to cause the OC curve to have any desired steepness (determined largely by n) and position (determined largely by c).

PROBLEMS

20·17 Using the values in Table 20·1, for $n = 10$, draw the OC curve for a sampling plan of $n = 10$, $c = 3$.

20·18 Construct the OC curve for $n = 5$, $c = 1$.

(a) Expand the binomial $(p + q)^5$ and consider only the terms that contain one or fewer p terms.

(b) Then give p different values from 0.1 to 0.9 and locate the points.

20·19   Using Table 20·1, construct the following OC curves:
(a) $n = 10$, $c = 0$
(b) $n = 10$, $c = 2$
(c) $n = 10$, $c = 4$

20·10   CRITERIA FOR A SAMPLING PLAN

Let us suppose that we desire a sampling plan which will accept lots that are less than 10 percent defective ($p' = 0.10$) and will reject lots that are poorer than this. The graphs in Fig. 20·7 show how different sampling plans approach

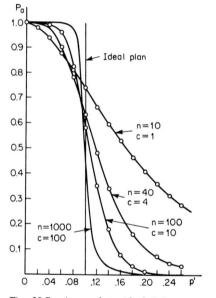

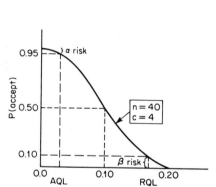

Fig. 20·7   Approach to ideal OC curve as n increases.

Fig. 20·8   Operating characteristic curve for a sampling plan, showing α and β risks.

the ideal, which would run vertically on the $p' = 0.10$ line. Observe that in these graphs the ratio of c to n is kept constant, here 1:10. The best approach to the ideal curve here involves a sample size of 1,000, with $c = 100$. A much larger sample size would be required to come appreciably closer to the ideal.

We select one of the curves, that for $n = 40$ and $c = 4$, to show the nature of the two risks that must be assumed; the curve is shown in Fig. 20·8.

20·11   CONSUMER'S (BUYER'S) RISK

The buyer wants to protect himself against purchasing a poor-quality lot having a sizable fraction of defectives. Occasionally he may accept such a lot on the basis of a sample which contains four or fewer defects. He must define what quality level (fraction defective) marks the lower limit of the rejectable quality. Material of this quality is called the *rejectable quality level* (RQL) and must have only a 10 percent (or other specified percentage) probability of acceptance. Here, the RQL would be approximately at 0.18 defective. The consumer's risk is often termed the beta (β) risk. That is, if a lot of 18 percent defective quality were submitted, the buyer would stand a 10 percent risk of accepting it when using this sampling plan.

20·12 PRODUCER'S (SELLER'S) RISK

The producer will surely find that on occasion some high-quality lots, with very few defective items, will be rejected on the basis of a sample which because of chance contained more than four defectives. In many industrial sampling plans, this risk is set at 5 percent, at a quality level he wants to be sure of accepting a high percentage of the time. This is called the *acceptable quality level*, or AQL. The producer wants a sampling plan that assures that if he submits material of this quality, it will have only a 5 percent probability of rejection and a 95 percent probability of acceptance. Here the AQL is at approximately 0.04 fraction defective. The seller's risk is often termed the alpha (α) risk. That is, if a lot of 4 percent defective quality were submitted, the producer would stand a 5 percent risk of having it rejected by the sampling plan.

20·13 DETERMINATION OF A SAMPLING PLAN

A producer and a consumer, in the course of negotiating the transfer of a product, should come to agreement on a sampling plan that will be fair to both. Let us suppose that the two would like to consummate the sale on the basis of maintaining at all times the quality of the product at fraction defective of 0.05. Since to maintain exactly this level would require the sampling plan to have an ideal OC curve, which is never attainable in practice, a more realistic approach to the problem is necessary.

The consumer must decide that, though he would like all submitted lots to be no poorer than $p' = 0.05$, he will be willing to accept lots of poorer quality (perhaps 0.07) if he can be assured that he will be buying such lots only infrequently; perhaps he is willing to accept only 10 percent of such poorer lots when by chance they fail to be detected by the plan. This decision establishes the RQL, in this case at $p' = 0.07$ and at $\beta = P = 0.10$. In effect, one point on the desired OC curve is now known (Fig. 20·9). Actually, the pressure will be on the producer to submit AQL quality or better or he will have high rejection, thus giving the consumer added assurance.

Similarly, the producer must decide that he is willing to agree that lots better than the desired $p' = 0.05$ level will have some small chance of being rejected by the sampling plan; conversely, he must agree that lots at, say, $p' = 0.03$ will have somewhat less than 100 percent chance of passing

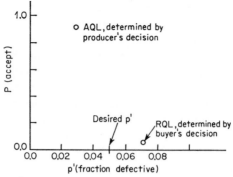

Fig. 20·9 Necessary information for the determination of a sampling plan operating characteristic curve.

inspection, perhaps only 95 percent. This decision establishes the AQL, at $p' = 0.03$ and $P = 100 - \alpha = 0.95$, and therefore determines a second point on the OC curve for the required sampling plan (Fig. 20·9).

What is now needed is an OC curve that will pass through these two points, and the corresponding n value and c value that will give this curve. Since we are here dealing with attribute sampling, and therefore with whole numbers only, it is usually not possible to find an OC curve that will pass through these two points exactly; the best we can do is to find a curve that will closely approach them.

We are here concerned with arriving at a single-sample attribute sampling plan that will have its OC curve pass through two specified points. The Peach-Littauer method[2] of obtaining such a sampling plan is one that illustrates the basic relationship of a sampling plan to its OC curve.

When the AQL and its associated α risk are set, along with the RQL and its associated β risk, it is possible to arrive at a value of n and c for a sampling plan which will provide the proper protection at the specified levels. Peach and Littauer have published tables to reduce the work required. An example is given in Table 20·3, where $\alpha = 0.05$ and $\beta = 0.10$. For other values of the two types of risks, other tables must be consulted.

Assume that it has been agreed to use a single sampling plan for attributes which will protect an AQL of 0.03 with an α risk of 0.05, and will also protect an RQL of 0.07 with a β risk of 0.10. We first form a ratio of RQL/AQL and then locate this value, or a just larger value in the fourth column of Table 20·3. Our ratio is $0.07 : 0.03 = 2.33$; in column four, we select 2.40 as the closest one on the high side of the ratio value 2.33. Reading to the extreme left of the same row, the value $c = 11$ becomes our acceptance number for the sampling plan.

To find the value of n, we use the second column, headed $p'n_{0.95}$. On the same line containing 11 we read the value 6.924. Since the value of our AQL

[2] *Annals of Mathematical Statistics*, vol. 17, no. 1, March, 1946.

Table 20·3 Data for Determining a Sampling Plan ($\alpha = 0.05$, $\beta = 0.10$)

c	$p'n_{0.95}$	$p'n_{0.10}$	$\dfrac{p'n_{0.10}}{p'n_{0.95}}$
0	0.051	2.30	45.10
1	0.355	3.89	10.96
2	0.818	5.32	6.50
3	1.366	6.68	4.89
4	1.970	7.99	4.06
5	2.613	9.28	3.55
6	3.285	10.53	3.21
7	3.981	11.77	2.96
8	4.695	12.99	2.77
9	5.425	14.21	2.62
10	6.169	15.41	2.50
11	6.924	16.60	2.40
12	7.690	17.78	2.31
13	8.464	18.96	2.24
14	9.246	20.13	2.18
15	10.04	21.29	2.12

NOTE: This technique for the calculation of a single sampling plan can be used only when the sample size is less than 10 percent of the lot size.

was 0.03, we substitute into the equation

$$np' = 6.924$$
$$n(0.03) = 6.924$$
$$n = 230.8 \quad \text{or} \quad 231$$

Thus the sampling plan based on $n = 231$, $c = 11$ will have its OC curve running very close to the (AQL, α), the first of the two agreed-upon points. The second point (RQL, β) will be approached when the sample size is determined by using the third column of the table. In the example being discussed, where c was found to be 11, on the same row of the table we read the value listed in the third column and form the following equation:

$$np'_{0.10} = 16.60$$
$$n(0.07) = 16.60$$
$$n = 237$$

Either of these two plans, $n = 231$, $c = 11$ or $n = 237$, $c = 11$, will have OC curves running very close to the two specified points. It is customary to use the AQL value to determine the sample size and then substitute this value of n in the np solution for the RQL and thus obtain a revised estimate of the RQL. There is, however, one important restriction. The sample size so determined must be less than 10 percent of the lot size; otherwise, some other type of sampling plan must be used. We can use the sampling plan $n = 231$, $c = 11$, provided that the lot size from which the sample is to be taken is more than 2,310 items.

It should be noted that the sampling plan here designed requires a large sample size and a large c value. The shape of the OC curve would therefore be very steep near the desired p' value of 0.05; and the sampling plan therefore discriminates well. This was forced by the choice of the p' values for the AQL and RQL, since they were relatively close together, with a ratio of only a little over 2. This sharp discrimination requires a large sample to fulfill the requirements placed upon it. Had the p' values for the AQL and RQL been widely different, we would have been using the upper levels of the table, where we would need considerably smaller values of n.

PROBLEMS

20·20   If $\alpha = 0.05$ and $\beta = 0.10$, what sampling plan should be used for the protections:
(a) AQL $= 0.02$; RQL $= 0.04$
(b) AQL $= 0.02$; RQL $= 0.08$
(c) AQL $= 0.02$; RQL $= 0.12$
(d) AQL $= 0.02$; RQL $= 0.20$

20·21   On the same graph paper, plot the points for the data in Prob. 20·20.
(a) Draw straight lines to connect the 2 points for each of the 4 sampling plans and indicate the sampling plan for each.
(b) What effect does increasing sample size have on the slope of the lines?
(c) Considering only the changes in the c value, how does an increase in the value c change the OC curve?

20·14   COMPUTING OC CURVES USING THE POISSON DISTRIBUTION

An operating characteristic curve should be drawn for every sampling plan for attributes, to show the ability of the plan to distinguish good lots from poor ones. The OC curve graphs the probabilities of acceptance of lots of materials at various quality levels for the particular sampling plan.

To illustrate the preparation of an OC curve using the Poisson distribution as an approximation to the binomial, we choose a single sampling plan for attributes which calls for $n = 100$, $c = 3$. We find on the horizontal axis of the graph (Fig. 20·10) a range of lot fractional defective values (p'); we find on the vertical scale P_a values which represent the probability of acceptance of the lot.

Table 20·4 is prepared, assuming various possible values of p' and a constant sample size of 100. The product $p'n$ is used to enter Table A·8; in the column headed $c = 3$, we find the P_a values required for plotting the OC curve. The chosen values of p' are plotted on the abscissa of the graph, and the table values of P_a are used as ordinates.

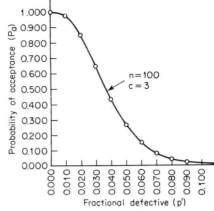

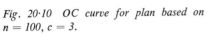
Fig. 20·10 OC curve for plan based on n = 100, c = 3.

From the OC curve (Fig. 20·10) we can see that incoming lots which are 3 percent defective will be accepted by the sampling plan with a probability of acceptance of about 0.65; a lot which is 8 percent defective will be accepted only about 4 percent of the time.

The better the incoming lots of material are, the greater the chance they have of being accepted by this sampling plan. If the material becomes more defective, it has less chance of being accepted. Even though we have no idea of the quality of the material being submitted, we can feel sure that the sampling plan will give us the protection indicated by the OC curve. Thus this sampling plan (defined by $n = 100$, $c = 3$) will accept lots no poorer than 1 percent defective 98 percent of the time but will accept material as poor as 7 percent defective only 8 percent of the time.

When the value of np' is less than 10.0, we can safely use Table A·8 for the calculation of the OC curve. When np' is greater than 10.0, the Poisson approximation is not very good, and we should compute the values by the use of the binomial distribution. An alternative method is to approximate

Table 20·4

p'	np'	$P_a(c = 3)$
0.00	0.00	1.00
0.01	1.00	0.981
0.02	2.00	0.857
0.03	3.00	0.647
0.04	4.00	0.433
0.05	5.00	0.265
0.06	6.00	0.151
0.07	7.00	0.082
0.08	8.00	0.042
0.09	9.00	0.021
0.10	10.0	0.010

the binomial distribution by the use of areas under the normal curve. Since most sampling plans are used in situations where the fractional defective values are small and the sample size is less than 10 percent of the lot size, the Poisson distribution is of general use in calculating probabilities of acceptance, as described above.

Consider now the sampling plan arrived at earlier in this chapter where $n = 231$ and $c = 11$. To construct an OC curve for this sampling plan, we again assume various degrees of defective materials and from Table A.8, list the various probabilities of securing such np' values. Figure 20·11 is a sketch of the OC curve for $n = 231$, $c = 11$.

Note that the OC curve goes through the AQL point at the specified value of 0.95, which allows only 0.05 to remain above the curve. This means that if material of the quality of 0.03 is submitted to this sampling plan, in the long run the sampling plan will accept the material 95 percent of the time and will reject it only about 5 percent of the time.

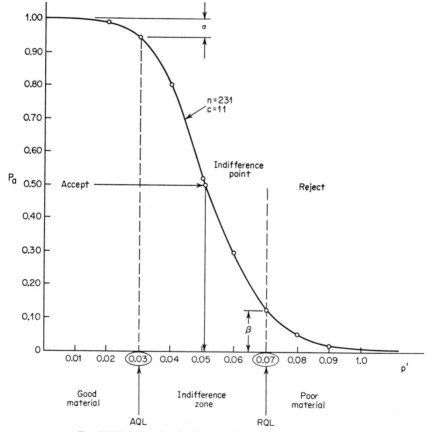

Fig. 20·11   Completed OC curve based on $n = 231$, $c = 11$.

The curve also goes through, or close to, the RQL value at its specified value of 0.10. The OC curve shows for each possible percentage of defective items in a lot the probability of accepting a lot of that quality. If the material submitted is better than the AQL, it has an even better chance of being accepted by the sampling plan, while material worse than the RQL has only a slight chance of being accepted.

There is a point on the OC curve that has a probability of 0.50 of acceptance. If material of this quality is submitted for appraisal, this sampling plan will accept it as often as it will reject it, hence the term *indifference point*. For this sampling plan the indifference point is at a quality of material with slightly more than 5 percent defective.

The area between the AQL and RQL is known as the *indifference zone*, as material of this quality will either be accepted or rejected on the basis of how closely it approaches the AQL or the RQL.

PROBLEMS

20·22   If the AQL is 2 percent with an α risk of 0.05 and the RQL is 6 percent with a β risk of 0.10, what values of the OC curve are known and could be plotted without using the Table A·8 values?

20·23   Using the values in Prob. 20·22, calculate a single sampling plan for attributes which will give the protection desired.

20·24   Calculate a single sampling plan for attributes which will protect an AQL of 0.01 with an α risk of 0.05 and an RQL of 0.05 with a β risk of 0.10.

20·15   DEFECTS-PER-UNIT ACCEPTANCE SAMPLING PLANS
In some cases where attribute inspection is used, a unit of material might contain one or more defects but still be considered as good product. If the number of defects per inspection unit, or per given area, becomes large, the product then becomes defective, but the mere presence of a single defect would not make it defective. The printing industry is well aware of the presence of small defects in most of the printed materials run on presses at high speeds. Some of these defects are so minor that the average person who uses the product would never know they existed. Even the trained press operator finds it hard to find all defects which are present in the finished copy. But this does not make the product unusable. As long as the number of defects per unit area are kept to a relatively small number, the sampling plan used should accept the product.

Here there is a need for a defects-per-unit sampling plan instead of a fraction defective plan. If the material submitted for inspection is rolls of film, paper, cloth, or similar product, or if it consists of large units such as radios, rockets, or automobiles, then a defects-per-unit sampling plan should be used.

Like the fraction defective plans, a single-sample defects-per-unit sampling plan specifies a sample size n and an acceptance number c. If the sample has a total number of defects that is less than or equal to c, the lot is accepted. If the sample contains a number of defects in excess of c, the lot is rejected. We use defects-per-unit sampling plans when the lot is large relative to the sample and when the distribution of units in the lot, with respect to the number of defects per unit, approximates the form of the Poisson distribution.

The Poisson distribution can be derived from the binomial by (1) letting the value of n approach infinity and (2) letting the value of p approach 0. Both n and p must change in such a way as to keep the value of np equal to a constant value, say λ. The Poisson distribution is then given by

$$P(r) = \frac{e^{-\lambda}\lambda^r}{r!} \qquad \text{for } r = 0, 1, 2, \ldots \qquad (20\cdot5)$$

The parameter λ must be positive. The mean of the distribution is $\mu = \lambda$, with variance also equal to λ. The formula will produce the probability of any value of r and the distribution for a given λ can be calculated. For a value of $\lambda = 1.0$, the distribution is shown in Fig. 20·12. The probabilities for the various value of r are:

$P(0) = 0.3679$	$P(5) = 0.0031$
$P(1) = 0.3679$	$P(6) = 0.0005$
$P(2) = 0.1839$	$P(7) = 0.0001$
$P(3) = 0.0613$	$P(8) = 0.0000$
$P(4) = 0.0153$	

Examples of uses of the Poisson distribution are the number of tire failures per week for a fleet of trucks or buses, the number of defects in the finish of manufactured pans, the number of typing errors per page, or as a first approximation to failure date such as the number of failures after a specified time. Whenever the inspection unit is large and the number of defects is small (relative to the lot size) the Poisson distribution is the one to use in the calculation of the ordinates of the OC curve.

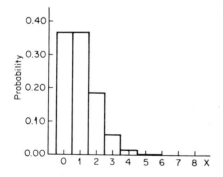

Fig. 20·12 Graph of the Poisson distribution for $\lambda = 1.0$.

Suppose, for example, that a sampling plan has $n = 100$ and $c = 17$. If a lot has an average of 0.15 defect per unit, the probability of acceptance of the lot may be found by using Table A·8 with a value of $\lambda = 100 \, (0.15) = 15.0$ and noting the probability of getting 17 or fewer defects. The table value is about 0.75; so the probability of acceptance of the specified lot using the given sampling plan is 0.75.

Some sampling plans, such as Military Standard 105D, will refer to defects per 100 units. Others will refer to defects per unit. In either case, the OC curve for a defects-per-unit plan with sample size n and acceptance number c is approximately the same as the OC curve for fraction defective. In the case of defects per unit, the abscissa scale will read "number of defects per unit" or "number of defects per hundred units." What has been said about fraction defective sampling plans can also be said about defects-per-unit plans by replacing "number of defectives" with "number of defects" and by replacing "lot fraction defective" with "lot number of defects per unit."

20·16 LOT QUALITY DETERMINATION

Where we are to use sampling to determine the quality of a single lot, we are often faced with making such a determination on a small sample size. Answers obtained from small sample sizes in the case of acceptance sampling are very often not too specific, as indicated by the values listed in Table 20·5.

If, from a specific lot, we take a sample of $n = 50$ and find there are two defectives $(r = 2)$, by using Table 20·5 we can find the two listed values 0 and 14. Thus the lot from which the sample was taken contains between 0 percent defective and 14 percent defective items. Such a statement is made with 95 percent confidence, since all values in the table are actually 95 percent confidence limits for binomial distributions.

If the sample size is 250, and 5 defectives were found, then the ratio r/N is first computed and the value 5/250 is used to enter the table. Looking down the r/N column until the ratio value of 0.02 is found, and moving across the row to the column 250, we find the two values 1 and 5. This means that the lot being sampled contains between 1 percent defective and 5 percent defective items.

Consider the following question: How large a sample should we take to be 95 percent sure that the sample fraction defective p_0 is within 5 percent of the true but unknown lot quality? We can use the techniques of working with the normal curve to approximate the answer. Recalling the formula $t = (X - \mu)/\sigma_X$ we can now write it as $t = (p_0 - p')/\sigma_P$. As we wish to have the sample value within 5 percent of the true value p', we know that $|p_0 - p'| \leq 0.05$. We do not know what σ_p is; so we will estimate it from the sample. We will use the formula $s_p = \sqrt{p_0(1 - p_0)/n}$. The value of this standard deviation can then be computed as a result of finding p_0 from a sample from the lot. But the value of s_p does not change a great deal regardless

Table 20·5 Ninety-five percent Confidence Interval (percent) for Binomial Distribution

Number observed r	Size of sample, N 10		15		20		30		50		100		Fraction observed r/N	Size of sample 250		1,000	
0	0	31	0	22	0	17	0	12	0	7	0	4	.00	0	1	0	0
1	0	45	0	32	0	17	0	17	0	11	0	5	.01	0	4	0	2
2	3	56	0	40	1	31	1	22	0	14	0	7	.02	1	5	1	3
3	7	65	4	48	3	38	2	27	1	17	1	8	.03	1	6	2	4
4	12	74	8	55	6	44	4	31	2	19	1	10	.04	2	7	3	5
5	19	81	12	62	9	49	6	35	3	22	2	11	.05	3	9	4	7
6	26	88	16	68	12	54	8	39	5	24	2	12	.06	3	10	5	8
7	35	93	21	73	15	59	10	43	6	27	3	14	.07	4	11	6	9
8	44	97	27	79	19	64	12	46	7	29	4	15	.08	5	12	6	10
9	55	100	32	84	23	68	15	50	9	31	4	16	.09	6	13	7	11
10	69	100	38	88	27	73	17	53	10	34	5	18	.10	7	14	8	12
11			45	92	32	77	20	56	12	36	5	19	.11	7	16	9	13
12			52	96	36	81	23	60	13	38	6	20	.12	8	17	10	14
13			60	98	41	85	25	63	15	41	7	21	.13	9	18	11	15
14			68	100	46	88	28	66	16	43	8	22	.14	10	19	12	16
15			78	100	51	91	31	69	18	44	9	24	.15	10	20	13	17
16					56	94	34	72	20	46	9	25	.16	11	21	14	18
17					62	97	37	75	21	48	10	26	.17	12	22	15	19
18					69	99	40	77	23	50	11	27	.18	13	23	16	21
19					75	100	44	80	25	53	12	28	.19	14	24	17	22
20					83	100	47	83	27	55	13	29	.20	15	26	18	23
21							50	85	28	57	14	30	.21	16	27	19	24
22							54	88	30	59	14	31	.22	17	28	19	25
23							57	90	32	61	15	32	.23	18	29	20	26
24							61	92	34	63	16	33	.24	19	30	21	27
25							65	94	36	64	17	35	.25	20	31	22	28
26							69	96	37	66	18	36	.26	20	32	23	29
27							73	98	39	68	19	37	.27	21	33	24	30
28							78	99	41	70	19	38	.28	22	34	25	31
29							83	100	43	72	20	39	.29	23	35	26	32
30							88	100	45	73	21	40	.30	24	36	27	33
31									47	75	22	41	.31	25	37	28	34
32									50	77	23	42	.32	26	38	29	35
33									52	79	24	43	.33	27	39	30	36
34									54	80	25	44	.34	28	40	31	37
35									56	82	26	45	.35	29	41	32	38
36									57	84	27	46	.36	30	42	33	39
37									59	85	28	47	.37	31	43	34	40
38									62	87	28	48	.38	32	44	35	41
39									64	88	29	49	.39	33	45	36	42
40									66	90	30	50	.40	34	46	37	43
41									69	91	31	51	.41	35	47	38	44
42									71	93	32	52	.42	36	48	39	45
43									73	94	33	53	.43	37	49	40	46
44									76	95	34	54	.44	38	50	41	47
45									78	97	35	55	.45	39	51	42	48
46									81	98	36	56	.46	40	52	43	49
47									83	99	37	57	.47	41	53	44	50
48									86	100	38	58	.48	42	54	45	51
49									89	100	39	59	.49	43	55	46	52
50									93	100	40	60	.50	44	56	47	53

Reproduced by permission from George W. Snedecor, "Statistical Methods," 5th ed., Iowa State University Press, Ames, Iowa, 1956.

of the value of p_0. We here use the maximum possible value 0.5. For $p_0 = 0.5$, the value of s_p is $0.5/\sqrt{n}$, which we shall use as the approximation to σ_p.

As we wish to be 95 percent sure that we will be within 5 percent of p', we divide the α value into two equal parts. Thus we cause the confidence interval to be equally spaced about the value p'. In attribute sampling, the sample sizes run considerably larger than they do in variable sampling; when $n > 30$, it is not unreasonable to use the probabilities of the normal distribution. This means that if α is 0.05, $t_{0.025} = 1.96$. Returning to the formula, and substituting in all the known or given values we have

$$t_{\alpha/2} = \frac{p_0 - p'}{\sigma_p}$$

$$t_{0.025} = \frac{0.05}{0.5/\sqrt{n}}$$

$$1.96 = 0.05\sqrt{n}/0.5$$

Solving for n, we have $\sqrt{n} = 19.6$; hence $n = $ approximately 385 items. It can be readily seen that the quality determination of a single lot can be expensive when the approximations must be fairly good.

20·17 MILITARY STANDARD 105D SAMPLING PLANS

The single sampling plans for attributes which we have discussed in this chapter were designed to protect a given AQL at a specified α risk and to protect also a given RQL at its specified β risk. Military Standard 105D (Mil-Std-105D) sampling plans, on the other hand, were designed to assure specified AQL values using specified sample sizes, with the RQL being undefined, though obtainable from the OC curves. Though the consumer would appear to be unprotected by the failure to specify the RQL, the plans do protect him in the sense that the producer who supplies a product at a quality poorer than the AQL will suffer more frequent rejections and with this a potential loss of contract.

Since the basic concept of acceptance sampling is that sampling plans cannot assure 100 percent perfection, it is necessary to decide on the quality level that is considered acceptable as a process average. In attribute sampling the acceptable quality level (AQL) is expressed in percent defective. Mil-Std-105D is a sampling plan which will provide average lot-to-lot protection. We are concerned with overall quality level, not just single-lot quality, when we set up sampling plans to be used for acceptance of materials on a repetitive lot basis.

Having specified the AQL value to be maintained, and knowing the lot size to be sampled, the Mil-Std-105D tables can be used to secure the proper sampling plan required. The tables include data for several types of sampling plans: single sampling plans, double sampling plans, and multiple sampling plans. In a single sampling plan, the decision (to accept the lot or to reject it) is made after one sample of n items from the lot has been inspected. In double sampling plans, two samples are taken, n_1 and n_2. After examining the first sample n_1, we make one of three decisions. The lot could be accepted without further sampling, the lot could be rejected without further sampling, or the second sample n_2 must be inspected prior to making a final decision. In double sampling, a relatively small sample is taken and examined. The plan then provides two sample values with $n_1 = n_2$. An acceptance number and a rejection number are specified. If, in the n_1 samples examined, the number of defectives found is less than the acceptance number, there is no need to sample again and the lot is accepted. Should there be more defectives than the rejection number, then again sampling is stopped and the entire lot is rejected. The acceptance number and the rejection number associated with each n_1 value are always more than one value apart. For example, for $n_1 = 100$, the acceptance number for a double sampling plan is $c_1 = 5$, and the rejection number $c_2 = 12$. If the number of defectives found in the 100 items constituting the first sample is between 5 and 12, then another sample, n_2, also of 100 items, must be taken and examined.

The acceptance number and the rejection number associated with the second sample will always be just one value apart; so in double sampling a decision is forced after both samples have been examined. The advantage of double sampling is that a decision often can be made after only the first sample has been examined. This would cut down considerably on the number of samples required. If the material being sampled is very good or very poor, then double sampling is profitable, as many decisions will be made after the first sample. Where marginal material is submitted for examination, both samples will be required to make the decision. In such cases, the total number of items will increase over that required if single sampling had been employed.

Multiple sampling plans are extensions of double sampling plans, requiring the examination of a series of even smaller samples until a final decision is forced with the last sample. The size of each sample taken in multiple sampling is smaller than that of single or double sampling plans. Associated with each n_i is an acceptance number and a rejection number. The average amount of inspection using multiple sampling plans is smaller than for single or double plans. The administration of the sampling program for multiple plans, however, may be harder.

The Military Standard plan calls for shifting from normal inspection to tightened or reduced inspection as the process average appears to shift to poorer or better levels. This is another feature which protects the consumer. A low-level percent defective may result in reduced inspection, cutting costs to the supplier. A high level will result in tightened inspection, adding to his cost.

The effect of using a sampling plan for acceptance sampling is to force the supplier to submit lots of such quality that a small percentage of the lots is rejected. Mil-Std-105D sampling plans are classified by the AQL's. If the AQL is the same for all suppliers of the same product, they are all forced to operate at very nearly the same level of quality. Once the suppliers know the AQL, they have the major portion of the information required to plan production and inspection, and to set prices in bids for contracts.

The Mil-Std-105D is a positive approach to acceptance sampling in that it sets the quality level to be obtained, and thereby a target for the supplier.

20·18 DODGE-ROMIG SAMPLING PLANS

Mil-Std-105D sampling plans were constructed to maintain an AQL at specified α values and are classified by their AQL's. In the type of sampling plan developed by Harold F. Dodge and Harry G. Romig, the major concern is assuring that the customer will be protected against poor-quality lots. The RQL (or lot tolerance percent defective—LTPD) is specified. To minimize inspection costs, in addition to setting limits on the outgoing quality level, defective items must be sorted out of rejected lots by 100 percent examination. The latter is called *rectifying inspection*. With 100 percent inspection of rejected lots (detailing) we can consider the average quality of the material turned out by the combination of sampling and 100 percent inspection. This outgoing quality is called AOQ, average outgoing quality, and is the result of sorting all rejected lots and replacing all defectives with good items.

As the quality of the manufactured items changes, the sampling plan being used will accept more submitted lots or will start to reject a greater number of lots should the change be in the wrong direction. As poorer lots are subjected to the sampling plan, more lots will be rejected and as a result more lots will be screened. This will require more work but will cause the outgoing lots to be of much higher quality then that of the original lots, since defects are replaced by good items.

As better quality is submitted to the same plan, less sorting and screening will be called for as more lots will be accepted. The outgoing quality will, in this case, be very close to the quality of the incoming material. If the rejected lots were not subjected to detailing, only the accepted lots would leave the plant and they would leave with the same quality level as at production. (We are assuming a constant defect level, some lots being accepted, some rejected.) A graph of this relationship would be a straight line, with a slope of 1.0, as shown in Fig. 20·13. Good quality, with a detailing type of sampling plan, would be about the same quality when it was shipped, but poor quality would be much better quality after going through rectifying inspection. In fact, it is possible to have every lot rejected and require 100 percent inspection on all material. This would greatly change the quality level of the

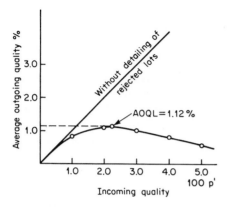

Fig. 20·13 Graph of the AOQL value for sampling plan based on $n = 75$ and $c = 1$, with detailing of rejected lots.

poor lots, and the curve of relationship on the graph in Fig. 20·13 would approach the horizontal line of 0 percent defective lots.

Somewhere between the two minima of outgoing quality is a maximum for each sampling plan. This maximum is a measure of the worst possible quality of material turned out by the plan and is known as the average outgoing quality limit, or briefly the AOQL. The AOQL is an average value over many lots. It may be exceeded in a particular instance, but it is a long-run indication of the worst quality that can be expected with a particular sampling plan.

The Dodge-Romig plans are especially concerned with the RQL, or lot tolerance percent defective (p_t). Either the sampling plans are set up to protect a specific p_t, or they are set up to protect a specific AOQL value. In either case, both values are listed in the table so the user will know them for any plan being used.

The information required to use the Dodge-Romig tables is the lot size and the desired RQL (p_t) or AOQL. The tables then supply the n and c values that will fulfill the objectives of the plan.

The value of AOQ is found by the following formula:

$$\text{AOQ} = \frac{P_A p'(N - n)}{N} \tag{20·6}$$

assuming replacement and detailing of rejected lots. If there is no detailing of rejected lots, the AOQ is the same as p'. Suppose that we are inspecting lots of $N = 10,000$, and we use the sampling plan $n = 75$, $c = 1$. Then, since the value of n is much smaller than N, we can ignore n in Eq. (20·6), which reduces to $\text{AOQ} = P_A p'$, where P_A is the probability of acceptance and p' is the lot fraction defective.

To illustrate how the value of AOQL can be calculated, we select different values of p', and from Table A·8 we find the value of P_A for a lot having the selected p' value. For example, if the lot has a value of $p' = 0.05$, and we use the sampling plan $n = 75$, $c = 1$, there will be a probability of acceptance of the lot of $P_A = 0.112$. Then by using the reduced equation for AOQ, we multiply p' by P_A. In this case, AOQ = (5 percent) (0.112) = 0.560. Table 20·6 shows the values for other selected p' values.

Table 20·6   Average Outgoing Quality for $n = 75, c = 1$

$p', \%$	P_A	AOQ
1.0	0.827	0.827
2.0	0.558	1.116
3.0	0.343	1.029
4.0	0.199	0.796
5.0	0.112	0.560

Figure 20·13 shows a graph of the results from Table 20·6. Incoming quality is graphed on the horizontal axis, and outgoing quality, or AOQ, is located on the vertical axis. The maximum point of the curve is the value of the AOQL for the sampling plan $n = 75, c = 1$.

The value of the average total inspection for each p' value, or average total inspection number (ATIN) is given by the formula

$$\text{ATIN} = nP_A + (1 - P_A)N \qquad (20\cdot7)$$

For $p' = 0.01$, ATIN = (75)(0.827) + (0.173)(10,000) = 1,792.

20·19   OTHER TYPES OF SAMPLING PLANS

A type of sampling plan similar to the multiple sampling plans explained in Mil-Std-105D is the sequential sampling plan. Items are taken singly, in order of production, and inspected. If, after a certain number have been examined, fewer than a specified number of defectives have been found, the whole lot is accepted; if more than a larger number of defectives have been found, the whole lot is rejected; if the number of defectives lies between these two values, inspection is continued. This method may be conveniently expressed by a graph, as in Fig. 20·14. Sequential inspection methods may

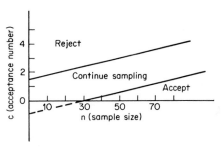

Fig. 20·14   Sequential sampling plan chart.

be especially cost-saving, inasmuch as a decision to reject or accept may be made on the basis of a very small sample; this is particularly true if the lot from which the samples are taken is exceptionally good or exceptionally bad.

There are many special attribute sampling plans available for special situations. More are appearing in the literature all the time. To include even brief descriptions of the more common plans now being used in industry would require many pages. Only a few are mentioned here.

Dodge developed sampling plans for continuous production. His CSP-1 technique can be summarized as follows:

1   At the start, inspect 100 percent of the units of production and continue inspection until i units in succession are found without defects.
2   When i units have been found free of defects, stop 100 percent inspection and inspect only a fraction f of the units by selecting them one at a time from the flow of production.
3   When a defective unit is found, revert to 100 percent inspection and continue until i units are found without defects.

Dodge set forth the values of i and f required for any desired AOQL.[3]

Two modifications of CSP-1 were developed later and are referred to as CSP-2 and CSP-3. In CSP-2, once sampling inspection is started, 100 percent inspection is not required when each defective is found, but it is required when a second defect occurs in the next k or fewer sample units. That is, if two defects are found during sampling, separated by k or fewer good inspected units, then 100 percent inspection is required. CSP-3 is a refinement of CSP-2 to provide greater protection against a sudden run of poor quality.

Other types of continuous sampling AOQL plans were developed by Wald, Abraham, and Wolfowitz. These plans start with sampling inspection in place of 100 percent inspection.[4]

Most acceptance sampling is by attributes, but statistical quality control techniques have led to an increase in the industrial use of acceptance sampling by variables. The most serious limitation on the use of sampling by variables is the fact that acceptance criteria must be applied separately to each quality characteristic. The great advantage in their use is that more information is obtained about the quality characteristic being considered.[5]

QUESTIONS

20·1   How many times does the letter "e" appear in the first paragraph of this chapter? After you have recorded the count, make a recount.

[3] *Industrial Quality Control*, vol. 4, no. 3, pp. 5–9, November, 1947.

[4] The plan by Wald et al. appeared in *The Annals of Mathematical Statistics*, vol. 16, pp. 30–49, March, 1945.

[5] See A. H. Bowker and H. P. Goode, "Sampling Inspection by Variables," McGraw-Hill Book Company, New York, 1952, and Military Standard 414, the latter being extensively used.

(a) Did you get the same number of occurrences of "e" each time you counted?

(b) Would this be an example of sampling inspection, or of 100 percent inspection of a "lot"?

20·2 Randomly select 5 lines in the first paragraph of this chapter and count the number of times that the letter e is found in the 5 lines. Count the number of lines contained in the first paragraph and calculate from the sample 5 lines how many times "e" probably occurs in the entire paragraph.

20·3 What is meant by the term attribute?

20·4 Expand the expression $(H + T)^5$ (H = heads, T = tails). In how many different ways can 5 coins be arranged with respect to heads and tails?

20·5 From the information in Question 20·4, calculate the probability of the following:

(a) Getting 1 head and 4 tails when tossing 5 honest coins

(b) Getting 2 heads and 3 tails

(c) Getting 2 heads or fewer

20·6 If the expansion of the binomial $(p + q)^n$ will always produce the entire binomial distribution, why is the use of the binomial formula necessary?

20·7 If the binomial formula were used to calculate the answers to Question 20·5, how many times would it have to be used to arrive at the answer to part (c)?

20·8 Why do the values in the table for the binomial probability for $n = 10$ differ from the values in Table A·8 for $n = 10$, at any of the pn values?

20·9 What is the role played by the OC curve in explaining how good a sampling plan is?

20·10 Why do all OC curves go through the probability of acceptance value of 1.0, at $pn = 0$?

20·11 If a single sampling plan is given, why is it possible to use only the binomial formula to calculate the OC curve for the sampling plan?

20·12 How could the expansion of $(p + q)^n$ be used to construct the OC curve for a single sampling plan?

20·13 What values must be changed if the OC curve must be:

(a) made steeper

(b) moved to the left

(c) made steeper and moved to the right

20·14 Why is it impossible to have a sampling plan which will have an OC curve with a vertical line separating good quality from poor quality?

20·15 If the AQL and RQL values are relatively close to each other, what type of sampling plan will be necessary?

20·16 What is the difference between the α and the β risks?

20·17 When sampling plans are used to accept or reject whole lots, what are the two types of mistakes that might be made by the sampling plan?

20·18   From the answer to Question 20·17, which type of risk is associated with each type of error?

20·19   The Military Standard 105D sampling plans are built to protect against making what type of error?

20·20   How do the Dodge-Romig sampling plans differ from the Mil-Std-105D sampling plans?

20·21   Under what conditions can the Poisson distribution be used as an approximation to the binomial distribution?

20·22   What does the symbol c stand for in sampling plans?

20·23   Explain how double sampling plans work.

20·24   Under what conditions would a double sampling plan lead to taking fewer samples than a single sampling plan would call for?

REFERENCES

1, 16, 24, 26, 28, 34, 35, 36, 40, 53, 54, 55, 76, 82, 84, 85, 92, 105, 112, 117

21 Shewhart Control Charts

The object of *product* control is to decide whether to accept or reject a lot on the basis of evidence from one or more samples drawn at random from the lot in question. The statistical tool used for making such decisions is the acceptance sampling plan.

On the other hand, the primary object of *process* control is to keep a process at a stable level. The major statistical tool used to do this is the control chart, which is a graphic method of presenting data based on a sequence of samples.

The quality-control chart as a graphical means of applying the statistical principle of significance to the control of production processes was first proposed by Dr. Walter A. Shewhart in 1924. His basic contribution to the development of statistical quality control is the concept of a *state of statistical control*. Past and current data are used as aids to infer the behavior of the process. Before reasonable inferences can be made, the process must be *in control*; that is, the process must be reproducible within limits. One of the primary functions of statistical quality control is finding the magnitude of these limits.

21·1 FUNDAMENTALS OF CONTROL CHARTS

Basic to control charts is the recognition of two types of variability. One type is the inherent or random component associated with the "noise" level. This is variability we have to live with. Such variability cannot be eliminated for economic or engineering reasons. The second type is variability associated with real changes in process level. There are usually assignable causes for this type of variation. This type includes the signals which we usually want to detect above the noise level.

A control chart is a graph of the variability of the response variable. It is much like a histogram with one important change: A time scale is added to the data plot. Thereby we retain the order of data collection and plot the results with respect to time. After we have gathered enough data, we compute the average and draw a line on the graph at the mean level. A measure of the inherent variability is also obtained from the data, and two lines (control-limit lines) are drawn parallel to the central line.

The data in Table 21·1 represent values typical of an industrial process. The values of the response variable were collected for samples of size $n = 4$. In Fig. 21·1 we have graphed the data and have added the upper and lower control limits (UCL and LCL). Two graphs are used to display the variability of the data. The upper graph is a plot of averages of the samples; the lower graph is a plot of the ranges of the samples. Both graphs use the same time scale.

Table 21·1 Data Typical of a Process in a State of Statistical Control

Time	X_1	X_2	X_3	X_4	ΣX	$\bar{X}$	R
8:30	19	23	19	10	71	17.75	13
8:31	16	18	25	22	81	20.25	9
8:32	15	22	16	16	69	17.25	7
8:34	21	23	22	21	87	21.75	2
8:35	26	21	15	22	84	21.00	11
8:37	20	13	21	20	74	18.50	8
8:38	16	13	26	21	76	19.00	10
8:39	21	25	23	25	94	23.50	4
8:40	22	19	15	18	74	18.50	7
8:41	15	20	21	21	77	19.25	6
8:42	16	21	17	19	73	18.25	5
8:43	19	19	12	22	72	18.00	10
8:44	16	21	26	25	88	22.00	10
8:45	17	21	19	13	70	17.50	8
8:46	22	19	23	26	90	22.50	7
8:47	17	19	17	22	75	18.75	5
8:48	14	20	17	13	64	16.00	7
8:49	21	23	19	26	89	22.25	7
8:50	21	23	23	24	91	22.75	3
8:51	20	18	22	21	81	20.25	3
8:52	22	20	18	16	76	19.00	6
8:53	20	22	20	23	85	21.25	3

$$\Sigma\bar{X} = 435.25 \qquad \Sigma R = 151$$
$$\bar{X} = 19.78 \qquad \bar{R} = 6.9$$

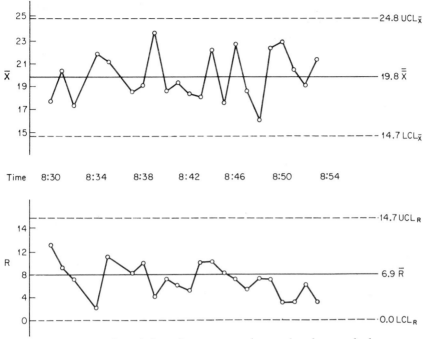

Fig. 21·1 Control charts for averages and ranges based on standards.

The control limits, as in Fig. 21·1, bound the region within which the noise is causing fluctuations. Beyond these limits, we assume that the signal is real and that an assignable cause has brought about a shift in process level. The limits are based on an estimate of the random component of variability σ. This is usually measured by dividing the data into "rational" subgroups (as shown in Table 21·1) that represent periods of time or product that should include only the random components. Note that σ in control-chart usage does not include any assignable cause components of variation.

Most Shewhart charts use 3σ limits. This means that 99.7 percent of the plotted points when a process is correctly centered will lie within limits. It also means that 0.3 percent of the points will lie outside, thus erroneously indicating that the process level is no longer correctly centered. By changing the number of σ units that the limits are located from the central line, the risk (α risk) of incorrectly calling a good process "out of control" can be changed. By increasing the sample size without changing the number of σ units, the noise belt, as defined by the control limits, can be reduced. This is because the σ we talk about in control-chart work is a generic term for the deviation of whatever statistic we are plotting. Thus for averages, used in Table 21·1, $\sigma_{\bar{x}}$ or $\sigma/\sqrt{n}$ would be a more accurate description. As n increases, $\sigma_{\bar{x}}$ becomes smaller. With a narrower noise belt, there will be more power to detect a real signal. As long as this is accomplished without changing the number of $\sigma_{\bar{x}}$ units separating the central line and the control limits, the α risk will remain unchanged. Of course, σ itself is also an important factor in determining the width of the noise belt. If the rational subgroups are incorrectly selected to include assignable causes of variation, then σ will be too large, the limits will be too wide, and the system will be too insensitive.

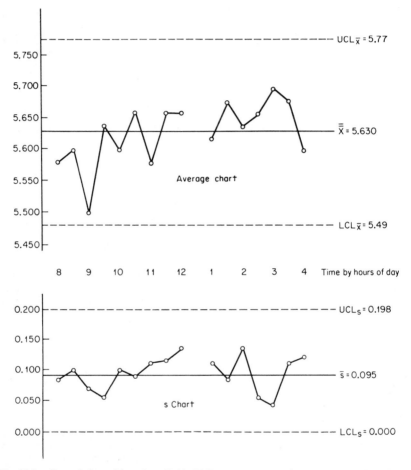

Fig. 21·2   Control charts (data from Table 21·9) for averages and sample standard deviations.

On the other hand, if all the sources of random variation are not allowed to enter into a rational subgroup, then σ will be too small, the limits will be too tight, and the system will be oversensitive.

In the Shewhart control-chart technique, it is assumed that there is a stable process operating at some standard level. By a stable process is meant one that yields product that is essentially uniform. The random variability, or noise level, is the variation we would encounter about the standard level in such a stable process. This variation is measured by the differences within the rational subgroups which we have formed from the data. These rational subgroups usually represent a short period of time during which assignable causes are not likely to occur, but in which all the sources of random error are free to have their effects. Rational subgroups are discussed in more detail in Sec. 21·3.

When a point plots outside the control limits, we must act on the basis that a signal has been received indicating that the process is "out of control," or, in other words, that there has been a shift in process level. Action is called for whenever a point exceeds the limits. The choice of the number of σ units (3σ, or 2σ, etc.) that the control limits are located away from the standard line determines how often we will draw an erroneous conclusion that there has been a shift in process level even though the process is running at standard. For 3σ limits, the risk of this type of error is 0.3 percent. For 2σ it is 5 percent. These probabilities must not be interpreted to mean that we can discount α percent of the "out of control" points. This interpretation might be fine for a historical review but will not work in a control program. Failure to act on these points would in effect widen the zone associated with the noise level and make it far more difficult for us to recognize a legitimate signal.

If we have limits that are too loose we will not detect process shifts. If we have limits that are too tight, we will be unable to run the process because of a continuous search for nonexisting trouble. If we are willing to assume that there exists only one standard level and that we want to know whenever there is a sizable shift from that level, the Shewhart control chart is an excellent tool.

PROBLEMS

21·1 Mark 200 chips, of equal kind and size, as indicated by the values of population A from Table 21·2. Place the chips in a container and mix them well. Draw one chip at a time, and after recording its value, return it to the container and repeat. After obtaining 50 such results, graph the data obtained showing the average of the samples, and control-limit lines at plus and minus three standard deviations as computed from sample data. (This will be a control chart for individuals.)

(a) Compare the sample average and sample standard deviation with the population values as listed in Table 21·2.

(b) What should be considered as the normal range of variability of the samples as drawn randomly from population A?

(c) Using the same 200 chips, collect a new set of 50 samples and graph the results. Compare the results of the second set of samples with the first set.

Table 21·2 Frequency Distributions for Sampling Experiments, and for Exercises in Control-chart Theory

Chip value	Population A	B	C	D
21	...	...	...	1
20	...	...	1	1
19	...	...	1	1
18	...	...	1	3
17	...	1	3	5
16	...	3	5	8
15	1	10	8	12
14	3	23	12	16
13	10	39	16	20
12	23	48	20	22
11	39	39	22	23
10	48	23	23	22
9	39	10	22	20
8	23	3	20	16
7	10	1	16	12
6	3	...	12	8
5	1	...	8	5
4	...	...	5	3
3	...	...	3	1
2	...	...	1	1
1	...	...	1	1
0	...	...	1	
Population	200	200	201	201
Mean value	10.0	12.0	10.0	11.0
Standard deviation	1.7	1.7	3.5	3.5

21·2 Collect from population A 20 sets of samples of size $n = 4$, by using 4 consecutive chips drawn at random, with replacement after each selection, and calculate the average and range of each sample of $n = 4$. [Note: $\sigma_X = \bar{R}/2.059$; $\text{UCL}_R = (2.28)\bar{R}$, $\text{LCL}_R = 0.0$.]

(a) Construct a control chart of the sample averages and ranges.
(b) Compare the results of the chart for averages with the chart for individuals.
(c) How does the chart for ranges give important information about the data collected?

21·3 Mark 200 chips with the values of population B as indicated in Table 21·2. After mixing the chips, select 20 random samples of $n = 4$, as was done in Prob. 21·2, and graph the results.

21·4 (a) Compare the results of the chart for averages from population B with the results from population A.

(b) Compare the results of the chart for ranges from the two populations.

21·5 Construct one large graph which contains the results of the 20 samples from Prob. 21·2, followed by the results of the 20 samples from Prob. 21.3.

(a) Discuss any change that takes place in the chart for averages.

(b) Should any change be noticeable in the chart for ranges? Explain.

21·6 Mark a set of chips to form population C and another set to form population D as listed in Table 21·2.

(a) After drawing 25 random samples of $n = 4$ from each population, C and D, construct graphs for the sample averages and ranges for each population.

(b) Compare the results from the two populations C and D.

(c) Compare the results from population B with the results from populations C and D.

21·7 On the same graph, plot the averages and ranges from samples of $n = 4$, using first the 20 samples from population A, followed by the samples from the other 3 populations.

(a) Comment on how the chart for averages changes.

(b) Comment on how the chart for ranges changes.

21·2 TO WHAT LEVELS DO WE CONTROL?

We use control charts for one of the following two distinctly different purposes:

1 To determine whether a process has achieved a state of statistical control. For this purpose, we collect data and test them against computed limits.
2 To tell whether current process is still at the base level. For this purpose, we compare current values against limits calculated from past experience or from a given standard.

It is rare that we can be satisfied with a process that is operating at any level whatever; in most cases the process must be set up to run to a standard. An exception is, for example, a novel process of chemical manufacture for which we have no previous record and for which we cannot set a standard of expected yield. Even here we will usually have pilot-plant studies which give some information about what to expect. In the pilot-plant operation we may want a consistent level and may not care what that level is as long as it is repeatable.

Standard values may be based on experience and prior data. Alternatively, we may set standards as aim points and base them on economic considerations of need and cost of production. Standards very often come from contract specifications and tolerances. Whatever may be the origin of the standards, we set values for μ and σ. From these set values, we determine in advance the numbers for the central line and limit lines for the control chart. For such a chart to be most successfully applied to a process, the value of the population standard deviation should be reliably known.

21·3 RATIONAL SUBGROUPS

A subgroup is a small group of measurements, often four or five, usually taken at nearly the same time. A rational subgroup is one in which assignable causes are not likely to be present but which contains all the effects of sources of random error. In general, subgroups should be chosen in a way that is likely to give the maximum chance for the measurements in each subgroup to be alike and the maximum chance for the subgroups to differ one from the other.

The most obvious rational basis for subgrouping is the order of production. Time need not be the governing criterion. For example, a rational subgroup in a batch-processing operation might be associated with each batch prepared. In this case, the sources of inherent variability, or within-batch effects, could include minor local fluctuations in concentration due to imperfect mixing, the presence of small amounts of impurities, uneven removal of decomposition products, plus the various components of measurement error. The assignable causes should be associated with batch-to-batch (subgroup-to-subgroup) variations. Mixing errors, changes in raw materials, different operators, and varying starting conditions are some of the assignable causes that might produce shifts between batches.

By choosing a rational subgroup that is so homogeneous that only a relatively few of the sources of random error can come into play, σ will be small. By choosing a rational subgroup that is large and includes not only all sources of random error but also some assignable causes, σ will be relatively large. The first method leads to excessive control, the second to insensitive control.

Many assignable causes produce effects that are real, but small and unimportant. In fact, it is often uneconomical to detect these effects, and much less economical to try to correct them. For example, a shift in agitation rate may cause an increase in the photographic density of film being processed in a particular machine. The increase may be 0.01 density units but only a shift as large as 0.10 may be photographically important. The 0.01 density shift is real and the cause is assignable to the change in agitation rate. But it is certainly not worth having a point plot out of control to inform us of this situation. No one would want to shut down the machine for a few hours to correct something that does not materially affect the final product. Hence, we often inflate the standard deviation by selecting a subgroup that includes some assignable causes it would not pay to eliminate. The width of the control limits may be regulated by the choice of the risk of looking for nonexistent trouble, the sample size, and the subgroups selected to furnish the estimate of σ.

21·4 SUBGROUP SELECTION

When we select subgroups from a population for preparing control charts, we must take care to choose the observations properly. We wish to obtain small lots of individuals which are representative of the population we intend to sample. The subgroups should be as homogeneous as possible but subject to all random variation we are not trying to eliminate. By selecting as subgroups lots with small within-group variation, and by basing control limits (where no standards are given) on this small variation, we will be able to detect long-term changes in the population. If the subgroups were instead to have large within-group variability, we would not be able except in extreme cases to detect the effects of assignable causes on the mean levels of the lots.

Examples of rational subgroups are:

1 In machine production of parts, a set of five items taken at a single time from the production line.
2 In the continuous production of a chemical, a set of four samples of the chemical taken from the line over a short period of time.
3 In a printing operation, three consecutive sheets from the press.
4 In the batch production of steel, four samples, each sample being taken at the same time from a different part of the batch container.

From this kind of sampling procedure, we can get two different kinds of information about the process. The means of the subgroups, compared with each other, tell us how the general level of the process is changing; the standard deviations (or the ranges) of the subgroups tell us how the within-group variability of the process is changing. Alternatively, a change in the subgroup means represents a long-time process change; a change in the subgroup standard deviations represents a short-time process change. We may decide to take, in a three-shift industrial operation, five samples during each shift. The *means* of the subgroups are related to the variation in production *between* shifts; the *standard deviations* of the subgroups are related to the variation in production within shifts.

In a situation where measurements are difficult to make, we may decide to select as subgroups sets of four analyses of the same sample; this is often done in chemical assay. Here a change in the *mean* of the subgroup of four estimates indicates a change in the sample itself; a change in the *range* of analyses indicates a change in the measurement process. In this application, it is essential that the several measurements made by the analyst should be independent. Often special techniques, such as "blind" analyses, are needed to provide this assurance.

To display the two kinds of information, control charts based on subgroups consist of two graphs. One is usually a plot of the averages of the subgroup values, called an $\bar{X}$ chart; the other is a plot of the standard deviations or ranges of the subgroups. The subgroup standard deviation is a preferable measure of the subgroup variability if the subgroup sample size is larger than 12; range becomes an inefficient estimate as sample size increases so that a cutoff at 12 is recommended. Since for large values of n the calculation of the range involves the use of only two numbers from the sample set, we run the risk of considerable error, and actually make use of only a small fraction of the available data. Hence if n is greater than 11, it is better to use the standard deviation.

21·5 DATA COLLECTION FOR CONTROL CHARTS BASED ON SUBGROUPS

In obtaining data for variables control charts based on rational subgroups, we need to make decisions about the following:

1 The characteristic of the process we are studying. The characteristics should be easily measurable and preferably significant as related to the end use of the product. We would like to obtain the data as rapidly as possible to avoid undue lag between the time of taking the sample and plotting the resulting observation. Thus we will be able to make changes in the process, if needed, so as to reduce the production of scrap.
2 The nature of the subgroups.
3 The sampling sequence. Ordinarily a periodic inspection schedule is adopted, based in part on inspection costs and experience with the process.
4 The measurement method, which should be clearly defined to reduce measurement errors.

We wish to make control charts for the operation of a printing press. We will use ink thickness as the response variable and estimate this by means of a reflectance measurement on a set of standard patches. We take 25 sheets from the press run, approximately 1 sheet per 100. We will make 5 measurements on each sheet. We can choose to make the subgroup a set of measurements (1) across the sheet; (2) along the sheet; (3) at random places on the sheet. Our decision about the selection of the subgroup depends on our estimate of the character of the significant within-sheet variation. We would select as a subgroup that set of measurements which would include only sources of variation we do *not* want to detect. The computed $\bar{X}$ values would indicate the average ink film thickness for each sheet; the R values would estimate within-sheet variation.

A filling machine for a drug product has three different outlets; volume is the response variable. We may make control charts based on:

1 A rational subgroup as one container from each of the three filling heads. $\bar{X}$ would estimate the time-to-time variation of the performance of the filling machine; R would estimate the difference among the outlets.

2 A rational subgroup as four successive containers from each outlet. In this case we would have a control chart for each outlet. $\bar{X}$ now would estimate the long-term average level for each outlet and R would estimate the short-term variation for each outlet.

Unless we have reason to believe that the three filling heads are very similar and might not become clogged individually, we would probably obtain a more realistic picture of the machine operation by using the second method of choosing the subgroups.

In a study of the variation in chemical analysis results of 10 different chemical laboratories, the same material was sent to each for analysis. (Note in this example that we can use control charts to deal with problems from conventional industrial processes.) Among the possible alternative methods of selecting subgroups, one which involves several analysts rather than one would give a better picture of the working of each laboratory.

PROBLEMS

21·8 Assign values of 1 through 13 to the cards in a deck of playing cards. Shuffle the deck. Determine the average of successive sets of 4 cards. Repeat 10 times, and plot a frequency distribution of the means. What do you conclude? Did the process involve randomization?

21·9 What would be logical subgroups in each of the following:
(a) 1,000-ft roll of paper 10 ft wide
(b) 100 rolls of paper as described in (a)
(c) an open-hearth (batch) process of steel manufacture
(d) a furnace making steel on a continuous basis
(e) 4 machines producing nails

21·10 Consider a loom where cloth is produced as a long continuous web and is wound into a large roll.
(a) What type of control would be involved in taking the rational subgroups along the length of the roll?
(b) What variability would be under investigation if the rational subgroups were taken across the roll?

21·11 If we were concerned with a bottle-filling machine that utilized 5 filling heads at the same time, explain the type of control we would be investigating if we obtained our rational subgroups by:
(a) Using 1 bottle from each filling head ($n = 5$)
(b) Taking 3 consecutive bottles from head number 1 ($n = 3$)

21·12   Coal is delivered from the mine by a conveyor belt to coal cars for transportation to the breaker. Discuss the methods of obtaining rational subgroups from the coal, and the type of variation being investigated by each type of sampling.

21·13   In a test of the consistency of measurements of paper burst strength, 2 different, and supposedly uniform, sheets of aluminum foil were tested on each of 4 machines by each of 5 operators. Each operator tested each sample 20 times. Suggest various ways in which rational subgroups could be chosen. Discuss the implications associated with each way.

21·6 FINDING CONTROL-CHART LINES

In Table 21·3 we give formulas for locating the central line and three-sigma limit lines for six different variables control charts. We have arranged the formulas according to:

1   Whether or not standards are given
2   Whether R or s is to be used as a measure of subgroup variability
3   Subgroup sample size

To use this table, a decision must be made first on these three questions. After the appropriate data have then been collected, the control-chart lines are found from the formulas given in the table, by the use of values for the coefficients given in Tables A·6 and A·7. The table coefficients depend on the value of the subgroup sample size n; they are conversion factors which convert the population parameter (or sample statistic) to give limit lines at plus and minus three standard deviations from the mean.

When we find limit lines from the formulas as given in Table 21·3, we assume an α risk of 0.003. Much experience with control charts indicates that limit lines based on three sigma are practical in that they represent a desirable compromise in the effort to minimize both types of error—namely, that of failing to detect a process change and that of making unnecessary investigation of a process. In some circumstances, we will wish to assume other α risks and set the limit lines at different values of sigma.

If we want to assume $\alpha = 0.05$, we will use two-sigma control limits. For the chart of X or $\bar{X}$ values, we multiply the proper A or E factor by 2/3 and substitute this value in the required formula. To find two-sigma limits for a range or standard deviation chart, we use the following rules: For ranges,

$$\mathrm{LCL}_R = \bar{R} - 2/3(D_4\bar{R} - \bar{R})$$
$$\mathrm{UCL}_R = \bar{R} + 2/3(D_4\bar{R} - \bar{R})$$

For sample standard deviations,

$$\mathrm{LCL}_s = \bar{s} - 2/3(B_4\bar{s} - \bar{s})$$
$$\mathrm{UCL}_s = \bar{s} - 2/3(B_4\bar{s} - \bar{s})$$

Similar multiples can be used to obtain any desired set width for the limit lines. Whatever the width of the limits, the central line position is unchanged.

Table 21·3   Formulas for Control Charts for Variables

I   Standards given (μ and σ known)

Type of chart	Central line	UCL	LCL
1 $\bar{X}$ and R, based on small subgroups	For $\bar{X}$: μ For R: $d_2\sigma$	$\mu + A\sigma$ $D_2\sigma$	$\mu - A\sigma$ $D_1\sigma$
2 $\bar{X}$ and s, based on small subgroups	For $\bar{X}$: μ For s: $c_3\sigma$	$\mu + A\sigma$ $B_6\sigma$	$\mu - A\sigma$ $B_5\sigma$
3 $\bar{X}$ and s, based on large subgroups	For $\bar{X}$: μ For s: σ	$\mu + \dfrac{3\sigma}{\sqrt{n}}$ $\sigma + \dfrac{3\sigma}{\sqrt{2n}}$	$\mu - \dfrac{3\sigma}{\sqrt{n}}$ $\sigma - \dfrac{3\sigma}{\sqrt{2n}}$

II   No standard given (μ and σ unknown)

Type of chart	Central line	UCL	LCL
4 $\bar{X}$ and R, based on small subgroups	For $\bar{X}$: $\bar{\bar{X}}$ For R: $\bar{R}$	$\bar{\bar{X}} + A_2\bar{R}$ $D_4\bar{R}$	$\bar{\bar{X}} - A_2\bar{R}$ $D_3\bar{R}$
5 $\bar{X}$ and s, based on small subgroups	For $\bar{X}$: $\bar{\bar{X}}$ For s: $\bar{s}$	$\bar{\bar{X}} + A_3\bar{s}$ $B_8\bar{s}$	$\bar{\bar{X}} - A_3\bar{s}$ $B_7\bar{s}$
6 $\bar{X}$ and s, based on large subgroups	For $\bar{X}$: $\bar{\bar{X}}$ For s: $\bar{s}$	$\bar{\bar{X}} + \dfrac{3\bar{s}}{\sqrt{n}}$ $\bar{s} + \dfrac{3\bar{s}}{\sqrt{2n}}$	$\bar{\bar{X}} - \dfrac{3\bar{s}}{\sqrt{n}}$ $\bar{s} - \dfrac{3\bar{s}}{\sqrt{2n}}$

Note from the formulas given in Table 21·3 that control limits for sample means are equally spaced about the central line, but control limits for sample ranges and standard deviations are unsymmetrical. These observations are related to the Central Limit theorem, which applies to sample means but not to estimates of sample variability. Furthermore, we cannot possibly obtain negative ranges or negative standard deviations; for these numbers the lower limit is zero.

We proceed now to apply some of these formulas to different situations.

21·7 CONTROL TO STANDARDS

For a papermaking machine, we have a long history of satisfactory operation. Among the data we have on the process are values for the mean and standard deviation, for which the coded values are $\mu = 50$ and $\sigma = 5$. We set up another machine of the same kind as the first, and we want to bring this machine into control. We intend that the new machine will operate at the same level as the first, and with the same variability; thus in this situation we have standard values which we wish the new machine to meet.

We will prepare $\bar{X}$ and R charts based on a subgroup of four observations representing measurements across the sheet of paper. We can prepare the charts before making any measurements on the new process. The formulas are found in Table 21·3 as type 1. The central line for the $\bar{X}$ chart will be at $\mu = 50$. For ± 3-sigma limits and $n = 4$ we find the coefficient A in Tables A·6 and A·7 to be 1.50. Thus the control limits for $\bar{X}$ will be placed at $\mu \pm A\sigma = 50 \pm 1.50 \times 5 = 50 \pm 7.5$, or at 57.5 and 42.5. The central line for the R chart will be at $d_2\sigma = 2.06 \times 5 = 10.3$. The coefficient d_2 is a factor which converts σ_X to $\bar{R}$. UCL for the range chart is placed at $D_2\sigma$. LCL for the range chart is placed at $D_1\sigma$. Tables A·6 and A·7 give $D_2 = 4.698$ and $D_1 = 0$. Using these coefficients, we find that the $\text{UCL}_R = 4.698 \times 5 = 23.5$ and $\text{LCL}_R = 0$. The control charts with the calculated lines should be drawn by the student.

Control charts based on standards are used to help us decide whether or not the process is doing what we want it to do. After we have prepared the control charts as in the preceding, we obtain data about the process and plot the corresponding points. Thus we take subgroups of 4 from the new process and plot the values of $\bar{X}$ and R. If the first sample gives $\bar{X} = 46$, $R = 17$, both points fall within the limit lines. The inference is that we have no reason to suppose that the process is operating improperly. If $\bar{X}$ had been 60, we would have taken this value as indicating that the process was operating at too high a level, and we would have made an adjustment to the process. If R had been 24, we would have inferred that the across-the-sheet variability of the process was excessive and we would have looked for the cause.

If a series of points remains within the limit lines for the $\bar{X}$ chart, we will apply the methods of Chap. 18 and test for a run of significant length. If we find from such a test that the process is operating at too low a level, we will ask the operator to make a process change. The intent is to cause the plotted data to vary randomly about the aim. Similarly, if we find that a series of points in sequence plots above $\bar{R}$, we will test for significance here also. If we reject the null hypothesis we will look for the cause of excessive subgroup variability. When the cause is found, we will correct it. Our desire is to have the two control charts give evidence of a state of statistical control at the proper level. When we accomplish this result, we will have evidence that the new process is operating at the preset standards.

PROBLEMS

21·14 A manufacturer of large castings attempted to maintain an aimed-at distribution of quality for a certain operating characteristic. The coded

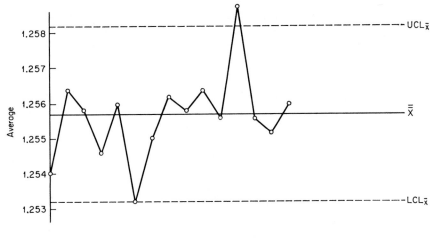

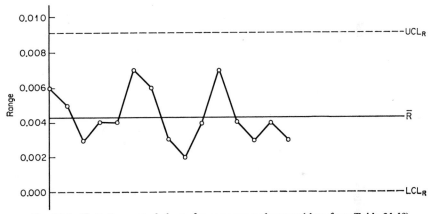

Fig. 21·3   Tentative control charts for averages and ranges (data from Table 21·10).

target values were $\mu = 35.00$ lb and $\sigma = 4.20$ lb. Table 21·4 gives observed values of $\bar{X}$ and s for daily samples of $n = 50$ observations each for 10 consecutive days.

(a) Construct a graph of the data and calculate the control limits, using 3-sigma limits.

(b) Comment on the past data, and discuss the possibility of being in control at the proper level.

21·15   It is desired to control the diameter of a product with respect to sample variations during each day. Samples of $n = 10$ are taken at definite intervals each day. The desired levels are $\mu = 0.200000$ in., with $\sigma = 0.00300$ in. Table 21·5 gives observed values of $\bar{X}$ and s for samples of 10 each, taken 10 times during a single day.

Table 21·4   Observed Values of $\bar{X}$ and s for 10 Consecutive Days

Sample number	Sample size, n	Average, $\bar{X}$	Standard deviation, s
1	50	35.2	5.20
2	50	36.4	4.78
3	50	33.3	3.80
4	50	34.8	4.50
5	50	33.5	4.05
6	50	34.0	4.35
7	50	34.5	4.98
8	50	33.6	5.00
9	50	32.8	3.29
10	50	34.6	3.77

Table 21·5   Observed Values of $\bar{X}$ and s for a Single Day

Sample number	Sample size, n	Average, $\bar{X}$	Standard deviation, s
1	10	0.19888	0.00350
2	10	0.20026	0.00310
3	10	0.19868	0.00330
4	10	0.20051	0.00335
5	10	0.20045	0.00300
6	10	0.20035	0.00315
7	10	0.19885	0.00325
8	10	0.20090	0.00380
9	10	0.19895	0.00390
10	10	0.20015	0.00355

(a) Construct charts of the data.
(b) Calculate the 3-sigma control limits and add them to the graphs.
(c) Is the process in control at the proper level? Discuss.

21·16   A manufacturer wishes to control variations in quality from lot to lot by taking a small sample from each lot. The target values are $\mu = 35.00$ lb and $\sigma = 4.20$ lb. Table 21·6 gives observed values of $\bar{X}$ and R for samples of $n = 5$, selected from 10 consecutive lots.
(a) Construct an $\bar{X}$ and R chart from the data.
(b) Place 3-sigma control limits on the charts, and discuss the process.

21·17   If the objective standards for greatest dimension of cereal flakes are $\mu = 35/64$ in. and $\sigma = 3/64$ in., is the process which produced the data in Table 21·7 in control at the proper level? Explain.

Table 21·6   Observed Values of $\bar{X}$ and R

Lot number	Sample size, n	Average, $\bar{X}$	Range, R
1	5	36.0	6.6
2	5	31.4	0.5
3	5	39.0	15.1
4	5	36.6	8.4
5	5	37.2	6.3
6	5	36.7	9.4
7	5	37.3	10.6
8	5	36.9	7.5
9	5	35.8	20.6
10	5	37.4	21.7

Table 21·7   Length of Greatest Dimension of a Cereal Flake, as Recorded in Order of Production, 4 Flakes per Sample, Recorded in $1/64$ in.

$\bar{X}$	R
33.2	13
31.2	14
33.0	15
35.0	4
32.0	6
34.0	4
35.2	7
32.3	10
32.0	4
33.4	6
35.3	7
34.8	6
34.7	14
32.3	6
35.0	10
35.2	14
35.0	7
34.7	4
33.0	12
33.0	11
32.5	8
32.7	16
34.0	5
34.6	8
34.5	9
32.7	6
32.0	8
31.6	6
32.0	9
32.5	2

21·18   The standard values for 10-point slugs should be 0.1400 in. with a standard deviation of 0.0008 in. Using the data contained in Table 21·8, set up control charts with 3-sigma limits, and discuss the process which produced the data.

Table 21·8   Measurement of 10-point Linotype Slugs Produced by One Machine for a Period of 2 Hr, Samples of Size n = 5, Selected at Random in Order of Production (10-point Slug Should be 0.1400 in.)

Subgroups	$\bar{X}$	R
1	0.1438	0.0020
2	0.1446	0.0041
3	0.1438	0.0025
4	0.1446	0.0021
5	0.1447	0.0035
6	0.1452	0.0022
7	0.1453	0.0004
8	0.1447	0.0029
9	0.1444	0.0029
10	0.1453	0.0035
11	0.1456	0.0023
12	0.1440	0.0048
13	0.1436	0.0012
14	0.1441	0.0027
15	0.1445	0.0022
16	0.1435	0.0029
17	0.1435	0.0035
18	0.1448	0.0022
19	0.1448	0.0017
20	0.1439	0.0024
21	0.1443	0.0035
22	0.1447	0.0023
23	0.1443	0.0037
24	0.1441	0.0032
25	0.1437	0.0035

21·8   CONTROL CHARTS—NO STANDARDS GIVEN

An entirely new chemical manufacturing process is being put into operation. Thus no previous data are available, and no standards can be applied to this operation. The question is this: Is this process in a state of statistical control; i.e., do data obtained from the process give us reason to believe that a constant cause system is at work?

We take 5 samples every $\frac{1}{2}$ hr from the process. The coded response variable is yield in percent. The data for an 8-hr shift were as given in Table 21·9. We intend to make a control chart based on the subgroup values of $\bar{X}$ and s; we have listed these values in the table. The average of the subgroup

Table 21·9 Yield in Percent of Chemical Manufacturing Process

Time	Yield					$\bar{X}$	s
8:00	5.7	5.5	5.5	5.6	5.6	5.58	0.084
8:30	5.5	5.5	5.6	5.7	5.7	5.60	0.100
9:00	5.5	5.4	5.6	5.5	5.5	5.50	0.071
9:30	5.6	5.6	5.6	5.7	5.7	5.64	0.055
10:00	5.7	5.7	5.5	5.5	5.6	5.60	0.100
10:30	5.7	5.5	5.7	5.7	5.7	5.66	0.089
11:00	5.7	5.7	5.5	5.5	5.5	5.58	0.110
11:30	5.6	5.6	5.8	5.7	5.6	5.66	0.114
12:00	5.6	5.6	5.5	5.8	5.8	5.66	0.134
1:00	5.6	5.6	5.6	5.8	5.5	5.62	0.110
1:30	5.6	5.6	5.7	5.7	5.8	5.68	0.084
2:00	5.5	5.8	5.5	5.7	5.7	5.64	0.134
2:30	5.7	5.7	5.7	5.6	5.6	5.66	0.055
3:00	5.7	5.7	5.7	5.7	5.8	5.72	0.042
3:30	5.6	5.6	5.8	5.8	5.6	5.68	0.110
4:00	5.6	5.5	5.5	5.8	5.5	5.60	0.122

averages $\bar{\bar{X}}$ is 5.63; the average of the subgroup standard deviations $\bar{s}$ is 0.095.

Since we have no available estimate of σ, we will base the control-chart limit lines on $\bar{s}$ and use the formulas in Table 21·3 for type 5. The $\bar{X}$ chart will have the central line at $\bar{\bar{X}} = 5.63$. The limit lines for this chart are at $\bar{\bar{X}} \pm A_3\bar{s}$. From Tables A·6 and A·7 we find for a sample of 5 that the value of A_3 is 1.427; thus the limit lines are placed at $5.63 \pm (1.427)(0.095) = 5.63 \pm 0.14$, or at 5.49 and 5.77. These limit lines are spaced at three sigma from the central line; the value of A_3 is based on the rule that the variability of a set of sample means changes according to the square root of the sample size, with a necessary correction for the bias associated with small values of n. For the s chart, the central line is placed at $\bar{s} = 0.095$. The limit lines are based on $\bar{s}$ and the coefficients B_7 and B_8, which are found to be 0.0 and 2.089 for $n = 5$ in Tables A·6 and A·7. Thus we find that $LCL_s = 0$ and $UCL_s = 0.198$. These limits are also at ± 3 sigma; the table coefficients are based on the probable values of a set of sample standard deviations when the mean standard deviation and the sample size are known.

21·9 CONTROL-CHART INTERPRETATION

In Fig. 21·2 we show the $\bar{X}$ and s charts for the data in Table 21·9. We note that all the points fall within the calculated limit lines; from this observation we have no reason to believe that the process is unstable. Beyond this, however, we observe the following: There is some indication of a gradual rise from the start of the day's operation through the afternoon, as shown by the plotted values of $\bar{X}$. Alternatively, there may be suggested a shift in process average after 9 A.M. If we were to find a similar pattern for data for succeeding days, we would probably look for the cause, since for this process we would like as high a yield as possible. The point we are making here is that control charts should be examined for every indication of unusual patterns. It is insufficient to be satisfied merely with the observation that "all points are in control." We see, as an additional example, that there is a suggestion of a trend in the s values from 9:30 to noon; we would look into this trend if we were to find it repeated as data from following days are added to the chart.

21·10 $\bar{X}$ AND R CHARTS—NO STANDARDS GIVEN

For subgroup samples of small n, i.e., less than 12, the range is a useful measure of within-lot variability; if n is large, the possible occurrence of extreme values of the response variable makes the range less useful. Since the range of a set of data is more easily calculated than the sample standard deviation, control charts are often based on the sample values of R. Such charts are constructed, in cases where no standards are applicable, by following the rules given for type 4 in Table 21·3.

For a new process of manufacturing microscope slides, we make a control chart using thickness as the response variable. A set of five consecutive slides is taken as a subgroup; a single measurement is made in a randomly chosen position on each slide. The short-time variability of the product will be shown by changes in the subgroup ranges; the long-time variability will be shown by changes in the subgroup means. We are assuming that for this new process no standards are available.

The results from measurements of the first subgroup were 1.255, 1.257, 1.253, 1.251, and 1.254. In recording the data as in Table 21·10 we have coded the data by subtracting from each observation the value 1.250 and multiplying the result by 1,000. Thus we can list the data as small whole numbers, facilitating both the recording operation and the calculations to follow. For a sequence of 15 subgroups, we show in Table 21·10 the observations, the $\bar{X}$ values, and the R values for each subgroup, and below the table the average $\bar{\bar{X}}$ of all the data and the average range $\bar{R}$.

We now calculate tentative upper and lower control limits for both plots, basing them on the value of $\bar{R}$. The basis of the calculations is the set of coefficients presented in Tables A·6 and A·7. The coefficients are derived from these considerations:

1 $\bar{R}$ estimates the variability of the process. The specific R values are related to the process variability and the subgroup sample size. Hence the control limits for the values of R are calculable from $\bar{R}$ by the use of the coefficients D_3 and D_4 in Tables A·6 and A·7. These coefficients are used as factors with $\bar{R}$ to give the lower and upper control limits for the range chart. Thus the table value for D_4 ($n = 5$) is 2.115; we multiply this coefficient by the uncoded value for $\bar{R}$, which is 0.0043, to get 0.0091 as

Sample	Thickness					$\bar{X}$	R
1	5	7	3	1	4	4.0	6
2	7	9	7	5	4	6.4	5
3	5	7	7	6	4	5.8	3
4	5	7	3	5	3	4.6	4
5	7	4	4	8	7	6.0	4
6	1	3	1	3	8	3.2	7
7	7	8	5	2	3	5.0	6
8	6	7	5	5	8	6.2	3
9	5	6	6	5	7	5.8	2
10	8	10	6	7	6	7.4	4
11	10	4	9	11	4	5.6	7
12	10	9	10	9	6	8.8	4
13	4	6	7	5	6	5.6	3
14	6	9	5	5	3	5.2	4
15	7	6	6	6	5	6.0	2

$\bar{\bar{X}} = 5.7$ $\bar{R} = 4.3$

Uncoded $\bar{\bar{X}} = 1.2557$ Uncoded $\bar{R} = 0.0043$

the UCL for the range chart. Since D_3 for $n = 5$ is zero, the value of the LCL for the ranges is also zero. The central line for the range chart we place at the value of $\bar{R}$.

2   We expect that the successive values of $\bar{X}$ will fluctuate in accordance with the process variability which is estimated by $\bar{R}$. The coefficient A_2 in Tables A·6 and A·7 is derived from the probable relationship between the expected standard deviation of a set of sample means and the average range of the set of samples. $A_2 \times \bar{R}$ approximates the three-sigma spacing for the upper and lower control limits from the value of $\bar{X}$. For this case, A_2 ($n = 5$) = 0.577. We multiply the value of the coefficient by $\bar{R}$ to get 0.0025. This last value is added to $\bar{\bar{X}}$ to obtain the value of UCL for the $\bar{X}$ chart; it is 1.2582. Similarly, LCL $= \bar{\bar{X}} - A_2\bar{R} = 1.2532$. We use the value of $\bar{\bar{X}}$ for the central line of the $\bar{X}$ chart.

In Fig. 21·3 we show the time plot of the data in Table 21·10, together with the tentative control-limit lines and the average lines. We calculated these lines on the assumption of a stable process. When we examine the plot, however, we see that two points are out of control—sample 6 falls on the LCL, and sample 12 plots above the UCL. We consider a sample which plots precisely on a limit line as being out of control, since we have in computing the values of the limits assumed that a value as far from the average as ± 3 sigma is an abnormal one.

The presence of 2 out-of-control values in a set of 15 indicates that we have not attained the necessary process stability. We have indication, in the two out-of-control values, that the process is afflicted with assignable causes of variation that must be found and corrected.

In the manufacture of glass slides, the temperature of the glass in the molten state has influence on the thickness of the finished slides. We check the temperature record and find that for the sample that plotted out of control on the high side the temperature was unusually low, and for the sample that plotted out of control on the low side the temperature was unusually high. We take the necessary steps to correct the temperature control. Supposing now that this assignable cause of abnormality has been corrected, we discard the data for the out-of-control points and recalculate $\bar{X}$ and $\bar{R}$ for the remaining 13 subgroups and the control-limit lines as well.

The new values are $\bar{R} = 0.0041$; $\bar{\bar{X}}$ is unchanged at 1.2557; $\text{UCL}_R = 0.0087$; $\text{LCL}_R = 0$; $\text{UCL}_{\bar{x}} = 1.2581$; $\text{LCL}_{\bar{x}} = 1.2533$. A replot of the data, using the new values, is shown in Fig. 21·4. As compared with Fig. 21·3,

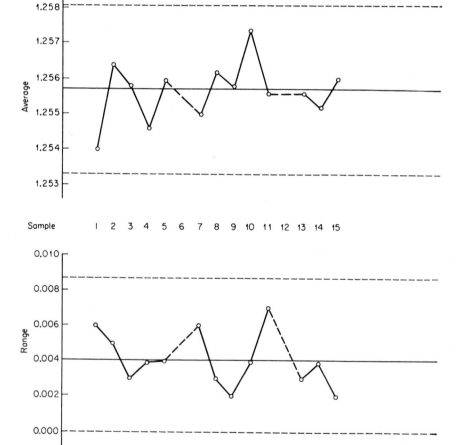

Fig. 21·4   Replot of data (Table 21·10) after detection of assignable causes for two points.

the control limits for both plots are slightly narrowed. The major difference is in the reduced variability of the $\bar{X}$ chart caused by the detection and elimination of an assignable cause of variation.

21·11 REVISION OF CONTROL LIMITS

Having now no evidence of process instability or abnormality, we extend forward the control-limit lines we have calculated and plot additional data as the production of microscope slides continues. In Table 21·11 are the $\bar{X}$ and R values for 15 more samples of $n = 5$ taken from the process.

Sample	$\bar{X}$	R
16	6.2	3
17	5.8	3
18	6.4	5
19	5.2	4
20	5.8	2
21	6.0	2
22	5.2	1
23	5.6	3
24	5.0	0
25	5.8	2
26	5.4	2
27	5.8	2
28	5.4	2
29	6.2	1
30	5.8	1

Table 21·11 Additional Data for Process Manufacturing Microscope Slides

$\bar{\bar{X}} = 5.7;$
$\bar{R} = 2.2$

We see that the process average is unchanged but that the value of $\bar{R}$ has fallen to 0.0022. This result we attribute to the better temperature control which has reduced process variability; thus we have in effect new process conditions. In Fig. 21·5 we plot the new data.

Based on the diminished value of $\bar{R}$ for the added set of 15 pieces of data, we calculate new values of the control-limit lines. These are substantially closer together than those we found from the original data. Such a result is typical of an improved process.

We see that the $\bar{X}$ chart gives evidence of good lot-to-lot consistency. There are, however, two points out of control on the R chart: sample 18 has an abnormally high range; sample 24 has a range of zero. Our next step should be to attempt to find the cause for both these points. We want to know the cause of the great range for sample 18 in order to reduce the within-lot variability. We want to find the cause for the production of a lot

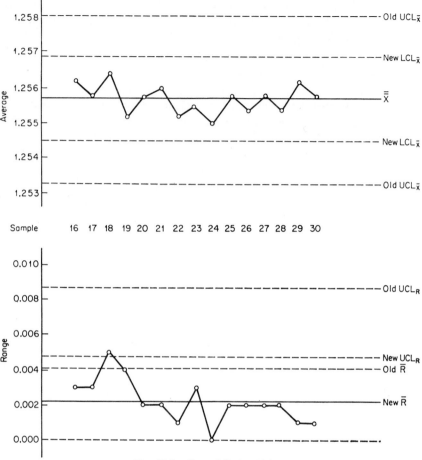

Fig. 21·5   Control-limit revision.

such as sample 24 with no variation whatever (although technically this point is not out of control in this case) because there may have been carelessness in the measurement of the slide thickness, or because there may have been a genuinely excellent lot and we wish to maintain the condition which led to this excellence.

This example is intended to make the point that, once control limits have been established for a stable process, the end of the study of the process is not yet in sight. Continual examination and reexamination of the process should be undertaken.

PROBLEMS

21·19   Successive lots of a crude organic solvent were analyzed for the concentration of a critical impurity. To check on the variability of the manufacturing process, each lot of solvent was sampled and three determinations were made on each sample. The coded results are given in Table 21·12.

Table 21·12   Impurity in Organic Solvent

Batch	Analysis		
	1	2	3
1	0.81	0.72	0.70
2	0.60	0.87	0.60
3	0.77	0.60	1.00
4	0.80	0.97	1.32
5	1.00	0.89	1.00
6	0.85	0.94	0.82
7	0.84	0.99	0.94
8	0.70	0.75	0.78
9	0.97	0.92	0.92
10	0.71	0.87	0.85
11	0.92	0.90	0.89
12	0.61	0.60	0.67
13	0.57	0.55	0.61
14	0.81	0.77	0.87
15	0.97	1.05	0.98
16	0.72	0.78	0.76
17	1.09	1.17	1.05
18	0.91	0.96	0.95
19	0.87	0.86	0.85
20	0.81	0.89	0.83

(a) Calculate and plot the values of $\bar{X}$ and R.
(b) Find tentative control limits for the data.
(c) Are there indications of instability in the process? If the answer is "yes," describe them.
(d) What information about the process is contained in the range chart? in the average chart?
(e) If the batch is a true solution, is one sample per batch sufficient? Explain.
(f) If the batch were not a true solution, should more than one sample be taken per batch? How should the sampling be done?
(g) If several samples were taken per batch, how should the data be plotted?
(h) For any points lying outside the tentative control limits, eliminate these from the data and calculate new limit lines. Plot these lines.
(i) The practice suggested in (h) is risky. Explain.

21·20   In the manufacture of a transistor for radio sets, measurements of depth in 0.001 in. were made on 5 items per hour. The results of the first

20 hr of production are given in the top part of Table 21·13, and the results of the next 7 hr are given in the lower part of the table.

Table 21·13   Measurements of Transistor Depth

Sample time	Measurement on each item of 5 items per hour					Average, $\bar{X}$	Range, R
1	40	43	37	34	35	37.8	9
2	38	43	43	45	46	43.0	8
3	39	33	47	48	39	41.2	15
4	43	41	37	38	40	39.8	6
5	42	42	45	35	36	40.0	10
6	36	44	43	36	37	39.2	8
7	42	47	37	42	38	41.2	10
8	43	37	45	37	38	40.0	8
9	41	42	47	40	40	42.0	7
10	42	37	45	40	32	39.2	13
11	37	47	42	37	35	39.6	12
12	37	46	42	42	40	41.4	9
13	42	42	39	41	42	41.2	3
14	37	45	44	37	40	40.6	8
15	44	42	43	35	44	41.6	9
16	40	32	44	45	41	40.4	13
17	37	37	42	43	41	40.0	6
18	37	42	42	45	43	41.8	8
19	42	42	43	40	35	40.4	8
20	36	42	40	39	37	38.8	6
21	42	44	40	38	43		
22	39	46	43	40	39		
23	40	45	42	39	37		
24	34	47	43	41	42		
25	38	45	41	37	41		
26	40	45	43	44	38		
27	45	45	37	38	40		

(a) Set up a control chart for averages and range, based on the results of the first 20 hr of production.

(b) If the process is in a state of statistical control, extend the control limits forward for future production.

(c) Plot the results of the next 7 hr of production to see if the process has remained the same as it was prior to control limits being set.

21·21   A manufacturer wishes to determine if his product exhibits a state of control. Table 21·14 gives the observed values of $\bar{X}$ and s for daily samples of $n = 50$ for 10 consecutive days. Does the process exhibit a state of control?

Table 21·14 Observed Values of $\bar{X}$ and s for 10 Days

Date	Sample size, n	Average, $\bar{X}$	Standard deviation, s
March 1	50	35.1	5.35
March 2	50	34.6	4.73
March 3	50	33.2	3.73
March 4	50	34.8	4.55
March 5	50	33.4	4.00
March 6	50	33.9	4.30
March 7	50	34.4	4.98
March 8	50	33.0	5.30
March 9	50	32.8	3.29
March 10	50	34.8	3.77

21·22   To determine whether there existed any assignable causes of variation in quality of a given product being shipped, samples were taken from 10 shipments. Table 21·15 gives the results of the inspection of the 10 shipments, using unequal sample sizes. Calculate 3-sigma control limits for the data and plot the data on an $\bar{X}$ and s chart using varying control limits.

Table 21·15 Inspection of 10 Shipments

Shipment	Sample size, n	Average, $\bar{X}$	Standard deviation, s
1	50	55.7	4.35
2	50	54.6	4.03
3	100	52.6	2.43
4	25	55.0	3.56
5	25	53.4	3.10
6	50	55.2	3.30
7	100	53.3	4.18
8	50	52.3	4.30
9	50	53.7	2.09
10	50	54.3	2.67

21·23   Using the data contained in Table 21·16, construct a control chart with 3-sigma limits.
(a) Is there evidence of a state of control with respect to average age at death in the given city?
(b) Is it reasonable to expect that a constant cause system might be operating with such a response variable? Explain.

	Table 21·16	Average Age at Death as Listed in Obituary of Large City. Samples of Size n = 6, Randomly Selected	

Date	Average age, $\bar{X}$	Range, R
Sept. 1	77.4	25
Sept. 2	72.0	27
Sept. 3	46.7	86
Sept. 4	72.0	31
Sept. 5	68.0	18
Sept. 8	64.5	63
Sept. 9	57.9	36
Sept. 10	53.6	67
Sept. 11	52.6	16
Sept. 12	73.0	43
Sept. 15	69.7	26
Sept. 16	70.6	63
Sept. 17	72.1	40
Sept. 18	69.8	28
Sept. 19	57.4	66
Sept. 22	66.0	48
Sept. 23	53.4	48
Sept. 24	72.6	24
Sept. 25	65.0	47
Sept. 26	89.3	2

21·24 Using the data from Table 21·17, determine whether there exist any assignable causes of variation in time required to do the selected task
(a) Use 3-sigma limits.
(b) Use 2-sigma limits.
(c) Discuss the results of the graphs in answer to parts (a) and (b).

21·25 Using the data contained in Table 21·18, determine if a constant cause system is in operation for the hours that the data were collected.

21·26 When a new four-spindle bottle-filling machine was put into operation, the amount of overfill was measured for the purpose of setting each of the spindles for correct filling. The number of milliliters of overfill is given in Table 21·19, for the first 10 sets of 4 bottles filled.
(a) Make a control chart, with tentative limits, for each spindle.
(b) Select as a subgroup each set, and make a control chart, with tentative control limits, for the entire set of data. (When we use such a method, we are dealing with "across-the-machine" variability.)
(c) What is the average amount of overfill for each spindle?
(d) Which is greater, the variability across the spindles or the variability within the spindles?
(e) What action should be taken to reduce all spindles to an average overfill of 0.00 milliliters?

Table 21·17   *Time, in Seconds, Required to Paste a Card Pocket into a Library Book. Samples of n = 3, Taken Approximately Every 10 Minutes (Each Subgroup Consists of Three Consecutive Books)*

Time	$\bar{X}$	R
10:00	23.3	3
10:10	21.0	3
10:20	22.3	5
10:30	22.3	2
10:40	24.3	3
10:50	23.0	4
11:00	21.3	4
11:10	22.0	4
11:20	23.3	1
11:30	23.6	1
11:40	23.3	4
11:50	25.0	2
2:00	23.0	2
2:10	22.3	2
2:20	21.3	3
2:30	22.6	4
2:40	23.6	2
2:50	23.6	3
3:00	22.6	5
3:10	24.6	4

Table 21·18   *Number of Cars Waiting for Traffic Light to Change, at Given City Intersection, Four Consecutive Light Changes Considered as Subgroup*

Time of day, P.M.	$\bar{X}$	R
4:30	1.8	3
4:40	3.0	5
4:50	2.0	3
5:00	2.8	2
5:10	2.0	2
5:20	3.5	5
5:30	2.0	2
5:40	2.5	3
5:50	2.8	2
6:00	2.5	4
6:10	3.0	4
6:20	3.2	4
6:30	2.0	3
6:40	1.2	1
6:50	2.0	2
7:00	1.2	3
7:10	1.5	1
7:20	1.2	1
7:30	1.5	3
7:40	1.8	3

Table 21·19 Amount of Overfill

Set number	Spindle			
	1	2	3	4
1	7.2	10.0	6.0	14.5
2	9.2	9.1	4.7	7.5
3	8.5	8.2	7.2	9.1
4	8.7	10.0	7.0	8.8
5	8.4	9.4	7.0	9.8
6	8.2	10.1	8.0	10.8
7	8.7	9.7	7.8	12.0
8	8.4	9.9	6.5	9.4
9	8.0	9.7	6.5	10.2
10	8.2	9.4	6.4	9.6

(f) After the machine has been readjusted, how should samples be taken to maintain a control chart on the filling operation?

QUESTIONS

21·1 If digits were selected from a table of random numbers and plotted in the form of a control chart for individuals, what pattern would you expect to find?

21·2 How would the following effects be demonstrated by a control chart:

(a) Progressive tool wear in an automatic lathe operation.

(b) A sudden change in the position of the tool in an automatic lathe operation.

(c) Oscillations in the temperature-regulating device in a chemical manufacturing process.

(d) An operator overcontrolling a filling-machine operation.

(e) An operator undercontrolling a filling-machine operation.

21·3 Since the presence of a single point outside control limits indicates trouble, how often would we be making the mistake of looking for trouble when there were none if we used the following limits: (a) 1-sigma; (b) 1½-sigma; (c) 2-sigma; (d) 2½-sigma; (e) 3-sigma?

21·4 Under what conditions would a control chart be valueless, or nearly so, in recording data from a process?

21·5 Classify the following situations according to whether they require attribute or variable control charts:

(a) taste test on a food

(b) tests of tire wear

(c) color in a printed label

(d) performance of an electric toaster

(e) tests of tailoring quality of men's suits. Briefly, justify your decision for each situation.

21·6   In graphing the results of sample data from a process, why is time of production an important factor in the analysis of the data?

21·7   Under what conditions might it be advisable to construct a frequency distribution out of the control-chart data?

21·8   Why is a long history of a process necessary in setting up a control chart on the process?

21·9   When graphing a time series of a process, what is to be considered as the "noise" of the system?

21·10   What constitutes a "signal" in the plotting of a process by control-chart techniques?

21·11   What is meant by a "constant cause system"?

21·12   What is the difference between "assignable" causes and "chance" causes?

21·13   What inference should be made when a single point on a control chart is found to fall outside the control limits?

21·14   When all points on a control chart fall inside the control limits, does it mean that there are no assignable causes present? Explain.

21·15   Whenever a pattern of points is detected in a control chart, the process is said to be "out of control." Explain.

21·16   What conditions must be met before a control chart for variables can be prepared?

21·17   What is the significance of the central line in a control chart for individuals?

21·18   Under what circumstances can a point fall outside 3-sigma limits when the point contributes to the limit calculations?

21·19   Under what circumstances should limits be set at other than 3-sigma spacing?

21·20   The control chart is a major tool in the control of a process. Why must the process be a repetitive one?

21·21   The first step in the preparation of a control chart is the collection of data to produce a history of the process. Why is this step necessary?

21·22   If a long time is required to obtain the information needed to plot a point on a control chart for individuals, what techniques might be used to strengthen the signal given by the control chart?

21·23   What is the meaning of the phrase "in a state of statistical control"?

21·24   As compared with plotting individuals, what advantages and disadvantages are associated with plots of averages?

21·25   What pattern of distribution must the parent population display if a control chart for averages is to be used? Explain.

21·26   What information is displayed by a control chart for averages? by a control chart for ranges?

21·27 When rational subgroups are used in control charts, two points are plotted for each sample. Explain.

21·28 As compared with a plot of sample standard deviations, what advantages and disadvantages are associated with plots of ranges?

21·29 Why should s, rather than R, be used for large sample sizes? Why is not the likelihood of inappropriately wide control limits about the same for both s and R?

21·30 Why should specification limits not be placed on a control chart for averages?

21·31 How would you proceed to calculate limits for 2-sigma rather than 3-sigma limits for a control chart based on $\bar{R}$? Under what circumstances would such a chart be preferred?

REFERENCES

7, 16, 18, 22, 26, 27, 28, 53, 61, 62, 82, 83, 86, 93, 98, 99, 101, 102, 109, 120

22 Acceptance Control Charts

For the control charts which we discussed in Chap. 21, we have assumed that there is only a single standard level about which a process varies. The central line defines this single process standard, and the control limits specify the region within which the process is expected to operate when only random causes of variation are present. When measurements of the process indicate that a point falls outside these limits, we assume that an assignable cause is operating and that we should take action to change the process conditions.

There are, however, many situations in which shifts in process level are unavoidable and are indeed to be expected since it would be uneconomic to try to remove their assignable causes. For example, when a new tool is placed in an automatic lathe, it is to be expected that there will be a change in process level. A similar result is likely when a new batch of raw materials is introduced into a chemical process. We have in such processes not a single process level but instead a family of acceptable process levels.

We can, if we like, calculate conventional control-chart limits to include these assignable causes of variation within the limit lines. The principle of the control chart is, however, violated by this procedure, and in widening the region between the limit lines we may overlook changes in the process other than those caused by the expected changes. In other situations, and these are very common, there is a single level at which the process is aimed, but tolerances permit acceptance of material falling within a reasonable distance of the aim level. For these situations, the concept of a single average line should also be replaced by the concept of a zone of acceptable quality.

22·1 ACCEPTABLE PROCESS LEVEL (APL)

Acceptance control charts may be considered as a combination of the Shewhart control charts and acceptance sampling for variables. Here we take a sample from the process and use it to test the hypothesis that there has been no shift in the process level which has been producing satisfactory quality. The alternative hypothesis is that the process has shifted to a level of unsatisfactory quality. We must therefore define what is meant by satisfactory quality. In other words, we will want to know how far the process level is allowed to shift and still be considered as being acceptable so that we would not reject the process too often. At what point is a shift in the process level to be considered as an unsatisfactory quality level? The answers to these questions should be found by referring to the tolerance statements in the specifications.

As very slight shifts in the process level will still allow for the manufacture of satisfactory quality, we will find it more reasonable to use a *band* of acceptable process levels rather than a single acceptable process level. Each edge of this band marks an acceptable process level (APL). A process which is centered on this edge of the band will have only an α risk of being rejected.

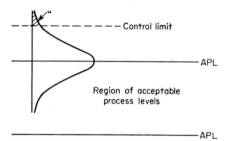

Region of acceptable
process levels

Fig. 22·1   Schematic of a process centered at the edge of the band of acceptable process levels.

If the process is centered within the region of the band, the risk of being rejected is even smaller. Figure 22·1 shows schematically such a band of acceptable process levels and the risk of rejection of the process when it is centered on the upper edge of the band.

The distribution curve about the APL, as in Fig. 22·1, is similar to that of the Shewhart system, except that it is centered on the edge of the acceptable process zone and not on the standard line. The distance from the APL (edge of the zone) to the control limit is $t_\alpha \sigma / \sqrt{n}$. So far we have located the process level associated with a quality level we wish to accept almost all $(1 - \alpha)$ the time it is being run.

22·2 REJECTABLE PROCESS LEVEL (RPL)

An acceptance control chart is also concerned with the process level which should be rejected almost all the time $(1 - \beta)$ it is run because it is associated with undesirable quality level. This is called the rejectable process level. A process centered at this level of quality should have a high probability of being rejected. Processes that are centered beyond the RPL will have an even greater probability of being rejected. This is as it should be, for as the process quality level grows poorer, there should be a greater probability of detection of poor quality. Figure 22·2 shows a schematic of the RPL lines added to the acceptance control technique.

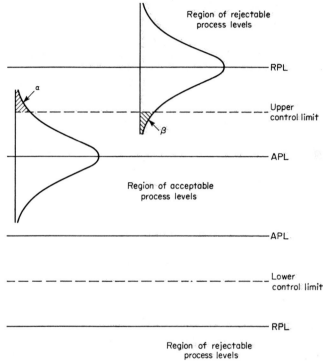

Fig. 22·2   Schematic of acceptance control chart, showing locations of APL and RPL.

The distance from the control-limit line to the RPL is $t_\beta \sigma / \sqrt{n}$. Just as there is an α value associated with the APL, there is a β value associated with the RPL. The curve centered about the RPL, in Fig. 22·2, is not considered in the Shewhart system. Yet, often the β risk of accepting poor quality is more important than the α risk of rejecting good quality. It is in the consideration of both types of risks that the acceptance control chart approaches the technique of acceptance sampling for variables.

When the process level lies somewhere between the APL and the RPL, in the indifference zone, the process is producing borderline quality (see Fig. 22·3). As long as the process level is in the indifference zone, the quality is not so good as we would like it nor so bad as that which we would call

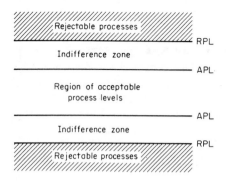

Fig. 22·3 APL and RPL lines in relation to process levels for acceptable, rejectable, and indifferent quality.

rejectable or poor material. The width of the indifference zone is related to the particular process and the risks we are willing to take in connection with the process. If we wish to make the indifference zone narrower, that is, have APL and RPL move closer to each other, we will have to use a larger sample size. As in any sampling system, there is no single sharp border between good and poor quality. The indifference zone between good and poor quality is an area of indifferent quality and can be made as small as we wish if we are willing to use large enough sample sizes.

Located between the APL and the RPL is the acceptance control limit (ACL), which is the action criterion. When the acceptance control chart is displayed for use, only the ACL lines are placed on the chart, along with the sample size to be used. The operator of the process need not be concerned with the APL and RPL; these are used by the engineer to select the proper control system. The acceptance control chart will be as easy to use, on the part of the operator, as the Shewhart charts, but the engineer will gain better insight into the process through its use.

22·3 PUTTING THE CHARTS TO USE

Like acceptance sampling plans, acceptance control charts involve four values:

1 An APL with its associated α risk
2 An RPL with its associated β risk
3 An ACL which is the signal for action to be taken
4 The sample size

To establish the chart limits, we need to know the variability of the process. This variability is measured by the subgroup standard deviation, which we will know either from the history of the process or from a special study of the process. (When sigma is not known, it can be estimated by the usual $\bar{R}$ or $\bar{s}$ techniques.)

In Fig. 22·4 we show the model for acceptance control charts, with the first three items in the list above represented. The subgroup sample size will determine the spread of the distributions sketched in the figure.

In relationship to the standard value, the locations of the APL, the RPL, and the ACL are fixed by the values of the normal distribution for the specified values of α, β, and sample size. Note that these are all one-tailed values of *t*, if the acceptable process zone has any reasonable width, since

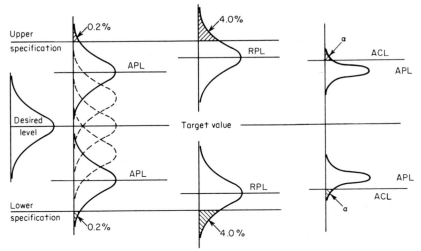

Fig. 22·4   Model for acceptance control charts.

we are invariably interested in an area under only one tail of any distribution. The acceptance control chart is defined by any two of the four values in the list above (plus the subgroup standard deviation), and the other two can then be calculated.

22·4 CONTROL CHARTS BASED ON APL AND RPL

A packaging machine is to be adjusted to make the net weight of the contents 32.0 ± 0.5 oz. The process is acceptable if less than 0.2 percent of the packages fall outside the specifications. The process is rejectable if more than 4.0 percent of the packages fall outside the specifications. We set α at 0.005 and β at 0.10.

We have in the preceding statements specified the APL and the RPL. If sigma is known to be 0.08, we have sufficient information to prepare the acceptance control chart. Four steps are required in the calculations; the results of each step are shown in Fig. 22·5.

1　The specification limits USL and LSL are set by the statement of the problem at 32.5 and 31.5.

2　The APL are determined by the requirement that at most 0.2 percent of the weights may be outside specifications. From Table A·1 we find $t_{0.002}$ to be 2.878. This factor, used with the known standard deviation, gives 0.23. The upper APL will be 0.23 *below* the upper specification limit; the lower APL will be 0.23 *above* the lower specification limit.

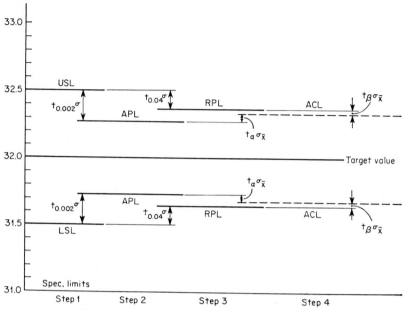

Fig. 22·5   Steps in the preparation of acceptance control charts.

We have found the limits on the process mean that will ensure that only 0.2 percent of the total distribution will fall beyond specifications. The upper APL = 32.27; the lower APL = 31.73.

3   The RPL are determined in a manner similar to that for the APL, except that we use $t_{0.04}$ because of the problem statement that the rejectable level is based on the production of more than 4 percent of out-of-specification weights. The table value for $t_{0.04}$ is 1.75; this, multiplied by the standard deviation, gives 0.14. Thus each RPL is placed 0.14 within the specification limit. The upper RPL = 32.36; the lower RPL = 31.64.

4   We find the ACL from the specified α and β risks and from the results in steps 2 and 3 above. We calculate the term

$$\text{APL} + \frac{t_\alpha}{t_\alpha + t_\beta} (\text{RPL} - \text{APL}) =$$

$$\text{APL} + \frac{2.57}{2.57 + 1.28} (32.36 - 32.27) = 32.27 + 0.06 = 32.33$$

This value determines the distance that the ACL should be placed outside the previously determined APL limits. This equation is derived (as shown in Fig. 22·5) from the fact that the difference between the RPL and APL is $t_\alpha \sigma_{\bar{x}} + t_\beta \sigma_{\bar{x}}$, while the distance between the ACL and the APL is only $t_\alpha \sigma_{\bar{x}}$.

We have determined the limit lines. We have yet to find the sample size needed to give us protection of the APL and the RPL at the values of α and β we have agreed upon. We find n from

$$n = \left[\frac{(t_\alpha + t_\beta)\sigma}{\text{RPL} - \text{APL}}\right]^2$$

$$= \left[\frac{(1.28 + 2.57)(0.08)}{0.09}\right]^2 = 11.6$$

Thus we would use a sample size of 12. This solution comes about from $\text{RPL} - \text{APL} = t_\alpha\sigma_{\bar{x}} + t_\beta\sigma_{\bar{x}} = (t_\alpha + t_\beta)\sigma/\sqrt{n}$.

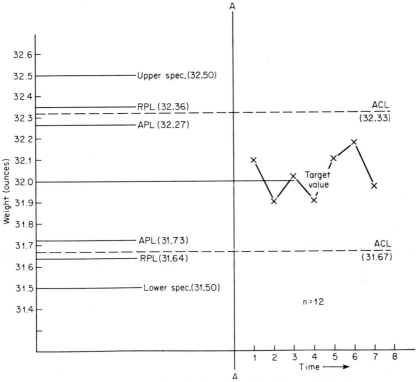

Fig. 22·6   Acceptance chart based on specified APL and RPL values.

Figure 22·6 shows the placement of the APL and the RPL lines in the determination of the location of the ACL lines. The part of the control chart in Fig. 22·6 which falls to the left of line AA is shown only to illustrate the method of obtaining the ACL values. This part of the chart is not shown to the chart user, the operator of the process upon which the chart is being placed. The operator will see the chart only as it appears to the right of the line AA in Fig. 22·6. All the user of the chart needs to control his process is a time scale to tell him where to plot the results of his samples, and the action lines (ACL) to indicate when the process level has shifted too far. When an average plots outside the ACL lines, the process is rejected and the operator or engineer must take some action to return the process average to its desired level.

A control chart for the ranges (or standard deviations) of each sample should be kept along with the acceptance control chart to observe the inherent stability of the process, since an underlying assumption is that the process is in control at each level.

22·5 CONTROL CHART BASED ON SPECIFIED APL AND n

In certain manufacturing operations such as the coating of photographic film, the making of paper, or the coating of a metal with a copper layer, the uniformity of the coating may be of more importance than the actual thickness. When a product has a uniform thickness, within reasonable units of a standard thickness, the operational steps which follow the coating operation can be adjusted to the coating thickness. On the other hand, variations in thickness of a single setting are difficult to adjust for, and cause problems in the various stages of manufacture which follow.

In a continuous papermaking process, we are concerned with paper thickness. The sources of variability are variation across the width of the roll, along the length of the roll, and in the measurement of thickness. The variation in thickness across the roll is much larger than along the roll. Therefore, we shall slit the paper into long narrow strips in the direction of the length of the roll. A rational subgroup is a single lengthwise strip of paper. We subdivide each strip into four small test squares for measuring purposes. After keeping results of such tests for a reasonably long period of time, we find the inherent variability to be $\sigma = 0.02$. (This value will include along-the-roll and measurement variability.) We now agree, based on requirements of the paper, that strips having a mean deviation from the grand average for the roll of less than ± 0.03 would be accepted. We agree to take an α risk of 0.05 of rejection at this level of uniformity. We have now fixed the APL (and its associated risk) and the sample size. The APL will be ± 0.03 from the grand average of thickness, and the sample size will be $n = 4$.

The ACL can now be found by using the formula

$$\text{ACL} = \text{APL} + \frac{t_\alpha \sigma}{\sqrt{n}}$$

The upper ACL line is found by using APL $= +0.03$ and adding the positive term $t_\alpha \sigma / \sqrt{n}$, while the lower ACL is found by using APL $= -0.03$ and adding the negative term.

$$\text{Upper ACL} = \text{APL} + \frac{t_\alpha \sigma}{\sqrt{n}} = +0.03 + \frac{(1.645)(0.02)}{\sqrt{4}} = +0.05$$

$$\text{Lower ACL} = \text{APL} - \frac{t_\alpha \sigma}{\sqrt{n}} = -0.03 - \frac{(1.645)(0.02)}{\sqrt{4}} = -0.05$$

The RPL values can also be determined once we have set the value of the β risk to be associated with the RPL. If we agree to use 0.05 for the β risk, then the values of RPL can be found as follows:

$$\text{Upper PRL} = \text{ACL} + \frac{t_\beta \sigma}{\sqrt{n}} = +0.05 + \frac{(1.645)(0.02)}{4} = +0.07$$

$$\text{Lower RPL} = \text{ACL} - \frac{t_\beta \sigma}{\sqrt{n}} = -0.05 - \frac{(1.645)(0.02)}{4} = -0.07$$

If the distance between APL and RPL (the indifference zone) is considered too great, then a larger sample size might be used in place of $n = 4$. In our example, the width of the indifference zone is $0.07 - 0.03 = 0.04$ units. If we were to use a sample size of $n = 16$, we could cut this distance in half. If a more uniform thickness is desired, we could move the values of APL closer to 0.00, say 0.01, and then with $n = 4$, the value of ACL would be ± 0.03 and RPL would be ± 0.05. This interval could also be decreased by using a larger sample size.

The control chart would then contain the same information as when APL and RPL were specified, that is, the central line and the control-limit lines. The central line in this example would be the value 0.00 (the grand average of the thickness the roll of paper coded to 0.00), and the plotted points would be the size of the average deviation from the grand mean for the sample of $n = 4$.

22·6 CONTROL CHART BASED ON SPECIFIED RPL AND n

There are situations where the size of the sample to be taken is determined on the basis of a management decision. We are told to take samples of $n = 4$ from a process where we are concerned with overall length determinations. This might be an automatic lathe, a wire-cutting device, or a book-stitching machine. We find that we must change the cutting tool every so often because the tool becomes dull with use. There will be variation due to resetting the new tool, from the raw material, from the machine, and from measurement. (Assume $\sigma = 0.02$.)

The specifications on the length of the material being cut read 6.50 ± 0.13 units. When a new tool is placed in the cutter, we are told to have only a 0.01 risk of acceptance of a situation which would permit 0.5 percent or more of the production units falling outside the specification limits. We have already given the sample size n and the β risk to be used, and enough information to locate RPL. The values of RPL are found as follows:

$$\text{RPL} = \text{specifications} \pm t_{0.005}\sigma$$
$$\text{RPL} = (6.50 \pm 0.13) \mp (2.576)(0.02)$$
$$\text{Upper RPL} = 6.58 \quad \text{and} \quad \text{lower RPL} = 6.42$$

The ACL values can be obtained, since we know that $\beta = 0.01$, by moving in from the location of the RPL lines by the appropriate amount. This will be $\text{RPL} - t_\beta\sigma/\sqrt{n}$.

$$\text{ACL} = \text{RPL} \pm \frac{t_{0.01}(0.02)}{\sqrt{4}}$$

$$\text{Upper ACL} = 6.58 - \frac{(2.326)(0.02)}{\sqrt{4}} = 6.56$$

$$\text{Lower ACL} = 6.42 + \frac{(2.326)(0.02)}{\sqrt{4}} = 6.44$$

If we assume an α risk of 0.05, the location of the APL can be found by moving in from the ACL lines by an amount equal to $1.645\sigma/\sqrt{n}$. In our example,

$$\text{Upper APL} = \text{upper ACL} - \frac{(1.645)(0.02)}{\sqrt{4}} = 6.56 - 0.02 = 6.54$$

$$\text{Lower APL} = \text{lower ACL} + \frac{(1.645)(0.02)}{\sqrt{4}} = 6.44 + 0.02 = 6.46$$

As before, the control chart as set up for the operator to use would contain only the center line and the control-limit lines. The location of the APL and RPL lines are for the added insight of the quality-control man or the engineer on the process.

22·7 COMMENTS ON ACCEPTANCE CONTROL CHARTS

The Shewhart control-chart system was based on the assumption that there is a stable process operating at some standard level. The control limits were designed to indicate the presence of an assignable cause and thereby show that the process level has shifted. The Shewhart charts were not designed to be used as acceptance tools but were to indicate a state of statistical control or the lack of control. Acceptance control charts are designed to apply to situations in which the control chart is used to decide whether to accept or reject a process on the basis of whether or not the product will meet specifications.

Whereas the Shewhart charts call for a single stable process level, acceptance charts make use of a zone around the central line which allows for small shifts in the process level so long as the process will yield acceptable product. As long as the process level does not change to a point where it is producing some undesirable percentage of product exceeding the product specifications, the process is considered as "in control."

The acceptance control chart is essentially a variables sampling plan when it uses the ACL lines to determine whether to accept or reject a process for a given sample size and standard deviation. The location of the ACL lines is related to the value of the APL and RPL, along with their associated risks. It has control features in that a chart of standard deviations or ranges is also kept, and the averages of each sample are plotted at regular intervals of production rather than being allowed to accumulate into inspection lots.

For further discussions of acceptance control charts, the references listed at the end of this chapter should be consulted. The articles by Richard A. Freund are especially useful. They give examples of applications to batch and continuous process operations.

PROBLEMS

22·1   The specifications for a paper-cutting machine call for the cut lengths to be 5.0 in. $\pm \frac{1}{16}$ in. As long as less than 0.1 percent of the cut lengths fall outside the specification limits, the process is to be considered acceptable, but whenever 5 percent or more of the cut lengths fall outside the specification limits, the process is to be considered as incorrectly set and adjustments made.

(a) Using an α risk $= 0.05$ and a β risk $= 0.05$, calculate the acceptance control limits and the sample size required, if the standard deviation of the process is 0.009.

(b) Using an $\alpha = 0.05$ and a $\beta = 0.10$, calculate the acceptance control limits and sample size for a process whose standard deviation is also 0.009.

(c) Using an $\alpha = 0.05$, and a $\beta = 0.05$, calculate the acceptance control limits and sample size for a process whose standard deviation is 0.02.

(d) Compare the answers to the above parts of the problem. What conclusions do you draw?

22·2   In the manufacture of metal pins, the specification for length of the pins is given as 1.1250 ± 0.0625. Samples of size $n = 4$ are taken every 30 min. The inherent variability of the process is found to be $\sigma = 0.0039$. If we wish to have a β risk of 0.01 of acceptance of the process when 0.5 percent or more of the individual pins fall outside specification limits, where should the acceptance control limits be set? (Assume an α risk of 5 percent.)

22·3   A process is to operate with the following specifications for fill control: 54.0 ± 0.6 grams. If less than 0.4 percent of fills fall outside specifications the process is to be considered satisfactory. The process is to be rejected if more than 6.0 percent of the fills fall outside specifications. If sigma is known to be 0.10, and we select α of 0.05 and β of 0.05,

(a) Where will the APL lines be placed?

(b) Where will the RPL values fall?

(c) Where should the ACL lines be placed?

(d) What size sample should be used?

QUESTIONS

22·1 Why might slight shifts in the process level be overlooked in some control-chart situations?

22·2 List the four values which must be specified in acceptance sampling procedures.

22·3 What is meant by the "indifference zone" in the acceptance control chart?

22·4 What is the underlying assumption in the Shewhart control-chart system?

22·5 Is the acceptable process level a line or a zone? Explain.

22·6 Why is the value of α stated for the APL?

22·7 How does the value of α change as the process level falls on either side of the APL?

22·8 Does the Shewhart control-chart system involve both α and β risks? Explain.

22·9 What is the distance from the control limits to (a) the APL; (b) the RPL?

22·10 If sigma for a given process is not known, how might it be obtained?

22·11 What is the role played by the sample size in acceptance control charts?

22·12 What is the usual sample size in the Shewhart control chart for variables?

22·13 Why is a control chart for ranges or standard deviations kept along with the acceptance control chart?

22·14 There are four requisites for acceptance control charts. How many of the four must be specified to define the acceptance control-chart system completely, if sigma is known?

22·15 How does the acceptance control chart resemble a variables sampling plan?

22·16 If we specify that which is good quality and that which is poor quality, do we have enough information to set up an acceptance control chart? Explain.

22·17 How do we know if the process level has shifted to a new position which will produce poor quality?

22·18 How can we control, or influence, the width of the indifference zone?

REFERENCES

7, 22, 27, 35, 49, 50, 53, 62

23 Attribute Control Charts

One limitation of control charts for variables is that they are charts for quality characteristics that can be measured and expressed in numbers. Many quality characteristics can be observed only as *attributes*, i.e., by classifying each item inspected into one of two classes, either conforming or nonconforming to the specifications. We talk about defectives and non-defectives because this is the most commonly used classification, but the techniques we are about to discuss apply for any such dichotomization.

When we are dealing with quality characteristics which have been observed as attributes, and we wish to chart the inspection results as control charts, we are to prepare an *attribute* control chart. The concept of percent defective is one with which most production people are already familiar.

In attribute inspection, a production unit is considered defective when it is qualitatively unsatisfactory. It may be a workable unit but contain a major defect, or it may contain too many minor defects to allow it to be sold. Items may have a measurable characteristic but may still be classified as defective if they are too large or too small with respect to a testing gauge such as a "go no-go" gauge.

When we use a control chart for defectives rather than an $\bar{X}$ and R chart, we utilize only the information that the item is or is not within a specified range of values, and not its actual value. For this reason, the sample size must be larger to provide a test of the same power. Information for an attribute chart is, however, easier to obtain and may still be economical.

23·1 TYPES OF ATTRIBUTE CONTROL CHARTS

There are many somewhat different types of attribute control charts. Most of them, however, can be classified as being one of two kinds:

1 Control charts based on *fraction defective* (*p* charts). Fraction defective charts are used when we score each inspected item as good or bad, without regard to different degrees of defectiveness. Examples are: a container leaks or does not; in a drop test a bag breaks or withstands the test; a package contains the correct weight or not. Thus the use of *p* charts implies that there are only two possibilities of the results of inspection— good or bad. Furthermore, it assumes there is a different probability of occurrence of bad and good items. If *p* represents the probability of encountering a defective item, then $1 - p$, or *q*, represents the probability of encountering a satisfactory item. For any specified number of items in a lot, there is an expected binomial distribution described by the expression $(p + q)^n$, as discussed in Chap. 20.

2 Control charts based on a count of *defects per unit* (*c* charts). This kind of control chart is used where the items being inspected are complex in nature, so that numerous defects are possible. In inspecting the operation of a printing press, for example, we may examine a single sheet from the press. An infinite number of defects is conceivable, of different types. Misregistration, dirt specks, color and tone variations, smears, etc., may occur in innumerable places on the sheet. Here we cannot easily classify a sample merely as defective or not. The reasons are that the occurrence of a perfect sheet—one with no defects whatever—is very improbable and that there are many degrees of defectiveness. Other examples of similar kinds of production involve automobiles, fabrics, books, radios, etc. The pattern which underlies this kind of situation is the Poisson distribution; it applies when the probability of occurrence of a defect is small in comparison with the total number of possible defects that might in theory occur. In an automobile, the number of different defects that *might* occur is very large indeed; the number that actually occurs is small in comparison.

PROBLEM

23·1 Each of the following involves attribute inspection. Determine whether a *p* chart or a *c* chart would be appropriate:

(a) blemishes in finished furniture
(b) taste of canned soup
(c) seams in garments
(d) spots in photographic film
(e) quality of color photographic prints
(f) breakdowns of insulation in electrical wire

(g) fit of pins in motor connecting rods

(h) knots in lumber

(i) performance of television set

23·2 COLLECTING DATA FOR p CHARTS

In the accumulation of data for the preparation of p charts, a systematic method of recording information is needed. A reasonable history of the process is necessary. As we take sample lots from the process, we record

NO. IN SAMPLE				QUALITY CONTROL $\bar{p}$ DATA SHEET						PART NO.	
SUB. GROUP	PART NAME									MACHINE NO.	
PCS. PER HOUR	OPERATION NO. AND DESCRIPTION									DEPT. NO.	SHEET NO.

Fig. 23·1   Form for the collection of data for p charts.

time of taking the sample; sample size n; number of defectives d; the fraction defective d/n; the kind of defect observed in each defective item. A form such as that shown in Fig. 23·1 is convenient for making the record.

23·3 CONTROL LINES FOR p CHARTS

The average fraction defective $\bar{p}$ is an estimate of the population fraction defective p'. The value of $\bar{p}$ can be found by averaging the p values for all the sample lots if the sample size is held constant. If the value of n varies, we find $\bar{p}$ by dividing the total number of defective items for all the lots by the total number of inspected items.

Trial-limit lines are based on the standard deviation of the expected distribution of the p values. This standard deviation s_p is given by

$$s_p = \sqrt{\frac{\bar{p}(1 - \bar{p})}{n}} \tag{23·1}$$

where $\bar{p}$ is the average fraction defective found in the inspected samples during the base period, and n is the sample size for the particular sample being plotted. Note that this means that there actually exists a family of control limits, each having the sample $\bar{p}$ value but dependent on the sample size. We may use shortcuts to keep this dependency from becoming unwieldy. Sample sizes can be grouped and represented by an average value for each group. Otherwise each observation may be divided by its own standard deviation $\sqrt{\bar{p}(1 - \bar{p})/n}$ to give a standardized value (p/σ_p) that may be plotted in terms of the number of standard deviations away from the central line with $\pm 3\sigma_p$ as control limits.

A cherry-pitting machine is used in a canning operation. We wish to make a p chart to describe the performance of such a machine. We instruct the inspector to select random lots of 200 cherries each as they come from the machine and to count the number of unpitted cherries in each lot of 200. Here each unpitted cherry is a defective item. The data are recorded in Table 23·1.

Table 23·1   Data Sheet for p Chart, Cherry-pitting Machine

Sample number	Time	Sample size	Number defective	Fraction defective, %
1	8:00	200	6	3.0
2	8:30	200	3	1.5
3	9:00	200	0	0.0
4	9:30	200	7	3.5
5	10:00	200	11	5.5
6	10:30	200	8	4.0
7	11:00	200	10	5.0
8	11:30	200	7	3.5
9	12:00	200	5	2.5
10	1:00	200	4	2.0
11	1:30	200	10	5.0
12	2:00	200	2	1.0
13	2:30	200	9	4.5
14	3:00	200	7	3.5
15	3:30	200	6	3.0
16	4:00	200	8	4.0
17	4:30	200	4	2.0
18	5:00	200	0	0.0
19	5:30	200	10	5.0
20	6:00	200	8	4.0
Total		4,000	125	

The control-chart lines are calculated as follows:

1   Find the average fraction defective $\bar{p}$. Since the sample lots are all the same size, we can divide the total number of defectives (125) by the total

number inspected (4,000) and get 0.0313 (3.13 percent). This value is used for the central line of the chart.

2 From Eq. (23·1) find the estimate of the standard deviation of p:

$$s_p = \sqrt{\frac{\bar{p}(1 - \bar{p})}{n}} = \sqrt{\frac{0.0313 \times 0.9687}{200}} = 0.0125$$

3 Decide on the distance between the central line and the control-limit lines. Here we will use $\pm 3s$; therefore, the limit lines will be at $\pm(3 \times 0.0125)$ from the mean. The upper control limit is then 0.0375 above the mean, or at 0.688. When we similarly calculate the lower limit at 0.0375 below the mean, we obtain a negative number. Since a negative value for unpitted cherries is meaningless, we place the lower control limit at zero. In using such a control chart, we can never find a point out of control on the low

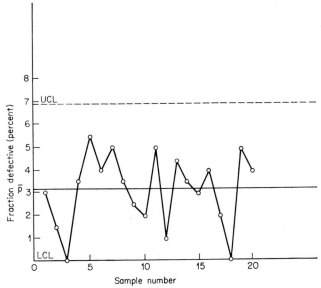

Fig. 23·2 Control chart for fraction defective (cherry-pitting machine).

side—not a fortunate situation. We would like to be able to detect unusually good performance of the machine and will not be able to do so. It is probable that an increase in sample size would have given a value other than zero for the lower control limit.

4 Prepare the control chart as in Fig. 23·2.

23·4 INTERPRETING p CHARTS

The principles involved in interpreting control charts were described in Chap. 21; these principles apply equally here. As data are added, we examine the chart for points out of control, for runs, and for trends. If a point falls out of control on the high side, the inspection indicates unusually poor quality; the cause should be found and corrected.

If, in collecting data for a p chart, a few very high points are found, and *if an assignable cause for these points is discoverable*, it is reasonable to ignore these points in calculating $\bar{p}$ and the limit lines. Furthermore, if assignable causes can be found for out-of-control points as the charting continues, and if these causes can be eliminated, the control-limit lines and $\bar{p}$ should be recalculated from time to time. Thus we can revise the control chart in the direction of better and more uniform performance as we identify and correct the causes of poor production. Before making the limits too tight, however, we ask the same type of questions asked for the acceptance control chart for variables. What are the specification requirements? Is there meaning in setting a zone of acceptable quality? The control limit is in effect the acceptance number of an attribute sampling plan, so that an APL and RPL can be readily determined.

A point out of control on the low side deserves as much attention as one out of control in the other direction. A point which plots below the lower control limit represents unusually good production. It is natural to want to continue such production; therefore, the cause of this out-of-control sample should be discovered so that the process may be improved. Alternatively, an abnormally low value of p may indicate a faulty inspection process which also should be corrected.

We should examine p charts for runs and trends in order to associate nonrandomness with other information about the process. In a three-shift operation, for example, runs or trends may be associated with a specific shift, or with the beginning or end of the operation of each shift. Such evidence is valuable in attempting to improve the process. If one shift produces an unusually good or an unusually poor product, the mere knowledge of this fact often encourages an improvement in performance.

PROBLEMS

23·2 Construct a p chart from the data in Table 23·2 (sample size was 50 throughout).

23·3 Find the control limits for a p chart (based on $\pm 3s$) for the following situations:
(a) number of lots $= 15$; $n = 75$; total defectives $= 82$
(b) number of lots $= 30$; $n = 75$; total defectives $= 164$
(c) number of lots $= 15$; $n = 75$; total defectives $= 102$

23·4 From your answers to Prob. 23·3,
(a) How are the control limits affected by the number of sample lots for the same fraction defective?
(b) How are the control limits affected by an increase in the number of defective items for the same number of lots?

Table 23·2 Data for p Chart Preparation

Sample number	Number defective	Sample number	Number defective
1	2	11	4
2	4	12	1
3	4	13	3
4	3	14	0
5	7	15	4
6	3	16	3
7	5	17	1
8	1	18	6
9	3	19	5
10	4	20	0

23·5 p CHARTS WHEN n VARIES

The situations discussed so far in this chapter have involved a constant value for sample size n. In some inspection problems, however, it may be impossible or inconvenient to maintain a fixed sample size. Production or inspection rates may vary in such a way as to necessitate selecting samples of different sizes.

Such situations present a real problem in p chart calculation, since Eq. (23·1) includes n as a factor in the determination of the placement of the limit lines, even though $\bar{p}$ is constant (or assumed constant). Since n appears in Eq. (23·1) in the denominator of the fraction, as n increases s decreases; therefore, as n increases the limit lines move closer together. Conversely, the use of smaller sample sizes causes the limit lines to move farther apart. A p chart is therefore more sensitive with larger sample sizes; furthermore, we should really calculate new limit lines for every new value of n. To do this, we rewrite Eq. (23·1) as

$$s_p = \frac{\sqrt{\bar{p}(1-\bar{p})}}{\sqrt{n}} \qquad (23·2)$$

and calculate the numerator independent of the value of $\sqrt{n}$. Since the value of $\bar{p}$, in Table 23·3, is 0.0905, the value of Eq. (23·2) becomes $0.288/\sqrt{n}$.

It is now easy to calculate the control limits for the various sample sizes, as only the denominator changes as sample size changes. If we are using ± 3 standard deviations as the control limits, we then multiply each value of s_p by 3 and plot the control limits on the graph as indicated in Fig. 23·3.

Figure 23·3 contains the values of p for the data from Table 23·3. It also contains a scale of control-limit values for various sample sizes. As each value of p is graphed, the location of the control limits for the sample size used is obtained from the vertical scale on the left side of the graph. Note that the lower control-limit line is 0.0 in all cases, because $\bar{p} - 3s_p$ is negative

Table 23·3   Data for p Chart (Variable
Sample Size)

Sample number	Sample size	Number defective	Fraction defective
1	40	4	0.100
2	30	2	0.067
3	60	4	0.067
4	60	7	0.117
5	50	5	0.100
6	40	2	0.050
7	10	3	0.300
8	50	6	0.120
9	40	3	0.075
10	40	2	0.050
Total	420	38	

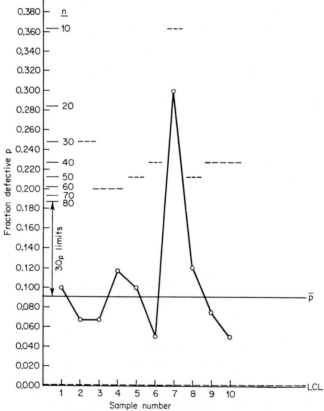

Fig. 23·3   Control chart with varying sample size.

for all values of n used. As negative values of p are meaningless, we set the control limit at 0.00.

As new points are added to the control chart, we can place the control-limit lines on the plotted point by merely knowing the sample size used and referring to the scale of values on the left side of the chart. As long as the sample sizes used remain multiples of 10, the control-limit lines are easily located.

If the sample size varies in such a way that any value of n might be selected, then we may use a zone system for figuring control-limit lines. We employ an average value of n over a range of sample sizes. We then use Eq. (23·2), the value of the denominator changing with the $\bar{n}$ which best fits the sample size. In Table 23·4 we show how to group various sample sizes and the value of $\bar{n}$ to be used for each case.

Table 23·4 Zone System of Grouping Sample Sizes for Use with p Charts of Varying Sample Size

Sample size	Use $\bar{n}$ of	$\sqrt{n}$
20–29	25	5
30–41	36	6
42–55	49	7
56–79	64	8
80–119	100	10
120–169	144	12
170–234	196	14
235–344	289	17
345–479	400	20
480–689	576	24
690–999	841	29
1,000–1,469	1,225	35
1,470–1,999	1,764	42
2,000–2,999	2,500	50

Since we assume that $\bar{p}$ remains constant, the value of the numerator in Eq. (23·2) need be computed only once. Then as the value of n changes, this constant can be divided by the appropriate $\sqrt{n}$. By this method the value of s_p for any sample size is quickly computed, and the relation of any point to its appropriate control limits is easily seen. This approximate method of dealing with variable sample sizes is especially useful when values of n are large, since then the control-limit lines do not vary so much as with small samples. If the sample sizes are generally small, we do better to calculate control limits for every sample.

We have illustrated the use of p charts in production control. They have, however, other uses, especially in connection with the description of the quality of lots of material coming into a plant. Inspection by sample lots, followed by plotting the data in p chart form, provides an exceptionally clear representation of what is being supplied. If different suppliers are involved, a separate *p* chart for each supplier gives a basis for a meaningful comparison of the material from different suppliers. The form in which the data appear on the chart (percent or fraction defective) is easily understood even by those who have had no training in statistical quality control. *p* charts are therefore especially useful in reporting to management personnel.

PROBLEMS

23·5 Plastic bags are sealed on 3 sides in an automatic machine which turns out several thousand bags per hour. As the bags are delivered to the shipping department, a handful is taken and inspected for leaks. If a bag has a single leak, it is considered defective. Prepare a *p* chart from the data in Table 23·5.

Table 23·5 Results of Inspection of Plastic Bags

Sample number	Bags inspected	Bags defective
1	15	1
2	17	2
3	14	0
4	16	1
5	10	0
6	17	2
7	15	0
8	11	3
9	16	1
10	17	0
11	15	0
12	13	1

23·6 In the production of vacuum tubes, the results of inspection for 9 production days were as given in Table 23·6. Construct a *p* chart from these data, using 2-sigma limits.

23·7 A photographic finishing plant kept records of the number of prints that had to be reprinted because they did not meet company standards. Each rejected print was considered as defective. Table 23·7 is the record for 1 day's production.

(a) Using the average sample size as a basis, prepare a *p* chart for the day's production, with the limit lines set at ± 2 sigma.

(b) Assume that all points out of control have led to the discovery of an assignable cause and that corrective action has been taken to prevent the recurrence of the effects of the cause. Recalculate the limits.

Table 23·6   Results of Inspection of Vacuum Tubes

Day	Inspected	Defective
1	2,000	64
2	1,500	37
3	1,750	40
4	2,000	53
5	2,000	45
6	1,700	38
7	1,500	40
8	1,500	25
9	1,750	21

Table 23·7   Results of Inspection of Photographic Prints

Hour	Prints inspected	Prints rejected
1	45	1
2	37	0
3	52	4
4	47	3
5	36	0
6	59	2
7	38	0
8	42	2
9	33	9
10	38	3
11	57	2
12	57	2
13	60	3

(c) Recalculate the control limits for all points which fall close to the control limits.

(d) Discuss the level of quality with respect to the time of day.

23·6 c CHARTS

The second major kind of attribute control chart requires that the inspector make a count of the *number of defects per unit*. In this context, a unit may be a single item, or it may be group of several items. The inspection may be for a single kind of defect or for several kinds of defects.

The data sheet for recording the inspection results should be similar to that shown in Fig. 23·1, with the provision of space for the identification of the type of defect. It is well to make the unit size sufficiently large so that the lower control limit, once it is calculated, is appreciably greater than zero; thus it will be possible to detect out-of-control points in the low direction. On the other hand, the use of an excessively large unit size may allow too long a trial period before we can detect the presence of trouble. The effect of a very large unit size is not unlike increasing the value for the acceptance number and working at a higher acceptable quality level.

The following are involved in the preparation of c charts:

1 Number of defects in each lot c.
2 Average number of defects $\bar{c}$; this value is found by dividing the total number of defects in all the units by the number of units. It is computed for the base period.
3 The standard deviation of the set of c values s_c; it is equal to $\sqrt{\bar{c}}$. Control-limit lines are customarily placed at a distance of $3s_c$ from the average line at $\bar{c}$.

In the manufacture of photographic film, defects are caused by the presence of dirt particles, radioactive fallout, and nonuniformities in the coating process. Since we cannot classify a film sample as merely "defective," a count of defects is made and a c chart prepared. The unit size may be a 100-ft length of film. Each unit is uniformly exposed and processed, and examined for defects. The c value for each unit is the total number of defects in that unit; the record would be as in Table 23·8. (Note that here each inspection unit is 100 ft of film and that the defects are per unit, not per foot, of film. Note further that in c charts there exists no concept of sample size as

Table 23·8   Data for the Preparation of c Chart

Lot number	Number of defects per lot
1	52
2	31
3	27
4	64
5	89
6	72
7	40
8	32
9	78
10	54
11	57
12	46
13	58
14	67
15	71

such but merely an agreed-upon unit of inspection. For this reason there is no variation of control chart limits with size. If, however, the results of several units are averaged together to form a point, we must use $s_c = \sqrt{\bar{c}/n}$, where n = number of units being averaged.) From the total number of defects (838) we find the value of $\bar{c}$ by dividing by the number of lots (15): $\bar{c} = 62.6$. The value of $s_c = \sqrt{\bar{c}} = \sqrt{62.6} = 7.9$. The control limits would be placed at $\pm 3 \times 7.9$ from 62.6, or at 86.3 and 38.9. Inspection of the data in Table 23·8 shows that lots 2, 3, and 8 will fall out of control on the low side, and lot 5 out of control on the high side. The process cannot, from these data, be considered to be stable, and an effort must be made to improve the process.

23·7 DEMERIT CHARTS

It often happens that a simple count of defects per lot is not very descriptive of the quality of the lot. Such a situation occurs where all defects are not of the same seriousness. In book production, minor misprints are defects and should be kept under control; a binding defect, however, may make the book unusable, and an error in collation may be only slightly less damaging.

In this type of production, it is useful to weight the observed defects according to some reasonable scale of values and then to prepare control charts based on the weighted values. Such control charts are called *demerit* control charts.

The choice of weighting values is to some extent arbitrary. It must be based on nonstatistical judgments related to estimates of how serious the defect will seem to the buyer. Some examples of weighting methods in use are

1   *Minor defect*: the customer will always accept the item, weighting 1.
    Major defect: the customer will complain about the product but will accept it, weighting 3.
    Critical defect: the customer will reject the product, weighting 6.
2   *Very minor defect*: weighting 1.
    Minor defect: weighting 3.
    Serious defect: weighting 10.
    Major defect: weighting 20.
3   *Not serious defect*: weighting 1.
    Moderately serious defect: weighting 10.
    Serious defect: weighting 50.
    Very serious defect: weighting 100.

Probably the best method of arriving at weighting values for demerit control charts is to examine customers' complaints and rejected items, and by trial to devise a demerit system that agrees well with the past history of the product. A system that does not weight the very serious defects considerably more than the minor ones will mask the appearance of the former, since only a few minor defects score as heavily as a critical one. On the other hand, a system that puts too much weight on a critical defect will hide the effect of the less serious defects, since a single appearance of the former shifts the score so violently.

The data sheet for collecting demerit records should provide space for sample number; type and number of defects; total demerits per lot. As in the preparation of c charts, the sample unit size is invariable, is chosen by agreement, and is of convenient size.

In demerit control charts, D specifies the number of demerits per inspection unit; the average number of demerits $\bar{D}$ is the central line on the chart; the estimate of the standard deviation of the D values is

$$s_D = \sqrt{w_1{}^2 c_1 + w_2{}^2 c_2 + \cdots + w_k{}^2 c_k}$$

where w_i stands for the weight used for each classification of defect and c_i the average number of defects per unit of each classification.

Suppose, for example, that we are concerned with four types of defects and we assign values to the defects as $w_1 = 1$, $w_2 = 5$, $w_3 = 50$, and $w_4 = 100$. After inspecting a given unit of production, we find that the average numbers of defects of each type are $\bar{c}_1 = 100$, $\bar{c}_2 = 20$, $\bar{c}_3 = 2$, and $\bar{c}_4 = 1$. The central line on the demerit chart will be $\bar{D} = w_1 c_1 + w_2 c_2 + w_3 c_3 + w_4 c_4$. This will be $1(100) + 5(20) + 50(2) + 100(1)$ and will equal a $\bar{D}$ of 400. Then

$$s_D = \sqrt{(1)^2(100) + (5)^2(20) + (50)^2(2) + (100)^2(1)}$$
$$= \sqrt{15{,}600}$$
$$= 125$$

The control limits, if we select $+$ three standard deviation limits, will then be placed at $400 \pm 3(125)$.

As in any control chart, the central line and the control-limit lines are indications of what the process is doing. If the process is in control, but the level of quality is not as desired, then some change must be made which affects process quality. Data must then be newly collected and a new control chart prepared, based on the newly obtained data.

PROBLEMS

23·8   In the manufacture of clear plastic hose, bubbles appear from time to time. Although they do not affect the performance of the hose, they are considered as defects since customers believe them to be weak spots. The hose is inspected in 10-ft lengths, and the number of bubbles counted. Use the data contained in Table 23·9 to construct a c chart. Is there evidence of a constant cause system?

Table 23·9   Defects in Hose

Footage of plastic	Number of defects
1–10	0
11–20	1
21–30	1
31–40	2
41–50	1
51–60	4
61–70	3
71–80	3
81–90	5
91–100	4
101–110	4
111–120	6

23·9   Assembled transistor radios are tested for loose connections, scratches, nicks, smears, etc., and the number of defects and their types are recorded. For one production line, the number of defects per radio was recorded as in Table 23·10. Construct a 3-sigma control chart for defects and comment on the possibility of a constant cause system operating.

23·10   Because of the high number of defects found in the radios inspected in Prob. 23·9, a demerit chart was prepared for the same radios. Demerits were assigned as follows:

(a) Minor defect (can be removed in cleaning and does not affect performance): 1 demerit.

(b) Major defect (harms the appearance but not the operation of the set): 3 demerits.

(c) Critical defect (makes the radio poor in performance): 9 demerits.

Table 23·10   Defects in Radios

Radio sample number	Defects found
1	11
2	10
3	9
4	12
5	7
6	4
7	13
8	2
9	0
10	11
11	9
12	8
13	11
14	9
15	7

Using the data contained in Table 23·11, prepare a demerit chart on the radios and compare the results with the answer to Prob. 23·9.

Table 23·11   Demerit Chart for Radios

Radio number	Total defects	Minor defects	Major defects	Critical defects
1	11	9	2	0
2	10	9	1	0
3	9	9	0	0
4	12	11	1	0
5	7	5	1	1
6	4	4	0	0
7	13	13	0	0
8	2	1	0	1
9	0	0	0	0
10	11	9	2	0
11	9	9	0	0
12	8	7	1	0
13	11	10	1	0
14	9	9	0	0
15	7	5	1	1

23·8 SUMMARY OF THREE-SIGMA LIMITS FOR ATTRIBUTE CONTROL CHARTS

See summary presented in tabular form on facing page.

Type of control chart	Central line	LCL	UCL
Fraction defective			
1. Standard given	p'	$p' - 3\sqrt{p'(1-p')/n}$	$p' + 3\sqrt{p'(1-p')/n}$
2. No standard given	$\bar{p}$	$\bar{p} - 3\sqrt{\bar{p}(1-\bar{p})/n}$	$\bar{p} + 3\sqrt{\bar{p}(1-\bar{p})/n}$
Number defective			
1. Standard given	$p'n$	$p'n - 3\sqrt{p'n(1-p')}$	$p'n + 3\sqrt{p'n(1-p')}$
2. No standard given	$\bar{p}n$	$\bar{p}n - 3\sqrt{\bar{p}n(1-\bar{p})}$	$\bar{p}n + 3\sqrt{\bar{p}n(1-\bar{p})}$
Number of defects			
1. Standard given	c'	$c' - 3\sqrt{c'}$	$c' + 3\sqrt{c'}$
2. No standard given	$\bar{c}$	$\bar{c} - 3\sqrt{\bar{c}}$	$\bar{c} + 3\sqrt{\bar{c}}$
Number of defects per unit area.			
1. Standard given	c'/n	$c'/n - (3/n)\sqrt{c'}$	$c'/n + (3/n)\sqrt{c'}$
2. No standard given	$\bar{c}/n$	$\bar{c}/n - (3/n)\sqrt{\bar{c}}$	$\bar{c}/n + (3/n)\sqrt{\bar{c}}$
Demerit chart	$\bar{D}$	$\bar{D} - 3\sqrt{w_1^2 c_1 + w_2^2 c_2 + \cdots + w_k^2 c_k}$	$\bar{D} + 3\sqrt{w_1^2 c_1 + w_2^2 c_2 + \cdots + w_k^2 c_k}$

n = number of units averaged per point.

NOTE: If 2-sigma limits are required, use 2 in place of 3 in the above formulas. In these formulas, p' and c' symbolize fraction defective and number of defects which are known to be characteristic of the process from past history. If they are considered as specifications instead, these values may be used, according to the formulas, for the construction of specification charts.

23·1   How do attribute control charts help to keep a process at a specified level of quality?

23·2   What do we mean when we say "process is in control"?

23·3   When a point falls outside the upper control limit on the attribute control chart, we should investigate the cause. Why?

23·4   When a point falls outside the lower control limit on the attribute chart, we should investigate the cause. Why?

23·5   After data have been collected, control limits are calculated on the past performance of the process. Under what conditions might we be able to revise the limits prior to projecting them ahead for control of production as it is being run?

23·6   If the sample sizes were not all the same in a p chart, how might it influence the control chart?

23·7   If the sample sizes were about the same size but not exactly the same size, what procedure might you follow in calculating limits?

23·8   Make a list of 5 situations where you believe a p chart would be the right type of control chart to use.

23·9   What is the probability distribution which underlies the p chart?

23·10   What is the probability distribution which underlies the c chart?

23·11   Under what conditions would the c chart be the proper type of control chart to use in controlling a process?

23·12   Make a list of 5 situations where a c chart should be used as the control chart.

23·13   When reading a p chart or a c chart, what do we look for?

23·14   Under what conditions should the demerit chart be used in place of the c chart?

REFERENCES
16, 27, 28, 49, 53, 69, 82, 98, 99, 120

24 Modifications of Control Charts for Variables

In Chap. 21, 22, and 23 we discussed the control charts which are fundamental and commonly employed. We shall now consider more recent developments of control charts and show the relationship of these to the Shewhart chart. Some modern methods make use of rational subgroups. Others estimate sigma using adjacent points rather than subgroups. Two of the techniques to be discussed will use methods other than arithmetic averages and are intended to detect small variations.

We have already described in Chap. 22 the acceptance control chart method, which is applicable to processes for which a single operating level is hardly to be expected and also relates control to specification requirements. This technique is designed to accept or reject the process, something the Shewhart technique was not designed to do. Whereas Shewhart charts permit us to determine whether a state of statistical control exists, the acceptance charts use the appropriate sample size to let us accept or reject the process as its output relates to specifications, with specified α and β risks.

24·1 CONTROL CHARTS BASED ON A MOVING AVERAGE

It is often true that taking rational subgroups from a process is impracticable. In some situations the time required to measure a single observation is so great that repeat observations cannot be considered. Therefore, the use of a rational subgroup as previously defined is not possible. Often the manufacture of a single item takes hours or days, as in the production of cameras, radios, or automobiles. In other situations the analysis of the product takes a long time, as in the evaluation of photographic film.

In still other situations the use of rational subgroups, based on measurements within a narrow time interval, is hardly sensible because the variation of the process over a short time, or within a process batch, is very small. Control limits based on the short-time (or within-batch) range may be impossibly tight for the estimation of the long-time (or between-batch) variation of the process. In the mass processing of photographic prints, it is characteristic of the process that short-time variation is small. The process is, however, likely to show wide swings in long-time performance; it is this kind of variation that must be kept within reasonable bounds. Conventional Shewhart $\bar{X}$ and R control chart limits cannot sensibly be based on ranges found from short-time variations because these are much smaller than the random variations encountered over extended periods of time. Such a situation is common in the chemical manufacturing industries.

Table 24·1   Calculations of Control Limits for Individuals

Batch number	Coded observations, X	Range of adjacent values of X
1	2	
2	1	1
3	4	3
4	3	1
5	1	2
6	3	2
7	2	1
8	4	2
9	1	3
10	2	1
11	2	0
12	3	1
13	5	2
14	4	1
15	6	2
Total	43 $\bar{X} = 2.9$	22 $R = 1.57$

For $n = 2$, $s_X = \bar{R}/d_2 = 1.57/1.128 = 1.4$.
Uncoded values: $\bar{X} = 98.9$; $s_X = 1.4$.
Control chart limit lines (2-sigma limits):

$$\text{UCL}_X = 98.9 + 2(1.4) = 101.7$$
$$\bar{X} = 98.9$$
$$\text{LCL}_X = 98.9 - 2(1.4) = 96.1$$

In processes like these, control charts may be based on a *moving* average. Single observations are made on the process, and the observations are arranged in the order of their acquisition: $X_1, X_2, X_3, \ldots, X_i$. We then find the averages of successive sets of two (or three or more) X values. For moving averages of two items, we find the average of X_1 and X_2, then of X_2 and X_3, etc. For moving averages of three items, we first find the averages of X_1, X_2, and X_3, then of X_2, X_3, X_4, etc.

From a sufficient number of observations, preferably 30 or more, we find $\overline{X}$, the average of the observations. This value—$\overline{\overline{X}}$—we use as the central line of the control chart. We place the limit lines at $\pm 3s_{\bar{x}}$, i.e., at a distance from the central line of three times the estimate of the standard deviation of the $\overline{X}$ values from their mean. On the chart, we plot the successive $\overline{X}$ values. (The average of the $\overline{X}$'s need not equal the average of the observations as all points are not used the same number of times—the first and last being used less often than the rest.)

Such control charts have two important characteristics, as compared with conventional control charts: (1) The plotted points are not independent of each other, since each point reflects the influence of two or more observations: the interobservation influence extends over a greater distance when the average involves many observations. (2) Temporary fluctuations in the process are likely to be overlooked, since the averaging process dampens out the variations.

The following data represent estimates of the potency of different batches of a drug: 98, 97, 100, 99, 97, 99, 98, 100, 97, 98, 98, 99, 101, 100, 102. We will prepare a control chart based on (1) a plot of individuals and (2) a plot of moving averages. Since we have only one sample from each batch of drugs, we have no alternatives to these methods. For ease in calculating, we code the data by subtracting 96 from each observation.

1 To obtain the control-limit line positions for plotting individuals, we summarize the calculations in Table 24·1. We will use 2-sigma limits for the chart; therefore, the limit lines are at ± 2.8 from the central line located at 98.9.

2 To obtain the control-limit lines for plotting moving averages of $n = 2$, we summarize the calculations in Table 24·2. We find s_X to be equal to 1.4, and since we are using averages of $n = 2$, we calculate $s_{\bar{x}}$ to be $s_X/\sqrt{2}$, or 1.0. As we are using 2-sigma limits for the chart, we will place the control limits at ± 2.0 from the central line located at 98.8.

In Fig. 24·1 we have plotted the two control charts, both for individuals (dotted line) and for moving averages of $n = 2$ (solid line). The first individual observation has been plotted at batch 1; the first moving-average value is plotted at batch 2, since this average value cannot be obtained until the second batch has been evaluated.

When we compare the two charts, we see that the use of the moving

Batch number	X values	$\bar{X}$	Range
1, 2	2, 1	1.5	1
2, 3	1, 4	2.5	3
3, 4	4, 3	3.5	1
4, 5	3, 1	2.0	2
5, 6	1, 3	2.0	2
6, 7	3. 2	2.5	1
7, 8	2, 4	3.0	2
8, 9	4, 1	2.5	3
9, 10	1, 2	1.5	1
10, 11	2, 2	2.0	0
11, 12	2, 3	2.5	1
12, 13	3, 5	4.0	2
13, 14	5, 4	4.5	1
14, 15	4, 6	5.0	2
Total		39.0	22

Table 24·2   *Calculations of Control Limits for Moving Averages of n = 2*

Coded values: $\bar{\bar{X}} = 39.0/14 = 2.8$, $\bar{R} = 1.57$ $s_X = \bar{R}/d_2$
Uncoded values: $\bar{\bar{X}} = 98.8$, $\bar{R} = 1.57$
Control chart limit lines (2-sigma limits):

$\text{UCL} = 98.8 + 2(1.4)/\sqrt{2} = 98.8 + 2.0 = 100.8$

$\text{LCL} = 98.8 - 2(1.4)/\sqrt{2} = 98.8 - 2.0 = 96.8$

(Note $s_{\bar{x}} = s_X/\sqrt{n}$.)

average has the effect of reducing the noise of the system, but the major patterns of the two charts are similar. That the control-limit lines are different for the two methods implies no severer control for averages than for individuals, but rather a more powerful tool as whenever we use averages. We expect that the variability of averages will be less than that of individuals; hence the limit lines are necessarily different and give us great power to detect

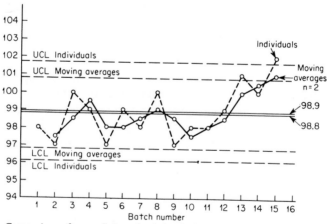

Fig. 24·1   *Comparison of control chart for individuals with control chart for moving average of n = 2.*

changes. Note that on moving-average charts, a problem observation will appear in more than one average until it is dropped.

If we were to use moving averages of $n = 4$, we would find that the limits were cut in half, and we would then gain even more power in detecting changes.

24·2 CONTROL CHARTS BASED ON A MOVING RANGE

A preferred method of preparing control charts for individuals involves the use of the *moving range*. When the observations from the process are arranged in order, we find a set of ranges as the difference between successive observations when we are using a moving range of two (as is seen in Table 24·1 or 24·2). Only the absolute differences are of interest; hence we ignore mathematical signs. From this set of ranges we find the average range $\bar{R}$. To find control limits for the X chart, we select from Table A·7 the value of the factor E_2 corresponding to the value $n = 2$ (in the case of moving ranges of two). Three-sigma limits for X are placed at $E_2\bar{R}$ from the central line at $\bar{X}$. We use the same data as plotted in Fig. 24·1 to prepare Table 24·3.

Table 24·3 Calculation of Control Limits for Individuals from Moving Range

Batch number	X	Moving range
1	98	
2	97	1
3	100	3
4	99	1
5	97	2
6	99	2
7	98	1
8	100	2
9	97	3
10	98	1
11	98	0
12	99	1
13	101	2
14	100	1
15	102	2

$\bar{X} = 98.9$, $\bar{R} = 1.57$.

Control-limit lines: For X chart:

$\text{UCL} = \bar{X} + E_2\bar{R} = 98.9 + 2.66 \times 1.57 = 103.0$
$\bar{X} = 98.9$
$\text{LCL} = \bar{X} - E_2\bar{R} = 98.9 - 2.66 \times 1.57 = 94.7$

For R chart:

$\text{UCL} = D_4\bar{R} = 3.267 \times 1.57 = 5.1$
$\bar{R} = 1.57$
$\text{LCL} = D_3\bar{R} = 0 \times 1.57 = 0.0$

If we wish to use the same data and gain more power through the distribution of averages of larger n values we may use moving ranges of three or more

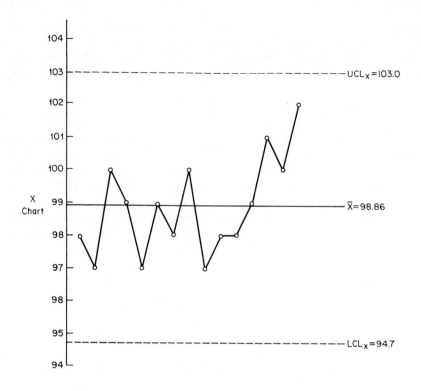

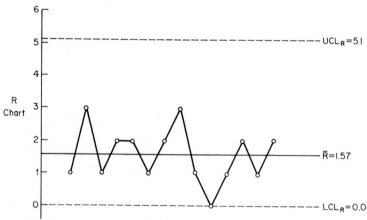

Fig. 24·2   Control chart for individuals, with moving range, n = 2.

observations. For moving ranges of three, we consider each successive set of the observations as a group, and determine the range for that group. Thus we first examine X_1, X_2, and X_3; the difference between the largest and smallest values in this group is R_1. Next we drop X_1 and annex the next observation X_4 to X_2 and X_3. We find R_2 for this new set of three observations. Continuing thus, we find R values for all the data at hand, and from these ranges we find $\bar{R}$. We calculate control limits from a new factor E_2, now for $n = 3$. A similar method is used for four or more observations treated as a group.

In addition to making a plot of the individual observations, we may plot moving ranges as well. We thus prepare a second chart, with central line at $\bar{R}$ and with limit lines at $D_4\bar{R}$ and $D_3\bar{R}$. The coefficients D_4 and D_3 are found in Table A·6.

The calculations for the preparation of such control charts are given in Table 24·3; the charts are shown in Fig. 24·2.

24·3 CONTROL CHARTS BASED ON MOVING AVERAGE AND MOVING RANGE

In general we will plot both moving averages and moving ranges. Thus we prepare two control charts—one for $\bar{X}$ values and one for R values—as in the charts described in Chap. 21. To find the three-sigma control limits we use the average of the set of moving ranges $\bar{R}$ and the table coefficients A_2 and D_3 and D_4, as was described in the earlier part of this chapter. We select the coefficients on the basis of the value of n used in the calculation for moving average and moving range.

Since moving-average and moving-range charts usually are for small sample sizes (especially when $n = 2$) and are thus still not too powerful, it is often useful to use tighter limits then the normal three-sigma limits. If limits at ± 2 sigma are desired, the control lines are, for the $\bar{X}$ chart,

$$\bar{X} \pm \tfrac{2}{3}(A_2\bar{R})$$

For the R chart the 2-sigma limits are

$$\bar{R} \pm \tfrac{2}{3}(D_4\bar{R} - \bar{R})$$

With moving-average and range charts, several points on the charts may be out of control because of only a single abnormal observation. The $\bar{X}$ chart, it is true, tends to show long-time trends and runs. An out-of-control point on this chart indicates that an abnormally high or low level of operation existed during at least part of the time that the two (or three or more) observations were collected. Out-of-control points on the R chart indicate the presence of abnormal short-time variability within the time period during which the data were collected. The interpretation of these charts must therefore be somewhat different than for the regular Shewhart charts and must be more carefully considered.

24·1   In the manufacture of small springs, one spring was taken from the process every 15 min and measured for compression strength. The results of such tests are contained in Table 24·4.

Table 24·4   Compression Strength of Springs

Time of sample	Observation, X	Time of sample	Observation, X
8:15	28	2:00	28
8:30	26	2:15	29
8:45	30	2:30	28
9:00	29	2:45	30
9:15	28	3:00	33
9:30	26	3:15	28
9:45	27	3:30	27
10:00	28	3:45	28
10:15	26	4:00	29
10:30	30	4:15	30
10:45	28	4:30	27
11:00	29	4:45	29
11:15	28	5:00	27
11:30	29	5:15	25
11:45	31	5:30	29
12:00	28	5:45	28
12:15	27	6:00	31
12:30	26	6:15	30
12:45	28	6:30	23
1:00	28	6:45	27
1:15	31	7:00	26
1:30	25	7:15	29
1:45	27		

(a) Construct a control chart for individuals, using a moving-range chart for moving ranges of 2.
(b) Construct a control chart for moving average of 2 observations.
(c) Construct a chart for a moving average of 3 observations.
(d) Compare the results in answer to (b) with the answer to part (c) and comment on the differences.
(e) Find the standard deviation of the individuals, and compare it with that of moving averages of two observations.
(f) Plot moving averages and ranges for $n = 5$, and place 2-sigma control limits on the data.

24·2   In a distilling plant batch lots of denatured alcohol were blended in a large tank. The percentage of methanol was to be controlled. As the variability of sampling within a single lot was found to be very small, only one observation per lot was taken. Control limits were based on the moving range of successive lots. Table 24·5 gives the results of the methanol content for 26 consecutive lots of denatured alcohol. Set up control charts for individuals X and moving range R of two observations.

Table 24·5   Methanol Content of Alcohol

Lot number	Percentage methanol, X	Moving range, R	Lot number	Percentage methanol, X	Moving range, R
1	4.6		14	5.5	0.1
2	4.7	0.1	15	5.2	0.3
3	4.3	0.4	16	4.6	0.6
4	4.7	0.4	17	5.5	0.9
5	4.7	0.0	18	5.6	0.1
6	4.6	0.1	19	5.2	0.4
7	4.8	0.2	20	4.9	0.3
8	4.8	0.0	21	4.9	0.0
9	5.2	0.4	22	5.3	0.4
10	5.0	0.2	23	5.0	0.3
11	5.2	0.2	24	4.3	0.7
12	5.0	0.2	25	4.5	0.2
13	5.6	0.6	26	4.4	0.1

24·4 CHARTS BASED ON A GEOMETRIC MOVING AVERAGE

In many continuous processes we dare not base our judgment about the stability of the process on single points. We have seen that the use of arithmetic averages of adjacent points increases our ability to detect smaller process changes. Arithmetic averages give equal weight to present and past data. It is often, however, reasonable to use a combining method that gives greater weight to more recent data, and less weight to data accumulated in the relatively remote past. Such a method involves the use of a *geometric moving average*. Such an average uses, instead of a simple arithmetic average, a calculating method by which a factor is applied to each observation. This factor is larger for recent data than for older data.

The geometric moving average weights each observation by a fraction r to find a term Z_t which is defined by

$$Z_t = rX_t + (1 - r)Z_{t-1} \qquad (24\cdot1)$$

X_t is the current observation from the process; Z_{t-1} is the value of Z calculated for the previous observation. The standard deviation of the values of Z_t is closely approximated by

$$\sigma_{Z_t} = \sigma\sqrt{\frac{r}{2 - r}} \qquad (24\cdot2)$$

where σ is the standard deviation of the X values from which Z_t is calculated.

CONTROL CHARTS FOR VARIABLES   531

Table 24·6 *Calculations for Control Chart Based on a Geometric Moving Average (Process Assumed at Standard with Z_0 at 6.0)*

Sample number	Response, X	$\frac{4}{5} Z_{t-1}$	$\frac{1}{5} X_t$	Z_t
1	5	4.80	1.00	5.80
2	7	4.64	1.40	6.04
3	4	4.83	0.80	5.63
4	8	4.50	1.60	6.10
5	6	4.88	1.20	6.08
6	7	4.86	1.40	6.26
7	9	5.01	1.80	6.81
8	4	5.45	0.80	6.25
9	3	5.00	0.60	5.60
10	7	4.48	1.40	5.88
11	6	4.72	1.20	5.92
12	9	4.74	1.80	6.54
13	5	5.23	1.00	6.23
14	4	4.98	0.80	5.78

Table 24·6 illustrates the method. The X values are data from a process with mean of 6.0 and standard deviation of 2.0. We take r as $\frac{1}{5}$. Each entry in the Z_t column is $\frac{1}{5}$ of the X value in the row plus $\frac{4}{5}$ of the Z_t value in the row immediately above. As we assume that the process is set to operate at the standard value 6.0, the value of Z_0 is 6.0. The central line of the chart is placed at the mean of the distribution, the standard value 6.0. The standard deviation of the Z_t values is found from Eq. (24·2) to be $2\sqrt{(0.2)/(2 - 0.2)}$, which turns out to be $\frac{2}{3}$. If we place the customary ± 3-sigma limits on the chart, the UCL is located at 8.0 and the LCL is at 4.0. The completed chart is shown in Fig. 24·3.

If we were to find the average of the 14 values of X as listed in Table 24·6, we would find it to be 6.0. This means that the average has not shifted during

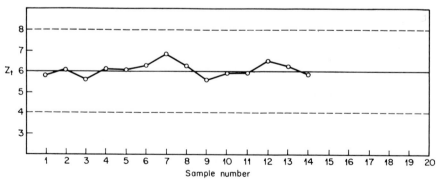

Fig. 24·3 Geometric-moving-average control chart.

the time the data were gathered. The chart (Fig. 24·3) indicates that no shift in the process average has occurred, as the plotted points do cluster about the standard value of 6.0. If a small shift in the process average were to take place, say in an upward direction, the values of X would cause the plotted points to start moving upward in the direction of the upper control limit. Although this is also true for the Shewhart charts, the length of time the process would have to run to detect the small shift in the mean would be less for the geometric-moving-average chart then it would for the Shewhart chart.

The geometric-moving-average chart is far more sensitive than the Shewhart chart in the case of small or moderate shifts in the process level. For larger shifts of the process level, the Shewhart chart is faster at picking up the shift in level.

The length of time the process must run, on the average, before the control chart will indicate a shift in the process level, when one has occurred, is called the *average run length* (ARL). For small shifts in level, the geometric-moving-average chart will have a smaller ARL than will the Shewhart chart. The selection of the proper value for the weighting value r can best be done by an examination of the ARL curves for various r values. Control Chart Tests Based on Geometric Moving Averages[1] includes such ARL curves.

PROBLEMS

24·3   Using the data contained in Table 24·6, construct a geometric-moving-average chart with a value of $r = 0.5$. (Use a 3-sigma chart.)

24·4   Using the data from Table 24·6, construct a 3-sigma chart for geometric moving averages with a value of $r = 0.9$.

24·5   Compare the answers from Probs. 24·3 and 24·4 with the charts in Fig. 24·3. Comment on the role played by the value of r.

24·6   Add a constant value of 2.0 to each value of X in Table 24·6. Use the resulting data to represent the results of additional sample numbers 15 to 28, with process level shifted a distance of one standard deviation. Continue plotting of the geometric-moving-average chart in Fig. 24·3.

(a) Use $r = \frac{1}{5}$.

(b) Use $r = 0.5$ and complete the chart in answer to Prob. 24·3 with the newly generated data.

24·5   CUMULATIVE-SUM CHARTS

A method similar to that for geometric moving averages but simpler in calculations plots the sum of the deviations of the observations from the aim point for the process. The control limits are prepared in such a way that they effectively give less weight to data as it becomes older.

[1] S. W. Roberts, *Technometrics*, vol. 1, no. 3, August, 1959.

As long as the observations cluster about the desired value, a cumulative-sum chart displays a graph that is essentially horizontal. The level of the line is immaterial. When, however, a change occurs in the process, the graph rises or falls. If the change has merely affected the process level by a fixed increase or decrease, the graph will show a straight-line rise or fall. The slope of the line suggests the severity of the effect. If the process drifts away from the aim point with a progressive increase in deviation, the rising or falling line will curve. Thus control limits for these charts will necessarily be other than horizontal lines.

A graphic method of determining whether or not the chart indicates cause for action requires the use of a movable mask, as shown in Fig. 24·4. The

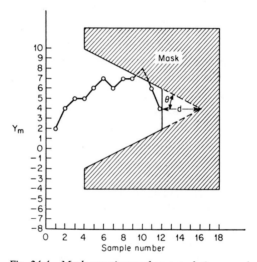

Fig. 24·4   Mask superimposed on cumulative-sum chart.

vertex of the mask is placed a distance d ahead of the last observation. The process is considered to be performing satisfactorily as long as all previously plotted points are visible and not obscured by the mask. When any point is hidden by the mask, a change in the process has been found. In Fig. 24·4 the point for sample 10 is covered by the mask. It was not until the mask was placed at sample 12 that sufficient evidence has been found to indicate a process shift. That the line slopes downward from sample 10 suggests that the shift began after that observation.

The cumulative-sum chart is sensitive to small shifts in process level. It has the additional advantage of showing graphically the approximate time of the occurrence of an out-of-control situation. The slope of the line indicating the shift suggests the magnitude of the shift, since it represents the cumulative deviation divided by the number of time intervals after we suspect that the shift has occurred.

The two significant characteristics of the mask are the distance d and the angle θ of the cut-out portion. The mask may be prepared by plotting cumulative-sum data from a known process with a long history, cut-and-try masks of various design until one seems to give the desired signals. An alternative method permits the calculation of d and θ from these data:

1 The standard deviation of the process. This is best based on observations of rational subgroups and is therefore usually $\sigma_{\bar{x}}$.
2 The difference D in the mean which the chart should detect.
3 The desired α risk.

To find the values of d and θ, we first find the value of δ from

$$\delta = \frac{D}{\sigma_{\bar{x}}} \qquad (24\cdot3)$$

This expression specifies the deviation we wish to detect in units of the standard deviation of the process subgroups. Now

$$\theta = \tan^{-1}\frac{\delta}{2} \qquad (24\cdot4)$$

that is, the angle θ is the angle which has a tangent equal to half the permitted deviation.

The value of d is found from

$$d = \frac{-2}{\delta^2}\ln\alpha \qquad (24\cdot5)$$

where ln represents the natural logarithm, and α is the desired α risk.

Equations (24·4) and (24·5) give the mask dimensions where the sums of the deviations from the aim point are plotted in units of $\sigma_{\bar{x}}$ and where these units occupy the same dimension on the graph as the spacing between observations on the horizontal axis of the chart. If a different scale is used for the vertical axis, the value of d is unchanged; the formula for θ, however, becomes

$$\theta = \tan^{-1}\frac{\delta}{2k} \qquad (24\cdot6)$$

where k is the ratio of $\sigma_{\bar{x}}$ to the desired plotting interval. It is well to scale the vertical axis so that a range of $\pm6\sigma_{\bar{x}}$ is available, so as to avoid having points fall off the scale.

Comparisons based on representative data indicate that cumulative-sum charts are very sensitive to small changes in level, especially when the desired α risk is very small (0.003 and smaller). The advantages of cumulative-sum charts over conventional charts decrease as the α risk is increased.

A process was known to operate at a mean of 50.0 and a standard deviation of 2.0. A cumulative-sum chart was used with the intention of detecting a shift in the process average equal to one standard deviation or more. Table 24·7 shows the results of the first 20 samples collected from the process after the control chart was started. The first 10 values are with the process in control at the proper level, and the second set of 10 values of $\bar{X}$ are the result of the process operating with a shift upward of 1.5 units. In other words, the process was operating at a level of 50.0 for the first 10 samples, and then a sudden shift in process level occurred. This shift in the process level was from a level of 50.0 to a level of 51.5. (Standard deviation of the process remained the same for all 20 samples.)

Table 24·7   Calculations for Cumulative-sum Chart

Sample number	$\bar{X}$	$\bar{X} - 50$	$Y_m = \Sigma\,(\bar{X}_i - 50)$
1	51.6	1.6	1.6
2	50.4	0.4	2.0
3	48.9	−1.1	0.9
4	49.3	−0.7	0.2
5	51.6	1.6	1.8
6	50.1	0.1	1.9
7	48.6	−1.4	0.5
8	50.8	0.8	1.3
9	49.8	−0.2	1.1
10	48.6	−1.4	−0.3
11	50.4	0.4	0.1
12	52.1	2.1	2.2
13	49.8	−0.2	2.0
14	52.4	2.4	4.4
15	51.9	1.9	6.3
16	51.1	1.1	7.4
17	49.9	−0.1	7.3
18	52.4	2.4	9.7
19	52.9	2.9	12.6
20	51.4	1.4	14.0

We show the necessary calculations in Table 24·7, and the plotted cumulative sums are graphed in Fig. 24·5. On the vertical scale of Fig. 24·5 the position 0 represents the assumed mean of the process. The scale factor k is 2, since the distance between time plots on the horizontal scale is equivalent to two units on the vertical scale.

For the preparation of the mask, we require two dimensions, d and θ. To find these, we must specify four values:

1   D, the size of the shift in process average which we wish to detect.
   $D = \sigma_{\bar{x}} = \sigma_x/\sqrt{n} = 2.0/\sqrt{4} = 1.0$.
2   $\delta = D/\sigma_{\bar{x}} = 1.0$.

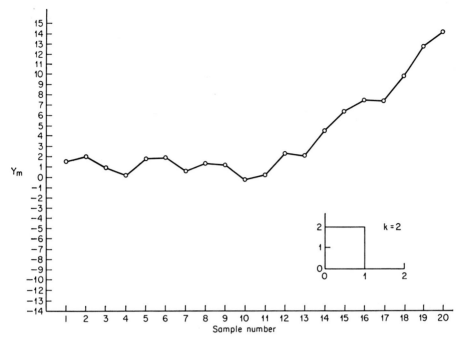

Fig. 24·5   Cumulative-sum chart.

3   We set the α risk at 0.00135.
4   The scale factor k is here 2.

To find θ we use Eq. (24·6):

$$\theta = \tan^{-1} \delta/2k = \tan^{-1} 1.0/4.0 = \tan^{-1} 0.25$$
$$\theta = 14°$$

To find d, we use Eq. (24·5):

$$d = (-2/\delta^2)\,(\ln \alpha) = (-2/1)(\ln 0.00135) = -2(-6.55)$$
$$d = 13.1$$

The mask is prepared as in Fig. 24·6. It is then placed over the chart of the data as in Fig. 24·7, aligning the last point plotted (or any point to be checked) with the arrow in the center of the V. In this example, if each plotted point is checked with the mask (in order of time), the first indication of a shift of process average as large as one standard deviation is found at the plot of the fifteenth point. When the mask is placed in proper position with the fifteenth point located at the arrow (distance d in front of the vertex of the V), the tenth and eleventh points are found to be hidden behind the mask. Since we have already pointed out that a shift occurred after the tenth sample (information we would not have in an actual situation), we can see that a series of five samples had to be taken from the process before the

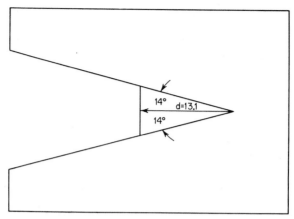

Fig. 24·6   Mask for cumulative-sum chart.

shift was detected. This is the run-length time required to make the detection of the shift.

A correction should have been made in the process at the first sign of a shift in the process level. In our example, this correction would have been indicated when the fifteenth sample was taken and the information plotted and checked with the mask. Since no action was taken, placing the mask at any point after the fifteenth will also show that there has been a shift in process average.

When we find evidence of a process shift in level, if we are using the cumulative-sum chart, we can often obtain a very good estimate of when the shift first occurred. By looking at the plotted points (see Fig. 24·5) it is possible to draw in a line on the upward trend of points. Where this straight line crosses the horizontal location of the process mean, the start of the

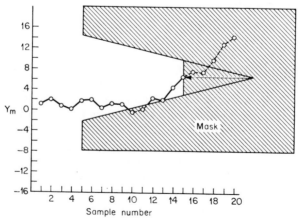

Fig. 24·7   Cumulative-sum chart with superimposed mask.

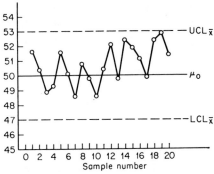

Fig. 24·8 Conventional control chart of same data as plotted in Fig. 24·5.

trouble is indicated. The straight line which best fits the upward climb of the points in our example will indicate the start of the shift in process level at about 10, where we know it actually did occur.

Figure 24·8 shows for comparison a conventional chart of the same data. Although a change in process level might be detectable from this chart as well, it is unlikely that it would have been detected by the fifteenth point. Thus for these data the cumulative-sum chart is superior.

The larger the shift in the process level, the steeper will be the slope of the cumulative-sum points. This means that the mask might be used to indicate the size of correction necessary to correct for different-sized shifts in level. By marking off zones on the edges of the V mask to show how fast the process is deviating from the standard, and listing specific instructions on what to do if the points disappear behind the various zones, we can tell the operator how much correction is called for in the correction of the process. Figure 24·9 shows a typical example of how the mask might be divided into zones of action. If the shift in the process level has been large enough to be detected near the apex of the V, a large correction is required. This also

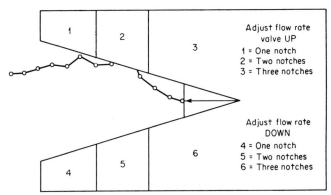

Fig. 24·9 Mask for cumulative-sum chart, with zones of action for correction of process.

suggests that lack-of-control points first signaled at a point a long time before the test point might be considered doubtful. If the point at which lack of control is first signaled is at a distance of more than $3d$ before 0, it would suggest doubt in the indication. With experience in the use of cumulative-sum charts, this doubt could be removed by restricting the overall length of the mask being used.

PROBLEM

24·7 A process is set to operate at a mean of 30 and has a standard deviation of 2.0. It is desirable to detect a shift in the process average of 1.0 standard deviation, in either direction from the mean of 30.0. From the process, subgroups of $n = 4$ were taken and the averages of the subgroups were recorded as follows:

31, 30, 29, 31, 31, 30, 31, 29, 30, 32, 31, 31, 28, 29, 29, 30, 30, 31, 31, 28, 29, 31, 31, 32, 31

(a) Calculate the cumulative sum of the deviations from the target value of 30.0, and construct a cumulative-sum graph.
(b) Using an $\alpha/2 = 0.00135$ (3 sigma equivalent) construct a V-shaped mask for the cumulative-sum chart.
(c) Using the mask with the cumulative-sum graph, determine if the process average has shifted by as much as one standard deviation.
(d) Change the last 10 averages of the above data by an increase of $+2$. Replot the cumulative-sum chart with the data changed and check with the mask to see if the process average has shifted by at least one standard deviation.
(e) Add a value of $+3$ to each of the last 10 averages of the given data, and after regraphing, check with the mask to see if the shift in average was detected sooner than in (d).
(f) Construct a conventional control chart from the same data as used in (d) and/or (e) and compare the results with the cumulative-sum chart. (Consider the range chart as being in control.)

QUESTIONS

24·1 Why is it not necessary always to have control-chart limits set at 3-sigma limits?

24·2 In the control charts for averages, how does the β risk enter into the control chart?

24·3 When should a control chart using a moving average be considered?

24·4 What does a moving average do to the signal of the control chart as revealed by the chart for individuals?

24·5 If a moving average of $n = 3$ is used, what will be the relationship of the 3-sigma limits as compared with the chart for individuals?

24·6   When plotting a control chart for individuals, if a moving-range chart is used for $n = 2$, how might we use the $\bar{R}$ value to calculate the 3-sigma limits of the chart for individuals?

24·7   If the control-chart limits are set at 3 sigmas when plotting individuals, how are the limits changed when subgroups of 4 are taken and the averages plotted?

24·8   What makes the geometric-moving-average chart different from the moving-average chart?

24·9   In the geometric-moving-average chart, are the plotted values considered to be independent of each other? Explain.

24·10   How does changing the fraction r in the geometric-moving-average chart influence the plotted values?

24·11   In a geometric-moving-average chart, if present data are much more important than the past data, what value should be given the fraction r?

24·12   In the cumulative-sum chart, the target value of the process need not appear on the chart. Explain.

24·13   Why is it necessary to include a scale factor in the calculation for the mask?

24·14   In cumulative-sum charts it is important to know the distance from the apex of the mask that the hidden point falls. Explain.

24·15   When plotting a cumulative-sum chart:
(a) What happens to the plotted points when the average of the process remains the same
(b) If the process average changes very slightly, how will the plotted points on the graph appear
(c) If a large shift in the process average suddenly occurs, how will the points on the graph change

24·16   When does the cumulative-sum chart have advantages over the conventional charts?

REFERENCES
6, 9, 18, 35, 41, 48, 49, 50, 51, 61, 63, 66, 86, 87, 88, 89, 90, 91, 94, 110

References

1. Abruzzi, Adam: Single Sampling Attribute Plans, *Industrial Quality Control*, vol. 15, no. 7, pp. 16–18, January, 1959.
2. Adams, J. K.: "Basic Statistical Concepts," McGraw-Hill Book Company, New York, 1955.
3. Addelman, S.: Some Two-level Factorial Plans with Split Plot Confounding, *Technometrics*, vol. 6, no. 3, pp. 253–258, August, 1964.
4. Addelman, S., and S. Bush: A Procedure for Constructing Incomplete Block Designs, *Technometrics*, vol. 6, no. 4, pp. 389–404, November, 1964.
5. Anderson, R. L., and T. A. Bancroft: "Statistical Theory in Research," McGraw-Hill Book Company, New York, 1952.
6. Aroian, L. A., and H. Levene: The Effectiveness of Quality Control Charts, *Journal of the American Statistical Association*, vol. 45, no. 252, pp. 520–529, December, 1950.
7. ASTM: "Manual on Quality Control of Materials," American Society for Testing and Materials, Philadelphia, 1951.
8. Baker, Robert A.: Subjective Panel Testing, *Industrial Quality Control*, vol. 19, no. 3, pp. 22–28, September, 1962.
9. Barnard, G. A.: Control Charts and Stochastic Processes, *Journal of the Royal Statistical Society*, Series B, vol. 21, no. 2, pp. 239–271, 1959.
10. Bennett, C. A., and N. L. Franklin: "Statistical Analysis in Chemistry and the Chemical Industry," John Wiley & Sons, Inc., New York, 1954.

11. Bicking, C. A.: Primer on Experimentation, *Industrial Quality Control*, vol. 17, no. 11, pp. 20–25, May, 1961.
12. Bicking, C. A.: Some Uses of Statistics in the Planning of Experiments, *Industrial Quality Control*, vol. 10, no. 4, pp. 20–23, January, 1954.
13. Bicking, C. A., and R. H. Gillespie: Exploring with Experimental Design, *Industrial Quality Control*, vol. 19, no. 12, pp. 17–21, June, 1963.
14. Bingham, Richard S., Jr.: Design of Experiments, from a Statistical Viewpoint, *Industrial Quality Control*, vol. 15, no. 8, pp. 29–34, May, 1959.
15. Bingham, R. S., Jr.: Try EVOP for Systematic Process Improvement, *Industrial Quality Control*, vol. 20, no. 3, pp. 17–23, September, 1963.
16. Bowker, A. H., and G. J. Lieberman: "Engineering Statistics," Prentice-Hall, Inc., Englewood Cliffs, N.J., 1959.
17. Bradley, R. A.: Determination of Optimum Operating Conditions by Experimental Methods. Part I, Mathematics and Statistics Fundamental to the Fitting of Response Surfaces, *Industrial Quality Control*, vol. 15, no. 1, pp. 16–20, July, 1958.
18. Breunig, H. L.: Some Uses of Statistical Control Charts in the Pharmaceutical Industry, *Industrial Quality Control*, vol. 21, no. 2, pp. 79–86, August, 1964.
19. Brownlee, K. A.: "Industrial Experimentation," Chemical Publishing Company, Inc., New York, 1948.
20. Brownlee, K. A.: The Principles of Experimental Design, *Industrial Quality Control*, vol. 13, no. 8, pp. 12–20, February, 1957.
21. Burington, R. S., and D. C. May: "Handbook of Probability and Statistics with Tables," McGraw-Hill Book Company, New York, 1953.
22. Burr, I. W.: "Engineering Statistics and Quality Control," McGraw-Hill Book Company, New York, 1953.
23. Chambers, E. G.: "Statistical Calculations for Engineers," 2d ed., Cambridge University Press, New York, 1958.
24. Cocca, O. A.: Mil-Std-105D, an International Standard for Attribute Sampling, *Industrial Quality Control*, vol. 21, no. 5, pp. 249–253, November, 1964.
25. Cochrane, W. G., and G. M. Cox: "Experimental Designs," 2d ed., John Wiley & Sons, Inc., New York, 1957.
26. Cowan, Alan: "Quality Control for the Manager," The Macmillan Company, New York, 1964.
27. Cowden, D. J.: "Statistical Methods in Quality Control," Prentice-Hall, Inc., Englewood Cliffs, N.J., 1957.
28. Crow, E. L., F. A. Davis, and N. W. Maxfield: "Statistics Manual," Dover Publications, Inc., New York, 1960.
29. Croxton, F. E., and D. J. Cowden: "Applied General Statistics," 2d ed., Prentice-Hall, Inc., Englewood Cliffs, N.J., 1955.

30. Davies, O. L.: "Design and Analysis of Industrial Experiments," Hafner Publishing Company, Inc., New York, 1956.

31. Davies, O. L.: "Statistical Methods in Research and Production," Hafner Publishing Company, Inc., New York, 1960.

32. DeBusk, R. E.: Experience in Evaluating Operations of Tennessee Eastman Company, *Industrial Quality Control*, vol. 19, no. 4, pp. 15–20, October, 1962.

33. Dixon, W. J., and F. J. Massey: "Introduction to Statistical Analysis," 2d ed., McGraw-Hill Book Company, New York, 1957.

34. Dodge, H. F., and H. G. Romig: "Sampling Inspection Tables— Single and Double Sampling," John Wiley & Sons, Inc., New York, 1944.

35. Duncan, A. J.: "Quality Control and Industrial Statistics," 3d ed., Richard D. Irwin, Inc., Homewood, Ill. 1965.

36. Eaton, H. C.: Engineered Assembly Tolerance, *Industrial Quality Control*, vol. 4, no. 2, pp. 16–17, September, 1947.

37. Ehrenfeld, S., and S. B. Littauer: "Introduction to Statistical Method," McGraw-Hill Book Company, New York, 1964.

38. Eisenhart, C., M. W. Hastay, and W. A. Wallis: "Techniques of Statistical Analysis," McGraw-Hill Book Company, New York, 1947.

39. Eitelman, H. K., W. A. Koppus, and R. W. Traver: A New Approach to Process Capability, *Industrial Quality Control*, vol. 18, no. 10, pp. 24–26, April, 1962.

40. Ewan, W. D., and K. W. Kemp: Sampling Inspection of Continuous Processes with No Autocorrelation between Successive Results, *Biometrika*, vol. 47, pp. 363–380, December, 1960.

41. Ferrell, Enoch B.: Control Charts for Log-normal Universes, *Industrial Quality Control*, vol. 15, no. 2, pp. 4–6, August, 1958.

42. Fisher, R. A.: "The Design of Experiments," Hafner Publishing Company, Inc., New York, 1954.

43. Fisher, R. A.: "Statistical Methods for Research Workers," Oliver & Boyd Ltd., Edinburgh and London, 1954.

44. Freeman, H. A.: "Industrial Statistics," John Wiley & Sons, Inc., New York, 1954.

45. Freund, J. E.: "Mathematical Statistics," Prentice-Hall, Inc., Englewood Cliffs, N.J., 1962.

46. Freund, J. E., "Modern Elementary Statistics," Prentice-Hall, Inc., Englewood Cliffs, N.J., 1956.

47. Freund, J. E., P. E. Livermore, and Irwin Miller: "Manual of Experimental Statistics," Prentice-Hall, Inc., Englewood Cliffs, N.J., 1960.

48. Freund, R. A.: Acceptance Control Charts, *Industrial Quality Control*, vol. 14, no. 4, pp. 13–23, October, 1957.

49. Freund, R. A.: Graphical Process Control, *Industrial Quality Control*, vol. 28, no. 7, pp. 1–8, January, 1962.

50. Freund, R. A.: A Reconsideration of the Variables Control Chart, *Industrial Quality Control*, vol. 16, no. 11, pp. 35–41, May, 1960.

51. Goldsmith, P. L., and H. Whitfield: Average Run Lengths in Cumulative Chart Quality Control Schemes, *Technometrics*, vol. 3, no. 1, pp. 11–20, February, 1961.

52. Goulden, C. H.: "Methods of Statistical Analysis," 2d ed., John Wiley & Sons, Inc., New York, 1952.

53. Grant, E. L.: "Statistical Quality Control," 3d ed., McGraw-Hill Book Company, New York, 1964.

54. Greb, D. J.: Sequential Sampling Plans with AOQL as the Primary Function, *Industrial Quality Control*, vol. 19, no. 11, pp. 24–28, 47–48, May, 1963.

55. Grubbs, F. E.: On Designing Single Sampling Plans, *Annals of Mathematical Statistics*, vol. 20, pp. 242–256.

56. Guenther, W. C.: "Analysis of Variance," Prentice-Hall, Inc., Englewood Cliffs, N.J., 1964.

57. Hamaker, H. C.: Examples of Designed Experiments, *Industrial Quality Control*, vol. 17, no. 9, pp. 16–20, March, 1961.

58. Harrington, E. C., Jr.: The Desirability Function, *Industrial Quality Control*, vol. 21, no. 10, pp. 494–498, April, 1965.

59. Hicks, C. R.: "Fundamental Concepts in the Design of Experiments," Holt, Rinehart and Winston, Inc., New York, 1964.

60. Hicks, C. R.: Fundamentals of Analysis of Variance, *Industrial Quality Control*, vol. 13, no. 2, pp. 17–20, August, 1956; no. 3, pp. 5–8, September, 1956; no. 4, pp. 13–16, October, 1956.

61. Hillier, F. S.: $\bar{X}$ Chart Control Limits Based on a Small Number of Subgroups, *Industrial Quality Control*, vol. 20, no. 8, pp. 24–29, February, 1964.

62. Hirsch, W.: "Introduction to Modern Statistics," The Macmillan Company, New York, 1957.

63. Hove, W. G., and R. H. Morris: "Exponential Smoothing," Unpublished paper presented before the Rochester, N.Y., sections of ASQC and ASA, December, 1960.

64. Hrorni, J. D.: Augmentation of Fractional Factorial Designs, *Conference Transactions*, 17th Annual Quality Control Conference, Rochester Society for Quality Control, Rochester, N.Y., 1961.

65. Hunter, J. S.: Determination of Optimum Operating Conditions by Experimental Methods, Part II, Models and Methods, *Industrial Quality Control*, vol. 15, no. 6, pp. 16–24, December, 1958; no. 7, pp. 7–15, January, 1959; no. 8, pp. 6–14, February, 1959.

66. Johnson, N. L., and F. C. Leone: Cumulative Sum Control Charts: Mathematical Principles Applied to Their Construction and Use, Part I, *Industrial Quality Control*, vol. 18, no. 12, pp. 15–21, June, 1962;

Part II, vol. 19, no. 1, pp. 29–36, July, 1962; Part III, vol. 19, no. 2, pp. 22–28, August, 1962.

67. Johnson, N. L., and F. C. Leone: "Statistics and Experimental Design," vol. I, John Wiley & Sons, Inc., New York, 1964.

68. Johnson, N. L., and F. C. Leone: "Statistics and Experimental Design," vol. II, John Wiley & Sons, Inc., New York, 1964.

69. Juran, J. M.: "Quality Control Handbook," 2d ed., McGraw-Hill Book Company, New York, 1962.

70. Kempthorne, Oscar: "The Design and Analysis of Experiments," John Wiley & Sons, Inc., New York, 1952.

71. Kenney, J. F., and E. S. Keeping: "Mathematics of Statistics," 3d ed., D. Van Nostrand Company, Inc., Princeton, N.J., 1954.

72. Kenworthy, O. O.: Factorial Experiments with Mixtures Using Ratios, *Industrial Quality Control*, vol. 19, no. 12, pp. 24–26, June, 1963.

73. Li, J. C. R.: "Introduction to Statistical Inference," Edwards Brothers, Inc., Ann Arbor, Mich., 1957.

74. Lindgren, B. W., and G. W. McElrath: "Introduction to Probability and Statistics," The Macmillan Company, New York, 1961.

75. Locke, L. G.: Bayesian Statistics, *Industrial Quality Control*, vol. 20, no. 10, pp. 18–21, April, 1964.

76. Mandelson, Joseph: Sampling, *Industrial Quality Control*, vol. 19, no. 1, pp. 5–6, July, 1962.

77. McCall, C. H., Jr.: Linear Contrasts, Part I, *Industrial Quality Control*, vol. 17, no. 1, pp. 19–21, July, 1960. Linear Contrasts, Part II, *Industrial Quality Control*, vol. 17, no. 2, pp. 12–16, August, 1960. Linear Contrasts, Part III, *Industrial Quality Control*, vol. 17, no. 3, pp. 5–8, September, 1960.

78. Mode, E. B.: "Elements of Statistics," Prentice-Hall, Inc., Englewood Cliffs, N.J., 1958.

79. Mood, A. M., and F. A. Graybill: "Introduction to the Theory of Statistics," 2d ed., McGraw-Hill Book Company, New York, 1963.

80. Mosteller, F., E. K. Rourke, and G. B. Thomas, Jr.: "Probability with Statistical Applications," Addison-Wesley Publishing Company, Inc., Reading, Mass., 1961.

81. Natrella, M. G.: "Experimental Statistics," National Bureau of Standard Handbook 91, Government Printing Office, Washington, D.C., 1963.

82. Nevelle, A. M., and J. B. Kennedy: "Basic Statistical Methods for Engineers and Scientists," International Textbook Company, Scranton, Pa., 1964.

83. Olds, E. G.: Power Characteristics of Control Charts, *Industrial Quality Control*, vol. 18, no. 1, pp. 4–9, July, 1961.

84. Ostle, B., and J. M. Wiesen: An Acceptance Sampling Plan, *Industrial Quality Control*, vol. 15, no. 3, pp. 8–9, September, 1958.

85. Pabst, W. R., Jr.: Mil-Std-105D, *Industrial Quality Control*, vol. 20, no. 5, pp. 4–8, November, 1963.

86. Page, E. S.: Comparison of Process Inspection Schemes, *Industrial Quality Control*, vol. 21, no. 5, pp. 245–448, November, 1964.

87. Page, E. S.: Continuous Inspection Schemes, *Biometrika*, vol. 41, pp. 100–115, June, 1954.

88. Page, E. S.: Control Charts for the Mean of a Normal Population, *Journal of the Royal Statistical Society*, Series B, vol. 16, no. 1, pp. 131–135, 1954.

89. Page, E. S.: Control Charts with Warning Limits, *Biometrika*, vol. 42, pp. 243, 257, June, 1955.

90. Page, E. S.: Cumulative Sum Charts, *Technometrics*, vol. 3, no. 1, pp. 1–9, February, 1961.

91. Page, E. S.: On Problems in Which a Change in a Parameter Occurs at an Unknown Point, *Biometrika*, vol. 44, pp. 248–252, June, 1957.

92. Price, Sidney: ABC's of Receiving Sampling Inspection, *Industrial Quality Control*, vol. 18, no. 5, pp. 26–28, November, 1961.

93. Proschan, F., and I. R. Savage: Starting a Control Chart, the Effect of Number and Size of Samples on the Level of Significance at the Start of a Control Chart for Sample Means, *Industrial Quality Control*, vol. 17, no. 3, pp. 12–13, September, 1960.

94. Roberts, S. W.: Control Charts Based on Geometric Moving Averages, *Technometrics*, vol. 1, no. 3, pp. 239–250, August, 1959.

95. Rosander, A. S.: "Elementary Principles of Statistics," D. Van Nostrand Company, Inc., New York, Princeton, N.J., 1951.

96. Schlaifer, R.: "Probability and Statistics for Business Decisions," McGraw-Hill Book Company, New York, 1959.

97. Shah, B. K., and C. G. Khatri: A Method of Fitting the Regression Curve, *Technometrics*, vol. 7, no. 1, pp. 59–65, February, 1965.

98. Shewhart, W. A.: "Economic Control of Quality of Manufactured Product," D. Van Nostrand Company, Inc., Princeton, N.J., 1931.

99. Shewhart, W. A.: "Statistical Methods from the Viewpoint of Quality Control," edited by W. E. Deming, The Graduate School, Department of Agriculture, Washington, D.C., 1939.

100. Siegel, S.: "Nonparametric Statistics for the Behavioral Sciences," McGraw-Hill Book Company, New York, 1956.

101. Simon, L. E.: "An Engineer's Manual of Statistical Methods," John Wiley & Sons, Inc., New York, 1941.

102. Smith, E. S.: "Control Charts," McGraw-Hill Book Company, New York, 1947.

103. Smith, J. G., and A. J. Duncan: "Elementary Statistics and Applications," McGraw-Hill Book Company, New York, 1944.

104. Snedecor, G. W.: "Statistical Methods," 5th ed., Iowa State College Press, Ames, Iowa, 1956.

105. Spooner, L. W.: The Concept of Indifference Quality Level (IQL) in Designing a Sampling Plan, *Industrial Quality Control*, vol. 16, no. 10, pp. 6–10, April, 1960.

106. Sprowls, R. C.: "Elementary Statistics," McGraw-Hill Book Company, New York, 1955.

107. Tarver, M. G., and B. H. Ellis: Selection of Flavor Panels for Complex Flavor Differences, *Industrial Quality Control*, vol. 17, no. 12, pp. 22–26, June, 1961.

108. Teegarden, K. L.: Critical Σd^2 Values for Rank Order Correlations, *Industrial Quality Control*, vol. 16, no. 11, pp. 48–49, May, 1960.

109. Traver, R. W., and J. M. Davis: How to Determine Process Capabilities in a Development Shop, *Industrial Quality Control*, vol. 18, no. 9, pp. 26–29, March, 1962.

110. Truax, H. M.: Cumulative Sum Charts, *Industrial Quality Control*, vol. 18, no. 6, pp. 18–25, December, 1961.

111. U.S. Department of Commerce, "Fractional Factorial Experiment Designs for Factors at Two Levels," National Bureau of Standards, Applied Mathematics Series 48, Government Printing Office, Washington, D.C., 1957.

112. U.S. Department of Defense, "Military Standard 105D, Sampling Procedures and Tables for Inspection by Attributes," Government Printing Office, Washington, D.C., 1963.

113. Van Eck, L. F.: Evolutionary Operations: A Path to More Effective Use of Process Data, *Industrial Quality Control*, vol. 19, no. 2, pp. 8–10.

114. Virene, E. P.: Nonparametric Method in Operating Life Testing, *Industrial Quality Control*, vol. 21, no. 11, pp. 560–562, May, 1965.

115. Volk, W.: "Applied Statistics for Engineers," McGraw-Hill Book Company, New York, 1958.

116. Wade, P. F.: An Introduction to the Use of Statistical Methods in Industrial Experimentation, *Industrial Quality Control*, vol. 18, no. 2, pp. 5–9, August, 1961.

117. Wald, A.: "Sequential Analysis," John Wiley & Sons, Inc., New York, 1947.

118. Walker, H. M., and J. Lev: "Statistical Inference," Holt, Rinehart and Winston, Inc., New York, 1953.

119. Walsh, J. E.: "Handbook of Nonparametric Statistics," D. Van Nostrand Company, Inc., Princeton, N.J., 1962.

120. Wescott, M. E.: Fundamental Control Concepts, *Rubber World*, December, 1958, January, 1959, March, 1959, May, 1959.

121. Whitney, D. R.: "Elements of Mathematical Statistics," Holt, Rinehart and Winston, Inc., New York, 1959.

122. Wilcoxon, Frank: "Some Rapid Approximate Statistical Procedures," American Cyanamid Company, New York, 1949.

123. Wine, R. L.: "Statistics for Scientists and Engineers," Prentice-Hall, Inc., Englewood Cliffs, N.J., 1964.

124. Winter, Robert F.: A Simplified Method for Solving Normal Equations, *Industrial Quality Control*, vol. 18, no. 4, pp. 14–16, October, 1961.

125. Wolf, Frank L.: "Elements of Probability and Statistics," McGraw-Hill Book Company, New York, 1962.

126. Wortham, A. M., and T. E. Smith: "Practical Statistics in Experimental Design," Charles E. Merrill, Inc., Englewood Cliffs, N.J., 1959.

127. Zelen, M., and W. S. Connor: Multi-factor Experiments, *Industrial Quality Control*, vol. 15, no. 9, pp. 14–17, March, 1959.

Appendix Tables

Table A·1 Areas under the Normal Curve

$t =$ $\dfrac{x - \mu}{\sigma}$	0.00	0.01	0.02	0.03	0.04	0.05	0.06	0.07	0.08	0.09
+0.0	0.5000	0.5040	0.5080	0.5120	0.5160	0.5199	0.5239	0.5279	0.5319	0.5359
+0.1	0.5398	0.5438	0.5478	0.5517	0.5557	0.5596	0.5636	0.5675	0.5714	0.5753
+0.2	0.5793	0.5832	0.5871	0.5910	0.5948	0.5987	0.6026	0.6064	0.6103	0.6141
+0.3	0.6179	0.6217	0.6255	0.6293	0.6331	0.6368	0.6406	0.6443	0.6480	0.6517
+0.4	0.6554	0.6591	0.6628	0.6664	0.6700	0.6736	0.6772	0.6808	0.6844	0.6879
+0.5	0.6915	0.6950	0.6985	0.7019	0.7054	0.7088	0.7123	0.7157	0.7190	0.7224
+0.6	0.7257	0.7291	0.7324	0.7357	0.7389	0.7422	0.7454	0.7486	0.7517	0.7549
+0.7	0.7580	0.7611	0.7642	0.7673	0.7704	0.7734	0.7764	0.7794	0.7823	0.7852
+0.8	0.7881	0.7910	0.7939	0.7967	0.7995	0.8023	0.8051	0.8079	0.8106	0.8133
+0.9	0.8159	0.8186	0.8212	0.8238	0.8264	0.8289	0.8315	0.8340	0.8365	0.8389
+1.0	0.8413	0.8438	0.8461	0.8485	0.8508	0.8531	0.8554	0.8577	0.8599	0.8621
+1.1	0.8643	0.8665	0.8686	0.8708	0.8729	0.8749	0.8770	0.8790	0.8810	0.8830
+1.2	0.8849	0.8869	0.8888	0.8907	0.8925	0.8944	0.8962	0.8980	0.8997	0.9015
+1.3	0.9032	0.9049	0.9066	0.9082	0.9099	0.9115	0.9131	0.9147	0.9162	0.9177
+1.4	0.9192	0.9207	0.9222	0.9236	0.9251	0.9265	0.9279	0.9292	0.9306	0.9319
+1.5	0.9332	0.9345	0.9357	0.9370	0.9382	0.9394	0.9406	0.9418	0.9429	0.9441
+1.6	0.9452	0.9463	0.9474	0.9484	0.9495	0.9505	0.9515	0.9525	0.9535	0.9545
+1.7	0.9554	0.9564	0.9573	0.9582	0.9591	0.9599	0.9608	0.9616	0.9625	0.9633
+1.8	0.9641	0.9649	0.9656	0.9664	0.9671	0.9678	0.9686	0.9693	0.9699	0.9706
+1.9	0.9713	0.9719	0.9726	0.9732	0.9738	0.9744	0.9750	0.9756	0.9761	0.9767
+2.0	0.9773	0.9778	0.9783	0.9788	0.9793	0.9798	0.9803	0.9808	0.9812	0.9817
+2.1	0.9821	0.9826	0.9830	0.9834	0.9838	0.9842	0.9846	0.9850	0.9854	0.9857
+2.2	0.9861	0.9864	0.9868	0.9871	0.9875	0.9878	0.9881	0.9884	0.9887	0.9890
+2.3	0.9893	0.9896	0.9898	0.9901	0.9904	0.9906	0.9909	0.9911	0.9913	0.9916
+2.4	0.9918	0.9920	0.9922	0.9925	0.9927	0.9929	0.9931	0.9932	0.9934	0.9936
+2.5	0.9938	0.9940	0.9941	0.9943	0.9945	0.9946	0.9948	0.9949	0.9951	0.9952
+2.6	0.9953	0.9955	0.9956	0.9957	0.9959	0.9960	0.9961	0.9962	0.9963	0.9964
+2.7	0.9965	0.9966	0.9967	0.9968	0.9969	0.9970	0.9971	0.9972	0.9973	0.9974
+2.8	0.9974	0.9975	0.9976	0.9977	0.9977	0.9978	0.9979	0.9979	0.9980	0.9981
+2.9	0.9981	0.9982	0.9983	0.9983	0.9984	0.9984	0.9985	0.9985	0.9986	0.9986
+3.0	0.99865	0.99869	0.99874	0.99878	0.99882	0.99886	0.99889	0.99893	0.99896	0.99900
+3.1	0.99903	0.99906	0.99910	0.99913	0.99915	0.99918	0.99921	0.99924	0.99926	0.99929
+3.2	0.99931	0.99934	0.99936	0.99938	0.99940	0.99942	0.99944	0.99946	0.99948	0.99950
+3.3	0.99952	0.99953	0.99955	0.99957	0.99958	0.99960	0.99961	0.99962	0.99964	0.99965
+3.4	0.99966	0.99967	0.99969	0.99970	0.99971	0.99972	0.99973	0.99974	0.99975	0.99976
+3.5	0.99977	0.99978	0.99978	0.99979	0.99980	0.99981	0.99981	0.99982	0.99983	0.99983

SOURCE: Eugene L. Grant, "Statistical Quality Control," 3d ed., McGraw-Hill Book Company, New York, 1964.

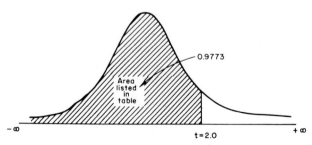

Table A·2 Critical Values of Student's t Distribution

$\bar{\alpha}$ ν	Two-tail Critical Values						
	0.50	0.25	0.10	0.05	0.025	0.01	0.005
1	1.00000	2.4142	6.3138	12.706	25.452	63.657	127.32
2	0.81650	1.6036	2.9200	4.3027	6.2053	9.9248	14.089
3	0.76489	1.4226	2.3534	3.1825	4.1765	5.8409	7.4533
4	0.74070	1.3444	2.1318	2.7764	3.4954	4.6041	5.5976
5	0.72669	1.3009	2.0150	2.5706	3.1634	4.0321	4.7733
6	0.71756	1.2733	1.9432	2.4469	2.9687	3.7074	4.3168
7	0.71114	1.2543	1.8946	2.3646	2.8412	3.4995	4.0293
8	0.70639	1.2403	1.8595	2.3060	2.7515	3.3554	3.8325
9	0.70272	1.2297	1.8331	2.2622	2.6850	3.2498	3.6897
10	0.69981	1.2213	1.8125	2.2281	2.6338	3.1693	3.5814
11	0.69745	1.2145	1.7959	2.2010	2.5931	3.1058	3.4966
12	0.69548	1.2089	1.7823	2.1788	2.5600	3.0545	3.4284
13	0.69384	1.2041	1.7709	2.1604	2.5326	3.0123	3.3725
14	0.69242	1.2001	1.7613	2.1448	2.5096	2.9768	3.3257
15	0.69120	1.1967	1.7530	2.1315	2.4899	2.9467	3.2860
16	0.69013	1.1937	1.7459	2.1199	2.4729	2.9208	3.2520
17	0.68919	1.1910	1.7396	2.1098	2.4581	2.8982	3.2225
18	0.68837	1.1887	1.7341	2.1009	2.4450	2.8784	3.1966
19	0.68763	1.1866	1.7291	2.0930	2.4334	2.8609	3.1737
20	0.68696	1.1848	1.7247	2.0860	2.4231	2.8453	3.1534
21	0.68635	1.1831	1.7207	2.0796	2.4138	2.8314	3.1352
22	0.68580	1.1816	1.7171	2.0739	2.4055	2.8188	3.1188
23	0.68531	1.1802	1.7139	2.0687	2.3979	2.8073	3.1040
24	0.68485	1.1789	1.7109	2.0639	2.3910	2.7969	3.0905
25	0.68443	1.1777	1.7081	2.0595	2.3846	2.7874	3.0782
26	0.68405	1.1766	1.7056	2.0555	2.3788	2.7787	3.0669
27	0.68370	1.1757	1.7033	2.0518	2.3734	2.7707	3.0565
28	0.68335	1.1748	1.7011	2.0484	2.3685	2.7633	3.0469
29	0.68304	1.1739	1.6991	2.0452	2.3638	2.7564	3.0380
30	0.68276	1.1731	1.6973	2.0423	2.3596	2.7500	3.0298
40	0.68066	1.1673	1.6839	2.0211	2.3289	2.7045	2.9712
60	0.67862	1.1616	1.6707	2.0003	2.2991	2.6603	2.9146
120	0.67656	1.1559	1.6577	1.9799	2.2699	2.6174	2.8599
∞	0.67449	1.1503	1.6449	1.9600	2.2414	2.5758	2.8070
ν α	0.25	0.125	0.05	0.025	0.0125	0.005	0.0025
	One-tail Critical Values						

SOURCE: E. S. Pearson, Critical Values of Student's t Distribution, *Biometrika*, vol. 32 pp. 168–181, 1941.

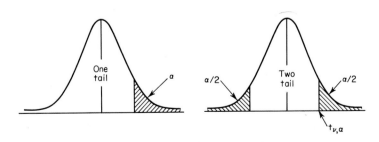

Table A·3 Critical Values of the χ^2 Distribution

α / ν	0.995	0.990	0.975	0.950	0.900	0.750
1	$392{,}704 \times 10^{-10}$	$157{,}088 \times 10^{-9}$	$982{,}069 \times 10^{-9}$	$393{,}214 \times 10^{-8}$	0.0157908	0.1015308
2	0.0100251	0.0201007	0.0506356	0.102587	0.210720	0.575364
3	0.0717212	0.114832	0.215795	0.351846	0.584375	1.212534
4	0.206990	0.297110	0.484419	0.710721	1.063623	1.92255
5	0.411740	0.554300	0.831211	1.145476	1.61031	2.67460
6	0.675727	0.872085	1.237347	1.63539	2.20413	3.45460
7	0.989265	1.239043	1.68987	2.16735	2.83311	4.25485
8	1.344419	1.646482	2.17973	2.73264	3.48954	5.07064
9	1.734926	2.087912	2.70039	3.32511	4.16816	5.89883
10	2.15585	2.55821	3.24697	3.94030	4.86518	6.73720
11	2.60321	3.05347	3.81575	4.57481	5.57779	7.58412
12	3.07382	3.57056	4.40379	5.22603	6.30380	8.43842
13	3.56503	4.10691	5.00874	5.89186	7.04150	9.29906
14	4.07468	4.66043	5.62872	6.57063	7.78953	10.1653
15	4.60094	5.22935	6.26214	7.26094	8.54675	11.0365
16	5.14224	5.81221	6.90766	7.96164	9.31223	11.9122
17	5.69724	6.40776	7.56418	8.67176	10.0852	12.7919
18	6.26481	7.01491	8.23075	9.39046	10.8649	13.6753
19	6.84398	7.63273	8.90655	10.1170	11.6509	14.5620
20	7.43386	8.26040	9.59083	10.8508	12.4426	15.4518
21	8.03366	8.89720	10.28293	11.5913	13.2396	16.3444
22	8.64272	9.54249	10.9823	12.3380	14.0415	17.2396
23	9.26042	10.19567	11.6885	13.0905	14.8479	18.1373
24	9.88623	10.8564	12.4001	13.8484	15.6587	19.0372
25	10.5197	11.5240	13.1197	14.6114	16.4734	19.9393
26	11.1603	12.1981	13.8439	15.3791	17.2919	20.8434
27	11.8076	12.8786	14.5733	16.1513	18.1138	21.7494
28	12.4613	13.5648	15.3079	16.9279	18.9392	22.6572
29	13.1211	14.2565	16.0471	17.7083	19.7677	23.5666
30	13.7867	14.9535	16.7908	18.4926	20.5992	24.4776
40	20.7065	22.1643	24.4331	26.5093	29.0505	33.6603
50	27.9907	29.7067	32.3574	34.7642	37.6886	42.9421
60	35.5346	37.4848	40.4817	43.1879	46.4589	52.2938
70	43.2752	45.4418	48.7576	51.7393	55.3290	61.6983
80	51.1720	53.5400	57.1532	60.3915	64.2778	71.1445
90	59.1963	61.7541	65.6466	69.1260	73.2912	80.6247
100	67.3276	70.0648	74.2219	77.9295	82.3581	90.1332
t_α	-2.5758	-2.3263	-1.9600	-1.6449	-1.2816	-0.6745

One-tail critical values

(For one-tail tests, $\alpha = \alpha_0$)
(For two-tail tests, $\alpha = \alpha_{0/2}$)

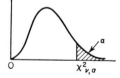

For $30 < \nu < 100$, linear interpolation where necessary will give four significant figures.
For $\nu > 100$, take $\chi^2_{\nu,\alpha} = \frac{1}{2}(t_\alpha + \sqrt{2\nu - 1})^2$.

Table A·3 Critical Values of the χ^2 Distribution (Continued)

α / ν	0.500	0.250	0.100	0.050	0.025	0.010	0.005
1	0.454937	1.32330	2.70554	3.84146	5.02389	6.63490	7.87944
2	1.38629	2.77259	4.60517	5.99147	7.37776	9.21034	10.5966
3	2.36597	4.10835	6.25139	7.81473	9.34840	11.3449	12.8381
4	3.35670	5.38527	7.77944	9.48773	11.1433	13.2767	14.8602
5	4.35146	6.62568	9.23635	11.0705	12.8325	15.0863	16.7496
6	5.34812	7.84080	10.6446	12.5916	14.4494	16.8119	18.5476
7	6.34581	9.03715	12.0170	14.0671	16.0128	18.4753	20.2777
8	7.34412	10.2188	13.3616	15.5073	17.5346	20.0902	21.9550
9	8.34283	11.3887	14.6837	16.9190	19.0228	21.6660	23.5893
10	9.34182	12.5489	15.9871	18.3070	20.4831	23.2093	25.1882
11	10.3410	13.7007	17.2750	19.6751	21.9200	24.7250	26.7569
12	11.3403	14.8454	18.5494	21.0261	23.3367	26.2170	28.2995
13	12.3398	15.9839	19.8119	22.3621	24.7356	27.6883	29.8194
14	13.3393	17.1170	21.0642	23.6848	26.1190	29.1413	31.3193
15	14.3389	18.2451	22.3072	24.9958	27.4884	30.5779	32.8013
16	15.3385	19.3688	23.5418	26.2962	28.8454	31.9999	34.2672
17	16.3381	20.4887	24.7690	27.5871	30.1910	33.4087	35.7185
18	17.3379	21.6049	25.9894	28.8693	31.5264	34.8053	37.1564
19	18.3376	22.7178	27.2036	30.1435	32.8523	36.1908	38.5822
20	19.3374	23.8277	28.4120	31.4104	34.1696	37.5662	39.9968
21	20.3372	24.9348	29.6151	32.6705	35.4789	38.9321	41.4010
22	21.3370	26.0393	30.8133	33.9244	36.7807	40.2894	42.7956
23	22.3369	27.1413	32.0069	35.1725	38.0757	41.6384	44.1813
24	23.3367	28.2412	33.1963	36.4151	39.3641	42.9798	45.5585
25	24.3366	29.3389	34.3816	37.6525	40.6465	44.3141	46.9278
26	25.3364	30.4345	35.5631	38.8852	41.9232	45.6417	48.2899
27	26.3363	31.5284	36.7412	40.1133	43.1944	46.9630	49.6449
28	27.3363	32.6205	37.9159	41.3372	44.4607	48.2782	50.9933
29	28.3362	33.7109	39.0875	42.5569	45.7222	49.5879	52.3356
30	29.3360	34.7998	40.2560	43.7729	46.9792	50.8922	53.6720
40	39.3354	45.6160	51.8050	55.7585	59.3417	63.6907	66.7659
50	49.3349	56.3336	63.1671	67.5048	71.4202	76.1539	79.4900
60	59.3347	66.9814	74.3970	79.0819	83.2976	88.3794	91.9517
70	69.3344	77.5766	85.5271	90.5312	95.0231	100.425	104.215
80	79.3343	88.1303	96.5782	101.879	106.629	112.329	116.321
90	89.3342	98.6499	107.565	113.145	118.136	124.116	128.299
100	99.3341	109.141	118.498	124.342	129.561	135.807	140.169
t_α	0.0000	+0.6745	+1.2816	+1.6449	+1.9600	+2.3263	+2.5758

SOURCE: E. S. Pearson, Tables of the Percentage Points of the χ^2 Distribution, *Biometrika*, vol. 32, pp. 188–189, 1941.

Table A·4 Critical Values of the F Distribution

One-tail critical values

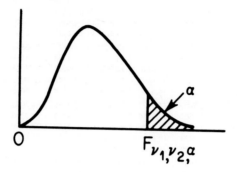

$$F_{\nu_1, \nu_2, \alpha}$$

(For one-tail tests, $\alpha = \alpha_0$)
(For two-tail tests, $\alpha = \alpha_{0/2}$)

Where interpolation is necessary, it should be carried out as a linear interpolation using reciprocals of the degrees of freedom involved instead of the actual degrees of freedom themselves.

Table A·4   Critical Values of the F Distribution (Continued)

$$\alpha = 0.10$$

ν_1 / ν_2	1	2	3	4	5	6	7	8	9
1	39.864	49.500	53.593	55.833	57.241	58.204	58.906	59.439	59.858
2	8.5263	9.0000	9.1618	9.2434	9.2926	9.3255	9.3491	9.3668	9.3805
3	5.5383	5.4624	5.3908	5.3427	5.3092	5.2847	5.2662	5.2517	5.2400
4	4.5448	4.3246	4.1908	4.1073	4.0506	4.0098	3.9790	3.9549	3.9357
5	4.0604	3.7797	3.6195	3.5202	3.4530	3.4045	3.3679	3.3393	3.3163
6	3.7760	3.4633	3.2888	3.1808	3.1075	3.0546	3.0145	2.9830	2.9577
7	3.5894	3.2574	3.0741	2.9605	2.8833	2.8274	2.7849	2.7516	2.7247
8	3.4579	3.1131	2.9238	2.8064	2.7265	2.6683	2.6241	2.5893	2.5612
9	3.3603	3.0065	2.8129	2.6927	2.6106	2.5509	2.5053	2.4694	2.4403
10	3.2850	2.9245	2.7277	2.6053	2.5216	2.4606	2.4140	2.3772	2.3473
11	3.2252	2.8595	2.6602	2.5362	2.4512	2.3891	2.3416	2.3040	2.2735
12	3.1765	2.8068	2.6055	2.4801	2.3940	2.3310	2.2828	2.2446	2.2135
13	3.1362	2.7632	2.5603	2.4337	2.3467	2.2830	2.2341	2.1953	2.1638
14	3.1022	2.7265	2.5222	2.3947	2.3069	2.2426	2.1931	2.1539	2.1220
15	3.0732	2.6952	2.4898	2.3614	2.2730	2.2081	2.1582	2.1185	2.0862
16	3.0481	2.6682	2.4618	2.3327	2.2438	2.1783	2.1280	2.0880	2.0553
17	3.0262	2.6446	2.4374	2.3077	2.2183	2.1524	2.1017	2.0613	2.0284
18	3.0070	2.6239	2.4160	2.2858	2.1958	2.1296	2.0785	2.0379	2.0047
19	2.9899	2.6056	2.3970	2.2663	2.1760	2.1094	2.0580	2.0171	1.9836
20	2.9747	2.5893	2.3801	2.2489	2.1582	2.0913	2.0397	1.9985	1.9649
21	2.9609	2.5746	2.3649	2.2333	2.1423	2.0751	2.0232	1.9819	1.9480
22	2.9486	2.5613	2.3512	2.2193	2.1279	2.0605	2.0084	1.9668	1.9327
23	2.9374	2.5493	2.3387	2.2065	2.1149	2.0472	1.9949	1.9531	1.9189
24	2.9271	2.5383	2.3274	2.1949	2.1030	2.0351	1.9826	1.9407	1.9063
25	2.9177	2.5283	2.3170	2.1843	2.0922	2.0241	1.9714	1.9292	1.8947
26	2.9091	2.5191	2.3075	2.1745	2.0822	2.0139	1.9610	1.9188	1.8841
27	2.9012	2.5106	2.2987	2.1655	2.0730	2.0045	1.9515	1.9091	1.8743
28	2.8939	2.5028	2.2906	2.1571	2.0645	1.9959	1.9427	1.9001	1.8652
29	2.8871	2.4955	2.2831	2.1494	2.0566	1.9878	1.9345	1.8918	1.8568
30	2.8807	2.4887	2.2761	2.1422	2.0492	1.9803	1.9269	1.8841	1.8490
40	2.8354	2.4404	2.2261	2.0909	1.9968	1.9269	1.8725	1.8289	1.7929
60	2.7914	2.3932	2.1774	2.0410	1.9457	1.8747	1.8194	1.7748	1.7380
120	2.7478	2.3473	2.1300	1.9923	1.8959	1.8238	1.7675	1.7220	1.6843
∞	2.7055	2.3026	2.0838	1.9449	1.8473	1.7741	1.7167	1.6702	1.6315

Table A·4   Critical Values of the F Distribution (Continued)

$$\alpha = 0.10$$

ν_1 / ν_2	10	12	15	20	24	30	40	60	120	∞
1	60.195	60.705	61.220	61.740	62.002	62.265	62.529	62.794	63.061	63.328
2	9.3916	9.4081	9.4247	9.4413	9.4496	9.4579	9.4663	9.4746	9.4829	9.4913
3	5.2304	5.2156	5.2003	5.1845	5.1764	5.1681	5.1597	5.1512	5.1425	5.1337
4	3.9199	3.8955	3.8689	3.8443	3.8310	3.8174	3.8036	3.7896	3.7753	3.7607
5	3.2974	3.2682	3.2380	3.2067	3.1905	3.1741	3.1573	3.1402	3.1228	3.1050
6	2.9369	2.9047	2.8712	2.8363	2.8183	2.8000	2.7812	2.7620	2.7423	2.7222
7	2.7025	2.6681	2.6322	2.5947	2.5753	2.5555	2.5351	2.5142	2.4928	2.4708
8	2.5380	2.5020	2.4642	2.4246	2.4041	2.3830	2.3614	2.3391	2.3162	2.2926
9	2.4163	2.3789	2.3396	2.2983	2.2768	2.2547	2.2320	2.2085	2.1843	2.1592
10	2.3226	2.2841	2.2435	2.2007	2.1784	2.1554	2.1317	2.1072	2.0818	2.0554
11	2.2482	2.2087	2.1671	2.1230	2.1000	2.0762	2.0516	2.0261	1.9997	1.9721
12	2.1878	2.1474	2.1049	2.0597	2.0360	2.0115	1.9861	1.9597	1.9323	1.9036
13	2.1376	2.0966	2.0532	2.0070	1.9827	1.9576	1.9315	1.9043	1.8759	1.8462
14	2.0954	2.0537	2.0095	1.9625	1.9377	1.9119	1.8852	1.8572	1.8280	1.7973
15	2.0593	2.0171	1.9722	1.9243	1.8990	1.8728	1.8454	1.8168	1.7867	1.7551
16	2.0281	1.9854	1.9399	1.8913	1.8656	1.8388	1.8108	1.7816	1.7507	1.7182
17	2.0009	1.9577	1.9117	1.8624	1.8362	1.8090	1.7805	1.7506	1.7191	1.6856
18	1.9770	1.9333	1.8868	1.8368	1.8103	1.7827	1.7537	1.7232	1.6910	1.6567
19	1.9557	1.9117	1.8647	1.8142	1.7873	1.7592	1.7298	1.6988	1.6659	1.6308
20	1.9367	1.8924	1.8449	1.7938	1.7667	1.7382	1.7083	1.6768	1.6433	1.6074
21	1.9197	1.8750	1.8272	1.7756	1.7481	1.7193	1.6890	1.6569	1.6228	1.5862
22	1.9043	1.8593	1.8111	1.7590	1.7312	1.7021	1.6714	1.6389	1.6042	1.5668
23	1.8903	1.8450	1.7964	1.7439	1.7159	1.6864	1.6554	1.6224	1.5871	1.5490
24	1.8775	1.8319	1.7831	1.7302	1.7019	1.6721	1.6407	1.6073	1.5715	1.5327
25	1.8658	1.8200	1.7708	1.7175	1.6890	1.6589	1.6272	1.5934	1.5570	1.5176
26	1.8550	1.8090	1.7596	1.7059	1.6771	1.6468	1.6147	1.5805	1.5437	1.5036
27	1.8451	1.7989	1.7492	1.6951	1.6662	1.6356	1.6032	1.5686	1.5313	1.4906
28	1.8359	1.7895	1.7395	1.6852	1.6560	1.6252	1.5925	1.5575	1.5198	1.4784
29	1.8274	1.7808	1.7306	1.6759	1.6465	1.6155	1.5825	1.5472	1.5090	1.4670
30	1.8195	1.7727	1.7223	1.6673	1.6377	1.6065	1.5732	1.5376	1.4989	1.4564
40	1.7627	1.7146	1.6624	1.6052	1.5741	1.5411	1.5056	1.4672	1.4248	1.3769
60	1.7070	1.6574	1.6034	1.5435	1.5107	1.4755	1.4373	1.3952	1.3476	1.2915
120	1.6524	1.6012	1.5450	1.4821	1.4472	1.4094	1.3676	1.3203	1.2646	1.1926
∞	1.5987	1.5458	1.4871	1.4206	1.3832	1.3419	1.2951	1.2400	1.1686	1.0000

$$\alpha = 0.05$$

v_1 / v_2	1	2	3	4	5	6	7	8	9
1	161.45	199.50	215.71	224.58	230.16	233.99	236.77	238.88	240.54
2	18.513	19.000	19.164	19.247	19.296	19.330	19.353	19.371	19.385
3	10.128	9.5521	9.2766	9.1172	9.0135	8.9406	8.8868	8.8452	8.8123
4	7.7086	6.9443	6.5914	6.3883	6.2560	6.1631	6.0942	6.0410	5.9988
5	6.6079	5.7861	5.4095	5.1922	5.0503	4.9503	4.8759	4.8183	4.7725
6	5.9874	5.1433	4.7571	4.5337	4.3874	4.2839	4.2066	4.1468	4.0990
7	5.5914	4.7374	4.3468	4.1203	3.9715	3.8660	3.7870	3.7257	3.6767
8	5.3177	4.4590	4.0662	3.8378	3.6875	3.5806	3.5005	3.4381	3.3881
9	5.1174	4.2565	3.8626	3.6331	3.4817	3.3738	3.2927	3.2296	3.1789
10	4.9646	4.1028	3.7083	3.4780	3.3258	3.2172	3.1355	3.0717	3.0204
11	4.8443	3.9823	3.5874	3.3567	3.2039	3.0946	3.0123	2.9480	2.8962
12	4.7472	3.8853	3.4903	3.2592	3.1059	2.9961	2.9134	2.8486	2.7964
13	4.6672	3.8056	3.4105	3.1791	3.0254	2.9153	2.8321	2.7669	2.7144
14	4.6001	3.7389	3.3439	3.1122	2.9582	2.8477	2.7642	2.6987	2.6458
15	4.5431	3.6823	3.2874	3.0556	2.9013	2.7905	2.7066	2.6408	2.5876
16	4.4940	3.6337	3.2389	3.0069	2.8524	2.7413	2.6572	2.5911	2.5377
17	4.4513	3.5915	3.1968	2.9647	2.8100	2.6987	2.6143	2.5480	2.4943
18	4.4139	3.5546	3.1599	2.9277	2.7729	2.6613	2.5767	2.5102	2.4563
19	4.3808	3.5219	3.1274	2.8951	2.7401	2.6283	2.5435	2.4768	2.4227
20	4.3513	3.4928	3.0984	2.8661	2.7109	2.5990	2.5140	2.4471	2.3928
21	4.3248	3.4668	3.0725	2.8401	2.6848	2.5727	2.4876	2.4205	2.3661
22	4.3009	3.4434	3.0491	2.8167	2.6613	2.5491	2.4638	2.3965	2.3419
23	4.2793	3.4221	3.0280	2.7955	2.6400	2.5277	2.4422	2.3748	2.3201
24	4.2597	3.4028	3.0088	2.7763	2.6207	2.5082	2.4226	2.3551	2.3002
25	4.2417	3.3852	2.9912	2.7587	2.6030	2.4904	2.4047	2.3371	2.2821
26	4.2252	3.3690	2.9751	2.7426	2.5868	2.4741	2.3883	2.3205	2.2655
27	4.2100	3.3541	2.9604	2.7278	2.5719	2.4591	2.3732	2.3053	2.2501
28	4.1960	3.3404	2.9467	2.7141	2.5581	2.4453	2.3593	2.2913	2.2360
29	4.1830	3.3277	2.9340	2.7014	2.5454	2.4324	2.3463	2.2782	2.2229
30	4.1709	3.3158	2.9223	2.6896	2.5336	2.4205	2.3343	2.2662	2.2107
40	4.0848	3.2317	2.8387	2.6060	2.4495	2.3359	2.2490	2.1802	2.1240
60	4.0012	3.1504	2.7581	2.5252	2.3683	2.2540	2.1665	2.0970	2.0401
120	3.9201	3.0718	2.6802	2.4472	2.2900	2.1750	2.0867	2.0164	1.9588
∞	3.8415	2.9957	2.6049	2.3719	2.2141	2.0986	2.0096	1.9384	1.8799

Table A·4   Critical Values of the F Distribution (Continued)

$$\alpha = 0.05$$

ν_2 \ ν_1	10	12	15	20	24	30	40	60	120	∞
1	241.88	243.91	245.95	248.01	249.05	250.09	251.14	252.20	253.25	254.32
2	19.396	19.413	19.429	19.446	19.454	19.462	19.471	19.479	19.487	19.496
3	8.7855	8.7446	8.7029	8.6602	8.6385	8.6166	8.5944	8.5720	8.5494	8.5265
4	5.9644	5.9117	5.8578	5.8025	5.7744	5.7459	5.7170	5.6878	5.6581	5.6281
5	4.7351	4.6777	4.6188	4.5581	4.5272	4.4957	4.4638	4.4314	4.3984	4.3650
6	4.0600	3.9999	3.9381	3.8742	3.8415	3.8082	3.7743	3.7398	3.7047	3.6688
7	3.6365	3.5747	3.5108	3.4445	3.4105	3.3758	3.3404	3.3043	3.2674	3.2298
8	3.3472	3.2840	3.2184	3.1503	3.1152	3.0794	3.0428	3.0053	2.9669	2.9276
9	3.1373	3.0729	3.0061	2.9365	2.9005	2.8637	2.8259	2.7872	2.7475	2.7067
10	2.9782	2.9130	2.8450	2.7740	2.7372	2.6996	2.6609	2.6211	2.5801	2.5379
11	2.8536	2.7876	2.7186	2.6464	2.6090	2.5705	2.5309	2.4901	2.4480	2.4045
12	2.7534	2.6866	2.6169	2.5436	2.5055	2.4663	2.4259	2.3842	2.3410	2.2962
13	2.6710	2.6037	2.5331	2.4589	2.4202	2.3803	2.3392	2.2966	2.2524	2.2064
14	2.6021	2.5342	2.4630	2.3879	2.3487	2.3082	2.2664	2.2230	2.1778	2.1307
15	2.5437	2.4753	2.4035	2.3275	2.2878	2.2468	2.2043	2.1601	2.1141	2.0658
16	2.4935	2.4247	2.3522	2.2756	2.2354	2.1938	2.1507	2.1058	2.0589	2.0096
17	2.4499	2.3807	2.3077	2.2304	2.1898	2.1477	2.1040	2.0584	2.0107	1.9604
18	2.4117	2.3421	2.2686	2.1906	2.1497	2.1071	2.0629	2.0166	1.9681	1.9168
19	2.3779	2.3080	2.2341	2.1555	2.1141	2.0712	2.0264	1.9796	1.9302	1.8780
20	2.3479	2.2776	2.2033	2.1242	2.0825	2.0391	1.9938	1.9464	1.8963	1.8432
21	2.3210	2.2504	2.1757	2.0960	2.0540	2.0102	1.9645	1.9165	1.8657	1.8117
22	2.2967	2.2258	2.1508	2.0707	2.0283	1.9842	1.9380	1.8895	1.8380	1.7831
23	2.2747	2.2036	2.1282	2.0476	2.0050	1.9605	1.9139	1.8649	1.8128	1.7570
24	2.2547	2.1834	2.1077	2.0267	1.9838	1.9390	1.8920	1.8424	1.7897	1.7331
25	2.2365	2.1649	2.0889	2.0075	1.9643	1.9192	1.8718	1.8217	1.7684	1.7110
26	2.2197	2.1479	2.0716	1.9898	1.9464	1.9010	1.8533	1.8027	1.7488	1.6906
27	2.2043	2.1323	2.0558	1.9736	1.9299	1.8842	1.8361	1.7851	1.7307	1.6717
28	2.1900	2.1179	2.0411	1.9586	1.9147	1.8687	1.8203	1.7689	1.7138	1.6541
29	2.1768	2.1045	2.0275	1.9446	1.9005	1.8543	1.8055	1.7537	1.6981	1.6377
30	2.1646	2.0921	2.0148	1.9317	1.8874	1.8409	1.7918	1.7396	1.6835	1.6223
40	2.0772	2.0035	1.9245	1.8389	1.7929	1.7444	1.6928	1.6373	1.5766	1.5089
60	1.9926	1.9174	1.8364	1.7480	1.7001	1.6491	1.5943	1.5343	1.4673	1.3893
120	1.9105	1.8337	1.7505	1.6587	1.6084	1.5543	1.4952	1.4290	1.3519	1.2539
∞	1.8307	1.7522	1.6664	1.5705	1.5173	1.4591	1.3940	1.3180	1.2214	1.0000

Table A·4 Critical Values of the F Distribution (Continued)

$$\alpha = 0.01$$

v_1 v_2	1	2	3	4	5	6	7	8	9
1	4052.2	4999.5	5403.3	5624.6	5763.7	5859.0	5928.3	5981.6	6022.5
2	98.503	99.000	99.166	99.249	99.299	99.332	99.356	99.374	99.388
3	34.116	30.817	29.457	28.710	28.237	27.911	27.672	27.489	27.345
4	21.198	18.000	16.694	15.977	15.522	15.207	14.976	14.799	14.659
5	16.258	13.274	12.060	11.392	10.967	10.672	10.456	10.289	10.158
6	13.745	10.925	9.7795	9.1483	8.7459	8.4661	8.2600	8.1016	7.9761
7	12.246	9.5466	8.4513	7.8467	7.4604	7.1914	6.9928	6.8401	6.7188
8	11.259	8.6491	7.5910	7.0060	6.6318	6.3707	6.1776	6.0289	5.9106
9	10.561	8.0215	6.9919	6.4221	6.0569	5.8018	5.6129	5.4671	5.3511
10	10.044	7.5594	6.5523	5.9943	5.6363	5.3858	5.2001	5.0567	4.9424
11	9.6460	7.2057	6.2167	5.6683	5.3160	5.0692	4.8861	4.7445	4.6315
12	9.3302	6.9266	5.9526	5.4119	5.0643	4.8206	4.6395	4.4994	4.3875
13	9.0738	6.7010	5.7394	5.2053	4.8616	4.6204	4.4410	4.3021	4.1911
14	8.8616	6.5149	5.5639	5.0354	4.6950	4.4558	4.2779	4.1399	4.0297
15	8.6831	6.3589	5.4170	4.8932	4.5556	4.3183	4.1415	4.0045	3.8948
16	8.5310	6.2262	5.2922	4.7726	4.4374	4.2016	4.0259	3.8896	3.7804
17	8.3997	6.1121	5.1850	4.6690	4.3359	4.1015	3.9267	3.7910	3.6822
18	8.2854	6.0129	5.0919	4.5790	4.2479	4.0146	3.8406	3.7054	3.5971
19	8.1850	5.9259	5.0103	4.5003	4.1708	3.9386	3.7653	3.6305	3.5225
20	8.0960	5.8489	4.9382	4.4307	4.1027	3.8714	3.6987	3.5644	3.4567
21	8.0166	5.7804	4.8740	4.3688	4.0421	3.8117	3.6396	3.5056	3.3981
22	7.9454	5.7190	4.8166	4.3134	3.9880	3.7583	3.5867	3.4530	3.3458
23	7.8811	5.6637	4.7649	4.2635	3.9392	3.7102	3.5390	3.4057	3.2986
24	7.8229	5.6136	4.7181	4.2184	3.8951	3.6667	3.4959	3.3629	3.2560
25	7.7698	5.5680	4.6755	4.1774	3.8550	3.6272	3.4568	3.3239	3.2172
26	7.7213	5.5263	4.6366	4.1400	3.8183	3.5911	3.4210	3.2884	3.1818
27	7.6767	5.4881	4.6009	4.1056	3.7848	3.5580	3.3882	3.2558	3.1494
28	7.6356	5.4529	4.5681	4.0740	3.7539	3.5276	3.3581	3.2259	3.1195
29	7.5976	5.4205	4.5378	4.0449	3.7254	3.4995	3.3302	3.1982	3.0920
30	7.5625	5.3904	4.5097	4.0179	3.6990	3.4735	3.3045	3.1726	3.0665
40	7.3141	5.1785	4.3126	3.8283	3.5138	3.2910	3.1238	2.9930	2.8876
60	7.0771	4.9774	4.1259	3.6491	3.3389	3.1187	2.9530	2.8233	2.7185
120	6.8510	4.7865	3.9493	3.4796	3.1735	2.9559	2.7918	2.6629	2.5586
∞	6.6349	4.6052	3.7816	3.3192	3.0173	2.8020	2.6393	2.5113	2.4073

Table A·4   Critical Values of the F Distribution (Continued)

$$\alpha = 0.01$$

v_2 \ v_1	10	12	15	20	24	30	40	60	120	∞
1	6055.8	6106.3	6157.3	6208.7	6234.6	6260.7	6286.8	6313.0	6339.4	6366.0
2	99.399	99.416	99.432	99.449	99.458	99.466	99.474	99.483	99.491	99.501
3	27.229	27.052	26.872	26.690	26.598	26.505	26.411	26.316	26.221	26.125
4	14.546	14.374	14.198	14.020	13.929	13.838	13.745	13.652	13.558	13.463
5	10.051	9.8883	9.7222	9.5527	9.4665	9.3793	9.2912	9.2020	9.1118	9.0204
6	7.8741	7.7183	7.5590	7.3958	7.3127	7.2285	7.1432	7.0568	6.9690	6.8801
7	6.6201	6.4691	6.3143	6.1554	6.0743	5.9921	5.9084	5.8236	5.7372	5.6495
8	5.8143	5.6668	5.5151	5.3591	5.2793	5.1981	5.1156	5.0316	4.9460	4.8588
9	5.2565	5.1114	4.9621	4.8080	4.7290	4.6486	4.5667	4.4831	4.3978	4.3105
10	4.8492	4.7059	4.5582	4.4054	4.3269	4.2469	4.1653	4.0819	3.9965	3.9090
11	4.5393	4.3974	4.2509	4.0990	4.0209	3.9411	3.8596	3.7761	3.6904	3.6025
12	4.2961	4.1553	4.0096	3.8584	3.7805	3.7008	3.6192	3.5355	3.4494	3.3608
13	4.1003	3.9603	3.8154	3.6646	3.5868	3.5070	3.4253	3.3413	3.2548	3.1654
14	3.9394	3.8001	3.6557	3.5052	3.4274	3.3476	3.2656	3.1813	3.0942	3.0040
15	3.8049	3.6662	3.5222	3.3719	3.2940	3.2141	3.1319	3.0471	2.9595	2.8684
16	3.6909	3.5527	3.4089	3.2588	3.1808	3.1007	3.0182	2.9330	2.8447	2.7528
17	3.5931	3.4552	3.3117	3.1615	3.0835	3.0032	2.9205	2.8348	2.7459	2.6530
18	3.5082	3.3706	3.2273	3.0771	2.9990	2.9185	2.8354	2.7493	2.6597	2.5660
19	3.4338	3.2965	3.1533	3.0031	2.9249	2.8442	2.7608	2.6742	2.5839	2.4893
20	3.3682	3.2311	3.0880	2.9377	2.8594	2.7785	2.6947	2.6077	2.5168	2.4212
21	3.3098	3.1729	3.0299	2.8796	2.8011	2.7200	2.6359	2.5484	2.4568	2.3603
22	3.2576	3.1209	2.9780	2.8274	2.7488	2.6675	2.5831	2.4951	2.4029	2.3055
23	3.2106	3.0740	2.9311	2.7805	2.7017	2.6202	2.5355	2.4471	2.3542	2.2559
24	3.1681	3.0316	2.8887	2.7380	2.6591	2.5773	2.4923	2.4035	2.3099	2.2107
25	3.1294	2.9931	2.8502	2.6993	2.6203	2.5383	2.4530	2.3637	2.2695	2.1694
26	3.0941	2.9579	2.8150	2.6640	2.5848	2.5026	2.4170	2.3273	2.2325	2.1315
27	3.0618	2.9256	2.7827	2.6316	2.5522	2.4699	2.3840	2.2938	2.1984	2.0965
28	3.0320	2.8959	2.7530	2.6017	2.5223	2.4397	2.3535	2.2629	2.1670	2.0642
29	3.0045	2.8685	2.7256	2.5742	2.4946	2.4118	2.3253	2.2344	2.1378	2.0342
30	2.9791	2.8431	2.7002	2.5487	2.4689	2.3860	2.2992	2.2079	2.1107	2.0062
40	2.8005	2.6648	2.5216	2.3689	2.2880	2.2034	2.1142	2.0194	1.9172	1.8047
60	2.6318	2.4961	2.3523	2.1978	2.1154	2.0285	1.9360	1.8363	1.7263	1.6006
120	2.4721	2.3363	2.1915	2.0346	1.9500	1.8600	1.7628	1.6557	1.5330	1.3805
∞	2.3209	2.1848	2.0385	1.8783	1.7908	1.6964	1.5923	1.4730	1.3246	1.0000

SOURCE: E. S. Pearson, Tables of Percentage Points of the Inverted Beta (F) Distribution, *Biometrika*, vol. 32, pp. 73–88, 1943.

Table A·5 Values of the Correlation Coefficient for Different Levels of Significance

ν	$\alpha = 0.10$	0.05	0.02	0.01
1	0.98769	0.996917	0.9995066	0.998766
2	0.90000	0.95000	0.98000	0.990000
3	0.8054	0.8783	0.93433	0.95873
4	0.7293	0.8114	0.8822	0.91720
5	0.6694	0.7545	0.8329	0.8745
6	0.6215	0.7067	0.7887	0.8343
7	0.5822	0.6664	0.7498	0.7977
8	0.5494	0.6319	0.7155	0.7646
9	0.5214	0.6021	0.6851	0.7348
10	0.4973	0.5760	0.6581	0.7079
11	0.4762	0.5529	0.6339	0.6835
12	0.4575	0.5324	0.6120	0.6614
13	0.4409	0.5139	0.5923	0.6411
14	0.4259	0.4973	0.5742	0.6226
15	0.4124	0.4821	0.5577	0.6055
16	0.4000	0.4683	0.5425	0.5897
17	0.3887	0.4555	0.5285	0.5751
18	0.3783	0.4438	0.5155	0.5614
19	0.3687	0.4329	0.5034	0.5487
20	0.3598	0.4227	0.4921	0.5368
25	0.3233	0.3809	0.4451	0.4869
30	0.2960	0.3494	0.4093	0.4487
35	0.2746	0.3246	0.3810	0.4182
40	0.2573	0.3044	0.3578	0.3932
45	0.2428	0.2875	0.3384	0.3721
50	0.2306	0.2732	0.3218	0.3541
60	0.2108	0.2500	0.2948	0.3248
70	0.1954	0.2319	0.2737	0.3017
80	0.1829	0.2172	0.2565	0.2830
90	0.1726	0.2050	0.2422	0.2673
100	0.1638	0.1946	0.2301	0.2540

For a total correlation, ν is 2 less than the number of pairs in the sample; for a partial correlation, the number of eliminated variates also should be subtracted.

Source: R. A. Fisher, "Statistical Methods for Research Workers," Oliver & Boyd Ltd., Edinburgh and London, 1954.

Table A·6 Factors for Computing Control-Chart Lines

Sample Size n	Chart for Averages — Factors for 3σ Control Limits				Chart for Individuals — Factors for 3σ Control Limits		Charts for Ranges — Factors for 3σ Control Limits				Factors for Central Line		Charts for Sample Standard Deviation (s) — Factors for 3σ Control Limits				Sample Size n
	A	A_1	A_2	A_3	E_2	E_3	D_1	D_2	D_3	D_4	d_2	c_3	B_5	B_6	B_7	B_8	
2.	2.121	3.760	1.880	2.659	2.660	3.760	0	3.686	0	3.267	1.128	0.7979	0	2.606	0	3.267	2.
3.	1.732	2.394	1.023	1.954	1.772	3.385	0	4.358	0	2.575	1.693	0.8862	0	2.276	0	2.568	3.
4.	1.500	1.880	0.729	1.628	1.457	3.256	0	4.698	0	2.282	2.059	0.9213	0	2.088	0	2.266	4.
5.	1.342	1.596	0.577	1.427	1.290	3.192	0	4.918	0	2.115	2.326	0.9400	0	1.964	0	2.089	5.
6.	1.225	1.410	0.483	1.287	1.184	3.153	0	5.078	0	2.004	2.534	0.9515	0.029	1.874	0.030	1.970	6.
7.	1.134	1.277	0.419	1.182	1.109	3.127	0.205	5.203	0.076	1.924	2.704	0.9594	0.113	1.806	0.118	1.882	7.
8.	1.061	1.175	0.373	1.099	1.054	3.109	0.387	5.307	0.136	1.864	2.847	0.9650	0.179	1.751	0.185	1.815	8.
9.	1.000	1.094	0.337	1.032	1.010	3.095	0.546	5.394	0.184	1.816	2.970	0.9693	0.232	1.707	0.239	1.761	9.
10.	0.949	1.028	0.308	0.975	0.975	3.084	0.687	5.469	0.223	1.777	3.078	0.9727	0.276	1.669	0.284	1.716	10.
11.	0.904	0.973	0.285	0.927	0.946	3.076	0.812	5.534	0.256	1.744	3.173	0.9754	0.313	1.637	0.321	1.679	11.
12.	0.866	0.925	0.266	0.886	0.921	3.069	0.924	5.592	0.284	1.716	3.258	0.9776	0.346	1.610	0.354	1.646	12.
13.	0.832	0.884	0.249	0.850	0.899	3.063	1.026	5.646	0.308	1.692	3.336	0.9794	0.374	1.585	0.382	1.618	13.
14.	0.802	0.848	0.235	0.817	0.881	3.058	1.121	5.693	0.329	1.671	3.407	0.9810	0.399	1.563	0.406	1.594	14.
15.	0.775	0.816	0.223	0.789	0.864	3.054	1.207	5.737	0.348	1.652	3.472	0.0923	0.421	1.544	0.428	1.572	15.

Averages

$\overline{\overline{X}} \pm A\sigma'$
$\overline{\overline{X}} \pm A_1\overline{\sigma}$
$\overline{\overline{X}} \pm A_2\overline{R}$
$\overline{\overline{X}} \pm A_3\overline{s}$

Individuals

$\overline{X} \pm 3\sigma'$
$\overline{X} \pm E_1\overline{\sigma}$
$\overline{X} \pm E_2\overline{R}$
$\overline{X} \pm E_3\overline{s}$

Ranges

LCL $= D_1\sigma'$
UCL $= D_2\sigma'$
LCL $= D_3\overline{R}$
UCL $= D_4\overline{R}$

Standard Deviations

LCL $= B_5\sigma'$
UCL $= B_6\sigma'$
LCL $= B_7\overline{s}$
UCL $= B_8\overline{s}$

Central Lines

Ranges: $d_2\sigma'$
Sample Standard Deviations: $o_8\sigma'$
Note: σ' is used to denote a Standard or known value.

Table A·8 Summation of Terms of Poisson's Exponential Binomial Limit

1,000 × probability of c or less occurrences of event that has average number of occurrences equal to c' or np'

c' or np'	0	1	2	3	4	5	6	7	8	9
0.02	980	1,000								
0.04	961	999	1,000							
0.06	942	998	1,000							
0.08	923	997	1,000							
0.10	905	995	1,000							
0.15	861	990	999	1,000						
0.20	819	982	999	1,000						
0.25	779	974	998	1,000						
0.30	741	963	996	1,000						
0.35	705	951	994	1,000						
0.40	670	938	992	999	1,000					
0.45	638	925	989	999	1,000					
0.50	607	910	986	998	1,000					
0.55	577	894	982	998	1,000					
0.60	549	878	977	997	1,000					
0.65	522	861	972	996	999	1,000				
0.70	497	844	966	994	999	1,000				
0.75	472	827	959	993	999	1,000				
0.80	449	809	953	991	999	1,000				
0.85	427	791	945	989	998	1,000				
0.90	407	772	937	987	998	1,000				
0.95	387	754	929	984	997	1,000				
1.00	368	736	920	981	996	999	1,000			
1.1	333	699	900	974	995	999	1,000			
1.2	301	663	879	966	992	998	1,000			
1.3	273	627	857	957	989	998	1,000			
1.4	247	592	833	946	986	997	999	1,000		
1.5	223	558	809	934	981	996	999	1,000		
1.6	202	525	783	921	976	994	999	1,000		
1.7	183	493	757	907	970	992	998	1,000		
1.8	165	463	731	891	964	990	997	999	1,000	
1.9	150	434	704	875	956	987	997	999	1,000	
2.0	135	406	677	857	947	983	995	999	1,000	

c' or np'	0	1	2	3	4	5	6	7	8	9
2.2	111	355	623	819	928	975	993	998	1,000	
2.4	091	308	570	779	904	964	988	997	999	1,000
2.6	074	267	518	736	877	951	983	995	999	1,000
2.8	061	231	469	692	848	935	976	992	998	999
3.0	050	199	423	647	815	916	966	988	996	999
3.2	041	171	380	603	781	895	955	983	994	998
3.4	033	147	340	558	744	871	942	977	992	997
3.6	027	126	303	515	706	844	927	969	988	996
3.8	022	107	269	473	668	816	909	960	984	994
4.0	018	092	238	433	629	785	889	949	979	992
4.2	015	078	210	395	590	753	867	936	972	989
4.4	012	066	185	359	551	720	844	921	964	985
4.6	010	056	163	326	513	686	818	905	955	980
4.8	008	048	143	294	476	651	791	887	944	975
5.0	007	040	125	265	440	616	762	867	932	968
5.2	006	034	109	238	406	581	732	845	918	960
5.4	005	029	095	213	373	546	702	822	903	951
5.6	004	024	082	191	342	512	670	797	886	941
5.8	003	021	072	170	313	478	638	771	867	929
6.0	002	017	062	151	285	446	606	744	847	916

c' or np'	10	11	12	13	14	15	16
2.8	1,000						
3.0	1,000						
3.2	1,000						
3.4	999	1,000					
3.6	999	1,000					
3.8	998	999	1,000				
4.0	997	999	1,000				
4.2	996	999	1,000				
4.4	994	998	999	1,000			
4.6	992	997	999	1,000			
4.8	990	996	999	1,000			
5.0	986	995	998	999	1,000		
5.2	982	993	997	999	1,000		
5.4	977	990	996	999	1,000		
5.6	972	988	995	998	999	1,000	
5.8	965	984	993	997	999	1,000	
6.0	957	980	991	996	999	999	1,000

c *c′ or np′*	0	1	2	3	4	5	6	7	8	9
6.2	002	015	054	134	259	414	574	716	826	902
6.4	002	012	046	119	235	384	542	687	803	886
6.6	001	010	040	105	213	355	511	658	780	869
6.8	001	009	034	093	192	327	480	628	755	850
7.0	001	007	030	082	173	301	450	599	729	830
7.2	001	006	025	072	156	276	420	569	703	810
7.4	001	005	022	063	140	253	392	539	676	788
7.6	001	004	019	055	125	231	365	510	648	765
7.8	000	004	016	048	112	210	338	481	620	741
8.0	000	003	014	042	100	191	313	453	593	717
8.5	000	002	009	030	074	150	256	386	523	653
9.0	000	001	006	021	055	116	207	324	456	587
9.5	000	001	004	015	040	089	165	269	392	522
10.0	000	000	003	010	029	067	130	220	333	458

	10	11	12	13	14	15	16	17	18	19
6.2	949	975	989	995	998	999	1,000			
6.4	939	969	986	994	997	999	1,000			
6.6	927	963	982	992	997	999	999	1,000		
6.8	915	955	978	990	996	998	999	1,000		
7.0	901	947	973	987	994	998	999	1,000		
7.2	887	937	967	984	993	997	999	999	1,000	
7.4	871	926	961	980	991	996	998	999	1,000	
7.6	854	915	954	976	989	995	998	999	1,000	
7.8	835	902	945	971	986	993	997	999	1,000	
8.0	816	888	936	966	983	992	996	998	999	1,000
8.5	763	849	909	949	973	986	993	997	999	999
9.0	706	803	876	926	959	978	989	995	998	999
9.5	645	752	836	898	940	967	982	991	996	998
10.0	583	697	792	864	917	951	973	986	993	997

	20	21	22
8.5	1,000		
9.0	1,000		
9.5	999	1,000	
10.0	998	999	1,000

Table A·8   Summation of Terms of Poisson's Exponential Binomial Limit
(Continued)

c c' or np'	0	1	2	3	4	5	6	7	8	9
10.5	000	000	002	007	021	050	102	179	279	397
11.0	000	000	001	005	015	038	079	143	232	341
11.5	000	000	001	003	011	028	060	114	191	289
12.0	000	000	001	002	008	020	046	090	155	242
12.5	000	000	000	002	005	015	035	070	125	201
13.0	000	000	000	001	004	011	026	054	100	166
13.5	000	000	000	001	003	008	019	041	079	135
14.0	000	000	000	000	002	006	014	032	062	109
14.5	000	000	000	000	001	004	010	024	048	088
15.0	000	000	000	000	001	003	008	018	037	070

	10	11	12	13	14	15	16	17	18	19
10.5	521	639	742	825	888	932	960	978	988	994
11.0	460	579	689	781	854	907	944	968	982	991
11.5	402	520	633	733	815	878	924	954	974	986
12.0	347	462	576	682	772	844	899	937	963	979
12.5	297	406	519	628	725	806	869	916	948	969
13.0	252	353	463	573	675	764	835	890	930	957
13.5	211	304	409	518	623	718	798	861	908	942
14.0	176	260	358	464	570	669	756	827	883	923
14.5	145	220	311	413	518	619	711	790	853	901
15.0	118	185	268	363	466	568	664	749	819	875

	20	21	22	23	24	25	26	27	28	29
10.5	997	999	999	1,000						
11.0	995	998	999	1,000						
11.5	992	996	998	999	1,000					
12.0	988	994	997	999	999	1,000				
12.5	983	991	995	998	999	999	1,000			
13.0	975	986	992	996	998	999	1,000			
13.5	965	980	989	994	997	998	999	1,000		
14.0	952	971	983	991	995	997	999	999	1,000	
14.5	936	960	976	986	992	996	998	999	999	1,000
15.0	917	947	967	981	989	994	997	998	999	1,000

Table A·8 Summation of Terms of Poisson's Exponential Binomial Limit (Continued)

c' or np' \ c	4	5	6	7	8	9	10	11	12	13
16	000	001	004	010	022	043	077	127	193	275
17	000	001	002	005	013	026	049	085	135	201
18	000	000	001	003	007	015	030	055	092	143
19	000	000	001	002	004	009	018	035	061	098
20	000	000	000	001	002	005	011	021	039	066
21	000	000	000	000	001	003	006	013	025	043
22	000	000	000	000	001	002	004	008	015	028
23	000	000	000	000	000	001	002	004	009	017
24	000	000	000	000	000	000	001	003	005	011
25	000	000	000	000	000	000	001	001	003	006

	14	15	16	17	18	19	20	21	22	23
16	368	467	566	659	742	812	868	911	942	963
17	281	371	468	564	655	736	805	861	905	937
18	208	287	375	469	562	651	731	799	855	899
19	150	215	292	378	469	561	647	725	793	849
20	105	157	221	297	381	470	559	644	721	787
21	072	111	163	227	302	384	471	558	640	716
22	048	077	117	169	232	306	387	472	556	637
23	031	052	082	123	175	238	310	389	472	555
24	020	034	056	087	128	180	243	314	392	473
25	012	022	038	060	092	134	185	247	318	394

	24	25	26	27	28	29	30	31	32	33
16	978	987	993	996	998	999	999	1,000		
17	959	975	985	991	995	997	999	999	1,000	
18	932	955	972	983	990	994	997	998	999	1,000
19	893	927	951	969	980	988	993	996	998	999
20	843	888	922	948	966	978	987	992	995	997
21	782	838	883	917	944	963	976	985	991	994
22	712	777	832	877	913	940	959	973	983	989
23	635	708	772	827	873	908	936	956	971	981
24	554	632	704	768	823	868	904	932	953	969
25	473	553	629	700	763	818	863	900	929	950

	34	35	36	37	38	39	40	41	42	43
19	999	1,000								
20	999	999	1,000							
21	997	998	999	999	1,000					
22	994	996	998	999	999	1,000				
23	988	993	996	997	999	999	1,000			
24	979	987	992	995	997	998	999	999	1,000	
25	966	978	985	991	994	997	998	999	999	1,000

SOURCE: Eugene L. Grant, "Statistical Quality Control," 3d ed., McGraw-Hill Book Company, New York, 1964.

Table A·9   95 percent Confidence Belts for the Coefficient of Correlation

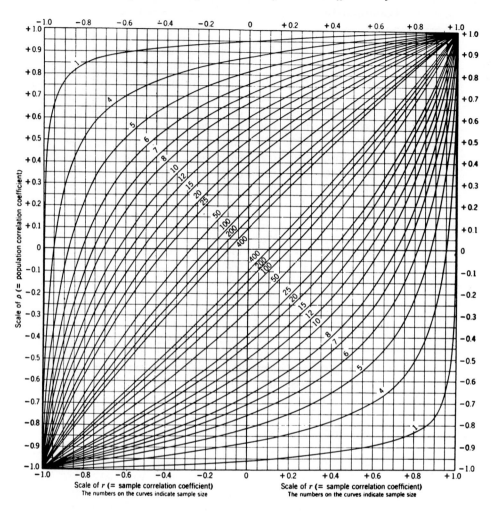

Scale of r (= sample correlation coefficient)
The numbers on the curves indicate sample size

Scale of r (= sample correlation coefficient)
The numbers on the curves indicate sample size

SOURCE: F. N. David, "Tables of the Correlation Coefficient," The Biometrika Office, University College, London, 1938. Reproduced by permission of Professor E. S. Pearson.

Table A·10   Criteria and Critical Values for Testing an Extreme Value

Statistic	Number of Obs., n	Critical Values	
		$\alpha = 0.05$	$\alpha = 0.01$
$r_{10} = \dfrac{X_2 - X_1}{X_n - X_1}$	3	0.941	0.988
	4	0.765	0.889
	5	0.642	0.780
	6	0.560	0.698
	7	0.507	0.637
$r_{11} = \dfrac{X_2 - X_1}{X_{n-1} - X_1}$	8	0.554	0.683
	9	0.512	0.635
	10	0.477	0.597
$r_{21} = \dfrac{X_3 - X_1}{X_{n-1} - X_1}$	11	0.576	0.679
	12	0.546	0.642
	13	0.521	0.615
$r_{22} = \dfrac{X_3 - X_1}{X_{n-2} - X_1}$	14	0.546	0.641
	15	0.525	0.616
	16	0.507	0.595
	17	0.490	0.577
	18	0.475	0.561
	19	0.462	0.547
	20	0.450	0.535
	21	0.440	0.524
	22	0.430	0.514
	23	0.421	0.505
	24	0.413	0.497
	25	0.406	0.489

SOURCE: W. J. Dixon, Processing Data for Outliers, *Biometrics*, vol. 9, pp. 74–89, 1953.

Table A·11 Number of Observations for t Test of Mean

The entries in this table show the numbers of observations needed in a t test of the significance of a mean in order to control the probabilities of errors of the first and second kinds at α and β, respectively

Level of t test

Value of $D = \dfrac{d}{s}$	0.01 (Single-sided α = 0.005, Double-sided α = 0.01)					0.02 (α = 0.01, α = 0.02)					0.05 (α = 0.025, α = 0.05)					0.1 (α = 0.05, α = 0.1)				
β →	0.01	0.05	0.1	0.2	0.5	0.01	0.05	0.1	0.2	0.5	0.01	0.05	0.1	0.2	0.5	0.01	0.05	0.1	0.2	0.5
0.05																				
0.10																				
0.15																				122
0.20	100									139					99					70
0.25					110	110				90				128	64			139	101	45
0.30		115	125	134	78		101	109	115	63	117	109	119	90	45	101	122	97	71	32
0.35		92	97	99	58		81	85	85	47	93	84	88	67	34	80	90	72	52	24
0.40		75	77	77	45		66	68	66	37	76	67	68	51	26	65	70	55	40	19
0.45			63	62	37			55	53	30		54	54	41	21		55	44	33	15
0.50				51	30				43	25			44	34	18		45	36	27	13
0.55	83	63	53	42	26	75	55	46	36	21	63	45	37	28	15	54	38	30	22	11
0.60	71	53	45	36	22	63	47	39	31	18	53	38	32	24	13	46	32	26	19	9
0.65	61	46	39	31	20	55	41	34	27	16	46	33	27	21	12	39	28	22	17	8
0.70	53	40	34	28	17	47	35	30	24	14	40	29	24	19	10	34	24	19	15	8
0.75	47	36	30	25	16	42	31	27	21	13	35	26	21	16	9	30	21	17	13	7

0.80	41	32	27	22	14	37	28	24	19	12	31	22	19	15	9	27	19	15	12	6
0.85	37	29	24	20	13	33	25	21	17	11	28	21	17	13	8	24	17	14	11	6
0.90	34	26	22	18	12	29	23	19	16	10	25	19	16	12	7	21	15	13	10	5
0.95	31	24	20	17	11	27	21	18	14	9	23	17	14	11	7	19	14	11	9	5
1.00	28	22	19	16	10	25	19	16	13	9	21	16	13	10	6	18	13	11	8	5
1.1	24	19	16	14	9	21	16	14	12	8	18	13	11	9	6	15	11	9	7	
1.2	21	16	14	12	8	18	14	12	10	7	15	12	10	8	6	13	10	8	6	
1.3	18	15	13	11	8	16	13	11	9	6	14	10	9	7	6	11	8	7	6	
1.4	16	13	12	10	7	14	11	10	9	6	12	9	8	7	6	10	8	7	5	
1.5	15	12	11	9	7	13	10	9	8	6	11	8	7	6	5	9	7	6		
1.6	13	11	10	8	6	12	10	9	7	5	10	8	7	6	6	8	6	6		
1.7	12	10	9	8	6	11	9	8	7		9	7	6	5	5	8	6	5		
1.8	12	10	9	8	6	10	8	7	7		8	7	6			7	6			
1.9	11	9	8	7	6	10	8	7	6		8	6	6			7	5			
2.0	10	8	8	7	5	9	7	7	6		7	6	5			6				
2.1	10	8	7	7		8	7	6	6		7	6				6				
2.2	9	8	7	6		8	7	6	6		7	6				6				
2.3	9	7	7	6		8	6	6	6		6	5				5				
2.4	8	7	7	6		7	6	6	6		6									
2.5	8	7	6	6		7	6	6	6		6									
3.0	7	6	6	5		6	5	5	5		5									
3.5	6	5	5			5														
4.0	6																			

SOURCE: O. L. Davies, "Design and Analysis of Industrial Experiments," Oliver & Boyd Ltd., Edinburgh and London, 1956.

The entries in this table show the number of observations needed in a t test of the significance of the difference between two means in order to control the probabilities of the errors of the first and second kinds at α and β, respectively

Level of t test

Value of $D = \frac{\delta}{s}$	0.01					0.02					0.05					0.1				
Single-sided test	$\alpha = 0.005$					$\alpha = 0.01$					$\alpha = 0.025$					$\alpha = 0.05$				
Double-sided test	$\alpha = 0.01$					$\alpha = 0.02$					$\alpha = 0.05$					$\alpha = 0.1$				
$\beta \rightarrow$	0.01	0.05	0.1	0.2	0.5	0.01	0.05	0.1	0.2	0.5	0.01	0.05	0.1	0.2	0.5	0.01	0.05	0.1	0.2	0.5
0.05																				
0.10																				
0.15																				
0.20																				137
0.25															124					88
0.30										123					87					61
0.35					110					90					64				102	45
0.40					85					70				100	50		108	78	35
0.45				118	68				101	55			105	79	39		108	86	62	28
0.50				96	55			106	82	45		106	86	64	32		88	70	51	23
0.55			101	79	46		106	88	68	38		87	71	53	27	112	73	58	42	19
0.60		101	85	67	39		90	74	58	32	104	74	60	45	23	89	61	49	36	16
0.65		87	73	57	34	104	77	64	49	27	88	63	51	39	20	76	52	42	30	14
0.70	100	75	63	50	29	90	66	55	43	24	76	55	44	34	17	66	45	36	26	12
0.75	88	66	55	44	26	79	58	48	38	21	67	48	39	29	15	57	40	32	23	11

0.80	77	58	49	39	23	70	51	43	33	19	59	42	34	26	14	50	35	28	21	10
0.85	69	51	43	35	21	62	46	38	30	17	52	37	31	23	12	45	31	25	18	9
0.90	62	46	39	31	19	55	41	34	27	15	47	34	27	21	11	40	28	22	16	8
0.95	55	42	35	28	17	50	37	31	24	14	42	30	25	19	10	36	25	20	15	7
1.00	50	38	32	26	15	45	33	28	22	13	38	27	23	17	9	33	23	18	14	7
1.1	42	32	27	22	13	38	28	23	19	11	32	23	19	14	8	27	19	15	12	6
1.2	36	27	23	18	11	32	24	20	16	9	27	20	16	12	7	23	16	13	10	5
1.3	31	23	20	16	10	28	21	17	14	8	23	17	14	11	6	20	14	11	9	5
1.4	27	20	17	14	9	24	18	15	12	8	20	15	12	10	6	17	12	10	8	4
1.5	24	18	15	13	8	21	16	14	11	7	18	13	11	9	5	15	11	9	7	4
1.6	21	16	14	11	7	19	14	12	10	6	16	12	10	8	5	14	10	8	6	4
1.7	19	15	13	10	7	17	13	11	9	6	14	11	9	7	4	12	9	7	6	3
1.8	17	13	11	10	6	15	12	10	8	5	13	10	8	6	4	11	8	7	5	
1.9	16	12	11	9	6	14	11	9	8	5	12	9	7	6	4	10	7	6	5	
2.0	14	11	10	8	6	13	10	9	7	5	11	8	7	6	4	9	7	6	4	
2.1	13	10	9	8	5	12	9	8	7	5	10	8	7	6	3	8	6	5	4	
2.2	12	10	8	7	5	11	9	7	6	4	9	7	6	5		8	6	5	4	
2.3	11	9	8	7	5	10	8	7	6	4	9	7	6	5		7	5	5	4	
2.4	11	9	8	6	5	10	8	7	6	4	8	7	6	5		7	5	4	4	
2.5	10	8	7	6	4	9	7	6	5	4	8	6	6	5		6	4	4	3	
3.0	8	6	6	5	4	7	6	5	4	3	6	5	4	4		5	4	4		
3.5	6	5	5	4	3	6	5	4	4		5	4	4	3		4	3	3		
4.0	6	5	4	4		5	4	4	3		4	4	3			4		3		

SOURCE: O. L. Davies, "Design and Analysis of Industrial Experiments," Oliver & Boyd Ltd., Edinburgh and London, 1956.

Table A·13 Number of Observations Required for the Comparison of a Population Variance with a Standard Value Using the χ^2 Test

The entries in this table show the value of the ratio R of the population variance σ_1^2 to a standard variance σ_0^2 which is undetected with frequency β in a χ^2 test at significance level α of an estimate s_1^2 of σ_1^2 based on ϕ degrees of freedom

ϕ	$\alpha = 0.01$				$\alpha = 0.05$			
	$\beta = 0.01$	$\beta = 0.05$	$\beta = 0.1$	$\beta = 0.5$	$\beta = 0.01$	$\beta = 0.05$	$\beta = 0.1$	$\beta = 0.5$
1	42,240	1,687	420.2	14.58	24,450	977.0	243.3	8.444
2	458.2	89.78	43.71	6.644	298.1	58.40	28.43	4.322
3	98.79	32.24	19.41	4.795	68.05	22.21	13.37	3.303
4	44.69	18.68	12.48	3.955	31.93	13.35	8.920	2.826
5	27.22	13.17	9.369	3.467	19.97	9.665	6.875	2.544
6	19.28	10.28	7.628	3.144	14.44	7.699	5.713	2.354
7	14.91	8.524	6.521	2.911	11.35	6.491	4.965	2.217
8	12.20	7.352	5.757	2.736	9.418	5.675	4.444	2.112
9	10.38	6.516	5.198	2.597	8.103	5.088	4.059	2.028
10	9.072	5.890	4.770	2.484	7.156	4.646	3.763	1.960
12	7.343	5.017	4.159	2.312	5.889	4.023	3.335	1.854
15	5.847	4.211	3.578	2.132	4.780	3.442	2.925	1.743
20	4.548	3.462	3.019	1.943	3.802	2.895	2.524	1.624
24	3.959	3.104	2.745	1.842	3.354	2.630	2.326	1.560
30	3.403	2.752	2.471	1.735	2.927	2.367	2.125	1.492
40	2.874	2.403	2.192	1.619	2.516	2.103	1.919	1.418
60	2.358	2.046	1.902	1.490	2.110	1.831	1.702	1.333
120	1.829	1.661	1.580	1.332	1.686	1.532	1.457	1.228
∞	1.000	1.000	1.000	1.000	1.000	1.000	1.000	1.000

Examples:

Testing for an increase in variance. Let $\alpha = 0.05$, $\beta = 0.01$, and $R = 4$. Entering the table with these values it is found that the value 4 occurs between the rows corresponding to $n = 15$ and $n = 20$. Using rough interpolation it is indicated that the estimate of variance should be based on degrees of freedom.

Testing for a decrease in variance. Let $\alpha = 0.05$, $\beta = 0.01$, and $R = 0.33$. The table is entered with $\alpha' = \beta = 0.01$, $\beta' = \alpha = 0.05$, and $R' = 1/R = 3$. It is found that the value 3 occurs between the rows corresponding to $n = 24$ and $n = 30$. Using rough interpolation it is indicated that the estimate of variance should be based on 26 degrees of freedom.

SOURCE: O. L. Davies, "Design and Analysis of Industrial Experiments," Oliver & Boyd Ltd., Edinburgh and London, 1956.

Table A·14 Number of Observations Required for the Comparison of Two Population Variances Using the F Test

The entries in this table show the value of the ratio R of two population variances σ_2^2/σ_1^2 which remains undetected with frequency β in a variance ratio test at significance level α of the ratio s_2^2/s_1^2 of estimates of the two variances, both being based on ν degrees of freedom

ν	$\alpha = 0.01$				$\alpha = 0.05$				$\alpha = 0.5$			
	$\beta = 0.01$	$\beta = 0.05$	$\beta = 0.1$	$\beta = 0.5$	$\beta = 0.01$	$\beta = 0.05$	$\beta = 0.1$	$\beta = 0.5$	$\beta = 0.01$	$\beta = 0.05$	$\beta = 0.1$	$\beta = 0.5$
1	16,420,000	654,200	161,500	4,052	654,200	26,070	6,436	161.5	4,052	161.5	39.85	1.000
2	9,801	1,881	891.0	99.00	1,881	361.0	171.0	19.00	99.00	19.00	9.000	1.000
3	867.7	273.3	158.8	29.46	273.3	86.06	50.01	9.277	29.46	9.277	5.391	1.000
4	255.3	102.1	65.62	15.98	102.1	40.81	26.24	6.388	15.98	6.388	4.108	1.000
5	120.3	55.39	37.87	10.97	55.39	24.51	17.44	5.050	10.97	5.050	3.453	1.000
6	71.67	36.27	25.86	8.466	36.27	18.35	13.09	4.282	8.466	4.284	3.056	1.000
7	48.90	26.48	19.47	6.993	26.48	14.34	10.55	3.878	6.993	3.787	2.786	1.000
8	36.35	20.73	15.61	6.029	20.73	11.82	8.902	3.438	6.029	3.438	2.589	1.000
9	28.63	17.01	13.06	5.351	17.01	10.11	7.757	3.179	5.351	3.179	2.440	1.000
10	23.51	14.44	11.26	4.849	14.44	8.870	6.917	2.978	4.849	2.978	2.323	1.000
12	17.27	11.16	8.923	4.155	11.16	7.218	5.769	2.687	4.155	2.687	2.147	1.000
15	12.41	8.466	6.946	3.522	8.466	5.777	4.740	2.404	3.522	2.404	1.972	1.000
20	8.630	6.240	5.270	2.938	6.240	4.512	3.810	2.124	2.938	2.124	1.794	1.000
24	7.071	5.275	4.526	2.659	5.275	3.935	3.376	1.984	2.659	1.984	1.702	1.000
30	5.693	4.392	3.833	2.386	4.392	3.398	2.957	1.841	2.386	1.841	1.606	1.000
40	4.470	3.579	3.183	2.114	3.579	2.866	2.549	1.693	2.114	1.693	1.506	1.000
60	3.372	2.817	2.562	1.836	2.817	2.354	2.141	1.534	1.836	1.534	1.396	1.000
120	2.350	2.072	1.939	1.533	2.072	1.828	1.710	1.352	1.533	1.352	1.265	1.000
∞	1.000	1.000	1.000	1.000	1.000	1.000	1.000	1.000	1.000	1.000	1.000	1.000

SOURCE: O. L. Davies, "Design and Analysis of Industrial Experiments," Oliver & Boyd Ltd., Edinburgh and London. 1956.

Table A·15 SSR, Multiple Range Test

v_2 \ g	Significance level 0.05						Significance level 0.01					
	2	3	4	5	6	7	2	3	4	5	6	7
1	18.0	18.0	18.0	18.0	18.0	18.0	90.0	90.0	90.0	90.0	90.0	90.0
2	6.09	6.09	6.09	6.09	6.09	6.09	14.0	14.0	14.0	14.0	14.0	14.0
3	4.50	4.50	4.50	4.50	4.50	4.50	8.26	8.5	8.6	8.7	8.8	8.9
4	3.93	4.01	4.02	4.02	4.02	4.02	6.51	6.8	6.9	7.0	7.1	7.1
5	3.64	3.74	3.79	3.83	3.83	3.83	5.70	5.96	6.11	6.18	6.26	6.33
6	3.46	3.58	3.64	3.68	3.68	3.68	5.24	5.51	5.65	5.73	5.81	5.88
7	3.35	3.47	3.54	3.58	3.60	3.61	4.95	5.22	5.37	5.45	5.53	5.61
8	3.26	3.39	3.47	3.52	3.55	3.56	4.74	5.00	5.14	5.23	5.32	5.40
9	3.20	3.34	3.41	3.47	3.50	3.52	4.60	4.86	4.99	5.08	5.17	5.25
10	3.15	3.30	3.37	3.43	3.46	3.47	4.48	4.73	4.88	4.96	5.06	5.13
11	3.11	3.27	3.35	3.39	3.43	3.44	4.39	4.63	4.77	4.86	4.94	5.01
12	3.08	3.23	3.33	3.36	3.40	3.42	4.32	4.55	4.68	4.76	4.84	4.92
13	3.06	3.21	3.30	3.35	3.38	3.41	4.26	4.48	4.62	4.69	4.74	4.84
14	3.03	3.18	3.27	3.33	3.37	3.39	4.21	4.42	4.55	4.63	4.70	4.78
15	3.01	3.16	3.25	3.31	3.36	3.38	4.17	4.37	4.50	4.58	4.64	4.72
20	2.95	3.10	3.18	3.25	3.30	3.34	4.02	4.22	4.33	4.40	4.47	4.53
30	2.89	3.04	3.12	3.20	3.25	3.29	3.89	4.06	4.16	4.22	4.32	4.36
40	2.86	3.01	3.10	3.17	3.22	3.27	3.82	3.99	4.10	4.17	4.24	4.30
∞	2.77	2.92	3.02	3.09	3.15	3.19	3.64	3.80	3.90	3.98	4.04	4.09

Index